D0461545

INTERNATIONAL TEXTBOOKS IN CIVIL ENGINEERING

Consulting Editor

RUSSELL C. BRINKER

Professor of Civil Engineering
New Mexico State University

Elementary Surveying

Elementary Surveying

FIFTH EDITION

RUSSELL C. BRINKER
Professor of Civil Engineering
New Mexico State University

INTERNATIONAL TEXTBOOK COMPANY

Standard Book Number 7002 2243 X

N-Q-RB

Printed in the United States of America by The Haddon Craftsmen, Inc., Scranton, Pennsylvania. Library of Congress Catalog Card Number: **69–16621.**

INTEXT EDUCATIONAL PUBLISHERS
666 FIFTH AVENUE
NEW YORK, N.Y. 10019

Preface

The fifth edition of ELEMENTARY SURVEYING continues the approach of the third and fourth editions which emphasized fundamental theory and procedures while still giving practical hints and illustrations not covered generally or at all elsewhere. Many remarks made in the fourth edition Preface apply also to this revision and are noted again.

Slight to major changes or additions have been made in every chapter including

1) Expanded information on government agencies and mapping.
2) Data on automatic reading and recording systems plus more suggestions on notekeeping.
3) Rewritten and updated material on the tellurometer, electrotape, and geodimeter with a new table on the characteristics of all known electronic distance measuring instruments.
4) Discussion of the laser and its applications in several chapters, and the airborne profile recorder.
5) New instrument illustrations.
6) An interlocking 2-loop level circuit adjustment with an example.
7) A new chart on the distribution of magnetic declinations.
8) Analysis of the number of instrument pointings to balance sighting and reading errors.
9) Description of the angles-to-the-right method.
10) An illustrative calculation of latitudes and departures using natural functions (but retaining the example with logarithms), and a Fortran program with output listing.
11) Five (instead of three) methods of traverse adjustment.
12) Additional material on contours.
13) Some rearrangement of instrument adjustment items.
14) Updated field astronomy example solutions with new technical and general points of interest.

15) Completely rewritten chapter on Photogrammetry.
16) A listing of slope staking steps and numerical example.
17) Circular curve layout by tangent offsets with a problem illustration.
18) Revision of the sight distance formula and examples to fit the changed height-of-eye specification.
19) Mention of the mass diagram.
20) Improved optical alignment equipment and utilization of the laser.
21) A new chapter on State Plane Coordinates, with numerical example, in Appendix A.
22) An increased number of problems at the end of each chapter and the data changed for old ones to provide over 1,000 homework problems (up from 653 and enough for many semesters without repeating). A solutions manual is available for the first time. Some problems are carried through several chapters so progressive steps in their completion can be followed.
23) Elimination of the log tables and enlargement of the natural function tables from 5 to 7 places for accuracy in handling desk calculator computations requiring 5 significant figures.

Stress in text and problems is placed upon the theory of errors and optics; correlation of theory and practical field methods; elimination of typical errors and mistakes; significant figures; use of basic values such as sin 1 min = 0.0003 and sin 1° = 0.01¾ in solving ordinary problems without tables or slide rule; and the interdependence of field, computation, and mapping requirements.

Engineers, architects, geologists, and foresters must be able to make measurements and analyze the precision and accuracy of the results obtained by other people. They should be qualified to properly locate and set machinery; to lay out houses, buildings, and other common structures; and to understand and prepare simple topographic maps. Each of these areas is discussed, and proper field procedures to obtain a desired precision are noted.

A few cost figures are introduced so that students will learn early in their college work to associate the three bases of engineering practice—theory, application, and costs. All surveying is a constant fight to eliminate or isolate errors and mistakes. In each chapter the student is reminded of this point, through lists of typical errors and mistakes.

Although the third, fourth, and fifth editions retain the title ELEMENTARY SURVEYING, the material goes beyond the elementary stage in length and scope. The large number of chapters, however, permits inclusion or omission of subjects to correspond with the class time available for students in civil engineering, other engineering curricula, architecture, geology, and forestry. The easy-reading qualities of the previous editions have been retained by keeping short lines for ready scanning and fewer lines per page.

Chapters are arranged in the order found most convenient at numerous colleges. Fundamental material is collected in the first sixteen chapters comprising Part I. Theory and use of the four fundamental surveying instruments—the tape, level, transit, and plane table—are described in detail, and new types of equipment noted. Any chapter following Chapter 11 can be omitted without loss of continuity, although many of them are short enough to be suitable for a single assignment.

Limited coverage of such subjects as photogrammetry, field astronomy, boundary surveys, and industrial applications of surveying methods is given in Part II to fit various programs offered. For example, the brief chapter on boundary surveys is intended to make students aware of a few problems involved in the survey and transfer of property, and the legal requirements of professional registration. Some instructors give broad survey-type courses and want their students to get an over-all view of the many surveying functions. It is believed that the arrangement in scope of material presented herein will meet that need also.

Taping, leveling, and transit work are taken up in or order because students find it easier to acquire some facility with the equipment in that sequence, and because this arrangement permits the start and continuation of field work with a minimum of lecture time. The suggested order of field assignments given in Appendix A makes it possible to begin effective computation and drafting-room problems after just a few periods in the field if bad weather is encountered.

The difficulty in getting through all of the preliminary material (basic concepts of the profession, history, theory of errors, and methods of notekeeping) before commencing field work during the first week is recognized. Nevertheless the author feels that these topics must precede the theory and use of instruments.

The subject of notes and noteforms—an important part of surveying and engineering—is discussed in a separate chapter. Most of the sample noteforms are collected in Appendix A instead of being scattered throughout the text.

Suggestions and criticism will be greatly appreciated.

RUSSELL C. BRINKER

University Park, New Mexico
January, 1969

Acknowledgments

The author wishes to acknowledge the use of material from the late Professors A. S. Cutler, O. S. Zelner, and L. F. Boon (University of Minnesota); Professor Emeritus C. B. Andrews (University of Hawaii); Professor (Retired) P. P. Rice (Rutgers University); and Mr. D. F. Griffin (formerly of the University of Southern California). Helpful suggestions, assistance, or pertinent material for this and/or previous editions were offered by Professor A. S. Chase (Auburn University); Professor L. Perez (Pennsylvania State University); Lt. Col. W. L. Baxter (formerly at the United States Military Academy); Professors L. G. Rich and J. P. Rastron (Clemson College); Professor D. V. Smith (Virginia Polytechnic Institute); Professor E. C. Wagner (University of Wisconsin); Professor H. E. Kallsen (formerly at Louisiana Polytechnic Institute); Professor G. B. Lyon (Cornell University); Professors W. Wintz, Jr., D. C. McKee, and J. M. DeMarche (Louisiana State University); W. Blakney (Auburn University); C. F. Meyer (Worcester Polytechnic Institute); Professor C. H. Drown (Sacramento State College); Professor J. R. Coltharp (University of Texas at El Paso); Professor Porter W. McDonnell, Jr. (Pennsylvania State University, Mont Alto Campus); Professor John O. Eichler (Georgia Institute of Technology); Professor J. L. Clapp (University of Wisconsin); Mr. W. C. Wattles of Glendale, California; Mr. R. B. Irwin of Los Angeles, and others.

Chapter 2 was prepared by Professor D. C. McNeese of the University of Washington; Chapter 18 and the Chapter on State Plane Coordinates in Appendix A by Professor Paul R. Wolf (University of California) who also provided many helpful suggestions throughout the entire book.

Illustrative material and other help has been freely supplied by the U. S. Bureau of Land Management, the U. S. Geological Survey, the U. S. Coast and Geodetic Survey, and the Army Map Service. Manufacturers of surveying equipment who furnished illustrations include the Keuffel and Esser Company, W. and L. E. Gurley, Kern Instruments, Inc., Wild Heerbrugg Instruments, Inc., Abrams Aerial Survey Corporation, Wallace and Tiernan, and Bausch and Lomb.

Contents

Part I

part I

chapter 1

Introduction

1–1. Definition of surveying. Surveying is the science and art of making the measurements necessary to determine the relative positions of points above, on, or beneath the surface of the earth, or to establish such points. The work of the surveyor, which largely consists of making such measurements, can be divided into four parts:

FIELD WORK—Making and recording measurements in the field.

COMPUTING—Making the necessary calculations to determine locations, areas, and volumes.

MAPPING—Plotting the measurements and drawing a map.

STAKEOUT—Setting stakes to delineate boundaries or to guide construction operations.

1–2. Importance of surveying. Surveying is one of the oldest arts practiced by man, because from the earliest times it has been found necessary to mark boundaries and divide land. It is now indispensable in all branches of engineering. For example, surveys are required prior to and during the planning and construction of highways, railroads, buildings, bridges, missile ranges, launching sites and tracking stations, tunnels, canals, irrigation ditches, dams, drainage works, water-supply and sewerage systems, pipelines and mine shafts. The use of surveying or surveying methods has become common in the layout of assembly lines and jigs, the fabrication of airplanes, the placement of equipment, to provide control for aerial surveys, and in many related tasks in aeronautical, agricultural, chemical, electrical, mechanical, and mining engineering, and in geology and forestry. Optical alignment represents an application of surveying in shop practice.

1–3. Training for all engineers. All engineers must know the limits of accuracy possible in construction, plant design and layout, and manufacturing processes, even though someone else does the actual surveying. This knowledge is best obtained by making measurements with the surveying equipment employed in practice which provides a true concept of the theory of errors, and small but recognizable differences in observed quantities.

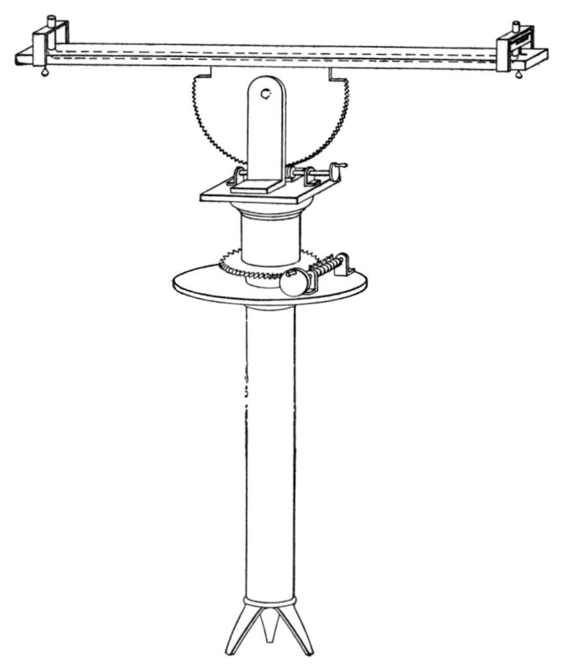

FIG. 1–1. The diopter.[1]

In addition to stressing the need for reasonable limits of accuracy, surveying emphasizes the value of significant figures. An engineer must know when to work to hundredths of a foot instead of to tenths or thousandths, or perhaps the nearest foot, and what precision in field data is necessary to justify carrying out computations to the desired number of decimal places. With experience, he learns how the equipment and personnel available govern procedures and results.

Neat sketches and computations are the mark of an orderly mind, which in turn is an index of sound engineering background and competence. Taking field notes under all sorts of conditions is excellent

[1] Figures 1–1, 2, 3, and 4 are shown through the courtesy of Professor Edward Noble Stone.

preparation for the kind of recording and sketching expected of engineers. Additional training having a carry-over value is obtained in arranging computations properly.

Engineers designing buildings, bridges, equipment, etc., are fortunate if their estimates of loads to be carried are correct within 5 per cent. Then a factor of safety of 2 or more is applied. But except for topographic work, only exceedingly small errors can be tolerated in surveying, and there is no factor of safety. Traditionally, therefore, surveying stresses mathematical precision.

1–4. History of surveying. The oldest historical records in existence today which bear directly on the subject of surveying state that this science had its beginning in Egypt. Herodotus says Sesostris (about 1400 B.C.) divided the land of Egypt into plots for the purpose of taxation. The annual floods of the Nile River swept away portions of these plots, and surveyors were appointed to replace the bounds. These early surveyors were called *rope-stretchers.* Their measurements were made by means of ropes with markers at unit distances.

As a consequence of this work the early Greek thinkers developed the science of geometry. Their advance, however, was chiefly along the lines of pure science. Heron stands out prominently for applying science to surveying, about 120 B.C. He was the author of several important treatises of interest to engineers, including one called *The Dioptra,* which related the methods of surveying a field, drawing a plan, and making calculations. This treatise also described one of the first pieces of surveying equipment recorded, the *diopter* (Fig. 1–1). For many years Heron's work was the most authoritative among Greek and Egyptian surveyors.

Real development in the art of surveying came through the practical-minded Romans, whose best-known treatise on surveying was by Frontinus. Although the original manuscript disappeared, copied portions have been preserved. This noted Roman engineer and surveyor, who lived in the first century, was a pioneer in the field and his treatise remained the standard for many years.

The engineering ability of the Romans was demonstrated by their extensive construction work throughout the Empire. The surveying necessary for this construction resulted in the organization of a surveyors' guild. Ingenious instruments were developed and used. Among these were the *groma* (Fig. 1–2), used for sighting; the *libella* (Fig. 1–3), an A frame with a plumb bob, used for leveling; and the

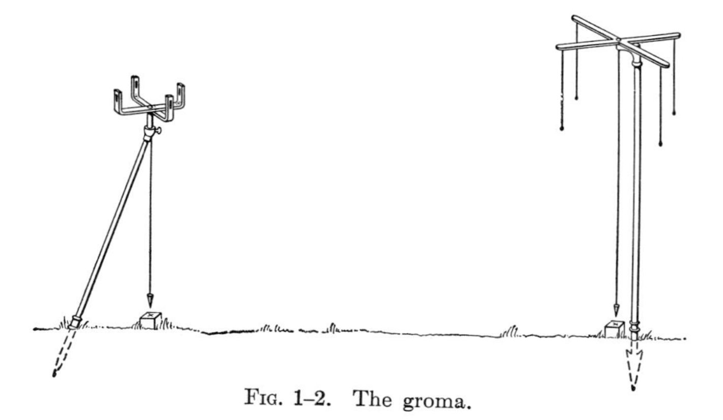

FIG. 1–2. The groma.

chorobates (Fig. 1–4), a horizontal straightedge about 20 ft long, with supporting legs and a groove on top for water to serve as a level.

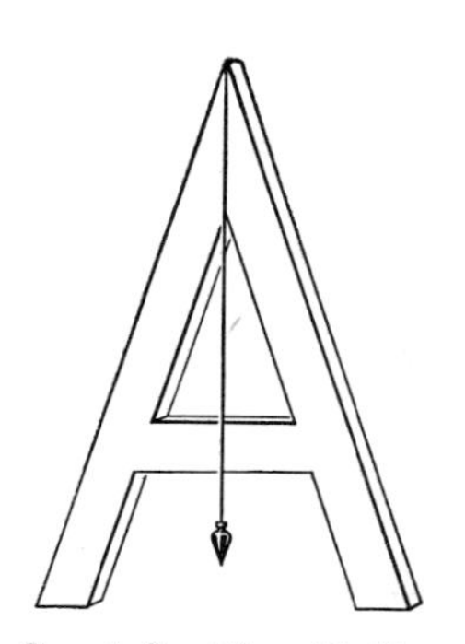

FIG. 1–3. The libella.

One of the oldest Latin manuscripts in existence is the *Codex Acerianus,* written about the sixth century. It contains an account of surveying as practiced by the Romans and includes several pages from Frontinus' treatise. The manuscript was found in the tenth century by Gerbert and served as the basis for his text on geometry, which was largely devoted to surveying.

During the Middle Ages, Greek and Roman science was kept alive by the Arabs. Little progress was made in the art of surveying, and the only writings pertaining to it were called "practical geometry."

In the thirteenth century Von Piso wrote *Practica Geometria,* which contained instructions on surveying. He also wrote *Liber Quadratorum,* dealing chiefly with the *quadrans,* a square brass frame having a 90° angle and other scales marked off on it. A movable pointer was used for sighting. Other instruments of the period were the *astrolabe,* a metal circle with a pointer hinged at its center and held by a ring at the top, and the *cross staff,* a wooden rod about 4 ft long with an adjustable crossarm at right angles to it. The known lengths of the arms of the cross staff permitted distances to be measured by proportion and angles.

In the eighteenth and nineteenth centuries the art of surveying advanced more rapidly. The need for maps and the location of national boundaries caused England and France to make extensive surveys requiring accurate triangulation. Thus geodetic surveying began. The United States Coast and Geodetic Survey was established by an act of Congress in 1807.

Increased land values and the importance of exact boundaries, along with the demand for public improvements in the canal, turnpike, and railroad eras, brought surveying into a prominent position. More recently, the large amount of general construction has entailed an augmented surveying program. Surveying is still the sign of progress.

During World Wars I and II, and the Korean and Viet Nam conflicts, surveying in its many branches played an important part because of the stimulus provided to improve instruments and speed the methods of making measurements and maps. As a result, a new era has opened for the surveying profession. The development of electronic devices for making distance measurements parallels the advances in electronics in other professions.

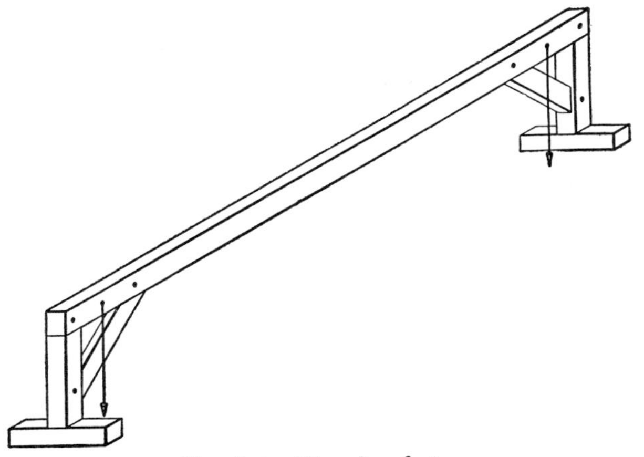

Fig. 1–4. The chorobates.

The basic surveying instruments of today—the transit, level, and steel tape shown in Fig. 1–5—are now frequently supplanted by the theodolite, self-leveling level, and electronic distance-measuring instruments. In the field of mapping, aerial surveying has replaced

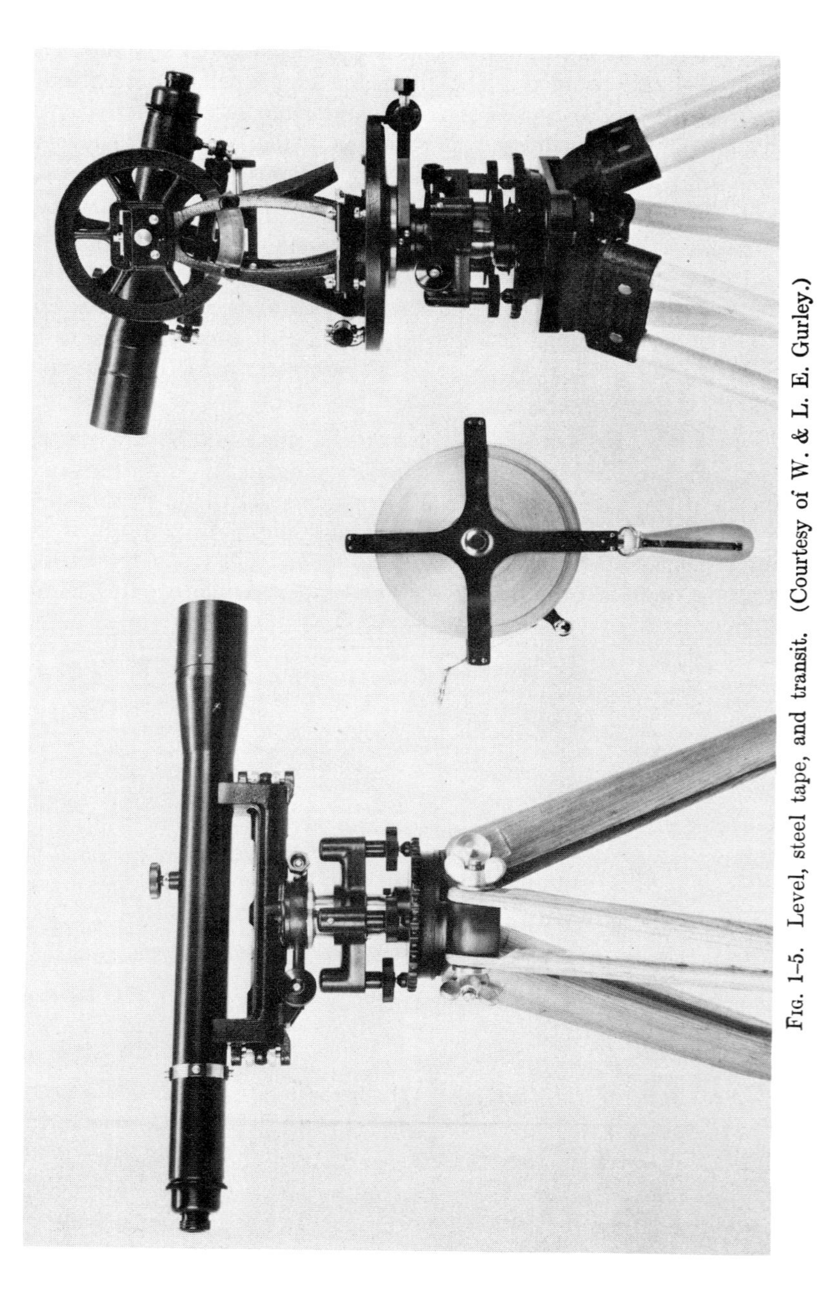

Fig. 1–5. Level, steel tape, and transit. (Courtesy of W. & L. E. Gurley.)

ground surveying on large-area projects; but the traditional ground surveys still are essential for establishing the locations of horizontal and vertical control points, and in land, construction, and small-area surveys.

1-5. Types of surveys. There are many types of surveys, each so specialized that a man proficient in one branch may have little contact with the other branches. The more-important classifications, described briefly, are:

Plane surveying. Surveying in which the curvature of the earth is neglected. It is applicable for small areas.

Geodetic surveying. Surveying in which the curvature of the earth is considered. It is applicable for large areas and long lines and is used to precisely locate basic points suitable for controlling other surveys.

Land or boundary surveys. Surveys to establish property corners and land lines. Usually closed surveys (surveys which start at one corner and end at the same corner).

Topographic surveys. Surveys made for the purpose of preparing maps showing locations of natural and artificial features, and elevations of points on the ground.

Route surveys. Surveys of and for highways, railroads, pipelines, transmission lines, canals, and other projects which do not close upon the starting point.

Hydrographic surveys. Surveys of lakes, streams, reservoirs, and other bodies of water.

Construction surveys. Surveys to provide locations and elevations of structures.

Photogrammetric surveys. Surveys in which photographs, either terrestrial or aerial, are used.

1-6. Present status of surveying. There is an increasing demand for good maps in the United States, and various governmental agencies are attempting to provide them, although such agencies are handicapped by insufficient funds and personnel. A common misconception is that the entire country has been adequately mapped by now. Actually, only about 75 per cent of the United States is covered by mapping at scales of one mile to the inch or larger. At the present rate of new map production and periodic revision of satisfactorily mapped areas, about a decade will be required to complete the job.

Accurate mapping cannot be done without good *basic control.* Two national survey nets for geodetic horizontal and vertical control are continually being extended over the country, primarily by the United States Coast and Geodetic Survey, for the control of nautical charts and topographic maps, and to provide coordinated position data for all surveyors. The horizontal-control survey net consists of arcs of first-order and second-order triangulation (see Sec. 4–2) and lines of first-order and second-order traverse (see Sec. 9–1). The data from this net are coordinated and correlated on the North American datum of 1927 extending through Alaska and Central America. The vertical-control survey net consists of over 327,000 kilometers of first-order and 475,000 kilometers of second-order spirit leveling which determine the elevations of 465,000 bench marks above a common datum (mean sea level).

The advent of missiles and satellites has caused a resurgence in geodesy. Electronic distance-measuring devices, and lasers to determine angles, distances, and elevations, have kept surveying abreast of rapid technological advances in other fields. The design of surveys introduces problems comparable with those in engineering. How many columns are needed at what spacing to support a certain building? How many triangulation stations in what relative locations will be required for alternate types of measurements in a particular control net or mapping project?

Three United States Government departments do mapping on a large scale:

a) The Coast and Geodetic Survey was organized to map the coast. Its activities now include triangulation, precise leveling, preparation of nautical and aeronautical charts, photogrammetric surveys, tide and current studies, collection of magnetic data, gravimetric surveys, and satellite triangulation. The basic control points established by this organization are the foundation for all large-area surveying.

b) The General Land Office, established in 1812, directed the public-lands surveys. Lines and corners have been set for most of the public lands. The Bureau of Land Management now has jurisdiction over the survey and sale of these lands.

c) The Geological Survey, established in 1879, will ultimately map the entire country. Its standard 7½- and 15-minute quadrangle maps show topographic and cultural features, and are suitable for general use and for a variety of engineering and scientific purposes. Over 7,500,000 copies are distributed each year. The price for single copies is 50 cents. Mail orders for maps covering areas east of the Mississippi River should be addressed to the Geological Survey, Washington, D.C. 20242, and for areas west of the Mississippi River to the Geological Survey, Federal Center, Denver, Colorado 80225.

In addition, units of the Corps of Engineers, U. S. Army, have made extensive surveys for emergency and military purposes. Some of these surveys provide data for engineering projects, such as those connected with flood control. The Army Map Service, which is an agency of the Corps of Engineers, is primarily concerned with the production and supply of military topographic maps, three-dimensional plastic relief maps, pictomaps, geodetic data, and related data for the entire Department of Defense. Additionally, it produces lunar and extraterrestrial maps for use in the United States space effort.

Extensive surveys have also been conducted for special purposes by the Forest Service, National Park Service, International Boundary Commission, Bureau of Reclamation, Tennessee Valley Authority, Mississippi River Commission, United States Lake Survey, and Naval Oceanographic Office.

1–7. The surveying profession. Surveying is classified as a learned profession because the modern practitioner needs a background of technical training and experience, and must exercise independent judgment. A registered (licensed) professional surveyor must have a knowledge of mathematics—particularly geometry and trigonometry—and of law as it pertains to land and boundaries. He should be accurate in computations and field operations, and be able to do neat drafting. Above all, he is governed by a professional code of ethics, and is expected to make reasonable minimum charges for his work.

The personal qualifications of a private surveyor are as important as his technical ability. He must be patient and tactful in dealing with clients and their hostile neighbors. Few people are aware of the painstaking search of old records that is required before field work is done. Diligent, time-consuming effort may be needed to locate corners on nearby tracts for checking purposes, and to find the marks for the property in question.

Permission to trespass on private property or to cut obstructing tree branches and shrubbery must be obtained through a proper approach. These privileges are not conveyed by employment in a state highway department (although a court order can be obtained if a property owner objects to necessary surveys) or by a surveying license.

All fifty states and the provinces of Canada now have registration laws for professional engineers, and most have separate registration for surveyors. In general, a surveyor's license is required for making boundary surveys, but construction and route surveying can be done by registered civil engineers. To qualify for registration as either a professional engineer or a surveyor, it is usually necessary to have a college degree (or an equivalent number of years of experience) plus four years of practical work, and to pass appropriate written examinations. The standard registration law makes it a misdemeanor "to practice or offer to practice land surveying" without a license. Technical considerations alone should discourage anyone other than an experienced surveyor from setting property corners. As in all professions, scholastic training is merely the first step toward the goal of every surveyor and engineer—true professional status denoted by registration.

PROBLEMS

NOTE: Answers for these problems can be obtained by consulting a dictionary and the succeeding chapters, from general experience, or by visiting the office of a surveyor, engineer, or land recorder.

1–1. List ten uses for surveying in activities other than land or boundary surveys.

1–2. How is surveying used in the missile and satellite programs?

1–3. How is surveying used in modern "dry" farming and contour plowing?

1–4. List some of the applications of surveying in geology, forestry, and mining.

1–5. Why is it necessary to make accurate surveys of underground mines?

1–6. What are patent surveys?

1–7. How is surveying used in connection with the construction of a highway through a mountain range?

1–8. What astronomical observations are made by surveyors?

1–9. How might a surveyor chart the shore lines and water depths of a winding river 800 ft wide?

1–10. What measurements are needed during the laying of a 36-in. city sewer? A 6-in. water main?

1–11. Sketch a means of using surveying methods applicable in the fabrication of a satellite launcher which is 160-ft long and which must have all dimensions correct to within 0.001 ft.

1–12. How might weather conditions affect the accuracy of surveys?

1–13. Devise a simple instrument which could be used for leveling a basement floor 30 ft × 30 ft prior to placing the concrete slab.

1–14. Describe a method for measuring a right angle to lay out the walls of an H-shaped ranch-type house 60 × 40 ft over all.

1–15. Is a vertical aerial photograph a map? Explain.

1–16. What organization in your state will furnish maps and surveying reference data to surveyors and engineers?

1–17. Why should the purchaser of a farm, a city lot, or a home require a survey before final payment is made?

1–18. Do the subdivision laws of your community specify the accuracy required for surveys made to lay out the subdivision? Why or why not?

1–19. List the legal requirements for registration as a land surveyor in your state.

1–20. What kinds of surveys would be classified as "engineering" surveys?

1–21. What basic mathematical and surveying principles are used in locating positions on the earth by means of satellite triangulation?

chapter 2

Theory of Measurements and Errors

DONALD C. MCNEESE
Associate Professor of General Engineering
University of Washington

2–1. Measurements in general. The process of making measurements in surveying requires a combination of human skill and mechanical equipment applied with the utmost judgment. Experience and good physical conditions improve the human factor; superior equipment enables good operators to do better work with more consistent results and in less time. The design of measurement programs, comparable with other engineering design, is now practiced. Matrix algebra and the electronic computer are two of the newer tools used to investigate and distribute errors in a large series of measurements.

2–2. Types of measurements made in surveying. It has been stated that surveying is the art and science of determining the relative positions of points on or near the earth's surface. A point may be located by using common rectangular or Cartesian coordinates. Distances are determined in the X-direction, called easterly or westerly, the Y-direction, called northerly or southerly, and the Z-direction, known as elevation. When points are located in this manner, the only unit involved is that of distance. For surveys covering small areas, these distances may be considered as lying on flat planes, without reference to the shape of the earth; thus the term "plane surveying" is used. For surveys of greater extent, the difference between a plane and the spherical earth becomes appreciable, and the geographic coordinates called latitude and longitude must be used.

Another method of locating a point is by some form of polar coordinates. In the *XY* plane, that is, the horizontal plane, an angle and a distance may be measured. A vertical distance, often referred to as elevation from some datum, may be measured by a direct method (taping, leveling) or may be determined by measuring an angle in the vertical plane and a slope or horizontal distance and applying trigonometry. Five kinds of measurements form the basis of all surveying: (a) horizontal angles, (b) horizontal distances, (c) vertical angles, (d) vertical distances, and (e) slope distances.

2–3. Units of measurement. The units of measurement in surveying are those for *length* and *angle.* The length unit in the English system, the *foot,* bears a definite relation to the *meter.* Originally the meter was defined as 1/10,000,000 of the earth's meridional quadrant. When the metric system was legalized for use in the United States in 1866, the meter was defined as the interval under certain physical conditions between lines on an International Prototype bar made of 90% platinum and 10% iridium, and was taken as equal to 39.37 in. A copy of the bar is held by the United States Bureau of Standards and is compared periodically with the international standard stored in France.

In October, 1960, the United States and other Meter-Treaty nations redefined the meter in terms of the wavelength of a certain kind of light. Now one meter is equal to the length of 1,650,763.73 waves of the orange-red light produced by burning gas of the element krypton (Kr 86). The inch, foot, yard, and other units of length, like the meter, should not actually change since the wavelength standard and the metal standard are in satisfactory agreement, but some apparently discrepant measurements are still being checked. The new definition will permit industries to make more accurate measurements and to check their own instruments without recourse to the standard meter-bar in Washington. The wavelength of orange-red krypton light is a true constant, whereas there is a risk of instability in the metal meter-bar. If the Meter-Treaty Conference had been held one year later, the laser might well have become the standard instead of krypton light.

In plane surveying, decimals of a foot are used extensively. In building construction, carpenters and builders use inches and fractions. The surveyor must be careful in making conversions, and certain that all measurements are properly labeled. Engineering, surveying, construction, and all science would benefit materially if the simpler, faster meter-decimal system (used by 90 per cent of the people of the world) replaced on a planned schedule the archaic and awkward yard, foot, inch, and fraction arrangement burdening theory and computations.

Units of length used in past surveys in the United States include the following:

1 foot	=	12 inches
1 yard	=	3 feet
1 meter	=	39.37 inches exactly = 3.2808 feet
1 rod	=	1 pole = 1 perch = $16\frac{1}{2}$ feet
1 vara	=	33 inches in California = $33\frac{1}{3}$ inches in Texas
1 Gunter's chain	=	66 feet = 100 links (lk.) = 4 rods
1 mile	=	5280 feet = 80 Gunter's chains (ch.)
1 nautical mile	=	6076.10 feet
1 engineer's chain	=	100 feet = 100 links

A common unit of area is the acre. Ten square chains (Gunter's) equal one acre. Thus, an acre contains 43,560 sq ft.

The unit of angle used in surveying is the *degree,* which is defined as $\frac{1}{360}$ the total angle around a point. One degree equals 60 minutes, and one minute equals 60 seconds.

Many methods have been used to divide a circle, for example, 400 *grads* and 6400 *mils.* The *degree-minute-second* system is supplanting these, mainly in the interest of standardization of angular measurement.

The *radian* is an angle subtended by an arc of a circle having a length equal to the radius of the circle. Obviously 2π radians = 360°, 1 radian = 57° 17′ 44.8″, and 0.01745 radians = 1 degree.

2–4. Significant figures. In recording measurements, an indication of the accuracy attained is the number of digits recorded. By definition, the number of significant figures in any value includes the positive (certain) digits plus one (*only one*) digit which is estimated, and therefore questionable. For example, a distance recorded as 873.52 ft is said to have five significant figures; in this case the first four digits are certain and the last digit is questionable. In order to be consistent with the theory of errors, it is essential that data be recorded with the correct number of significant figures. If a significant figure is dropped off in recording a value, the time spent in acquiring certain accuracy has been wasted. On the other hand, if data are recorded with more figures than those which are significant, false accuracy will be implied and time may be wasted in making computations.

The number of significant figures is often confused with the number of decimal places. Decimal places may have to be used to maintain the correct number of significant figures, but in themselves they do not indicate significant figures. Some examples follow:

Two significant figures: 24, 2.4, 0.24, 0.0024, 0.020
Three significant figures: 364, 36.4, 0.000364, 0.0240
Four significant figures: 7621, 76.21, 0.0007621, 24.00

Zeros at the end of an integral value may cause difficulty because they may or may not indicate significant figures. In the value 2400 it is not known how many figures are significant; there may be two, three, or four. One method of eliminating this uncertainty is to place a bar over the last significant figure, as in $240\bar{0}$, $24\bar{0}0$, or $2\bar{4}00$. Another method is to express the value in terms of powers of ten; the significant figures in the measurement are then written as a number between 1 and 10, including the correct number of zeros at the end, and the decimal point is placed by annexing a power of 10. As an example, 2400 becomes $2.400 \times (10)^3$ if both zeros are significant, $2.40 \times (10)^3$ if one is, and $2.4 \times (10)^3$ if there are only two significant figures.

In engineering computations it is imperative that calculations be consistent with the measured values. For addition or subtraction, retain as the last significant figure in the answer the digit found in the last column *full* of significant figures. Two examples are shown.

$$\begin{array}{r} 46.4012 \\ 1.02 \\ 375. \\ \hline 422. \end{array} \qquad \begin{array}{r} 57.301 \\ 1.48 \\ 629. \\ \hline 688. \end{array}$$

In multiplication the percentage error of the product is equal to the sum of the percentage errors of the factors. This percentage error can be maintained by having the number of significant figures in the answer the same as the least number of significant figures in any of the factors. For example, 362.56 × 2.13 will give 772.2528 when multiplied out; but there should be only three significant figures in the answer, which becomes 772.

In surveying, three types of problems relating to significant figures are encountered.

1. The field measurements are given to some specific number of significant figures, thus dictating that a corresponding number of significant figures should be shown in a computed value. In an intermediate calculation it is common practice to carry at least one more digit than required, and then to round off the answer to the correct number of significant figures.
2. There may be an implied number of significant figures. For instance, the length of a football field might be specified as 100 yards. But in laying out the field, such a distance would probably be measured to the nearest hundredth of a foot, not the nearest half-yard.
3. Each factor may not cause an equal variation. For example, if a steel tape 100.00 ft long is to be corrected for a change in temperature of 15 deg F, one of these numbers has five significant figures while the other has only two. A 15-deg variation in temperature changes the tape length by only 0.01 ft, however. Therefore an adjusted tape length to five significant figures is warranted for this type of data. Another example is the computation of a slope distance from horizontal and vertical distances, as in Fig. 2–1. The vertical distance V is given to two significant figures, and the horizontal distance H is measured to five significant figures. From these data the slope distance S can be computed to five significant figures. For small angles of slope, a considerable change in the vertical distance produces a relatively small increment in the difference between the slope and horizontal distances.

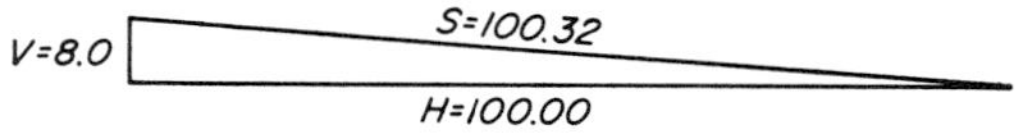

FIG. 2–1. Slope correction.

2–5. Rounding off numbers. Rounding off a number is the process of dropping one or more digits so that the answer contains only those digits which are significant or necessary in subsequent computations. When rounding off numbers to any required degree of accuracy in this text, the following procedure will be observed:

a. When the digit to be dropped is less than 5, the number is written *without* the digit. Thus 78.374 becomes 78.37.

b. When the digit to be dropped is exactly 5, the nearest *even* number is used for the preceding digit. Thus 78.375 becomes 78.38.

c. When the digit to be dropped is greater than 5, the number is written with the preceding digit *increased* by one. Thus 78.376 becomes 78.38.

The procedures in (a) and (c) are standard practice. When rounding off the value 78.375 in (b), however, some computers always take the next-higher hundredth, whereas others invariably use the next-lower hundredth. Using the nearest even digit produces better-balanced results in a series of computations. An exception occurs in the case of taped distances, where recorded measurements tend to be larger than the true values and therefore always rounding off to the next-lower digit is reasonable.

2–6. Direct and indirect measurements. Measurements may be made directly or indirectly. A *direct measurement* is obtained by applying a tape to a line, or by applying a protractor to an angle, or by turning an angle with a transit.

An *indirect measurement* is secured when it is not possible to apply the unit of measure directly to the distance or angle to be measured. The quantity is therefore determined by its relation to some other measured quantity. Thus the distance across a river can be found by measuring the length of a line on one side of the river and measuring the angle at each end of this line to a point on the other side. The desired distance is then computed by one of the standard trigonometric formulas. Since many indirect measurements are made in surveying, a thorough knowledge of geometry and trigonometry is essential.

2–7. Errors in measurements.. It can be unconditionally stated that (1) *no measurement is exact,* (2) *every measurement contains* *errors,* (3) *the true value of a measurement is never known,* and therefore (4) *the exact error present is always unknown.* These facts are demonstrated by noting the following two statements: No matter how large a number a person selects, there is always a larger one; regardless of how small a number is chosen, a still smaller one exists. When a distance is scaled with a rule divided into tenths of an inch, the distance can be read only to hundredths (by interpolation). If a better rule graduated in hundredths of an inch is available, however, the same distance might be estimated to thousandths of an inch. And with a rule graduated in thousandths of an inch, a reading to ten-thousandths is possible. Obviously, accuracy of measurements depends upon the division-size; the reliability of the equipment used; and upon human limitations in interpolating closer than about one-tenth of a scale division. As better equipment is developed (i.e., our number system is carried out further), recorded measurements will more closely approach their true values. Chapter 25 touches upon the attempts to split one-millionth of an inch into one ten-millionth for industrial gages. Note that *measurements,* not *counts* (of cars, bolts, buildings, or other objects), are under consideration herein.

Mistakes are caused by a misunderstanding of the problem, by carelessness, or by poor judgment. Large mistakes are often referred to as *blunders,* and are not considered in the succeeding discussion of errors. They are detected by systematic checking of all work, and must be eliminated by redoing part or all of a job. It is very difficult to detect small mistakes, because they merge with errors. When not detected, these small mistakes must therefore be treated as errors, and will *contaminate* the various types of errors.

2–8. Sources of errors in making measurements. Errors in measurements fall into three classes:

Natural errors. These are caused by variations in wind, temperature, humidity, refraction, gravity, and magnetic declination. For example, the length of a steel tape varies with changes in temperature.

Instrumental errors. These result from any imperfection in the construction or adjustment of instruments, and from the movement of individual parts. For example, the painted graduations on a rod may not be perfectly spaced, or the rod may be warped. The effect of most instrumental errors can be reduced by observing proper surveying procedures and by applying computed corrections.

Personal errors. These arise from limitations of the human senses of sight, touch, and hearing. For example, there is a small error in the measured value of an angle when the vertical cross hair in a transit is not aligned perfectly on the target sighted.

2–9. Types of errors. Errors in measurements are of two types: systematic errors and accidental errors.

Systematic errors. These errors conform to known mathematical and physical laws. They always have the same sign, but their magnitude may be constant or variable, depending upon conditions. Systematic errors, also known as *cumulative errors,* can be computed and their effects eliminated by applying corrections. For example, a 100-ft steel tape which is 0.02 ft too long introduces a plus 0.02-ft error each time it is used. The change in length of a steel tape resulting from a given temperature differential can be computed by a simple formula, and the correction easily made.

Accidental errors. These are the errors which remain after mistakes and systematic errors have been eliminated. They are caused by factors beyond the control of the observer, obey the law of probability, and are sometimes called *random errors.*

The magnitude and algebraic sign of an accidental error are matters of chance. There is no absolute way to compute accidental errors or to eliminate them. Accidental errors are also known as *compensating errors,* since they tend to partially cancel themselves in a series of measurements. For example, a tapeman interpolating to hundredths of a foot on a tape graduated only to tenths will estimate too high on some lengths and too low on others. Individual personal characteristics may nullify such partial compensation, however, since some people are inclined to interpolate high while others interpolate low. It will be shown later in the discussion on probability that *the number of accidental errors remaining after cancellation of some of the plus and minus values is the square root of the number of opportunities for error.*

2–10. Magnitude of errors. *Discrepancy* is the difference between two measured values of the same quantity. It is also the difference between the measured value and the known value of a quantity. A small discrepancy between two measured values indicates that probably there are no mistakes, and accidental errors are small. It does not reveal the magnitude of systematic errors, however.

Precision denotes relative or apparent nearness to the truth and is based upon the refinement of the measurements and the size of the discrepancies. The degree of precision attainable is dependent upon the sensitiveness of the equipment and upon the skill of the observer. In surveying, precision should not be confused with *accuracy,* which denotes absolute nearness to the truth. A survey may be precise without being accurate. As an illustration, if refined methods are employed and readings taken carefully, say to 0.001 ft, but there are errors in the measuring device (or in the procedures), the survey cannot be accurate. Also, a survey may appear to be accurate when rough measurements have been taken. For example, the angles of a traverse (see page 211) may be measured with a compass to only the nearest ¼ degree and yet produce a zero error of closure. On good surveys, precision and accuracy are consistent throughout.

Agreement between two values for the same quantity implies accuracy, as well as precision, but does not assure it. For example, two measurements of a distance made with a tape assumed to be 100.000 ft long but actually 100.020 ft long might give results of 453.270 and 453.272 ft. These values are precise, but they are not accurate since there is an error of approximately 0.090 ft in each measurement. The *apparent* precision obtained would be expressed as 0.002/453.271 = 1/220,000, which is excellent, but the measured distance is incorrect.

2–11. Minimizing errors. All field operations and office computations are governed by the constant fight to minimize or at least reduce the number of errors.

Mistakes can be corrected only if discovered. Comparing several measurements of the same quantity is one of the best ways to isolate mistakes and errors. Making a common-sense estimate and analysis is another.

As an example, assume that five measurements of a line are recorded as follows: 567.91, 576.95, 567.88, 567.90, and 567.93. The second value disagrees with the others, apparently because of a transposition of figures in reading or recording. This mistake can be eradicated by (a) repeating the measurement, (b) casting out the doubtful value, or (c) rectifying the questionable figure.

When a mistake is detected it is usually best to repeat the measurement. If a sufficient number of other measurements of the quantity are available and in agreement, as in the foregoing example, the widely divergent result may be discarded. Serious consideration must be given to the effect on the average before discarding a value. It is seldom safe to change a measurement, even though there appears to be a simple transposition of numbers. Tampering with physical data is always bad practice and will surely cause trouble, even though done infrequently.

Since systematic errors result from known causes, their values can be calculated and proper corrections applied to the measurements, or a procedure can be used which will automatically eliminate the errors. For example, the error due to the sag of a tape supported at the ends only can be computed and subtracted from each measurement. If, however, the tape is supported throughout its length or at short intervals, the sag error is zero or negligible. A leveling instrument which is out of adjustment gives incorrect differences of elevation unless all sights are made the same length to cancel the error of adjustment.

2–12. Scope of probability. At one time or another, everyone has had an experience with games of chance, such as flipping a coin, card games, or dice. In basic mathematic courses, laws of combinations and permutations are introduced. It is shown that things which happen by chance or by accident are governed by mathematical principles which are referred to as probability. These principles are applicable in many sociological and scientific measurements. In Sec. 2–9 it was pointed out that accidental errors exist in surveying measurements. The frequency and magnitude of these accidental errors are governed by the same general principles of probability.

For convenience, the term *error* will be used to mean only accidental error for the remainder of this chapter. Means of computing and correcting for systematic errors will be discussed in later chapters. It will be assumed that all systematic errors have been eliminated before the accidental errors are considered.

2–13. Occurrence of accidental errors. In making practically any type of physical measurement, it is necessary to record values of distances. This is true in reading scales, dials, gages, or any other devices which indicate measurements. It is characteristic of a measuring device that ***it cannot be read exactly.*** This is one of the reasons why there is no such thing as an exact measurement, although of course an exact *count* of the number of bolts in a box can be obtained.

One of the scales most frequently read by many engineers, the slide rule, will serve to illustrate a point. Some readings on a slide rule can be estimated to four figures, some to three, and others to only two figures. But no matter how precisely the indicator is set, or how accurately a reading is made, it is never possible (theoretically) to get an exact setting or reading.

To develop the principle of how accidental errors occur, assume that a measurement of 10.46 is made on a scale on which a reading can be estimated to 0.01 and is correct to ± 0.05. In this case the true value of the measurement is between 10.41 and 10.51; and to the nearest hundredth, it may be 10.41, 10.42, 10.43, 10.44, 10.45, 10.46, 10.47, 10.48, 10.49, 10.50, or 10.51. Thus, there are eleven possible values for the correct answer. For the purpose of this discussion it can be assumed that all these values have the same possibility of being correct. The probability of any one answer being correct is therefore 1/11 or 0.0909.

Let it be supposed that two adjacent measurements are made with this scale, each measurement having this same possible error. The correct value of the sum of the two measurements may be the sum of any pair of the eleven possibilities for each separate measurement, all having an equal chance of being correct. From mathematics, if one event can happen n ways and another event can happen r ways, the two events together can happen nr ways. For the assumed conditions, there are $(11)(11) = 121$ possibilities. The difference between the sum of the measurements and the true value will be between -0.10 and $+0.10$. Only one pair of possible values can give a difference of -0.10; that is the pair for which the difference in each measurement is -0.05. An error of -0.09 can be obtained in two ways; there may be a difference of -0.05 in the first reading and a difference of -0.04 in the second, or a difference of -0.04 in the first reading and a difference of -0.05 in the second. This analysis can be continued to obtain the results shown in Table 2–1.

If three adjacent measurements are taken in the same manner, then the range of error would be from —0.15 to +0.15. The possibility of a difference of —0.15 is one; i.e., all three measurements

TABLE 2–1
PROBABILITY FOR TWO MEASUREMENTS

Value of Error	Number of Possibilities	Probability	Probability, as Decimal
–0.10	1	1/121	0.0083
–0.09	2	2/121	0.0165
–0.08	3	3/121	0.0248
–0.07	4	4/121	0.0331
–0.06	5	5/121	0.0413
–0.05	6	6/121	0.0496
–0.04	7	7/121	0.0579
–0.03	8	8/121	0.0661
–0.02	9	9/121	0.0744
–0.01	10	10/121	0.0826
0.00	11	11/121	0.0909
0.01	10	10/121	0.0826
0.02	9	9/121	0.0744
0.03	8	8/121	0.0661
0.04	7	7/121	0.0579
0.05	6	6/121	0.0496
0.06	5	5/121	0.0413
0.07	4	4/121	0.0331
0.08	3	3/121	0.0248
0.09	2	2/121	0.0165
0.10	1	1/121	0.0083
			Total = 1.000

would have to be off by —0.05. Also, by the principles of mathematics, the total number of chances is $(11)(11)(11) = (11)^3 = 1331$. The complete development is listed in Table 2–2.

The values in Table 2–2 are plotted in Fig. 2–2 and show the typical bell-shaped probability curve. Each rectangle plotted has a width corresponding to an error of 0.01 and the height of each represents the probability of the error corresponding to the value at the center of its width. Therefore the area under the entire curve represents the sum of all the probabilities in the last column in Table 2–2, or *one*. It can also be stated that the total area between *any* two ordinates is equal to the sum of the partial areas or probabilities between the same two ordinates.

TABLE 2–2
PROBABILITY FOR THREE MEASUREMENTS

Value of error	Number of Possibilities	Probability
±0.15	1 each, or 2	0.0008
±0.14	3 each, or 6	0.0023
±0.13	6 each, or 12	0.0045
±0.12	10 each, or 20	0.0075
±0.11	15 each, or 30	0.0113
±0.10	21 each, or 42	0.0158
±0.09	28 each, or 56	0.0210
±0.08	36 each, or 72	0.0270
±0.07	45 each, or 90	0.0338
±0.06	55 each, or 110	0.0413
±0.05	66 each, or 132	0.0496
±0.04	75 each, or 150	0.0563
±0.03	82 each, or 164	0.0616
±0.02	87 each, or 174	0.0654
±0.01	90 each, or 180	0.0676
0.00	91 91	0.0684
	Sum: 1331	1.0000

If the same measurements had been taken with a smaller possible error (i.e., if a more precise measurement had been made), the probability curve would be similar to Fig. 2–3. This curve shows a greater percentage of measurements which have small errors and fewer measurements with large errors. For measurements taken less precisely, the opposite effect is produced and the probability

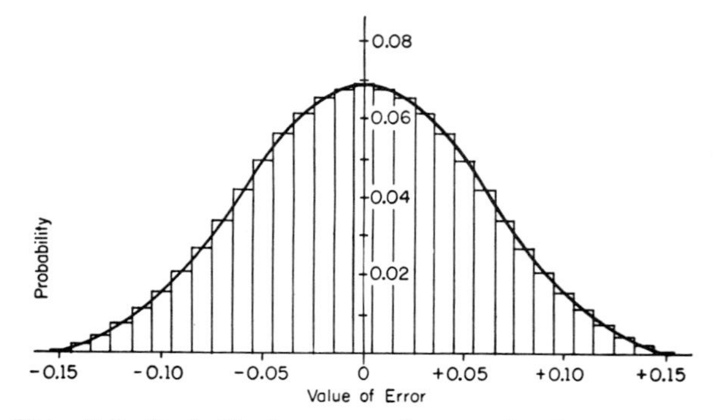

FIG. 2–2. Probable frequency of errors for three measurements, with maximum error of ±0.05 unit in each.

curve is shown in Fig. 2–4, where a greater percentage of the measurements have large errors. In all three cases, the curve has maintained its bell-shaped characteristic.

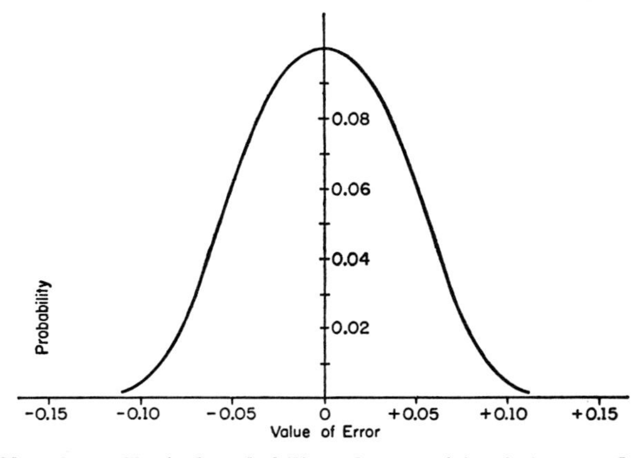

FIG. 2–3. Typical probability when precision is increased.

2–14. General laws of probability. From an analysis of the data in the preceding section and the curves in Figs. 2–2, 2–3, and 2–4, these general laws of probability can be stated:

a. Small errors occur more often than large ones; that is, they are more probable.

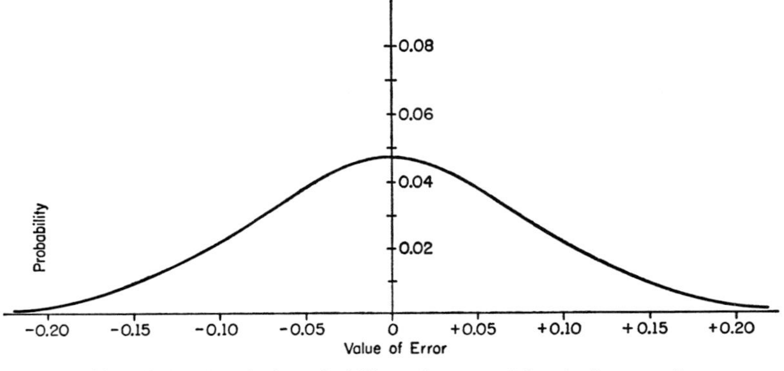

FIG. 2–4. Typical probability when precision is decreased.

b. Large errors occur infrequently and are therefore less probable; unusually large errors may be *mistakes* rather than accidental errors.

c. Positive and negative errors of the same size happen with equal frequency; in other words, they are equally probable.

2–15. The probability equation. In the example discussed in Sec. 2–13, it was assumed that accidental errors occurred in only three readings. Also it was implied that the equipment used was simple and the error was primarily due to lack of precision in estimating the reading. With more elaborate equipment there are many additional sources of accidental errors. A typical probability curve is shown in Fig. 2–5. The equation (mostly intuition and not

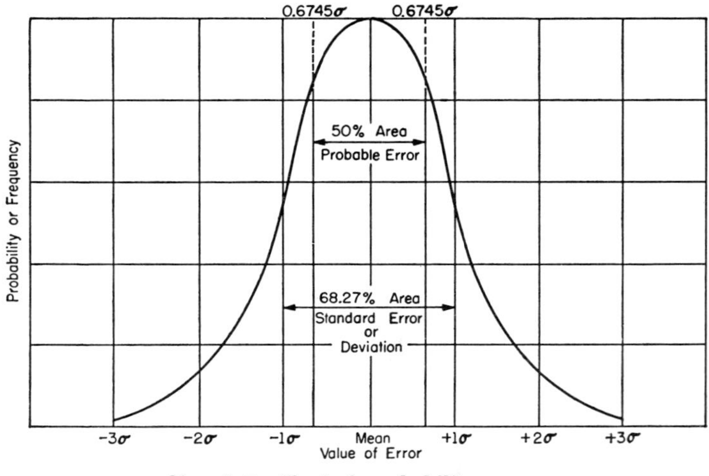

Fig. 2–5. **Typical probability curve.**

a rigid mathematical formula) which satisfies such a probability curve is based on two assumptions:

a. The accidental errors are infinite in number.

b. The increment of error of measurements is infinitesimal.

The following is one form of the equation for the probability curve:

$$y = \frac{1}{\sigma\sqrt{2\pi}} e^{-x^2/2\sigma^2} \tag{2–1}$$

where y = probability or frequency of a certain error;
x = error, or deviation from the mean;
e = 2.718;
σ = standard deviation (to be covered in Sec. 2–17).

2–16. The most probable value; the mean. In physical measurements the true value of any measurement is never known. Therefore it is not possible to say what error does exist in any measurement. According to one of the general laws of probability stated in Sec. 2–14, positive and negative errors of a certain magnitude happen with equal frequency. From this law, a basic assumption of probability is that *the most probable value of a group of repeated measurements is the average, or the arithmetic mean.* The detectable error of any particular measurement is the amount by which it deviates from the mean. This error, or deviation from the mean, is known as the *residual.*

It is not to be assumed that the mean is without error. The error of the mean will be discussed in Sec. 2–23.

2–17. The standard deviation or standard error. There are several methods of interpreting and evaluating the most probable value of a group of measurements. Whether the data be sociological, biological, psychological, or physical in nature, the most-used method of evaluation and comparison is based on the *standard deviation.*

The equation for standard deviation is

$$\sigma_s = \sqrt{\frac{\Sigma v^2}{n - 1}} \tag{2–2}$$

where σ_s = standard deviation of an individual observation;
v = residual of an individual observation;
n = number of observations.

In surveying, a deviation is called an error; hence, the term *standard error* will be used in this discussion. Also, it is common practice to omit the symbol $\pm$ from the equation, because it is known that an accidental error may be either plus or minus.

When the mean is an exact value, as it may be for some data other than physical measurements, then n is substituted for the term $n - 1$. This also may be done when n is a large value. Since neither of these two conditions tends to exist in surveying, the term $n - 1$ is used.

Figure 2–6 is a graph showing the percentage of the area of the probability curve corresponding to the range of error between equal positive and negative values. From the curve it can be seen that the area between errors of $+\sigma$ and $-\sigma$ represents 68.27 per cent of the total area under the probability curve, and hence this partial area represents the limits of errors which will occur 68.27 per cent of the time. This relation is shown more clearly on the typical probability curve in Fig. 2–5. In a corresponding manner

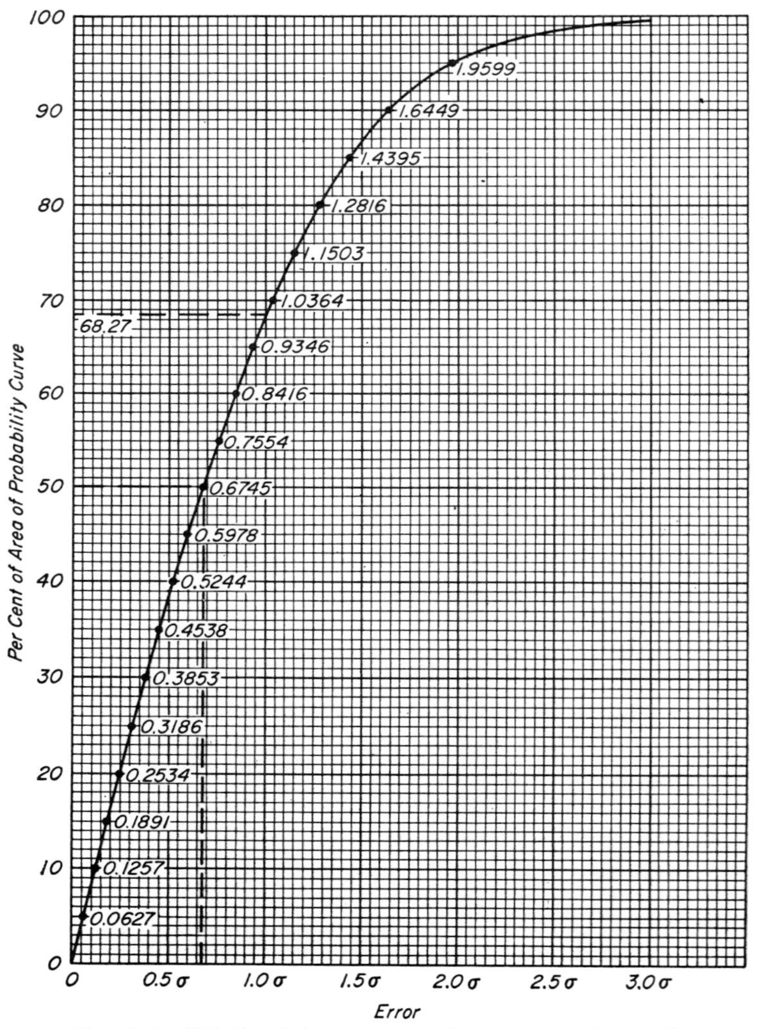

FIG. 2–6. Relation between error and percentage of area of probability curve.

the percentage of error for any proportion of the probability curve can be obtained from Fig. 2–6. Once the standard error has been found, the entire probability curve may be evaluated.

2–18. Interpretation of standard error. It has been shown that the standard error represents the error which will not be exceeded 68.27 per cent of the time; more specifically, it establishes the values of the limits within which the measurements will fall 68.27 per cent of the time. In other words, if a measurement were repeated ten times, it would be expected that approximately seven of the results would fall within the limits established by the standard error and, conversely, about three results would fall anywhere outside of these limits. Another interpretation is that if one additional measurement were made, it would have a 68.27 per cent chance of falling within the limits set by the standard error.

As an example to clarify the definitions and use of the equation for standard deviation, suppose that a line has been measured ten times with the results shown in the first column in illustration 2–1.

Illustration 2–1

STANDARD ERROR

Length ft	Residual v ft	v^2
1000.57	+0.12	0.0144
1000.39	−0.06	0.0036
1000.37	−0.08	0.0064
1000.39	−0.06	0.0036
1000.48	+0.03	0.0009
1000.49	+0.04	0.0016
1000.32	−0.13	0.0169
1000.46	+0.01	0.0001
1000.47	+0.02	0.0004
1000.55	+0.10	0.0100
1000.45 (Average)	$\Sigma - 0.01$	$\Sigma = 0.0579$

Eq. 2–2, $\sigma_s = \pm\sqrt{\dfrac{\Sigma v^2}{n-1}} = \pm\sqrt{\dfrac{0.0579}{9}} = \pm 0.08$ ft

Eq. 2–4, $PE_s = \pm 0.6745\,\sigma_s = \pm 0.6745\,(0.08) = \pm 0.05$ ft

It is assumed that these measurements have already been corrected for all systematic (cumulative) errors before the standard error (σ_s) and the probable error (PE_s) of a single measurement are computed. The following conclusions can be drawn:

a. The most probable length is 1000.45 ft.
b. The standard error of a single measurement is ±0.08 ft.
c. The normal expectation is that 68 per cent of the time the recorded length would lie between 1000.37 ft and 1000.53 ft; i.e., about 7 of the values would be between these limits. It so happens that 7 of them are in this range.

2–19. Probable error. By definition, the "probable error" establishes the limits within which the measurements will fall 50 per cent of the time. In other words a measurement will have the same chance of falling within these limits as it will have of falling any place outside of these limits. Care must be taken not to use the term "probable error" to mean just any probability. The word *probability* is a general term and may be used if the chances are small or large, but probable error can mean only the 50–50 chance. In the past, probable error has been used extensively in discussing accidental errors, but it is being replaced by standard error.

As shown in Fig. 2–6, 50 per cent of the area of the probability curve corresponds to 0.6745 standard error. Thus the equation for probable error is

$$PE = 0.6745\, \sigma_s \qquad (2\text{–}3)$$

or

$$PE = 0.6745 \sqrt{\frac{\Sigma v^2}{n - 1}} \qquad (2\text{–}4)$$

In the typical problem in illustration 2–1, the probable error of a single measurement is ±0.05 ft. Therefore it can be expected that one-half (5) of the values will lie between 1000.40 ft and 1000.50 ft. Actually 4 values do.

2–20. General probability. From the data given in Fig. 2–6, the probability of any error can be determined. The general equation is

$$E_p = C_p\, \sigma \qquad (2\text{–}5)$$

where E_p = the percentage error;
C_p = the numerical factor taken from Fig. 2–6.

For example, the 25 per cent error for the typical problem in illustration 2–1 would be

$$E_{25} = \pm\, 0.3186\, \sigma = \pm\, 0.3186\, (0.08) = \pm\, 0.03 \text{ ft}$$

The measurement should be between 1000.42 ft and 1000.48 ft one time out of four.

By use of the notation just described, the standard error and the probable error can be expressed as follows:

Standard error $= E_{68.27} = 1\sigma$.
Probable error $= E_{50} = 0.6745\sigma$.

The x-axis is an asymptote of the probability curve; so the 100 per cent error cannot be evaluated. In other words, no matter what error is found, a larger error can be expected. For this reason, the 90 per cent error is often used as a maximum value.

An economic factor is also involved. The degree of precision selected is based upon the equipment, time, and money available. It is generally assumed that doubling the accuracy increases the cost four times. Thus it may be desirable (and necessary) to establish a procedure which will produce acceptable results 90 per cent of the time rather than one which will give acceptable results more often.

The 90 per cent error for the problem in illustration 2–1 is

$$E_{90} = \pm 1.6449\ \sigma = \pm 1.6449\ (0.08) = \pm 0.13 \text{ ft}$$

All of the measurements in the illustration lie between 1000.45 ± 0.13, i.e., 1000.32 ft and 1000.58 ft.

2–21. The error of a sum. By analyzing the equation for standard error (Eq. 2–2), it can be observed that *accidental errors are a function of the square root of the sum of the squares of the individual errors.* Correspondingly, the equation for the error of a sum is

$$E_{\text{sum}} = \sqrt{E_a{}^2 + E_b{}^2 + E_c{}^2 + \cdots} \qquad (2\text{–}6)$$

where E represents any specified error, and a, b, and c are the separate measurements.

As an example, assume that a base line (discussed in Sec. 4–2) is measured in three sections, and that the probable errors of the sections equal ±0.012, ±0.028, and ±0.020 ft, respectively. Then the probable error of the total length is

$$E_{\text{sum}} = \pm \sqrt{0.012^2 + 0.028^2 + 0.020^2} = \pm 0.036 \text{ ft}$$

A similar computation applies to the error of any product and, thus, to the error of an area.

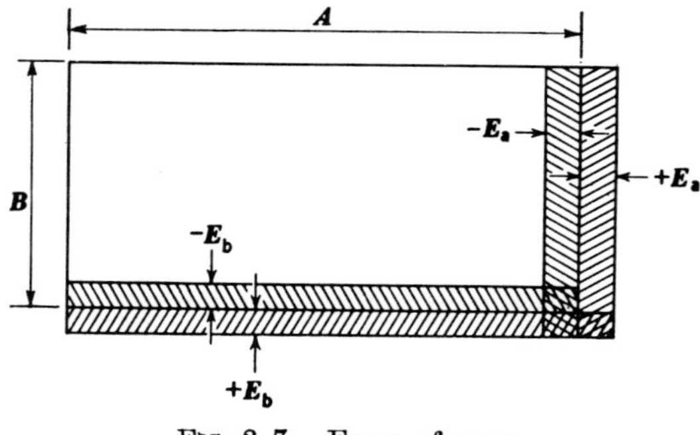

FIG. 2–7. Error of area.

In Fig. 2–7 the error in one direction (side A) is E_a and the error in the other direction (side B) is E_b. The error in the area due to E_a is BE_a, and the error in the area due to E_b is AE_b. Then the equation for the probable error in the area (the error in the product AB) is

$$E_{\text{prod.}} = \sqrt{A^2E_b{}^2 + B^2E_a{}^2} \tag{2–7}$$

As an example, assume a rectangular lot 50.00 ± 0.01 by 100.00 ± 0.02 ft. The probable error of the area is

$$\sqrt{50^2(0.02)^2 + 100^2(0.01)^2} = \pm 1.41 \text{ sq ft}$$

The area as computed might be (50.00) (100.00) = 5,000.0000 sq ft. However, the rule for significant figures (Sec. 2–4) states that there cannot be more significant figures in the answer than in any of the individual factors used. Accordingly, the area should be rounded off to 5000 sq ft. This analysis demonstrates that the area is 5000 ± 1.41 sq ft, that the last digit in 5000 is questionable, and verifies the number of significant figures to be used in the area.

2–22. The error of a series. Sometimes a series of similar quantities, such as the angles of a traverse, may be measured so that each measurement is in error by about the same amount. The total error in all the measured quantities of such a series is called the *error of the series,* designated as E_{series}. If it is assumed that there is the same error E in each measurement of a series and Eq. 2–6 is applied to determine the error of the series, the result is

$$E_{series} = \sqrt{E^2 + E^2 + E^2 + \cdots} = E\sqrt{n} \qquad (2\text{–}8)$$

where E represents the error in any individual measurement, and n represents the number of measurements.

This equation shows that when the same operation is repeated, the accidental errors tend to balance out and the remaining error of the series is proportional to the square root of the number of observations. The equation has extensive use. For instance, it is applied to determine the allowable error of closure for the angles of a traverse, which will be discussed in Chapter 9.

As an example involving the application of Eq. 2–8, assume that any distance of 100 ft can be taped with a probable error of ±0.02 ft if certain techniques are employed; and that it is desired to find the probable error in taping a mile by using these techniques. Since the number of 100-ft lengths in a mile is 52.80,

$$E_{series} = \pm E\sqrt{n} = \pm 0.02\sqrt{52.80} = \pm 0.15 \text{ ft}$$

As another application, assume that a distance of 1000 ft is to be measured with an error of not more than ±0.10 ft; and it is desired to determine how accurately each 100-ft length must be measured to insure that the error will not exceed the permissible limit. Since $E_{series} = E\sqrt{n}$, and since $n = 10$, the allowable error E in 100 ft is

$$E = \frac{E_{series}}{\sqrt{n}} = \frac{0.10}{\sqrt{10}} = \pm 0.03 \text{ ft}$$

Now for comparison suppose it is required to measure a length of 10,000 ft with an error of not more than ±0.10 ft. If 100 ft is again considered the unit length, $n = 100$ and the allowable error E in 100 ft is

$$E = \frac{0.10}{\sqrt{100}} = \pm 0.01 \text{ ft}$$

This analysis shows that a larger number of possibilities provides a greater chance for the errors to cancel out.

2–23. The error of the mean. In Sec. 2–16 it was stated that the most probable value of a group of repeating measurements is the arithmetic mean and that the mean itself is subject to error. Nevertheless, in many types of surveys, the error of the mean is commonly used for comparisons.

By applying Eq. 2–8, which is a special application of Eq. 2–6, it is possible to find the error of the sum of a series of measurements where each measurement has the same error. Since the sum divided by the number of measurements gives the average value of a measurement, the error of the mean, E_m, may be found by the relation

$$E_m = \frac{E_{\text{sum}}}{n}$$

from which

$$E_m = \frac{E_s\sqrt{n}}{n} = \frac{E_s}{\sqrt{n}} \tag{2-9}$$

where E_s = specified error of a single measurement;
E_m = some specified error of the mean;
n = number of observations.

The error of the mean may be applied to any of the other criteria which have been developed. The standard error of the mean is

$$\sigma_m = \frac{\sigma_s}{\sqrt{n}} = \sqrt{\frac{\Sigma v^2}{n(n-1)}} \tag{2-10}$$

The probable error of the mean is

$$\text{PE}_m = \frac{\text{PE}_s}{\sqrt{n}} = 0.6745\sqrt{\frac{\Sigma v^2}{n(n-1)}} \tag{2-11}$$

These equations show that *the error of the mean varies inversely as the square root of the number of repetitions.* For instance, to double the accuracy, that is to reduce the error by one-half, four times as many measurements must be taken.

In illustration 2–1, the standard error of the mean might have been the desired quantity. If so,

$$\sigma_m = \frac{\sigma_s}{\sqrt{n}} = \pm\frac{0.08}{\sqrt{10}} = \pm 0.025 \text{ ft}$$

Also,

$$\text{PE}_m = \pm 0.025\ (0.6745) = \pm 0.02 \text{ ft}$$

and

$$E_{90} = \pm 0.025\ (1.6449) = \pm 0.04 \text{ ft}$$

These values show the limits of several probabilities for the exact length of the line. For instance, it can be said that the exact length of the line has a 50–50 chance of being within ±0.02 of the mean and has the same chance of being outside of these limits.

2–24. Applications. In the sample problems it has been shown that the equations of probability are applied in two ways:

a. To analyze measurements which have already been made, for comparison with other results or with specification requirements.

b. To establish procedures and specifications in order that the required results will be obtained.

The application of the various probability equations must be tempered with judgment and caution. It should be remembered that one of the basic assumptions made was that an infinite number of errors is being considered. Frequently, in surveying, only a few observations are made—often from four to ten. If these observations are typical and representative, then the results will be accurate; if they are not, the results may be misleading.

2–25. Adjustments. In Sec. 2–7 it was emphasized that the true value of any measured quantity is never known. There are types of problems, however, in which the sum of several measurements must equal a fixed amount; for example the sum of the interior angles of a polygon must total $(n - 2)(180°)$. In practice, therefore, if the angles of a polygon are measured, they are adjusted to make them add up to the required value. Correspondingly, distances—either horizontal or vertical—may be adjusted to meet certain requirements. The methods employed will be explained in later chapters where the operations are taken up in detail. In making these adjustments, the principles of probability are most important.

For more-precise surveys, the adjustments are often made on the basis of *least squares.* The principle of least squares, which is developed from the law of probability, requires *adjusting the observed values so as to produce a minimum sum of the squares of the errors (residuals).*

It is evident that some measurements are more accurate than others because of the use of better equipment, improved techniques, and more-favorable field conditions. In making adjustments it may therefore be desirable to assign *weights* to individual observations.

As an example, suppose that four measurements of a distance are recorded as 482.16, 482.17, 482.20, and 482.18, and they are given weights of 1, 2, 2, and 4, respectively, by the survey-party chief. The weighted mean is found by multiplying each measurement by its weight, adding the products, and dividing the total by the sum of the weights. In this case the weighted mean is

$$\frac{482.16 + 482.17(2) + 482.20(2) + 482.18(4)}{1 + 2 + 2 + 4} = 482.18 \text{ ft}$$

As another example, assume that the measured angles of a certain triangle are: $A = 49°51'15''$, wt. 1; $B = 60°32'08''$, wt. 2; and $C = 69°36'33''$, wt. 3. The angles would be adjusted as in the accompanying tabulation.

	Angle	Wt.	Correction	Numerical Corr.	Rounded Corr.	Adjusted Angles
A	49°51′15″	1	$3x$	+2.18″	+2.2″ (or 2″)	49°51′17″
B	60°32′08″	2	$1.5x$	+1.09″	+1.1″ (or 1″)	60°32′09″
C	69°36′33″	3	x	+0.73″	+0.7″ (or 1″)	69°36′34″
Sum	179°59′56″	6	$5.5x$	+4.00″	+4.0″ 4″)	180°00′00″

$$5.5x = 4'' \text{ and } x = +0.73''$$

It must be emphasized that the method of least squares is valid only for observations of equal reliability or for those which can be weighted as to reliability. It is misleading and a waste of time to use the principles of probability and least squares if systematic errors and mistakes have not been eliminated by the employment of proper procedures, equipment, and calculations.

2–26. Methods of computation. Elementary surveying computations require only arithmetic, simple geometry, and trigonometry. Logarithms were universally used some years ago but have limited application now. Surveying offices employ calculating machines and natural functions of angles, for speed and accuracy.

When possible, longhand computations should be arranged to

take advantage of the process which normally produces the fewest mistakes. The preferable order is addition, subtraction, multiplication, and division.

Electronic computers have come into wide use in surveying computations. Where the work load is such that computers can be made available, repetitious operations which were long and tedious, even with the best desk calculators, can now be programmed to obtain results in a matter of seconds or minutes. Programs for common calculations are readily available and one will be illustrated in Chapter 10.

Traverse tables expedite certain computations. They consist of tabulations of the products of distances from 1 to 100 and the sines and cosines of angles from 0° to 90°. Table I in Appendix B duplicates two pages from a traverse table which gives values for every 15 min of angle.

Slide rules must be used with caution in making surveying computations. The number of significant figures required for certain types of work cannot be secured on a slide rule. If, however, the small error is used instead of the value to be corrected in a computation, satisfactory results can often be obtained. For example, suppose that a distance of 837.23 ft is recorded for a measurement made with a 100-ft tape which is actually 99.87 ft long, and that the distance corrected for the error in the tape is required. If the problem is set up as 837.23(99.87)/100, an adequate result cannot be secured on the slide rule. However, there is an error of 0.13 ft per tape length, and the correction amounting to 0.13(8.3723) = 1.09 ft is easily read on the slide rule. The corrected length is 837.23 − 1.09 = 836.14 ft. Slide-rule calculations are always a convenient and rapid method for checking purposes. Special slide rules manufactured for stadia-reduction work can be read to the normally required maximum of three significant figures.

Tables of logarithms of numbers and trigonometric functions, and tables of natural trigonometric functions should have one more place than the number of significant figures desired held in a computation. Consequently, with the elimination of the 6-place logarithm tables included in previous editions, the natural sin, cos, tan and cot tables have been enlarged from five to seven places.

PROBLEMS

A distance XY is measured electronically in meters. In problems 2–1 through 2–3 convert the lengths to feet and decimals of a foot.

2–1. 5,179.28 meters

2–2. 4,827.16 meters.

2–3. 6,542.87 meters.

For problems 2–4 through 2–6 compute the length in feet corresponding to the distance measured with a Gunter's chain.

2–4. 11 ch. 45 lk.

2–5. 20 ch. 06 lk.

2–6. 39 ch. 98 lk.

2–7. How many square feet are there in 7½ acres? How many square chains?

2–8. What is the length in feet and decimals for a distance of 27 feet 8⅜ inches shown on a building blueprint?

What is the area, in acres, of a rectangular parcel of land measured with a Gunter's chain if the recorded sides are as listed in problems 2–9 through 2–12?

2–9. 20 ch. 04 lk. by 19 ch. 80 lk.

2–10. 39 ch. 65 lk. by 40 ch. 08 lk.

2–11. 20.44 ch. by 20.86 lk.

2–12. 40.16 ch. × 40.12 lk.

Compute the area, in acres, of a triangular lot shown on a village plat if the right-angle sides are recorded as shown in problems 2–13 through 2–16.

2–13. 146.41 ft and 35.6 ft.

2–14. 199.98 ft and 1,000.0 ft.

2–15. 1 ch. and 2 ch. 12 lk.

2–16. 2 ch. 01 lk. and 3 ch. 15 lk.

A certain angle is expressed in grads. In problems 2–17 through 2–19 convert it to the degree-minute-second system.

2–17. 83 grads.

2–18. 37 grads.

2–19. 147 grads.

In problems 2–20 through 2–22 express the value in terms of powers of 10 to the indicated number of significant figures.

2–20. 825,000 to 6 sig. figs.

2–21. 72,521 to 4 sig. figs.

2–22. 3,125 to 4 sig. figs.

How many significant figures are there in the numbers of problems 2–23 through 2–29?

2–23. Sum of 725.108, 0.001752, 825, and 4.8.

2–24. Sum of 128.21, 0.0167, 9.28, and 0.36.

2–25. Product of 267.45 and 0.707.

2–26. Quotient of 1,768.23 divided by 2.3.

2–27. Square root of 62,300.
2–28. Square of 4.67.
2–29. Sum of (128.1 + 0.0167 + 9.28 + 0.36) divided by 4.27.

In problems 2–30 and 2–31 convert the adjusted angles of a triangle to radians and show the computational check.

2–30. 47°23′18″, 82°44′06″ and 49°52′36″.
2–31. 36°20′40″, 79°16′30″ and 64°22′50″.
2–32. Explain the difference between systematic and accidental errors.
2–33. List 3 kinds of accidental errors and 3 types of systematic errors that can occur in measuring a line with a steel tape.
2–34. Discuss the difference between precision and accuracy.
2–35. A pair of dice is tossed once. Prepare a table similar to Table 2–1 on probability except that the left-hand column in this case should be labeled "the value of the sum."

A distance XY is measured 10 times and the results in feet listed in problems 2–36 through 2–40. Calculate the most probable length of the line, the standard error of a single measurement, and the standard error of the mean.

2–36. 743.06, 742.92, 742.97, 742.91, 742.93, 743.01, 742.89, 742.98, 743.04 and 743.02.
2–37. Same as problem 2–36 except discard the measurement 742.89.
2–38. Same as problem 2–36 except discard the 742.89 and 742.91 measurements.
2–39. Same as problem 2–36 except include an additional measurement of 743.00.
2–40. Same as problem 2–36 except include two additional measurements of 743.00 and 743.03.

In problems 2–41 through 2–45 determine the limits within which the measured length can be expected to fall 50 per cent of the time, 90 per cent of the time, and the number of values that actually fit within these limits.

2–41. For the data of problem 2–36.
2–42. For the data of problem 2–37.
2–43. For the data of problem 2–38.
2–44. For the data of problem 2–39.
2–45. For the data of problem 2–40.

Two sets of measurements are made with similar equipment, personnel and field conditions with the results shown in problems 2–46 and 2–47. Which set is better? Why? Show computations.

2–46. Set A 427.62 427.60 427.64 427.63 427.61.
Set B 427.65 427.64 427.65 427.65 427.61.
2–47. Set A 724.14 724.17 724.15 724.16 724.18 724.19.
Set B 724.18 724.15 724.16 724.14 724.13 724.18.
2–48. What is the probable error for the data in problem 2–38? Give your interpretation of probable error.

In problems 2–49 and 2–50 calculate the most probable value of the angle, the standard error of a single measurement, and the standard error of the mean for the observations recorded.

2–49. 34°08′20″ 34°09′00″ 34°09′10″ 34°10′00″ 34°08′00″ 34°09′00″

2–50. Same as problem 2–49 with additional measurements of 34°09′40″ and 34°09′30″ included.

Levels are run between points *A* and *B* six times with the results listed in problems 2–51 and 2–52. If the elevation of point *A* is 678.217 ft and *B* is higher than *A*, what is the most probable elevation of *B*?

2–51. 20.200 20.206 20.201 20.197 20.202 20.203 ft.

2–52. 36.998 37.000 37.001 36.999 37.002 37.000 ft.

2–53. A field party is capable of making 100-ft taping measurements with a standard error of ± 0.01 ft. What standard error would be expected if a distance of one mile was taped by this party?

2–54. Same as problem 2–53 except for a standard error of ± 0.02 ft per 100-ft length.

2–55. A line of differential levels was run and required 16 setups of the instrument. If each backsight and foresight rod reading has a standard error of ± 0.005 ft, what is the standard error in the level line?

2–56. Same as problem 2–55 except for 20 setups.

In problems 2–57 and 2–58 a line is measured in 3 sections with the lengths and standard errors listed in feet. What is the standard error of the total length?

2–57. 892.68 ± 0.11 1639.79 ± 0.14 325.41 ± 0.06.

2–58. 741.05 ± 0.12 1346.62 ± 0.16 598.22 ± 0.10.

2–59. A line of differential levels is run from *BMA* to *BMB*, from *BMB* to *BMC*, and then from *BMC* to *BMD*. The results obtained, with their standard errors are, respectively; 28.37 ± 0.08, 83.76 ± 0.06, and 216.04 ± 0.09 ft. What is the standard error in the difference of elevation between *BMA* and *BMD*?

2–60. A distance *AB* was measured 4 times as 522.15, 522.03, 521.98, and 522.13. The measurements were given weights of 1, 2, 3, and 2 respectively by the observer. Calculate the weighted mean for the distance *AB*. What difference results if later judgment revises the weights to 1, 3, 2, 2?

2–61. Determine the weighted mean for the following angles and weights: 38°21′55″, wt. 1; 38°21′59″, wt. 2; 38°22′02″, wt. 3; and 38°21′57″, wt. 1.

2–62. Same as problem 2–61 except the weights are 1, 1, 2, 2 respectively.

What is the area of a rectangular field, and the error in the area, for the recorded values shown in problems 2–63 and 2–64?

2–63. 235.00 ± 0.10 ft by 440.00 ± 0.15 ft.

2–64. 50.14 ± 0.01 ft by 132.65 ± 0.03 ft.

2–65. Specifications for measuring the angles of an 8-sided figure limit the total angular error to one minute. How accurately must each angle be measured?

2–66. Same as problem 2–65 except for a 9-sided figure and total limiting angular error of 30 seconds.

2–67. A square parcel of land containing 5 acres is to be surveyed for a subdivision. What is the maximum error permissible in the measurement of a side if the area must be correct within ± 100 sq ft?

Adjust the following angles of a triangle *ABC* for which the values and respective weights are given in problems 2–68 and 2–69.

2–68. A = 55°32′16″ B = 39°11′46″ C = 85°15′51″ Wts. 1, 2, 4.

2–69. A = 52°31′10″ B = 57°24′36″ C = 70°04′18″ Wts. 3, 2, 1.

2–70. In problem 2–65, if the probable error of reading the transit is ± ½ minute, how many separate readings must be taken at each corner to insure that (a) the average at the corner will be between the required limits 50 per cent of the time, and (b) the mean will fall between the limits 90 per cent of the time?

2–71. If a value of ± 2.5σ covers an entire class of 84 students in surveying (with each grade having a range of one standard deviation), using Fig. 2–5, how many A, B, C, D, and F grades should be given?

chapter **3**

Surveying Field Notes

3–1. General. Surveying field notes are the permanent record of work done in the field. If the notes are incomplete, incorrect, or destroyed, much or all of the time spent in making accurate measurements may be lost. Hence, the notekeeper's job is frequently the most difficult one in the party. A field book containing information gathered over a period of weeks is worth thousands of dollars, assuming that it costs $150 to $200 per day to maintain a party of four men in the field. Every field book should therefore have the name and address of the owner lettered with India ink on the cover and inside.

The data in field notes are normally used by office personnel to make a map or computations. Accordingly, it is essential that notes be intelligible to others without verbal explanations. The Reinhardt system of slope lettering is generally employed, for clarity and speed.

Property surveys are subject to court review under some conditions, in which case the field notes become an important factor in the litigation. Also, because survey notes may be used as references in land transactions for generations, it is necessary to properly index and preserve them. The salable "good will" of a surveyor's business depends largely upon his library of field books. Cash receipts may be kept in an unlocked desk drawer, but field notes are stored in a fireproof safe.

Original notes are those taken while measurements are being made. All other sets are copies and must be so marked. Copied notes may not be accepted in court. They are always subject to suspicion because of the possibility of errors and omissions. The value of a distance or an angle placed in the field book from memory, half an hour after the observation, is definitely unreliable.

Students are tempted to scribble notes on scrap sheets of paper for later transference in neat form to the regular field book. This practice defeats the purpose of a surveying course, which is to provide experience in taking notes under job conditions. On practical work, a surveyor is not likely to spend his own time at night transcribing scribbled notes. Certainly his employer will not pay him for this evidence of incompetence.

Notes should be lettered with a pencil of at least 3-H hardness so that an indentation is made in the paper. Books so prepared will withstand damp weather in the field (or even a soaking) and still be legible, whereas graphite from a soft pencil will leave an undecipherable smudge under such circumstances.

Erasures are not permitted in field books. If a number has been recorded incorrectly, a line is run through it without destroying its legibility, and the correct value is noted above (see Plate A-3, page 520). If an entire page is to be deleted, diagonal lines are drawn from opposite corners and VOID is lettered prominently.

3–2. Requirements of good notes. Five points are considered in appraising a set of field notes:[1]

Accuracy. This is the most important quality in all surveying operations.

Integrity. A single omitted measurement or detail may nullify use of the notes for plotting or computing. If the field party is far from the office, it may be time-consuming and expensive to obtain the missing data. Notes should be checked carefully for completeness *before leaving the survey site,* and never "fudged" to improve closures.

Legibility. Notes can be used only if they are legible. A professional-looking set of notes is likely to be professional in quality of measurements.

Arrangement. Noteforms appropriate to the particular survey contribute to accuracy, integrity, and legibility.

Clarity. Advance planning and proper field procedures are necessary to insure clarity of sketches and tabulations, and to make errors and omissions more evident. Mistakes in drafting and computing are the end results of ambiguous notes.

3–3. Types of field books. Since field books contain valuable

[1] Some material in this chapter has been abstracted from *Engineers Field Notes,* by Roth and Rice, privately published in 1940 and now out of print. By permission of the authors.

data, must take hard wear, and must be permanent in nature, it is poor economy to use anything but the best book for practical work. Various types of field books are available, but bound and loose-leaf books are the most common.

The *bound book*, a standard for many years, has a sewed binding and a hard stiff cover of impregnated canvas, leather, or imitation leather, and contains 80 leaves.

The *bound duplicating book* permits copies of notes to be made through carbon paper. Alternate pages are perforated for easy removal.

The *loose-leaf book* has come into wide use because of many advantages, which include (a) assurance of a flat working surface, (b) simplicity of filing individual project notes, (c) ready transfer of partial sets of notes between field and office, (d) provision for carrying pages of printed tables and diagrams, (e) the possibility of using different rulings in the same book, (f) a saving in sheets (since none are wasted by filing partially filled books), and (g) lower total cost.

The *stapled, sewed, or spiral-bound books* are not suitable for practical work. They may be satisfactory for abbreviated surveying courses having only a few field periods, because of the limited service required and the low cost.

Special column and page rulings provide for particular needs in leveling, transit work, topographic surveying, and cross-sectioning.

New automatic reading and recording systems are being developed in surveying but the notekeeper's job will become more complicated rather than eliminated. In one recording method, numbers are dialed, as on a telephone, giving a visual check and simultaneously punching a tape for direct input to digital computers. Sketches cannot be handled, however.

3-4. Kinds of notes. Four types of notes are kept in practice: (a) sketches, (b) tabulations, (c) descriptions, and (d) combinations of these. The most common type is a combination form, but an experienced recorder selects the version best fitted to the job at hand. Appendix A contains a set of typical noteforms illustrating some of the field problems covered in this text.

In a simple survey, such as one for measuring the distances between hubs on a series of lines, a sketch showing the lengths is sufficient. In measuring the length of a line forward and backward, a tabulation properly arranged in columns is adequate, as on Plate A–1.[2] The proverb about one picture being worth 10,000 words might well have been written for notekeepers.

The location of a reference point may be difficult to identify without a sketch, but often a few lines of description are enough. Bench marks usually are so described, as on Plate A–3.

In notekeeping, this axiom is always pertinent: When in doubt about the need for any information, include it and make a sketch.

3–5. Arrangement of notes. The arrangement of notes depends upon departmental standards and individual preference. Highway departments, mapping agencies, and other organizations engaged in surveying furnish their field men with sample noteforms, similar to those in Appendix A, to aid in preparing uniform and complete records which can be checked quickly.

It is desirable for students to have an expertly designed set of noteforms covering their first field work, to set high standards and save time. Opportunities for them to devise appropriate variations are still present, particularly in topographic work.

The noteforms shown in Appendix A are a composite of several models. They stress the open style, in which some lines or spaces are skipped for clarity. Thus angles measured at a point X are placed *opposite* X in the notes, but distances measured between points X and Y on the ground are recorded on the line *between* points X and Y in the field book.

The left- and right-hand pages are practically always used in pairs and therefore they carry the same number. A complete title should be lettered across the top of the left page and may be extended over the right page. Titles may be abbreviated on succeeding pages for the same survey project. The location and type of work are placed beneath the title on the left or right page. Some surveyors prefer to condense the title on the left page, thereby leaving the top of the right page free for the date, party, weather, and other items. In practice this condensation may be necessary to reserve the right page for sketches and bench-mark descriptions. The arrangements shown in Appendix A have been

[2] Plates A-1 to A-13 are full-size pages of typical field notes and are grouped in Appendix A for convenience in referring to them in the office or in the field.

found to be eminently satisfactory and demonstrate to students the flexibility of practical noteforms.

The left page is ruled in six columns and normally is reserved for tabulations only. Column headings are placed between the first two horizontal lines at the top of the page, and should follow from left to right in the anticipated order of reading and recording.

The upper right-hand corner of the right page should contain four items:

a) *Date, time of day* (A.M. *or* P.M.), *starting and finishing time.* These entries are necessary to document the notes and furnish a time table, as well as to correlate different surveys. Precision, troubles encountered, and other facts may be gleaned from the time required for the survey.

b) *Weather.* Wind velocity, temperature, and other weather conditions, such as rain, snow, sunshine, and fog, have a decided effect upon accuracy in surveying operations. A chainman is unlikely to do the best possible work at a temperature of minus 20 F or with rain pouring down his neck. Hence weather details are important in reviewing field notes.

c) *Party.* The names and initials of members of the party, and their duties, are necessary for documentation and future reference. Jobs may be shown by symbols, such as ⊼ for instrumentman, ϕ for rodman, N for notekeeper, H.C. for head chainman, and R.C. for rear chainman. The party chief is frequently the notekeeper.

d) *Instrument type and number.* The type of instrument used, and its adjustment, affect the accuracy of a survey. Identification of the particular equipment employed aids in isolating errors in some cases.

To permit ready location of desired data, each field book must have a table of contents at the beginning, and this should be kept current daily. In practice, surveyors cross-index their notes on days when field work is impossible.

3–6. Suggestions on recording notes. Observing the points listed here will eliminate some of the common mistakes in notekeeping:

1) Use the Reinhardt system of lettering. Reserve uppercase letters for emphasis.
2) Letter the name and address of the owner on the cover and first inside page.

3) Use a hard pencil—at least 3-H or 4-H.
4) Begin a new day's work on a new page. For property surveys having complicated sketches, this rule may be waived.
5) Always record directly in the field book immediately following a measurement, rather than on a sheet of scrap paper for copying later.
6) Do not erase recorded data. Run a single line through an incorrect value (so that it is still legible), and place the correct value above or below it. Void an entire page by running diagonal lines to the corners of the page.
7) Carry a straightedge for ruling lines, and a small protractor for laying off angles.
8) Run notes down the page except in route surveys, where they progress upward to conform with the sketches.
9) Use sketches instead of tabulations when in doubt.
10) Make sketches to general proportions, rather than exactly to scale or without plan. Letter parallel with or perpendicular to the appropriate feature.
11) Exaggerate details on sketches if clarity is thereby improved.
12) Line up descriptions and sketches with corresponding numerical data. For example, the beginning of a bench-mark description should be placed on the same line as its elevation, as on Plate A–3 in Appendix A.
13) Avoid crowding. If it is helpful to do so, use several right-hand pages of descriptions and sketches for a single left-hand page of tabulation. Similarly, use any number of pages of tabulation for a single sketch.
14) Use explanatory notes when they are pertinent.
15) Employ conventional symbols and signs, for compactness.
16) Place north at the top, or left side, of all sketches, if possible. A meridian arrow must be shown.
17) Keep tabulated figures inside the column rulings, with the decimal points and digits in line vertically.
18) Make a mental estimate of all measurements before receiving and recording them, to eliminate large errors.
19) Repeat aloud values given for recording. For example, be-

fore recording a distance 124.68, call out "one, two, four, point, six, eight" for verification by the tapeman who submitted the measurement.

20) Place a zero before the decimal point for numbers less than one. For example, record 0.37 instead of .37.
21) Show the precision of measurements by recording significant zeroes. For example, show 3.80 instead of 3.8 if the reading was actually determined to hundredths.
22) Do not write one figure over another, or on column rulings, or on lines of sketches.
23) Make all possible arithmetic checks on the notes, and record them before leaving the field, as shown on Plate A–3.
24) Record essential computations made in the field, in order that they can be checked later.
25) Title, index, and cross-reference each new job or continuation of a previous one.
26) Compute all closures and ratios of error before leaving the field. On large projects where daily assignments are made for several parties, completed work is shown by the satisfactory closures.
27) Sign surname and initials in the lower right-hand corner of the right page on all original notes. Letter COPY in large letters diagonally across the pages of copied notes, but do not obscure or touch the sketch or any figures in so doing.

PROBLEMS

3–1. Why are sketches in field notes not generally drawn to exact scale?

3–2. Prepare a set of field notes from which the data of problem 2–36 might have been taken. See Plate A-1 in the Appendix.

3–3. Similar to problem 3–2 for the data of problem 10–4. See Plate A-8.

3–4. Using assumed data, prepare a highway-accident field sketch covering a truck-car side collision near the center of a 60° intersection of a main highway and a side road.

3–5. Prepare a set of field notes and a sketch covering measurements of the three angles of a triangle which total 180°00′45″, thus giving a closure of 0°00′45″. See Plate A-8.

3–6. Prepare a set of field notes and a sketch covering the layout of an H-shaped ranch-type house 50 ft on the parallel sides with a 40-ft cross line. See Plate A-12.

3–7. Measure the length of each diagonal of a TV screen six times, and tabulate your notes. Compute the most probable length of the diagonals and the probable error of the mean.

3–8. Measure all sides and projections of the house in which you live, and determine its position on the lot. Draw a sketch giving pertinent dimensions.

3–9. Draw an analogy between the signer of a set of field notes and the signer of a check.

3–10. The width of a stream (approximately 120 ft) is to be determined by measuring a "base line" *AB*, which is roughly parallel with one shore line, and the angles at *A* and *B* to a point *C* near the opposite shore line. The angle at *C* is also measured. Prepare a set of notes and a sketch, using practical values that are mathematically correct for this situation.

3–11. An east-west vehicular tunnel about 2000 ft long is to be built through a ridge and approximately 100 ft below its top. The slope of the tunnel will be 3.00% (3 ft drop per 100 ft horizontally) downward from east to west. It will be drilled from both ends and also in both directions from a vertical shaft near the center. Sketch your conception of the surveys required to properly locate the tunnel in plan and elevation.

3–12. Differentiate between notes that should run down the page and those which generally run up the page. Illustrate as necessary.

3–13. In general, what information should be included in a good set of field notes?

3–14. A theodolite with a digital readout of angles is used in an isolated area. What advantages and disadvantages might result?

3–15. A resurvey was made of a lot in a 100-year old subdivision but the date was omitted by mistake in the field book. What effect might this have on you as the new purchaser of the lot?

3–16. Describe two problems that might arise from omission in the field book of the first names and middle initials of members of a surveying field party.

3–17. State your idea of a sixth point to be considered in appraising a set of field notes.

3–18. List the types of field notes used in each of Plates A-1 to A-13 in Appendix A.

chapter 4

Linear Measurements

4–1. General. Linear measurement is the basis of all surveying. Even though angles may be read precisely with elaborate equipment, the length of at least one line must be measured to supplement the angles in locating points.

Standard taping procedure is so simple that beginners in professional survey parties are traditionally assigned this work. As a result, the importance of correct techniques may be overlooked. Nevertheless, an energetic head chainman is probably more important than a fast instrumentman in keeping a field party moving.

In plane surveying the distance between two points means the horizontal distance. If the points are at different elevations, the distance is the horizontal length between plumb lines at the points.

Lines may be measured directly by applying a unit of length to them. The unit generally used in plane surveying in the United States and Canada is the foot, decimally divided. In architectural and machine work it is the foot divided into inches and fractions of an inch. The meter is commonly employed in geodetic surveying. Chains, varas, rods, and other units have been, and still are, utilized in some localities and for special purposes.

4–2. Methods of making linear measurements. In surveying, linear measurements are obtained by (a) odometer readings, (b) pacing, (c) taping, (d) tachymetry (stadia), (e) subtense bar, and (f) electronic devices.

Triangulation and other methods of computing distances from known lengths and angles have been developed. In triangulation, as indicated in Fig. 4–1, all of the angles and one or more lines (the base lines) are measured accurately, and the remaining distances are computed. Triangulation systems may consist of chains of triangles, quadrilaterals, and/or central-point figures. The "chain" of two quadrilaterals and a central point figure in Fig.

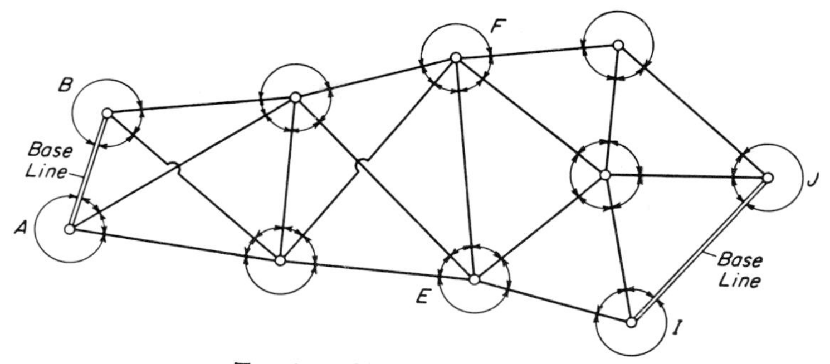

Fig. 4-1. Triangulation figures.

4-1 has measured base lines *AB* and *IJ* at the ends so that the computed length *EF* can be checked (and adjusted) by calculations carried forward from *AB* and backward from *IJ*.

In *trilateration*, all distances are measured and the coordinates of points usually computed. Angles larger than 30° give better results in triangulation, but smaller angles are suitable in trilateration. Electronic distance-measuring instruments now make trilateration a common and sometimes preferred system.

A method of *Airborne Control (ABC)* utilizes a helicopter with an optical plumb bob (hoversight). Electronic distance measurements (and/or angles) are taken to the helicopter as it hovers at a measured height over a ground station.

4-3. Odometer readings. Distances measured by an odometer on a vehicle are suitable for some preliminary surveys in route-location work. They also serve as a rough check on measurements made by other methods. A precision of approximately 1/200 is reasonable. Other types of measuring wheels are available and useful for determining short distances, particularly on curved lines.

4-4. Pacing. Distances obtained by pacing are sufficiently accurate for many purposes in surveying, engineering, geology, agriculture, forestry, and military field sketching. Pacing is also used to detect large errors which may occur in taping or in stadia readings.

The process of pacing consists of counting the number of steps or paces in a required distance. The length of an individual's pace must first be determined. This is best done by walking with natural steps back and forth over a measured level course at least 300 ft long, and averaging the number of steps taken. For short distances the length of each pace is needed, but the number of steps taken per 100 ft is desirable for checking long lines. Plate A–2 in Appendix A gives the notes for a field problem on pacing.

It is possible to adjust one's pace to an even 3 ft, but a person of average height finds such a step tiring if maintained very long. The length of an individual's pace shortens when going uphill, lengthens when going downhill, and changes with age. For long distances, a pocket instrument called a *pedometer* can be carried to register the number of paces.

Some surveyors prefer to count *strides*. A stride is the distance between the place where one foot is lifted and where it is put down again. A stride equals two paces.

Pacing is one of the most valuable things learned in surveying, since it has practical applications for everybody and requires no equipment. Experienced pacers can measure distances of 100 ft or longer with an accuracy of 1/50 to 1/100, if the terrain is open and reasonably level.

4–5. Taping. The most common method of measuring lengths in surveying is taping. Two types of problems arise: (a) measuring a distance between fixed points, such as two stakes in the ground, and (b) laying out a distance with only the starting mark in place. In either case the procedure consists of applying the tape length a number of times. An accuracy of 1/3000 to 1/10,000 can be obtained with properly controlled conditions.

Taping is performed in six steps: (a) lining in, (b) applying tension, (c) plumbing, (d) marking tape lengths, (e) reading the tape, and (f) recording the distance. The application of these steps in taping on level and sloping ground is detailed in Secs. 4–10 and 4–11.

Various types of equipment have been used in the United States to measure lengths. Early surveyors struggled with braced wooden panels, wood and metal poles (which resulted in the term *pole* as a unit of measurement), and other devices. The 100-ft steel tape and the 50-ft metallic tape are now standard, but other kinds of tapes and chains are important because of their present or past use.

Wires. Before thin flat steel could be produced efficiently, wires were utilized for measuring lengths. They still are practical in special cases, for example, hydrographic surveys.

Gunter's chain. The Gunter's chain was the best measuring device available to surveyors for many years in the United States and is referred to in old field notes and deeds. It is 66 ft long and has 100 *links*, each link equal to 7.92 in. The links are made of heavy wire, have a loop at each end, and are joined together by 3 rings, Fig. 4–2. The outside ends of the handles fastened to the

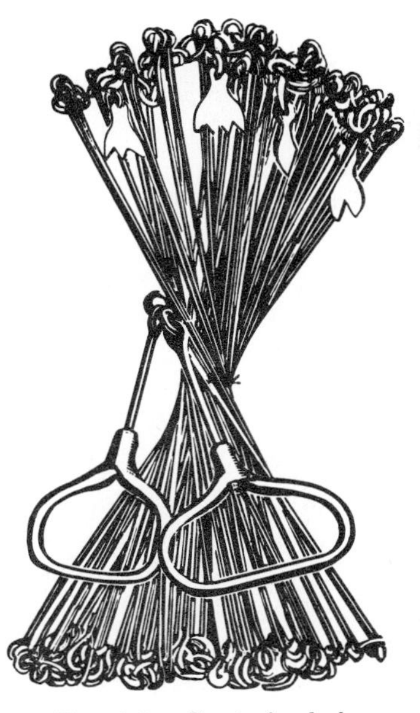

Fig. 4–2. Gunter's chain.

end links are the 0 mark and the 66-ft mark. Successive tags, having 1, 2, 3, or 4 teeth mark every tenth link from each end. The center tag is plain. With 600 or 800 connecting link and ring surfaces subject to frictional wear, hard use elongates the chain, and its length must then be adjusted by means of bolts in the handles.

Distances measured with chains are recorded either in chains and links or in chains and decimals of chains; for example, 7 ch. 94.5 lk. or 7.945 ch. Decimal parts of links are estimated. The 66-ft length of the Gunter's chain was selected because of its relevance to the mile and the relationship of the square chain to the acre. Thus 1 chain = $\frac{1}{80}$ mile, and 10 square chains = 10×66^2 = 43,560 sq ft = 1 acre.

Engineer's chain. The engineer's chain is constructed in the same manner as the Gunter's chain, but it is 100 ft long and each of its 100 links has a length of 1 ft.

Chains are used sparingly if at all today, although a steel tape graduated like a Gunter's chain is manufactured. Nevertheless, the many chain surveys on record oblige the modern practitioner to understand the limits of accuracy possible with this equipment. The term chaining continues to be used interchangeably with taping, even though tapes are employed exclusively.

Surveyor's and engineer's tapes. These are made of steel $\frac{1}{4}$ in. to $\frac{3}{8}$ in. wide and weigh 2 to 3 lb per 100 ft. They may be wound upon a reel, or done up in loops 5 ft long to make a figure 8, and then *thrown* into a circle having a diameter of about $9\frac{1}{2}$ in. For heavy-duty taping, *band chains* (also known as *chain tapes*) are available with cross sections $\frac{1}{8}$ to $\frac{5}{16}$ in. by 0.016 to 0.025 in. Lengths of 50, 100, 150, 200, 300, and 500 ft, and 30 or 50 meters, are standard.

Graduations placed at every foot are marked from 0 to 100. Some tapes have only the last foot at each end subdivided into tenths, or tenths and hundredths, of a foot. Others are graduated in feet, tenths, and hundredths throughout. Still others (*adding* tapes) have an extra graduated foot beyond the zero mark. In all cases, a metal ring or loop at each end allows a handle or leather thong to be attached. The National Bureau of Standards will issue a Report only, *not* a Certificate of Comparison, for tapes having the 0- and 100-ft points at the outer edges of the end loops instead of at line graduations on the band itself, and for tapes with graduations on babbitt-metal bosses bonded to the band but not an integral part of it.

Special-purpose tapes. Tapes having suitable cross sections, lengths, composition, and graduation arrangements are manufactured for special purposes, such as base-line measurements, city engineering work, oil riggers' and gaugers' use, and topographic surveys.

Builder's tapes. Builder's tapes have smaller cross sections and are lighter in weight than surveyor's tapes. Since most building plans prepared by engineers and architects carry dimensions in feet and inches, the builder's tape is graduated in those units.

Invar tapes. Invar tapes are made of a special nickel steel (35% nickel and 65% steel) to reduce length variations caused by differences in temperature. The thermal coefficient of expansion and contraction is only about 1/30 to 1/60 that of an ordinary steel tape. The metal is soft and somewhat unstable. This weakness of invar tapes, along with their cost of perhaps ten times that of ordinary tapes, makes them suitable only for precise geodetic work, and as a standard for comparison with working tapes.

Lovar tapes. A somewhat newer version, the lovar tape, has properties and a cost between those for steel and invar tapes.

Metallic tapes. Metallic tapes are actually made of high-grade linen $\frac{5}{8}$ in. wide with fine copper wires running lengthwise to give additional strength and to prevent excessive elongation. Metallic tapes commonly used are 50, 100, and 200 ft long and come in leather cases. Although not suitable for precise work, metallic tapes are convenient and practical for many purposes.

Glass-fiber tapes. Glass-fiber tapes can be used for the same types of work as metallic tapes.

Accessories. Chaining pins (surveyor's arrows) are used to mark tape lengths. Most chaining pins are made of number 12 steel wire and are 10 to 15 in. long. They are sharply pointed at one end, have a round loop at the other end, and are painted with alternate red and white bands. Sets of eleven pins carried on a steel ring are standard.

Hand level. This simple instrument, Fig. 5–14, is used to keep the tape ends at equal elevations when measuring over rough terrain.

Tension handles. Tension handles facilitate application of the exact standard tension. A complete unit consists of a wire handle, a clip to fit the end ring of the tape, and a spring balance reading up to 30 lb in $\frac{1}{2}$-lb calibrations.

Clamp handles. These are used to apply tension by a positive, quick grip using a scissor-type action on any part of a steel tape without damage to the tape or hands (see Fig. 4–10).

Pocket thermometer. Thermometers for field use are about 5 in. long, graduated from perhaps —30° to +120° F in 1° or 2° divisions, and kept in protective metal cases.

Tape-repair kits. A tape-repair kit contains sleeve splices to be placed over the two parts of a broken tape, hammered down, and fastened with eyelets by a combined hand puncher and riveter.

Poles. Range poles (flags or lining rods) are made of wood, steel, or aluminum, and are about 1 in. thick and 6 to 10 ft long. They are round or hexagonal in cross section and are marked with alternate red and white bands 1 ft long which can be used for rough measurements. A wooden range pole has a metal shoe at the base. The main utility of range poles is in marking the alignment.

Plumb bobs. Plumb bobs for taping should weigh a minimum of 8 oz and have a fine point. At least 6 ft of fish-line cord free of knots is necessary. Plumb-bob points are now standardized to simplify replacement.

Full equipment for a chaining party consists of one 100-ft steel tape; one 50-ft metallic tape; two range poles; 11 chaining pins on a ring; two plumb bobs; one hand level or clinometer; keel (colored crayon); and a field book. This equipment is shown in Fig. 4–3.

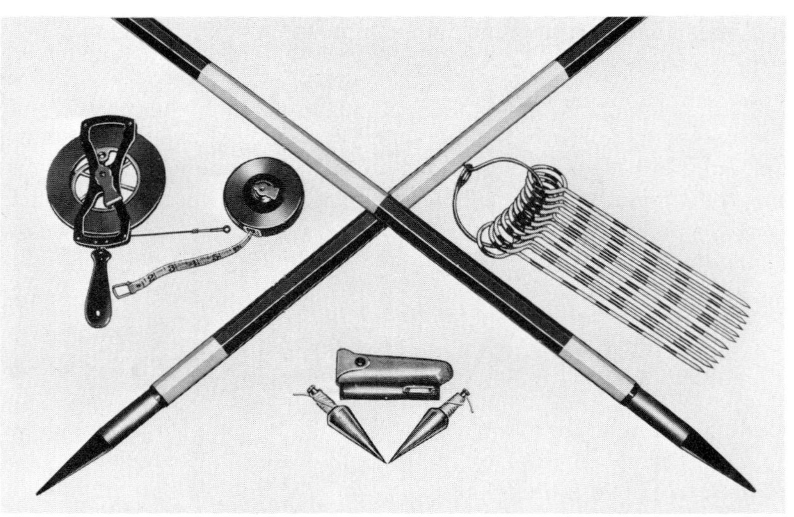

Fig. 4–3. Taping equipment for a field party. (Courtesy of W. & L. E. Gurley.)

Shop equipment. Shop equipment will be discussed in Chapter 25.

4–6. Tachymetry. Tachymetry is a surveying method used to quickly determine the distance, direction, and a relative elevation of a point with respect to the instrument station by a single observation on a rod or other object at the point. The best example of tachymetry in the United States (where the term is less familiar) is the stadia method.

Stadia measurements of horizontal and slope distances are obtained by sighting through a telescope equipped with two or more horizontal cross hairs at a known spacing, and noting the apparent intercepted length between the top and bottom hairs on a rod held vertically. The distance from telescope to rod is found by proportional relationships in similar triangles. A precision of 1/500 is achieved with reasonable care. A detailed explanation of the method is given in Chapter 12.

A range finder with an accuracy of 1/1000 has uses in surveying.

4–7. Subtense bar. Several other methods of determining distances indirectly have been developed in which the angle subtended by a known distance between endmarks on a horizontal rod or bar (such as a subtense bar) is read on a precise transit or theodolite.

The invar subtense bar shown in Fig. 4–4 (along with the appro-

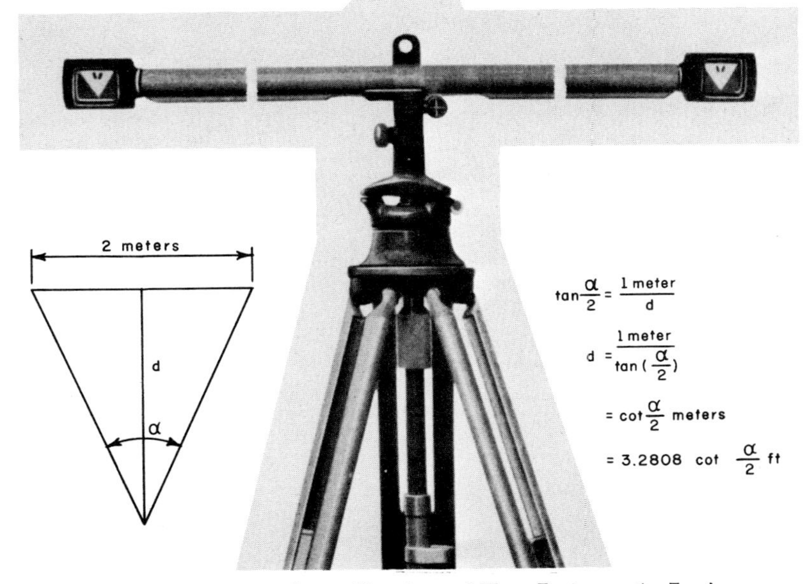

FIG. 4–4. Subtense bar. (Courtesy of Kern Instruments, Inc.)

priate theory) is set on a tripod and aligned perpendicular to the traverse line by means of a sighting device on top of the bar. Fixed targets near the ends of the bar are exactly 2 meters apart. The angle between the targets is measured with a theodolite reading to 1 sec of arc or less, and the horizontal distance is taken from a table, such as Table III in Appendix B. One important advantage of the subtense-bar method is that the ***horizontal distance is always obtained directly, even though an inclined sight is taken because*** *α* ***is measured in a horizontal plane.***

For sights of 500 ft or less, and using a 1-second theodolite, the accuracy should be equal to that for an ordinary tape traverse. Accuracy of 1/1,000 to 1/5,000 is attainable by taking several readings of the angle at both ends of the line for averaging.

4–8. Electronic devices. One of the major advances in surveying following World War II has been the development of electronic devices for measuring distance. The chief advantages of electronic surveying are the extremely high precisions obtained, and the ease and speed of measuring long distances (up to 50 or more miles) across inaccessible terrain. The methods have attained widespread use and acceptance by the U. S. Coast and Geodetic Survey for its precise control surveys.

While some airborne electronic equipment is capable of measuring distances of several hundred miles, emphasis in this text will be placed on the most common so-called "ground" devices, the *tellurometer, electrotape,* and *geodimeter.* The latter is an electro-optical instrument, the others use microwaves. The theory of operation of these instruments is rather complicated and will be treated only briefly herein. Their use in the field is relatively simple, however.

Tellurometer and *Electrotape.* The tellurometer and electrotape systems of electronic distance measurement are similar in theory and operation, and use the wavelengths of microwaves as their "yardstick" or measuring medium. Both systems consist of two portable identical units mounted on tripods. Each unit is a complete assembly and includes all components necessary for taking a measurement—a transmitter, receiver, antenna, measurement circuitry, and built-in communication system. Units are centered over the terminal points of a line to be measured by means of plummets. Figure 4–5 shows a tellurometer unit, Fig. 4–6 an electrotape.

In measuring with either of these microwave systems, one unit

Fig. 4–5. Tellurometer Model MRA–3 (Courtesy Tellurometer Inc.)

operates in the "master" mode, the other in the "remote" mode. The units are identical, and thus interchangeable, i.e., either may be made to operate as master or remote simply by pushing a switch. Readings are taken only in the master mode at both ends of the line. Since each unit contains its own temperature stabilized wavelength calibration, this procedure gives two independent measurements of the distance and hence a valuable check.

By a method called *phase comparison,* microwave systems measure the slope distance between terminal points of the line. A continuous *carrier* radio wave of high frequency (in the range of approximately 10,000 to 10,500 megacycles per second) is frequency modulated with lower frequency *pattern* frequencies (about 7.5 or 10 megacycles per second depending upon the particular model). This mod-

FIG. 4–6. Electrotape Model DM–20 (Courtesy Cubic Corp.)

ulated wave is transmitted from master unit to remote unit. The high frequency carrier wave, as its name implies, serves only as a carrier of the pattern frequencies. It is less susceptible to scattering and also reduces transmitting power requirements thereby keeping equipment light and portable. Pattern frequencies provide the necessary quality to get accurate measurements; are generated within the units themselves; and are precisely controlled by means of thermostatically regulated quartz crystals.

A transmitted signal is received at the remote unit and instantaneously re-transmitted to the master where the phase angle of the returned modulation is compared with the phase angle of the transmitted modulation. The difference in phases or the *phase change* is a measure of the double slope distance which the wave has traveled, and in newer models is converted internally to slope distance and read directly in digital form.

Microwaves traveling double the slope distance undergo a complete 360° phase change for each even multiple of exactly one-half the wavelength separating the instruments. Therefore, if the instruments are spaced precisely an even multiple of the half wavelength, the indicated phase change will be zero. When the two instruments are not exactly an even multiple of the half wavelength apart (the usual case), the fractional part of a wavelength is indicated by the phase change. Ambiguity of the unknown number of full cycles that the wave has undergone is resolved by transmitting lower frequency (longer wavelength) pattern frequencies.

To simplify the electronic circuitry and have portable equipment, the longer wavelength patterns are simulated by varying the modulation frequencies slightly. Phase changes of longer wavelengths are then obtained by comparing the phase changes of the slightly varied pattern frequencies. The tellurometer 7.5 mc/s MRA-3 will be analyzed more thoroughly to illustrate the principles of the microwave phase-shift method of distance determination. This particular device actually generates five pattern frequences—A, B, C, D, and E. Table 4–1 lists the frequencies of these pattern waves; the difference frequencies found by comparing the phase shift of the A pattern with the phase shifts of the B, C, D, E, and A–patterns; and finally the half wavelengths of the corresponding difference frequencies. (The A–pattern is equal in magnitude but opposite in sign from the A pattern).

TABLE 4–1

PATTERN AND DIFFERENCE FREQUENCIES OF THE TELLUROMETER MRA–3

Pattern	Frequency (mc/s)	Difference Combination	Difference Freq. (kc/s)	Half Wave-length (meters)
A	7.492377	A – B	1.498	100,000
B	7.490879	A – C	14.985	10,000
C	7.477392	A – D	149.848	1,000
D	7.342529	A – E	1,498.475	100
E	5.993902	A – (A–)	14,984.754	10*

* The wavelength may be computed from the equation $\lambda = V/f$, where λ is the wavelength in meters, V is the velocity of the wave in meters per second, and f is the frequency of the wave in cycles per second. The velocity of an electromagnetic wave is 299,792.5 km/s in a vacuum. For the A – (A–) difference combination which has a difference frequency of 14,984.754 kc/s, the corresponding half wavelength may be calculated as

$$\tfrac{1}{2}\lambda = \tfrac{1}{2}\left[\frac{299{,}792.5 \times 1{,}000}{14{,}984.754 \times 1{,}000}\right] = 10 \text{ meters}$$

With the tellurometer MRA-3, the phase angle of the A pattern is automatically compared with the phase angles of the A–,B, C, D, and E patterns and corresponding digital readings of distances obtained directly. For simplicity, these readings are referred to as the A, B, C, D, and E readings, and each indicates in decimal form, to three decimal places, the fractional part of the half wavelength. The rightmost digit of each reading therefore represents 1/1,000th of the half wavelength of the respective difference frequencies. The lowest or finest resolution of distance is obtained from the A readings as 1/1,000th of 10 meters or one centimeter so the A readings are called *fine readings*. The E, D, C, and B readings are *coarse readings* and serve only to successively resolve the unknown number of full wavelengths in the previous shorter wavelength difference frequency. Thus coarse readings need be noted to just two decimal places.

An error in the fine readings termed the *zero error* can occur within the instrument, and to eliminate it the A readings are taken both forward and reversed. This process is analogous to taking "circle right" and "circle left" readings when measuring angles with a theodolite. The zero error is of no consequence for coarse readings.

Reflection of microwaves from a ground or water surface can set up a condition designated as *ground swing* which affects the accuracy of readings. Experienced observers are able to detect this condition by examining fine readings obtained from carrier waves transmitted over a range of frequencies. Ground swing occurs in a cyclic manner and may be eliminated by taking a set of fine readings using perhaps a dozen or so different carrier frequencies, each differing from the previous one by some constant increment. The final A reading is best found by plotting fine readings versus carrier frequencies, connecting these plotted points with a best-fit sinusoidal curve, and then selecting the zero axis. Satisfactory results are generally secured, however, by just averaging the results. During the measuring procedure, the operators who may not be in sight of each other communicate by means of the built-in radio telephone.

In Table 4–1 it is seen that from the A−B difference frequency, unambiguous distances up to 100 kilometers can be measured. For those rare measurements greater than 100 kilometers, length to the nearest 100 kilometers is determined roughly from existing small-scale maps.

A distance read directly from the instrument is based upon an assumed average atmospheric condition. Since the velocity of microwaves varies somewhat with atmospheric pressure and humidity, small corrections are required. Operators carry aneroid barometers and wet- and dry-bulb thermometers to measure the ambient meteorological conditions. The final corrected slope distance is then obtained by multiplying the instrument reading and a constant taken from a nomogram to correct for the difference between ambient and assumed average atmospheric conditions. Lastly, the slope distance must be reduced to a horizontal or geodetic length.

To further clarify distance determination procedures, the results of an actual tellurometer survey using the 7.5 mc/s MRA-3 are presented in illustration 4–1 in standard note form. On the right-hand side of the illustration, A forward and reverse (fine) readings taken at various carrier frequencies (*Cav*) are tabulated, and on the left-hand side initial and final sets of coarse readings are listed. The final set is obtained at a different carrier frequency for a rough check on the first set. Note that the coarse readings (B, C, D, and E) are shown only to two decimal places. The three digits in the A block on the left side are A forward readings taken at the same carrier frequency as the coarse readings.

The coarse figure beneath the tabulated coarse readings is the rough distance in meters taken from the left-most digit of each of the B, C, D, and E readings and all three digits of the A reading. A finer resolution of the distance is secured from the average of all fine readings, 4.130, and entered in place of the last three digits of the coarse figure. The final slope distance uncorrected for meteorological conditions is 8,564.130 meters.

To get good results, the measured line should be relatively free of obstacles between instruments so the microwaves can travel unhampered. Stationary obstructions near the middle of the line are permissible if not large enough to cut off a considerable portion of the beam width. Haze, fog, light rain, and darkness have no adverse effect on the accuracy of readings but haze, fog, and rain do somewhat reduce range of the equipment. The beam has sufficient width (a few degrees on each side of center) to enable operators to first locate each other approximately. More precise pointings are then obtained by rotating the instruments to get maximum signal strength as in direction finding.

Competent surveyors and engineers can produce useful work with the tellurometer or electrotape after just a few days of training.

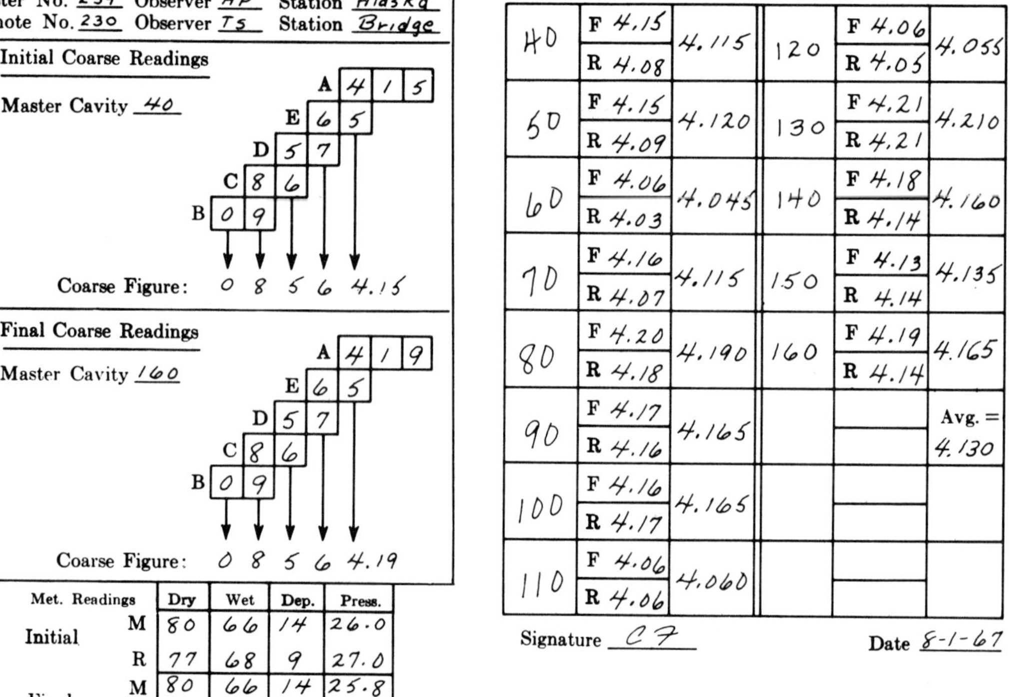

Illustration 4

(Courtesy Duff Surveys, San Francisco, Calif.) SURVEY NOTES FOR TELLUROMETER MRA-3

'Micro-distancer Model MRA3

Field Record

Master No. 239 Observer HP Station Alaska
Remote No. 230 Observer TS Station Bridge

Initial Coarse Readings

Master Cavity 40

A	4	1	5
E	6	5	
D	5	7	
C	8	6	
B	0	9	

Coarse Figure: 0 8 5 6 4.15

Final Coarse Readings

Master Cavity 160

A	4	1	9
E	6	5	
D	5	7	
C	8	6	
B	0	9	

Coarse Figure: 0 8 5 6 4.19

Met. Readings		Dry	Wet	Dep.	Press.
Initial	M	80	66	14	26.0
	R	77	68	9	27.0
Final	M	80	66	14	25.8
	R	80	69	11	27.0

Fine Readings

Cav.		Mean	Cav.		Mean
40	F 4.15	4.115	120	F 4.06	4.055
	R 4.08			R 4.05	
50	F 4.15	4.120	130	F 4.21	4.210
	R 4.09			R 4.21	
60	F 4.06	4.045	140	F 4.18	4.160
	R 4.03			R 4.14	
70	F 4.16	4.115	150	F 4.13	4.135
	R 4.07			R 4.14	
80	F 4.20	4.190	160	F 4.19	4.165
	R 4.18			R 4.14	
90	F 4.17	4.165			Avg. = 4.130
	R 4.16				
100	F 4.16	4.165			
	R 4.17				
110	F 4.06	4.060			
	R 4.06				

Signature CF Date 8-1-67

Measurements of the length of a line can be made in about 20 minutes including the time required to unpack, set up, dismantle, and repack the units. The electrotape model DM-20 is another popular portable instrument comparable with the tellurometer model MRA-3. Each weighs about 35 pounds and is capable of measuring lines from about 100 ft to 50 miles or more. The accuracy of both instruments has been well established at approximately 3/1,000,000 of the distance measured with an additional inherent error of about $\mp \frac{1}{2}$ inch.

Lengths up to nearly 100 miles have been measured with satisfactory results but on such long lines, meteorological conditions can vary considerably and affect the results. High precision is expected on lines of more than one mile but the inherent error of $\mp \frac{1}{2}$ inch, and sometimes the effect of ground swing, become more significant on short lengths. The tellurometer and electrotape have been used successfully in the tropics and in polar regions. Both instruments are powered by a standard 12-volt battery, and permit rapid coverage with multiple-station setups (for example, trilateration with a 3-station system), and by leap-frogging units.

Geodimeter. The geodimeter (*GEO*detic-*DI*stance-*METER*) functions on the energy of the visible spectrum, i.e., light. It has two basic parts, an *electrical unit* and a *prisim reflector,* each standing on a tripod as shown in Figs. 4–7 and 4–8. Centering of the units over a line's terminal points is by plummet.

Using the method of *phase comparison,* an electrical unit measures the double slope distance between terminal points—the distance from electrical unit to reflector and back. The electrical unit, powered by a battery or generator, transmits modulated pulses of light toward the prism reflector which returns the pulses to the electrical unit. The returned signal is converted to electrical current by a photo tube. At the exact instant each pulse is transmitted toward the reflector, the same signal is routed internally to the photo tube via delay circuitry within the unit.

The purpose of the delay circuitry is to advance or delay the internal signal and place it out of phase with the transmitted signal as it returns from the reflector so the two signals will exactly null each other. A null indicator dial provides a visual guide to the operator as he turns a knob to advance or delay the internal signal the required amount. This amount is a measure of the phase change that the transmitted light pulse has undergone in traveling

FIG. 4–7. Model 6 Geodimeter (Courtesy AGA Corp. of America.)

the double slope distance to the reflector and back. Phase change is a function of distance and is converted within the electrical unit directly to digits in centimeters for easy reading.

The frequency of the transmitted light is exactly controlled within the electrical unit. Since the velocity of light is precisely known, the wavelength corresponding to the transmitting frequency is likewise known. In traveling the double slope distance, the light pulse undergoes an integral number of half cycles or 180° phase changes if the one-way distance is an even multiple of the quarter wavelength of the transmitted pulse. The indicated phase change then is zero because the returning transmitted signal will null the internal signal with no advance or delay required. If the one-way slope distance is not an even multiple of the transmitted quarter wavelength (the usual case), advance or delay is required to null the two signals and the one-way distance is an unknown number of quarter wavelengths plus a fractional part of a wavelength as indicated by the advance or delay.

To eliminate ambiguity of an unknown number of quarter wave-

FIG. 4–8. Three-prism housings. (Courtesy of AGA Corp. of America)

lengths, two additional sets of light pulses at slightly different but also precisely known frequencies are transmitted and the phase changes determined as before. From these additional readings the exact distance is calculated. For the geodimeter, as for all electronic distance measuring devices, corrections due to variable atmospheric conditions must be computed and applied to the measured lengths which are then reduced to horizontal or geodetic distances.

The Model 6 is one of several geodimeters available. It is lightweight (35 pounds), well suited for general surveying work, and capable of results having average errors of ∓ 1/500,000 ∓ 0.03 ft. The unit has a daytime range of 2 miles and a night limit of about 10. Advantages of the geodimeter are that one man can operate the equipment, and humidity and ground swing do not affect accuracy.

A combined-function geodimeter is the Model 7T which can measure horizontal and vertical angles by estimation to ∓ 10 seconds, and daylight distances of 50 to 1600 ft with an accuracy of 10 mm. Weighing only 24 lb, it fills the need for an accurate short-length measuring device and one which can be used to lay off distances. Desired values of angles and/or lengths are set on scales and read while simultaneously watching the reflector as it is moved to the correct position, and the required point set. An optical plummet is built in.

The Laser-Geodimeter Model 8 has a receiving system with a filter which shuts out all light with a wavelength other than that of the signal. Consequently, the daylight and night ranges are very nearly the same—from 30 to 40 miles (minimum distance about 15 yards). The uncertainty about corrections for atmospheric conditions is lessened by using a light beam, such as the helium laser, for the measuring "tape." The larger geodimeters are being used to measure first-order baselines, one of the most exacting surveying jobs practiced.

Other systems. Various electronic distance-measuring systems are being used or developed for hydrographic and aerial surveying, and global and satellite navigation. Table 4–2 lists some of them with their ranges and frequencies.

TABLE 4–2

CHARACTERISTICS OF OTHER ELECTRONIC DISTANCE MEASURING EQUIPMENT

Name	Application	Frequency of Transmitted Energy	Range
Autotape	Precision hydrographic surveys	3000 mc/s (10-cm radar)	L.O.S.
Decca	Medium-range navigation and surveying	70–130 kc/s	300 miles
Distomat	Precise distance measurement	1000 mc/s (radio)	12 miles
E.O.S.	Precision base line measurement	60 mc/s (visible light)	L.O.S.
Hi-Fix	Medium-range hydro surveys	1700–2000 kc/s	150 miles
Lambda	Long range surveying	100–200 kc/s	300–500 mi.
Lorac	Medium-range hydro surveys	1700–2000 kc/s	250 miles
Loran-A	Long range navigation	1850–1950 kc/s	400–700 mi.
Loran-B	Medium range precise navigation	1850–1950 kc/s	
Loran-C	Extra long range navigation	100 kc/s	
Map	Short range surveying	radar	30 miles
Moran	Short range surveying	400 mc/s	25 miles
Omega	Extra long range navigation	10–14 kc/s	5000 miles
Raydist	Medium range hydro surveys	1600–3300 kc/s	150 miles
Shiran	Precision short range surveying	3000 mc/s	L.O.S.
Shoran	Short range surveying	200–300 mc/s	40 miles
Transit	Precise long range navigation	150–400 mc/s	signal from satellite

4–9. Care of taping equipment. The following points are pertinent in the care of tapes and range poles:

a. The cross-sectional area of a surveyor's steel tape is $\frac{1}{4}$ in. $\times$ $\frac{1}{50}$ in. $=$ 0.005 sq in. At a permissible stress of 20,000 lb per sq in., a pull of 20,000 $\times$ 0.005 $=$ 100 lb will do no damage. If the tape is kinked, however, a pull of less than 1 lb will break it. Always check to be certain that any loops and kinks are eliminated before tension is applied.

b. If a tape gets wet, it should be wiped first with a dry rag and then with an oily rag.

c. Tapes should either be kept on a reel or thrown, but should not be handled both ways. Double throwing is not recommended.

d. Each tape should have an individual number or tag to identify it.

e. Broken tapes can be mended by riveting and/or a sleeve device, but a mended tape should not be used on important work.

f. Range poles are made with the metal shoe and the point in line with the section above. This alignment may be lost if the pole is used improperly.

4–10. Taping on level ground. The six steps in the process of taping on level ground will now be described in some detail.

Lining in. The line to be measured should be definitely marked at both ends, and at intermediate points where necessary, to ensure unobstructed sight lines. Range poles are ideal for this purpose. The forward tapeman is lined in by the rear tapeman (or by the transitman, for greater accuracy). Directions are given by vocal or hand signals.

Applying tension. The 100-ft end of the tape is held over the first (rear) point by the rear tapeman while the forward tapeman, holding the zero end, is lined in. For accurate results the tape must be straight and the two ends held at the same elevation. A specified tension, generally 10, 12, or 15 lb, is applied. In order to maintain a steady pull, each tapeman wraps the leather thong at the end of the tape around his left hand, keeps his left forearm against his body, and faces at right angles to the line. In this position he is off the line of sight. Also, he need only tilt his body to hold, decrease, or increase the pull. Sustaining a constant tension with outstretched arms is difficult, if not impossible, for a pull of 15 lb or more. Good communication between head and rear tapemen will avoid jerking the tape, save time, and get better results.

Plumbing. Weeds, brush, obstacles, and surface irregularities usually make it undesirable to lay the tape on the ground. Instead, each end point of a measurement is marked by placing the plumb-bob string over the proper tape graduation and securing it with the right thumb. The rear tapeman continues to hold a plumb bob over the fixed point while the forward tapeman marks the length. In measuring a distance shorter than a full tape length, the forward tapeman moves the plumb-bob string to a point on the tape over the ground mark.

Marking tape lengths. When the tape has been lined in properly, tension has been applied, and the rear tapeman is over the point, he calls "stick," the forward tapeman places a pin from his set of ten (one is in the ground at the rear point) exactly opposite the zero mark of the tape, and calls "stuck." If the point is being plumbed and the ground is soft, the plumb bob is released by raising the left thumb. A chaining pin is then carefully set in the hole made by the plumb-bob point. The pin should form a right angle with the tape but approximately a 45° angle with the ground. The point where the pin enters the ground is checked by repeating the measurement.

After checking the measurement, the forward tapeman signals that he is finished, the rear tapeman pulls up the pin beside him, and they move ahead. The forward tapeman drags the tape, paces roughly 100 ft, and stops. Just before the 100-ft end reaches the point which has been set, the rear tapeman calls "tape" to notify the forward tapeman that he has gone 100 ft. The process is repeated until a partial tape length is needed at the end of the line.

When a surveyor is working on pavement, the plumb bob is eased to the surface, and the position of the point is marked by a scratch, a spike, keel, a nail in a bottle cap, or other means.

Reading the tape. There are two common styles of calibrations on surveyor's tapes. *It is necessary to identify the type being used before starting work,* in order to avoid making 1-ft mistakes repeatedly.

The more common type of tape is calibrated from 0 to 100 by full feet in one direction, and has an additional foot beyond the zero end graduated from zero to 1 foot in tenths (and perhaps hundredths) in the other direction, making the complete tape 101 ft long. With a full-foot graduation held by the rear tapeman at the last pin set, as the 87-ft mark in Fig. 4–9(a), the gradua-

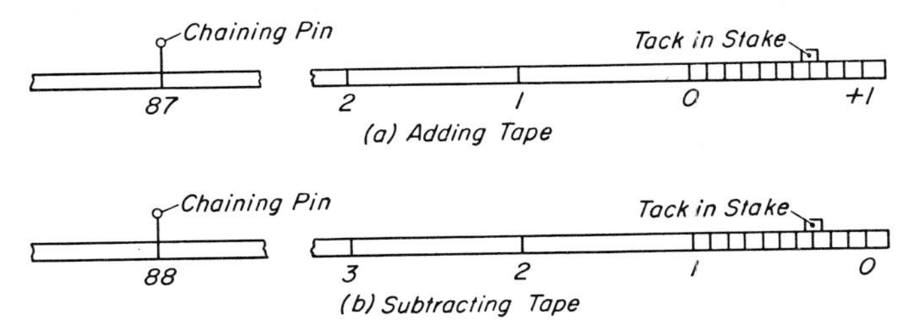

FIG. 4–9. Reading partial tape length.

tions between zero and the end of the tape should straddle the closing point. The head tapeman reads the additional length of 0.68 ft beyond the zero mark. To insure correct recording, the rear man calls "87." The head man repeats and adds his reading, calling "87.68" for the partial tape length. Since the part of a foot has been added, this type of tape is known as an *adding* tape.

The other kind of tape found in practice is calibrated from 0 to 100 by full feet, and the first foot at each end (from 0 to 1 and from 99 to 100) is graduated in tenths (and perhaps hundredths). Thus the complete tape is 100 ft long. With a full-foot graduation held at the last chaining pin set, the graduated section of the tape between the zero mark and the 1-ft mark should straddle the closing point, as indicated in Fig. 4–9(b), where the 88-ft mark is being held on the last chaining pin and the tack marking the end of the line is opposite 0.32 ft read from the zero end. The partial tape length is then 88.00 − 0.32 = 87.68 ft. The quantity 0.32 ft is said to be *cut off,* and this type of tape is called a *subtracting* or *cut* tape. To insure subtraction of a foot from the number at the full-foot graduation used, the following field procedure and calls are recommended: Rear tapeman calls "88"; forward tapeman says "0.32" (point three-two); rear tapeman answers "87.68"; forward tapeman replies "check."

Subtraction of the decimal of a foot is avoided if the forward tapeman reads (counts) 0.68 ft backward from the 1-ft graduation. Calls of 88, 0.68, 87.68, and check are made in this procedure.

The same routine should be used throughout all taping by a party, and the results tested in every possible way. A single failure to subtract one foot in the measurement of a partial tape length with a subtracting tape will destroy the precision of a hundred other measurements. For this reason, the adding tape is more nearly foolproof. The greatest danger arises when changing from one style to the other.

It is customary to have the 100-ft end of the subtracting tape ahead in route surveys, where stationing along the line is continuous. Some surveyors prefer this arrangement in other work also, when setting intermediate points or when measuring partial tape lengths.

Recording the distance. Accurate field work may be cancelled by careless recording. After the partial tape length is obtained at the end of a line, the rear tapeman determines the number of full 100-ft tape lengths by counting the pins he has collected from the original set of eleven. For distances longer than 1000 ft, a notation is made in the field book when the rear tapeman has ten pins and one remains in the ground. The forward tapeman starts out again with ten pins and the process is repeated.

Taping is a skill that can best be taught and learned by field demonstrations and practice.

4-11. Horizontal measurements on uneven ground. In taping on uneven or sloping ground, it is standard practice to hold the tape horizontal and to plumb at one or perhaps both ends. It is difficult to keep the plumb line steady for heights above the chest. Wind exaggerates this problem and may make accurate work impossible.

Where a 100-ft length cannot be held horizontal without plumbing from above shoulder level, shorter distances are measured and accumulated to total a full tape length. This procedure is called *breaking tape.* The technique used in breaking tape is illustrated in Fig. 4–10.

As an example of this operation, assume that when the 100-ft end of the tape is held at the rear point, the forward tapeman can advance only 30 ft without being forced to plumb from above his chest. A pin is therefore set beneath the 70-ft mark, as in Fig. 4–11. The rear tapeman moves ahead to this pin and holds the 70-ft graduation while another pin is set at, say, the 25-ft mark. Then, with the 25-ft graduation over the second pin, the full 100-ft distance is marked at the zero point.

Fig. 4–10. Breaking tape.

Note that the partial tape lengths are added mechanically to make a full 100 ft by holding the proper graduations. No mental arithmetic is required. The rear tapeman returns the pins set at the intermediate points to the forward tapeman, to keep the tally clear on the

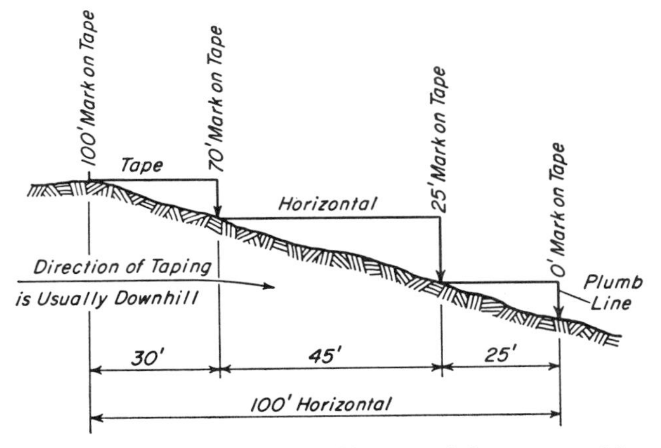

Fig. 4–11. Procedure for breaking tape (when tape used is not in a box or on a reel).

number of full tape lengths established. In all cases the tape is leveled by eye or by hand level, the tapeman keeping in mind the natural tendency to have the downhill end of the tape too low. Practice will develop the knack of holding the tape at right angles to the plumb-bob string.

Taping downhill is preferable to measuring uphill, since in taping downhill the rear point is held steady on a fixed object while the other end is plumbed. In taping uphill, the forward point must be set while the other end is wavering somewhat.

4–12. Slope measurements. In measuring the distance between two points on a steep slope, it may be desirable to tape along the slope and to determine the angle of inclination by means of a clinometer, rather than to break tape every few feet. Long tapes, 200 to 500 ft in length, are advantageous for measuring along slopes (as well as across rivers and ravines).

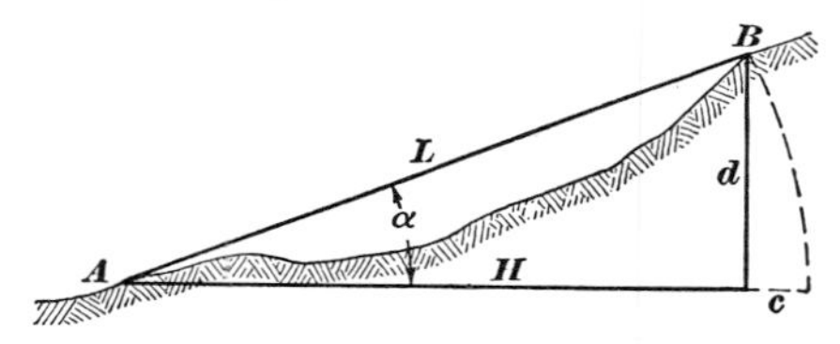

FIG. 4–12. Slope measurement.

In Fig. 4–12 the horizontal distance between points A and B can be computed from the relation

$$H = L \cos \alpha \tag{4–1}$$

where H = horizontal distance between the points;
L = slope distance between the points;
α = vertical angle from the horizontal.

The horizontal projection of the slope distance can also be computed by subtracting a correction, c, from the slope distance. This correction is obtained from the equation

$$c = L(1 - \cos \alpha) = L \text{ vers } \alpha \tag{4–2}$$

On base-line work, the difference in elevation, d, between the ends of the tape is found by leveling, and the horizontal projection is computed. Thus in Fig. 4–12

$$c = L - H$$

$$d^2 = L^2 - H^2 = (L - H)(L + H) = c(L + H)$$

and
$$c = \frac{d^2}{L + H} \tag{4-3}$$

or
$$c = \frac{d^2}{2L} \quad \text{[approximately]} \tag{4-4}$$

Formula 4–4 is preferred for making corrections since it is readily solved longhand, or by slide rule with sufficient accuracy. Rearranging the formula permits computation of d or L as the unknown instead of c.

The error in using the approximate formula for a 100-ft length grows with increasing slope but for inclinations up to 10 per cent the answer is correct to the nearest 0.001 ft. More precise results are obtained for slopes steeper than 10 per cent by using the partial series formula

$$c = \frac{d^2}{2L} + \frac{d^4}{8L^3} \tag{4-5}$$

4–13. Stationing. In route surveying, stationing is carried along continuously from a starting point designated as station 0 + 00. The term *station* is applied to each 100-ft length, where a stake is normally set, and also to any point whose position is given by its total distance from the beginning hub. Thus station 7 + 84.9 is a unique point 784.9 ft from the starting mark, this distance being measured along the survey line. The partial length beyond a *full* station, in this example 84.9 ft, is termed a *plus*.

Taping in stations with a subtracting tape is done most conveniently by carrying the 100-ft end ahead. Since stakes are driven at every change in direction (*angle point*) of a route survey as well as at each full station, it is necessary to follow each plus station with a full one. A special procedure for *determining the plus* at a stake and for *getting off a plus* with the least chance of making an arithmetic error will now be described.

To determine the plus station of an angle point, for example at station 7 + 84.9, the 100-ft end of the tape is pulled beyond the angle point and the forward tapeman then walks back and holds a full-foot graduation on the stake (in this case the 85-ft graduation). Meanwhile the rear tapeman reads the number of tenths of a foot from the 1-ft mark (in this illustration 0.9 ft).

To get off the plus and establish the next full station, 8 + 00, the rear tapeman holds the 84-ft graduation at the plus station, and the forward tapeman sets a pin 0.9 ft back from the 100-ft mark. Note that all subtractions are eliminated by holding the foot mark corresponding to the plus station at that station, and by reading the decimal part of the plus from the 100-ft graduation. This method, like others to be discussed, exemplifies the advantage of systematizing field procedures to reduce the possibility of errors.

4–14. Sources of error in taping. There are three sources of error in taping:

a. *Instrumental errors.* The tape may be different in length from its nominal length because of a defect in its manufacture or repair, or as a result of kinks.
b. *Natural errors.* The horizontal distance between the end graduations of a tape varies because of the effects of temperature, wind, and the weight of the tape itself.
c. *Personal errors.* The tapemen may be careless in setting pins, reading the tape, and/or manipulating the equipment.

Errors in taping may be classified another way, under the following nine headings:

1. Incorrect length of tape.
2. Temperature other than the standard 68 F.
3. Pull (tension) not consistent.
4. Sag due to weight and wind.
5. Poor alignment.
6. Tape not horizontal.
7. Improper plumbing.
8. Faulty marking.
9. Incorrect reading or interpolation.

Some of the nine classifications produce systematic errors; others, accidental errors.

4–15. Incorrect length of tape. Incorrect length of tape always increases, or always decreases, individual measurements and is the most important of any of the errors. It is a systematic error.

Tape manufacturers do not guarantee steel tapes to be exactly their nominal length, for example 100.00 ft. The true length of a tape is obtained by comparing it with a standard tape or distance. The National Bureau of Standards in Washington, D.C., will make such a comparison for a nominal fee and will certify the exact distance between the end graduations under given conditions of temperature, tension, and manner of support.

A 100-ft steel tape usually is standardized for each of two sets of conditions: for example, 68 F, a 10-lb pull, and the tape fully supported throughout; and 68 F, a 12-lb pull, and the tape supported at the ends only. Schools and surveying offices normally have at least one standardized tape which is used only to check other tapes subjected to wear.

A taping error due to incorrect length of tape occurs each time the tape is laid down. If the true length of a tape is known by standardization, and is not exactly equal to the nominal length of say 100.00 ft which has been recorded for each measured tape length, the correction factor to be applied can be determined from the formula

$$C_l = \frac{l - l'}{l'} \tag{4-6}$$

and the true length of a line can be found by the formula

$$T = L + \frac{l - l'}{l'} L = L + C_l L \tag{4-6a}$$

where C_l = correction factor to be applied to the measured (recorded) length to obtain the true length;
l = actual tape length;
l' = nominal tape length;
L = measured (recorded) length of line;
T = true length of line.

If the length of a tape is assumed to be 100.00 ft but when the tape is compared with a standardized tape its length is found to be 100.02 ft, then $C_l = (100.02 - 100.00)/100.00 = +0.0002$. A line which measured 565.75 ft with this tape has a true length of $565.75 + 565.75(0.0002) = 565.86$ ft. For a tape 99.98 ft long and a measured length of 565.75 ft, the true distance is $565.75 + 565.75(-0.0002) = 565.64$ ft.

From a practical standpoint, the effect of any error is to make the tape length incorrect. Note that the true (actual) distance equals the measured distance plus a correction, and the proper algebraic sign is "built-in." This is also true for the corrections discussed in succeeding sections.

4–16. Temperature. Steel tapes are standardized for 68 F in the United States. A temperature greater or less than this value causes a change in length which must be considered.

The coefficient of thermal expansion and contraction of steel is approximately 0.0000065 per unit length per degree F. For any tape

$$C_t = k(T_t - T)L \tag{4–7}$$

where C_t = correction in length of line due to nonstandard temperature;
k = coefficient of thermal expansion and contraction of tape;
T_t = temperature of tape at time of measurement;
T = temperature of tape when it has standard length;
L = measured (recorded) length of line.

The error due to temperature changes may be eliminated in either of two ways:

a) The correction to the measured length of the line may be calculated by formula 4–7. For example, assume that the recorded length of a line measured at 30.5 F with a steel tape that is 100.00 ft long at 68 F is 872.54 ft. The change in the recorded length of the line due to temperature is

$$0.0000065(30.5 - 68)872.54 = -0.21 \text{ ft}$$

The correct length of the line is

$$872.54 - 0.21 = 872.33 \text{ ft}$$

b) An invar tape whose coefficient of thermal expansion and contraction is 0.0000001 or 0.0000002, or a lovar tape with a coefficient of perhaps 0.0000022, can be used. The temperature effect on the length of an invar tape is negligible for most practical work.

Errors due to temperature changes may be either systematic or accidental, but they are more likely to be systematic. On a day when the temperature is always above 68 F, or always below 68 F, they will be systematic. If the temperature is above 68 F during part of the time occupied in measuring a long line, and below 68 F for the remainder of the time, the errors will be partially compensating, or accidental.

Temperature effects are the most difficult to correct for in taping.

The air temperature read from a thermometer may be quite different from the temperature of the tape to which the thermometer is attached. Sunshine, shade, wind, evaporation from a wet tape, and other conditions make the tape temperature uncertain. Field experiments prove that temperatures on the ground or in the grass may be 10 to 25 deg higher or lower than those at shoulder height because of a 6-in. "layer of weather" (microclimate) on top of the ground. Since a 1-deg error in the temperature of the tape is significant in measuring a long line, the importance of large variations is obvious.

Shop measurements made with steel scales and other devices likewise are subject to temperature effects. The precision required in fabricating a large airplane can be lost by this one cause alone.

4–17. Pull. When a steel tape is pulled with a tension greater than the standard amount, it elongates in an elastic manner. The modulus of elasticity of a material is the ratio of the unit stress to the unit elongation, or

$$E = \frac{\text{unit stress}}{\text{elongation per unit length}} = \frac{P/A}{e/L}$$

Also, the correction for pull is

$$C_p = e = (P_1 - P)\frac{L}{AE} \tag{4-8}$$

where C_p = e = total elongation in one tape length due to increase in pull, in feet;
P_1 = pull applied to the tape, in pounds;
P = standard pull for the tape, in pounds;
L = length of tape, in feet;
A = cross-sectional area of the tape, in square inches;
E = modulus of elasticity of steel, in pounds per square inch.

The average value of E is 29,000,000 lb per sq in. for the kind of steel used in tapes.

Errors due to incorrect pull may be either systematic or accidental. The pull applied by an experienced tapeman is sometimes greater than the desired amount, and sometimes less. An inexperienced person, particularly one who has not used a spring balance on a tape, is likely to apply less than the standard tension consistently.

Errors resulting from incorrect tension can be eliminated by three methods:

a) A spring balance can be used to measure and maintain the standard pull.

b) The elongation caused by pull can be calculated. Assume that a steel tape is 100.000 ft long under a pull of 12.0 lb, and has a cross-sectional area of 0.005 sq in. The increase in the length of the tape for a pull of 20.0 lb is

$$\frac{(20 - 12)100}{29{,}000{,}000 \times 0.005} = 0.0055 \text{ ft or } 0.006 \text{ ft}$$

A line measuring 872.54 ft with this tape and tension requires a correction of $8.7254 \times 0.0055 = 0.048$ ft.

c) Sag can be decreased by greater tension; therefore these factors may be adjusted to offset each other. The following formula can be solved by trial to obtain a balanced value:

$$P_t = \frac{0.2W\sqrt{AE}}{\sqrt{P_t - P}} \tag{4-9}$$

where P_t = pull, in pounds, for tension correction to offset sag correction;
P = pull for the standardized tape, in pounds (supported throughout and giving $L = 100.000$);
W = weight of the tape, in pounds;
A = cross-sectional area of the tape, in square inches;
E = modulus of elasticity of steel, in pounds per sq. in.

The cross-sectional area of a steel tape can be obtained by measuring its width and thickness with calipers (1 lb of steel occupies 3.526 cu in.), or by dividing the total weight of the tape by the unit weight of steel (490 lb per cu ft).

The pull required to balance sag for a tape having a cross-sectional area of 0.005 sq in. and a weight of 1.7 lb is found by trial to be 30.3 lb. Thus

$$30.3 = \frac{0.2(1.7)\sqrt{0.005 \times 29{,}000{,}000}}{\sqrt{30.3 - 12.0}}$$

The pull required to make the distance between end graduations exactly 100.00 ft under given conditions of temperature and end support is called the *normal tension.* Normal tension is not commonly used, because it may be too large for convenient application and it changes with temperature variations.

4–18. Sag. A steel tape not supported along its entire length sags in the form of a catenary. A good example of the catenary is the cable of a suspension bridge. Sag shortens the horizontal (chord) distance between end graduations, since the tape length remains the same, Fig. 4–13. Sag can be diminished (by greater tension) but not eliminated, unless the tape is supported throughout.

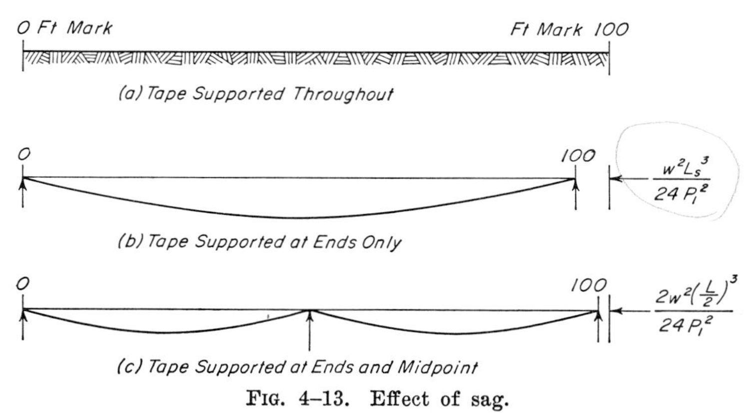

FIG. 4–13. Effect of sag.

The actual sag of a tape (for example, 6 in. below the horizontal) is not important. The reduced chord distance between the end graduations is the critical factor.

For a small deflection v at the center of the length of a tape, the equation of a parabola can be used to investigate the sag effect. Thus

$$L_s - d = \frac{8v^2}{3d}$$

Also, by mechanics,

$$P_1v = \frac{wd^2}{8}$$

Combining these two relations, and assuming that $L_s = d$ in order to simplify the result, the following equation is obtained:

$$C_s = -\frac{L_s}{24}\left(\frac{W}{P_1}\right)^2 = -\frac{W^2L_s}{24P_1^2} = -\frac{w^2L_s^3}{24P_1^2} \qquad (4\text{–}10)$$

where C_s = correction for sag (difference between length of curve and straight line from one support to the next), in feet;

L_s = unsupported length of tape, in feet;
d = chord distance between supports, in feet;
w = weight of tape per foot of length, in pounds;
$W = wL_s$ = total weight of tape between supports, in pounds;
P_1 = pull on tape, in pounds.

The correction C_s is to be added algebraically to L_s to obtain d, and its sign must always be negative. Therefore, errors due to sag alone are systematic. However, the error due to sag depends on the tension applied to the tape, and the error due to pull and sag combined may be either systematic or accidental.

Three methods can be used to eliminate errors caused by sag:

a. Support the tape at short intervals or throughout.
b. Increase the pull to make the tape stretch an amount equal to the sag correction required.
c. Calculate the sag correction for each measurement and apply it to the recorded length.

As an example of method c, assume that a steel tape 100.000 ft long weighs 1.50 lb and is supported at its ends only, as in Fig. 4–13b. If a 12.0-lb pull is applied, the correction for sag is $-100(1.50)^2/24(12.0)^2 = -0.065$ ft. The correction for sag is proportional to the cube of the unsupported length, since $W^2 = w^2L^2_s$ and w and P_1 are constants in a given case. If the 100-ft tape is supported at its center, as well as at its ends (see Fig. 4–13c), there are two unsupported lengths of 50 ft and the correction for sag in each such length would be $-0.065/2^3 = -0.008$ ft. The total correction for the tape length of 100 ft would then be $-0.008 \times 2 = -0.016$ ft.

On slopes where measurements less than 100 ft are used in breaking tape, it may be desirable to vary the pull or compute special sag corrections.

4–19. Alignment. If one end of a tape is off line, or if the tape is snagged on an obstruction, a systematic error is introduced. The size of this type of error can be calculated from the formula

$$C_a = -\frac{d^2}{2L} \tag{4–11}$$

where C_a = correction for offset from alignment;
d = distance the tape is off line;
L = length of tape involved.

If a pin marking the end of a 100-ft length is set 1.4 ft off line, the error in that measurement is $1.4^2/200 = -0.01$ ft. A similar error enters the next tape length when the succeeding pin is set correctly on line.

If the center of a 100-ft tape catches on brush and is 1.0 ft off line, the error produced in the two 50-ft lengths is

$$-\frac{2 \times 1.0^2}{2 \times 50} = -0.02 \text{ ft}$$

Errors resulting from poor alignment are systematic and always make the recorded length longer than the true length. They may be reduced (but never eliminated) by care in setting pins, lining in properly, and keeping the tape straight. Snapping the tape while applying tension will straighten it. A moderate amount of field practice enables a rear tapeman to keep the forward tapeman within much less than a foot of the correct line.

4–20. Tape not horizontal. The error caused by a tape being inclined in the vertical plane is similar to that caused by the tape being out of line in the horizontal plane. Its value can be determined by the formula

$$C_g = -\frac{h^2}{2L} \tag{4–12}$$

where C_g = grade correction;
h = difference in elevation between the ends of the tape;
L = length of tape.

Errors due to the tape not being horizontal are systematic and always make the recorded length longer than the true length. They are reduced by using a hand level to check elevations of the tape ends, or by running differential levels over the taping points. The errors cannot be completely eliminated since the tape is certain to be out of level on some measurements despite the best efforts of a tapeman.

4–21. Plumbing. Practice and steady nerves are necessary to hold a plumb bob still for a period long enough to mark a point or permit an instrument sight. The plumb bob moves around, even in calm weather. On very light slopes, and on smooth surfaces such as pavements, inexperienced tapemen obtain better results by laying the tape on the ground instead of plumbing. Experienced tapemen plumb most measurements.

Errors due to improper plumbing are accidental, since they may make distances either too long or too short. The errors would be systematic, however, when taping directly against or in the direction of a strong wind.

Touching the plumb bob on the ground, or steadying it with one foot, decreases its swing. Practice in plumbing will reduce errors.

4–22. Incorrect marking. Chaining pins should be set perpendicular to the taped line but inclined 45° to the ground. This position permits plumbing to the point where the pin enters the ground without interference from the loop.

Brush, stones, and roots deflect a chaining pin and may increase the effect of incorrect marking. Errors from these sources tend to be accidental and are kept small by carefully locating a point and then checking it.

4–23. Interpolation. The process of reading to hundredths on tapes graduated only to tenths is called interpolation. This process is readily learned and can be applied in many branches of engineering. Some individuals tend to interpolate high; others tend to interpolate low. Also, some people are inclined to favor certain digits; for example, they read 3 rather than 2 or 4, or perhaps 7 instead of 6 or 8.

Errors due to interpolation are accidental over the length of a line. They can be reduced by care in reading; by using a small scale to determine the last figure; and by correcting any disposition toward particular values. Tabulating the number of times each digit from 0 through 9 is interpolated in work covering a period of several days will expose any predilection for a few numerals.

4–24. Summary of effects of taping errors. An error of 0.01 ft is significant in many surveying measurements. Table 4–3 lists the nine sources of errors, classifies them as instrumental, natural or personal, and systematic or accidental, and gives the departure from normal that will produce an error of 0.01 ft in a 100-ft length. The summary verifies practical experience that recorded lengths of lines are more often too long than too short.

TABLE 4–3

TYPES OF ERRORS

Type and Class of Error	Systematic or Accidental	Departure from Normal to Produce 0.01-Foot Error for a 100-Foot Tape
Tape length I	Sys.	0.01 ft
Temperature .... N	Sys. or Acc.	15 F
Pull .... P	Sys. or Acc.	15 lb
Sag .... N, P	Sys.	$7\frac{3}{8}$ in. at center for 100-ft tape standardized by support throughout
Alignment .... P	Sys.	1.4 ft at one end of 100-ft tape or $8\frac{1}{2}$ in. at midpoint
Tape not level.... P	Sys.	1.4 ft
Plumbing .... P	Acc.	0.01 ft
Marking .... P	Acc.	0.01 ft
Interpolation ..... P	Acc.	0.01 ft

The accepted method of reducing errors on precise work is to make several measurements of the same line with various tapes, at different times of the day, and in opposite directions.

4–25. Mistakes or blunders. Careless manipulation of equipment results in personal errors so large that they are classified as mistakes. Examples are:

a. Reading the tape incorrectly.
b. Miscounting the number of full tape lengths.
c. Using the end mark as zero on an adding tape.
d. Transposing figures, or recording a distance improperly.
e. Failing to subtract one foot in making a measurement covering a partial tape length when using a subtracting tape (which does not have the extra graduated foot).

Mistakes are not compensating in nature. They are reduced or eliminated by standard field procedures, and by measuring lines in both directions.

4–26. Tape problems. All tape problems develop from the fact that a nominal 100-ft tape is longer or shorter than 100.00 ft because of manufacture, temperature changes, tension applied, or for some other reason. There are four versions of the problem: A line can be *measured* between two fixed points, or a distance *laid off* from one fixed point, with a tape which is either too long or too short. The solution of a particular problem is always simplified and verified by drawing a sketch.

Assume that the fixed distance *AB* in Fig. 4–14 is measured with a tape that is later found to be 100.03 ft long. Then (the conditions are greatly exaggerated) the first tape length would extend to point *1*; the next, to point *2*; and the third, to point *3*. Since the distance remaining from *3* to *B* is less than the correct distance from the 300-ft mark to *B*, the *recorded* length *AB* is too small and must be increased by a correction. If the tape had been too short, the *recorded* distance would be too large, and a correction would have to be subtracted.

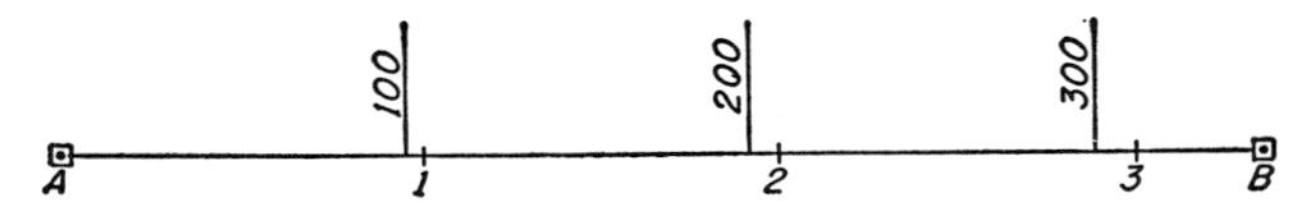

FIG. 4–14. Taping between fixed points, tape too long.

In laying out a given distance from one fixed point, the reverse is true. The correction must be subtracted from the desired length for tapes longer than the nominal value, and must be added for tapes shorter than the assumed length.

4–27. Laying out a right angle with a tape. Many problems arising in the field can be solved by taping. For example, a right angle is laid out readily by the 3–4–5 method. In Fig. 4–15a, to erect a perpendicular to *AD* at *A*, measure 30 ft along *AD* and set point *B*. Then with the zero graduation of the tape at *B* and the 100-ft mark at *A*, form a loop in the tape by bringing the 50- and 60-ft graduations together and pull each part of the tape taut to locate point *C*. One

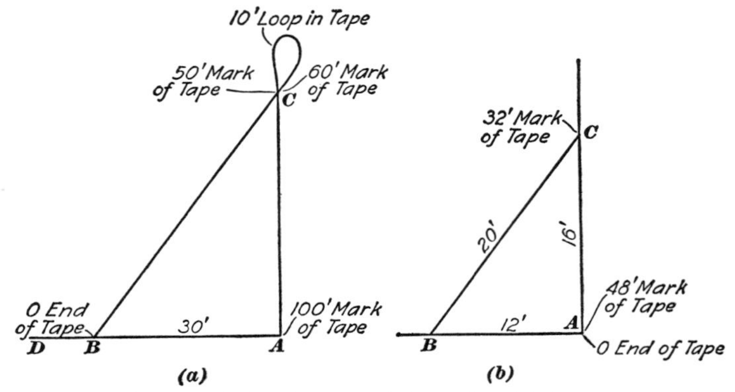

FIG. 4–15. Laying out right angle with tape.

person can make the layout alone by tying the tape thongs to stakes beyond A and B.

If a 50-ft metallic tape is used, a possible procedure is indicated in Fig. 4–15b. The zero mark can be held at A, the 12-ft mark at B, the 32-ft mark at C, and the 48-ft mark at A. Any other distances in the proportions of 3, 4, and 5 may be used.

4–28. Measuring an angle with a tape by the chord method. If all three sides of a triangle are known, the angles can be computed. To find angle A, Fig. 4–16, measure any definite lengths along AM and AN, such as AB and AC. Also measure BC. Then

$$\sin \tfrac{1}{2}A = \sqrt{\frac{(s-b)(s-c)}{bc}} \qquad (4\text{–}13)$$

where a, b, and c are the sides of the triangle ABC, and where $s = \frac{1}{2}(a + b + c)$.

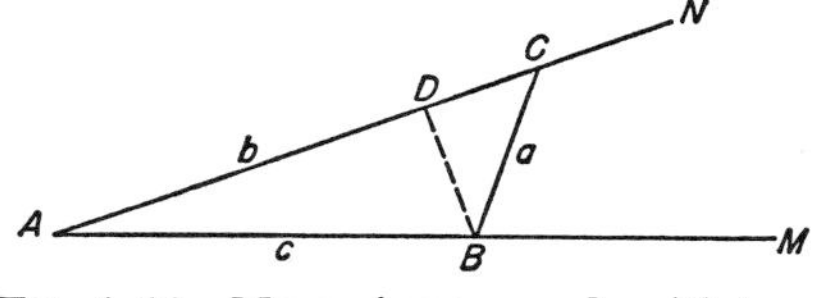

FIG. 4–16. Measuring an angle with tape.

For $b = 30.0$ ft, $c = 25.0$ ft, and $a = 12.5$ ft, angle A is found to be equal to 24°09′.

An isoceles triangle is formed by making AB equal to AC. Then

$$\sin \tfrac{1}{2}A = \frac{a}{2c} \qquad (4\text{–}14)$$

Selecting a value of 50 ft for AB and AC simplifies the arithmetic. Thus if $AB = AC = 50.0$ ft, and BC measures 20.90 ft, then $\sin \frac{1}{2}A = 0.2090$ and angle $A = 24°\ 08'$.

Angle A can also be found from the cosine law

$$a^2 = b^2 + c^2 - 2bc \cos A \text{ and } \cos A = \frac{b^2 + c^2 - a^2}{2bc} \qquad (4\text{–}15)$$

The arithmetic is simplified if b is made equal to c. Selecting 100 ft for b and c, formula 4–15 reduces to $\cos A = \dfrac{20{,}000 - a^2}{20{,}000}$. If a measures 41.80 ft, $\cos A = 0.9126$ and $A = 24°08'$.

4–29. Measuring an angle with a tape by the tangent method. If AD and a perpendicular BD are measured (Fig. 4–16), tan A = BD/AD. By making AD equal to 50 or 100 ft, the tangent can be easily computed. To illustrate, if AD = 100.00 ft and BD measures 44.80 ft, then tan A = 0.4480 and angle A = 24° 08′.

4–30. Laying off angles. An angle can be laid off by reversing the tangent method just described. Along the initial side of the angle, a unit distance of 10, 20, 50, or 100 ft is laid off, as AB in Fig. 4–17. A perpendicular BC is erected, and its length is made equal to 100 times the natural tangent of the desired angle A. Points A and C are connected to give the required angle at A. This method is used by draftsmen as well as by surveyors in the field.

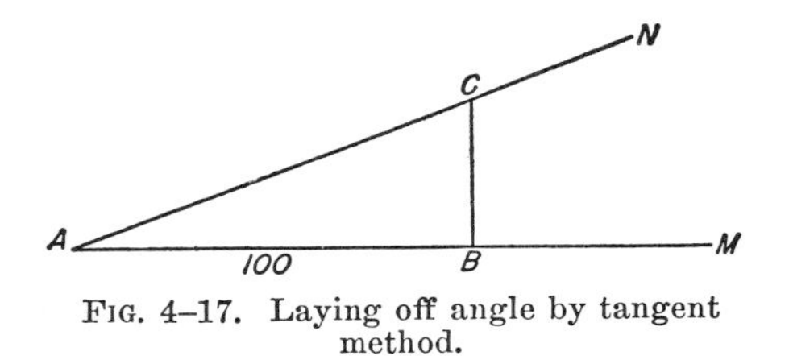

FIG. 4–17. Laying off angle by tangent method.

4–31. Tape survey of a field. A field may be completely surveyed by taping. In fact this was the only method available before instruments for measuring angles were built.

The procedure consists in dividing the area into a series of triangles and measuring the sides of each one. For small areas, one corner of the field is selected as the apex, and the distances to all other corners and the perimeter are measured. In Fig. 4–18, if corner G is chosen as the reference point, the distances GA, AB, BC, CD, DE, EF, and FG along the perimeter and the diagonal distances GB, GC, GD, and GE locate all corners of the field.

For larger areas it is better to establish a central point, such as P in Fig. 4–19, and to measure the perimeter and all lines radiating from P to the corners. The field can be plotted and the area then calculated from these data. The central-point method may appear to require more work, but the short lengths of the interior distances compensate for their greater number. Also, all corners are more likely to be visible from a selected point within the field.

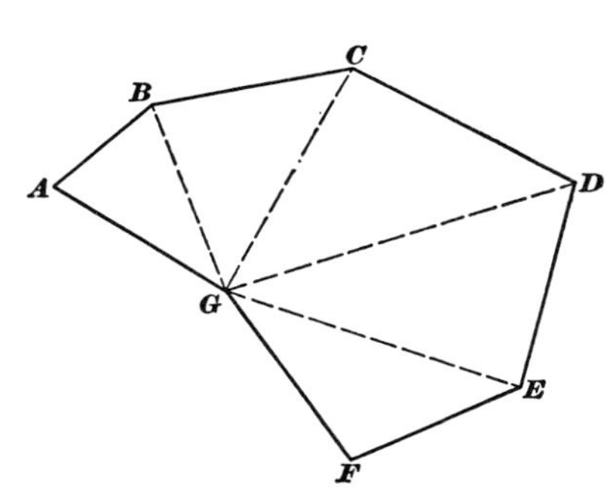

FIG. 4–18. Survey of field by tape.

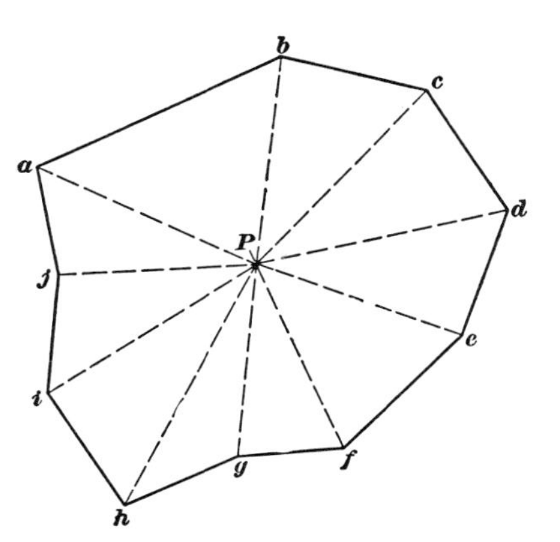

FIG. 4–19. Survey of field using central point.

4–32. Solution of a taping problem. The reduction of a typical base-line measurement will illustrate the method of making tape corrections. Data taken in the field are shown in the first four columns of Table 4–4. The tape ends were supported on *chaining bucks* (tripods).

TABLE 4–4

BASE-LINE TAPING ON BUCKS

Section	Measured (Recorded) Distance (ft)	Temperature (deg F)	Difference in Elevation (ft)	Inclination Correction (ft)
A–1	100.000	58	1.26	0.0079
1–2	100.000	58	0.98	0.0048
2–3	100.000	59	0.60	0.0018
3–4	100.348	59	0.81	0.0033
4–5	100.000	59	1.22	0.0074
5–6	100.000	60	2.06	0.0212
6–7	100.000	60	2.54	0.0323
7–8	100.000	60	2.68	0.0359
8–B	70.216	61	1.87	0.0249
Sum or avg.	870.564	59.3	....	0.1395

The properties of the tape are as follows:

Tape NBS 5269 (the number is the NBS record designation)

Standardization by the National Bureau of Standards when supported on a horizontal flat surface:

Tension, 15 lb Interval 0 mark to 100 mark, 100.008 ft

Standardization when supported at ends only:

Tension, 15 lb Interval 0 mark to 100 mark, 99.963 ft

Four corrections will be applied: (a) inclination (slope) correction, (b) temperature correction, (c) standard-length correction for full tape length, and (d) standard-length and sag corrections for partial tape length.

In geodetic work, a fifth correction is made to reduce the measurement to sea level. The formula used is $C_r = -Lh/R$, where C_r is the correction to be subtracted (unless the line is below sea level) from the measured length L at an average elevation h; and R is the radius of the earth = approximately 20,906,000 ft. Reduction to mean sea level makes all lengths throughout the country comparable, regardless of the elevations at which they are measured.

Inclination correction. Individual corrections for each tape measurement are shown in the last column of Table 4–2. Each is computed by formula 4–12, which is $C_g = h^2/2L$, and their total is -0.140 ft. The adjusted length becomes $870.564 - 0.140 = 870.424$ ft.

Temperature correction. Using the average temperature and substituting in formula 4–7, we find that the total correction is

$$C_t = (59.3 - 68) \times 0.0000065 \times 870.424 = -0.049 \text{ ft}$$

Standard-length correction for full tape lengths. The standard distance between the 0 mark and the 100-ft mark (tape supported at the ends only) differs from the nominal length by $100.000 - 99.963 = 0.037$ ft. For eight full tape lengths the correction is $8 \times 0.037 = -0.296$ ft.

Standard-length and sag correction for partial tape length. If the tape is supported throughout, the distance between the 0 mark and the 100-ft mark is 100.008 ft. The interval separating other graduations is assumed to be increased proportionately. The length correction for 70.216 ft is therefore $70.216 \times 0.008/100 = +0.006$ ft.

According to formula 4–10, the correction C_s for sag is proportional to L_s^3 when the weight per foot of tape and the tension are constant, as in this example. The sag correction from the standardization data is 99.963 − 100.008 = −0.045 ft. Then the correction for approximately a 70-ft span is $(70/100)^3 \times -0.045 = -0.015$ ft.

The true length of the base line is 870.564 − 0.140 − 0.049 − 0.296 + 0.006 − 0.015 = 870.564 − 0.494 = 870.070 ft.

Field data and corrections for this base-line computation have been carried out consistently. Ordinary taping precision does not justify working in thousandths of a foot, but the procedure is the same.

It is immaterial whether the inclination correction is made before or after the temperature adjustment. The temperature correction for the partial tape length should be computed separately if the temperature deviates considerably from the average.

PROBLEMS

4–1. List six methods of measuring distances. Give an advantage and a disadvantage of each.

4–2. Illustrate, by means of assumed values of measurements from *A* to *B* and from *B* to *A*, reasonable values of precision for (a) pacing, (b) taping, (c) stadia, (d) subtense bar, (e) tellurometer, and (f) geodimeter. Use distances of approximately 1200 ft.

The readings in problems 4–3 through 4–6 are taken on a 2-meter subtense bar with a one-second theodolite. Compute the horizontal distance from theodolite to subtense bar and check by means of Table III in the Appendix B.

4–3.	0°31′16″	0°31′18″	0°31′17″	0°31′17″
4–4.	0°33′25″	0°33′24″	0°33′25″	0°33′24″
4–5.	0°35′46″	0°35′45″	0°35′47″	0°35′46″
4–6.	0°38′52″	0°38′53″	0°38′53″	0°38′52″

4–7. If the microwaves of a tellurometer system travel 186,000 mi per sec under given conditions, what is the unit of distance corresponding to each millimicrosecond (mμs) of the time needed for the wave to travel from the master to the remote unit and return?

4–8. The speed of light through the atmosphere at standard barometric pressure of 29.92 in. of mercury is accepted as 186,218 mi per sec for measurements with a geodimeter. What time lag in the equipment will produce an error of 100 ft in the distance to a target 40 miles away?

In problems 4–9 through 4–11 compute the horizontal distance for the recorded slope distance *XY*.

4–9. $XY = 432.81$ ft, slope angle = 4°40′.

4–10. $XY = 528.96$ ft, difference in elevation of *X* and $Y = 12.1$ ft.

4–11. $XY = 615.54$ ft, grade = 3.00%.

4–12. Compute the slope corrections for a 100-ft tape length, using the appropriate approximate formula and an exact method for slopes of 10%,

15%, 20%, and 25%. Carry out the computations far enough to show differences in successive results, and tabulate your answers.

A 100-ft steel tape, NBS 421, cross-sectional area 0.0030 sq in., weight 1 lb, standardized at 68 F, is 100.012 ft between end marks when supported throughout under 12 lb pull, and 99.990 ft when supported at the ends only with a 15-lb pull. What is the true length of a recorded distance *AB* for the conditions given in problems 4–13 through 4–16?

	Recorded Dist. *AB*	Aver. Temp.	Means of Support	Tension
4–13.	**432.81** ft	**68** F	Throughout	12 lb
4–14.	**367.20** ft	**68** F	At ends only	15 lb
4–15.	**624.46** ft	**58** F	Throughout	12 lb
4–16.	**811.92** ft	**86** F	At ends only	15 lb

For the NBS 421 tape of problems 4–13 through 4–16, what is the true length of the recorded distance *BC* for the conditions shown in problems 4–17 through 4–20?

	Recorded Dist. *BC*	Aver. Temp.	Means of Support	Tension	Elev. Diff./100 Ft
4–17.	**342.18** ft	**53** F	Throughout	20 lb	1.4 ft
4–18.	**100.00** ft	**83** F	Ends only	18 lb	2.2 ft
4–19.	**297.75** ft	**105** F	Throughout	22 lb	3.5 ft
4–20.	**523.66** ft	**42** F	Ends only	18 lb	4.0 ft

In problems 4–21 through 4–26 determine the length *CD* to be laid out using a 100-ft steel tape, NBS 422, cross-sectional area 0.0060 sq in. and weight 2.0 lb, standardized at 68 F to be 100.006 ft between end marks when supported throughout with a 10-lb pull, and 99.961 ft if supported at the ends only under 15 lb tension.

	Reqd. Dist. *CD*	Aver. Temp.	Support	Tension	Elev. Diff./100 Ft
4–21.	**91.00** ft	68 F	Throughout	10 lb	0
4–22.	**87.68** ft	75 F	Ends only	15 lb	0
4–23.	**162.48** ft	83 F	Throughout	12 lb	2.4 ft
4–24.	**91.00** ft	48 F	Ends only	20 lb	3.6 ft
4–25.	**226.58** ft	34 F	Ends only	20 lb	2.6% grade
4–26.	**350.00** ft	104 F	Throughout	10 lb	1.5° slope

4–27. **The chaining pins for a line measured with a 100-ft steel (adding) tape which has an extra graduated foot are set by mistake at the 101-ft mark, and the length is recorded as exactly 6 full tape lengths.** What **is the correct distance?**

4–28. **The distance between two fixed points on a missile range in the desert was measured at a temperature of 112 F with a 100-ft steel tape which had been standardized at 68 F, and was recorded as 2208.56 ft.** What **is the distance corrected for temperature?** What **distance would have been recorded if the temperature at the time of measurement had been 20 F?** What precisions are obtained if the measurements at 112 F and **20 F are not corrected for temperature?**

4–29. A distance of 827.43 ft is recorded for the measurement of a line with a 100-ft steel tape supported at the ends only and under a 15.0-lb tension. The tape weighs 1.5 lb, and it is 100.00 ft long when supported throughout and under 15.0-lb tension. Compute (a) the correction for sag for a full tape length, (b) the correction for sag for the partial tape length, and (c) the total distance after correction for sag.

4–30. A chain tape $\frac{5}{16}$ by 0.025 in. is 100.08 ft long when supported throughout its length and under a tension of 15.0 lb. What is the length between end graduations if the same tension is applied but the tape is supported at the ends and at the middle point? If it is supported at the ends and at the quarter points?

4–31. What is the normal tension for a 100-ft steel tape having a cross-sectional area of 0.0050 sq in. when it is supported at the ends only at 68 F, if the tape is 99.992 ft long at the same temperature when supported at the ends only and subjected to a 15.0-lb pull?

4–32. Same as 4–31 except for a tape of cross-section 0.0060 sq in. and 100.004 ft at 68 F for a 15.0 lb pull.

In problems 4–33 through 4–36 what error results from the condition noted?

4–33. One end of a 70-ft length of tape is off line by 1.0 ft?

4–34. The zero end of a 55-ft length of tape is too high by 2.2 ft?

4–35. One end of a 100-ft tape is off line by 0.8 ft and too high by 1.2 ft?

4–36. The end of a 100-ft tape is off line by 1.4 ft and too low by 2.8 ft?

4–37. To determine the angle AOB between two intersecting fences without setting up a transit, convenient distances OA = 100.00 ft and OB = 87.50 ft are measured from the intersection along the fence lines. If the distance AB is 62.6 ft, what is the intersection angle?

4–38. Similar to problem 4–37, but assume that $OA = OB = 100.00$ ft and $AB = 76.4$ ft.

4–39. If in a 6-sided figure comparable with Fig. 4–18 except that sides EF and EG are omitted, $GA = 256.1$, $AB = 268.6$, $BC = 567.6$, $CD = 551.8$, $DE = 314.2$, $EG = 303.6$, $GB = 219.1$, $GC = 588.7$, and $GD = 298.3$ ft, compute the angles at each corner.

4–40. Compute the total area enclosed in the figure of problem 4–39.

4–41. A base line taped on chaining bucks has a recorded length of 1650.234 ft. The average temperature was 81.6 F, and the total inclination correction for the line is −0.174 ft. Standardization data for the tape used (in practice several tapes would be employed for checking purposes) are as follows: supported throughout and tension 18.0 lb, 0 to 100 mark = 100.016 ft; supported at ends only and tension 18.0 lb, 0 to 100 mark = 99.978 ft. Compute the sea level length of the base line, making all five corrections, if the average elevation of the line is 4200 ft.

4–42. Determine the most probable length of a line AB, the probable error, and the 90% error of a single measurement for the following series of measurements of AB made under the same conditions and by use of the same procedures: 628.52, 628.58, 628.55, 628.57, 628.56, and 628.58 ft.

4–43. Similar to problem 4–42, but assume that the distances recorded are as follows: 443.64, 443.63, 443.67, 443.61, 443.65, 443.65 ft.

4–44. The probable error of taping a 600-ft distance is ± 0.12 ft. Using the same procedures, what should it be for a 1200-ft distance?

4–45. A property line is to be measured with an accuracy of 1/5000 using a 100-ft steel tape. If the probable error of a single tape measurement is ± 0.01 ft, explain a suitable field procedure for the work.

4–46. In measuring a length in one direction with a 100-ft steel tape on a day when the temperature is about 70 F, and without a thermometer, hand level, spring balance, or transit for alignment, what maximum error might you expect in 500 ft? 1000 ft? 2500 ft? In 2500 ft if the 1000- and 1500-ft lengths were measured separately?

4–47. A student counted 102, 105, 103, 105, 103 and 104 paces in walking along a 300-ft course on level ground. He then took 177, 176, 174, 176, 178, and 175 paces in walking an unknown distance *AB*. What is the length of *AB?*

4–48. Discuss some of the advantages of electronic distance measurement.

4–49. Explain briefly how a distance can be measured by the method of phase comparison.

4–50. Prepare a table similar to Table 4–1 for the following pattern frequencies, except that the half wavelengths are to be tabulated in ft for the difference frequencies A-B, A-C, A-D, A-E and A-alone (i.e., the difference frequency for A-alone is 10.000 mcs). A = 10.000 mcs, B = 9.999 mcs, C = 9.990 mcs, D = 9.900 mcs, and E = 9.000 mcs.

4–51. Discuss what is meant by "ground swing." How is it eliminated?

4–52. In measuring a distance with an MRA-3 tellurometer, the following coarse readings were taken: $B = 17$, $C = 68$, $D = 71$, and $E = 12$. The average of all fine readings was 3.18. Calculate the slope distance uncorrected for meteorological conditions.

4–53. What are the basic differences between the Geodimeter and the microwave electronic distance measuring equipment?

4–54. An irregular field is measured with a steel tape 100.10 ft long, and the area found to be 36.472 acres. What is the true area?

4–55. If the 1/5 point (20-ft mark) of a 100–ft tape is snagged 1.0 ft off line but the tape is pulled tight, what error in the full tape length will be recorded?

4–56. In taping a 1000–ft long line on a property survey with a 100–ft standardized steel tape ¼ in. by 0.020 in., which of the following errors is most serious? (a) A constant temperature difference of + 5 deg from standard, (b) an alignment error of 0.1 ft on each tape length, (c) tape out of level 0.1 ft on each length, (d) a tension variation of +3 lb from standard on each tape length.

4–57. A line 6 miles long to be laid off must have an error less than 10 ft. Would standard taping procedures be satisfactory for this measurement? Discuss the reasons for your decision.

chapter 5

Leveling

5–1. General. Leveling is a relatively simple process but a vital one in all construction work, in setting machinery and equipment, and in the fabrication of large objects such as airplanes. Leveling has its own special terms and these will be defined in the following topic as an introduction to the subject.

5–2. Definitions. The basic terms in leveling defined below are illustrated in Fig. 5–1.

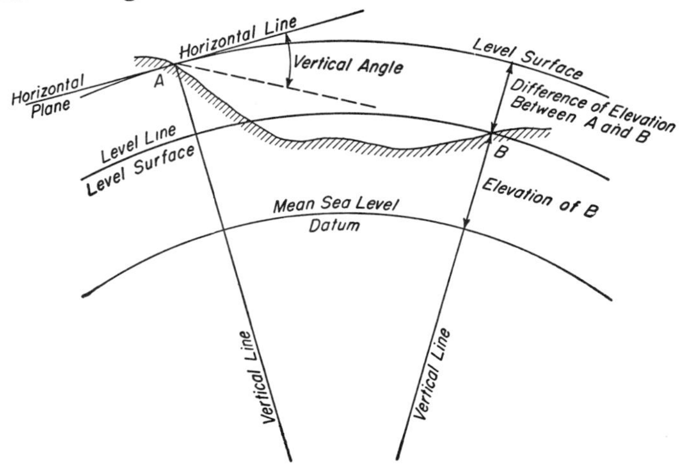

Fig. 5–1. Leveling terms.

Vertical line. A line to the center of the earth from any point. It is commonly considered to be the line defined by a plumb line.

Level surface. A curved surface which at every point is perpendicular to the plumb line (the direction in which gravity acts). Level surfaces are approximately spheroidal in shape. A body of still water is the best example. In the survey of a limited area, a level surface is sometimes treated as a plane surface.

Level line. A line in a level surface, therefore a curved line.

Horizontal plane. A plane perpendicular to the plumb line.

Horizontal line. A straight line perpendicular to the vertical.

Datum. Any level surface to which elevations are referred (for example, mean sea level). Also called datum plane, although not actually a plane.

Mean sea level. The average height of the surface of the sea for all stages of the tide over a 19-year period, usually determined from hourly height readings. The standard datum of elevations throughout the United States is referred to as the "Sea Level Datum of 1929." It was arrived at through a least-squares adjustment in 1929 of all first-order leveling in the United States and Canada. In this adjustment, mean sea level was held at zero for 26 tidal stations along the Atlantic and Pacific Oceans and the Gulf of Mexico. It is not likely that another such general adjustment will be made, although elevations of individual lines and marks may be revised as more data are accumulated.

Elevation. The vertical distance from a datum, usually mean sea level, to a point or object.

Bench mark. A relatively permanent material object, natural or artificial, bearing a marked point whose elevation above or below an adopted datum is known or assumed. Common examples are metal discs set in concrete, non-movable parts of fire hydrants, and curbs.

Leveling. The process of finding the difference in elevation between two points by measuring the vertical distance between the level surfaces through the points.

If the elevation of a point X is 802.56 ft, X is 802.56 ft above some datum. Mean sea level is the nation-wide reference surface made available for the use of local surveyors by the United States Coast and Geodetic Survey through its establishment of thousands of bench marks. Locations and elevations of these marks can be obtained by writing to its Director, Washington 25, D.C.

Many organizations have their own arbitrarily selected datums. Hence it is necessary to state the datum (and perhaps the year of adjustment) on which an elevation is based.

Vertical control. A series of bench marks or other points of known elevation established throughout a project. Also termed *basic control,* or *level control.* The basic vertical control for the topographic mapping of the United States consists of first- and second-order leveling. Less precise third-order (or even fourth-order) leveling is satisfactory for filling the gaps between second-order bench marks and for many projects.

5–3. Curvature and refraction. From the definitions of a level surface and a horizontal line, it is evident that a horizontal line departs from a level surface because of the *curvature* of the earth. In Fig. 5–2 the departure DB from a horizontal line through point A is expressed approximately by the formula

$$C = 0.667M^2 = 0.024F^2 \tag{5-1}$$

where C is the departure, in feet, of a level surface from a horizontal line, M is the distance in miles, and F is the distance in thousands of feet.

Since points A and B are on a level line, they have the same elevation. Curvature of the earth therefore causes a rod held on B to be read too high by an amount BD.

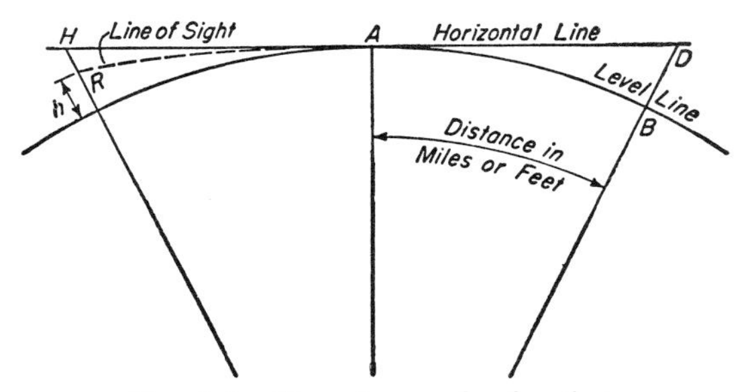

Fig. 5–2. Curvature and refraction.

Light rays passing through the earth's atmosphere are bent or refracted toward the earth's surface as shown in Fig. 5–3. Thus a theoretically horizontal line of sight, as AH in Fig. 5–2, is bent to the curved form AR. The result is that an object at R appears to be at H, and the reading on a rod held at R is diminished by the distance RH.

The effect of *refraction* in making objects appear higher than they really are (and thus rod readings smaller), can be remembered by considering what happens when the sun is on the horizon, as in Fig. 5–3. At the moment the sun has just passed below the horizon, it is seen just above the horizon. The sun's diameter of approximately

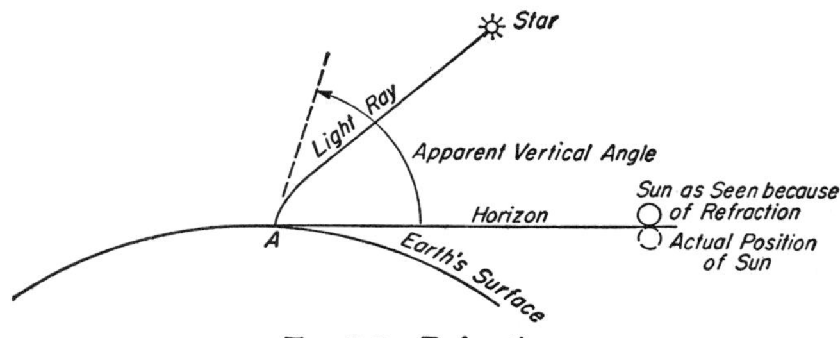

FIG. 5-3. Refraction.

32 min is roughly equal to the average refraction on a horizontal sight.

The angular displacement resulting from refraction is variable. It depends upon atmospheric conditions and the angle the line of sight makes with the vertical. For a horizontal sight it is expressed approximately by the formula

$$R = 0.09M^2 = 0.003F^2 \quad (5\text{-}2)$$

This is closely one-seventh of the effect of curvature of the earth, but in the opposite direction.

The formula for the combined effect of curvature and refraction is approximately

$$h = 0.574M^2 = 0.0206F^2 \quad (5\text{-}3)$$

For sights of 100, 200, and 300 ft, h equals 0.00021 ft, 0.00082 ft, and 0.0019 ft, respectively.

5-4. Methods of determining differences in elevation. Differences in elevation are determined directly by taping, by a leveling instrument or a barometer, and indirectly by trigonometric leveling.

Taping method. Application of a tape to the vertical line between two points is sometimes possible. This method is used in determining the depths of mine shafts and in the layout and construction of multistory buildings. When a water or sewer pipe is being laid, a graduated rod or pole may replace the tape.

Leveling instrument. A horizontal plane of sight tangent to the level surface at any point is readily established by means of a level vial. If a sighting device (telescope) and level vial are combined, vertical distances can be measured by observing on graduated rods. This is the method commonly used by engineers.

The recently developed laser alignment system provides a horizontal beam which permits vertical and horizontal measurements to 0.001 inch from 0 to 300 feet. The laser system is described in Chapter 25.

Barometer, or altimeter. **This instrument measures air pressure and converts the readings to elevations in feet. Readings taken at critical points provide relative elevations.**

Trigonometric leveling. The difference in elevation between two points can be determined by measuring (a) the inclined distance between them and (b) the vertical angle to one point from a horizontal plane through the other. Thus in Fig. 5–4 if the slope distance AB or DC and the vertical angle EDC are measured, then the difference in elevation between A and B is $EC = DC \sin EDC$.

Trigonometric leveling is commonly used in topographical work where inclined distances and vertical angles are obtained by means of a transit or alidade. It is a technique especially useful for leveling in very rugged terrain.

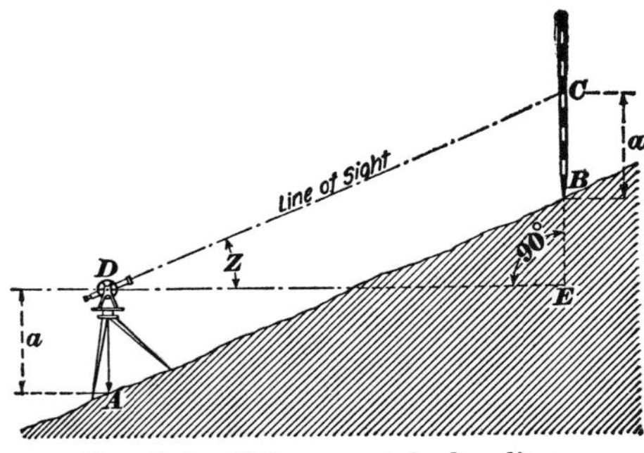

Fig. 5–4. Trigonometric leveling.

5–5. Types of leveling instruments. The types of levels used by most surveyors are the American-type engineer's level (dumpy and wye), shown in Figs. 5–7 and 5–9; the tilting level, Fig. 5–10; the European-type engineer's level, Fig. 5–11; and the self-leveling level, Fig. 5–12. For less-accurate work, the hand level, Fig. 5–14, is often used.

The principal parts of the engineer's level are the level vial, telescope, level bar, and leveling head. A hand level, as the name implies, is a hand-held instrument consisting of a telescope and level vial.

5–6. Level vials. A level vial is a glass tube which is sealed at both ends and contains a sensitive liquid and a small air bubble. The liquid must be nonfreezing, quick-acting, and relatively stable in length for normal temperature variations. Purified synthetic alcohol has generally replaced the mixture of alcohol and ether formerly used. Uniformly spaced graduations are etched on the exterior surface of the tube to show the exact position of the bubble. On vials now made, the divisions are generally 2 mm long, but 0.01-ft and 0.1-in. divisions have also been used.

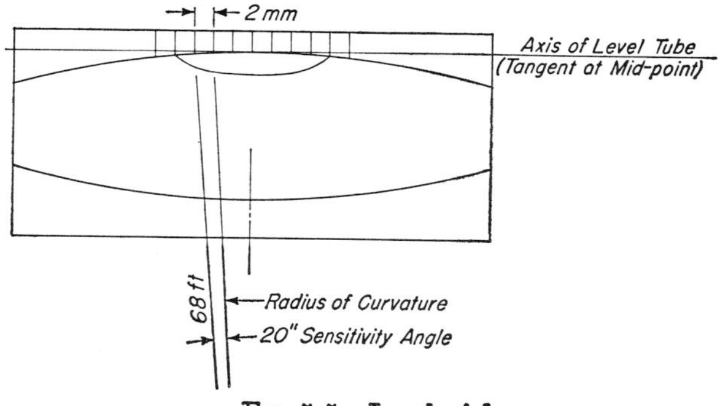

FIG. 5–5. Level vial.

The *axis* of a level vial is the longitudinal line tangent to the upper inside surface at the mid-point. When the bubble is in the center of its run, the axis should be a horizontal line, as in Fig. 5–5.

The sensitivity of a level vial is determined by the radius of curvature provided in the grinding. The larger the radius, the more sensitive the bubble. A highly sensitive bubble is necessary for precise work but may be a handicap in rough surveys because of the longer time required to center it.

A properly designed instrument has the sensitivity of the level vial correlated with the resolving power, or resolution, of the telescope. A slight movement of the bubble should be accompanied by a minute change in the observed rod reading at a distance of perhaps 200 ft. Sensitiveness of a level vial is expressed in two ways: (a) by the angle, in seconds, subtended by one division on the scale, and (b) by the radius of curvature of the tube.

If one division subtends an angle of 20 sec at the center, it is called a 20-sec bubble. Because of the variable lengths of divisions that have been used, this is not always a fair comparison. A 20-sec bubble on a vial having 2-mm divisions has a radius of 68 ft. The angles of level vials used on America-made instruments range from about 20 to 90 sec, the usual value for a level being approximately 20 sec for 2-mm divisions.

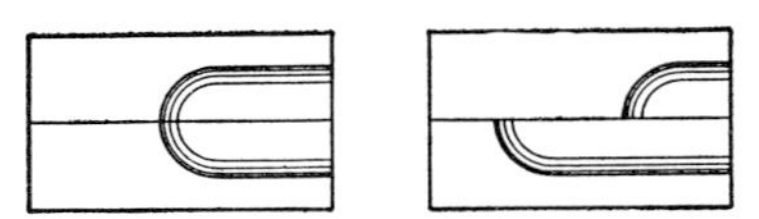

FIG. 5-6. Coincident type of level bubble. Correctly set in left view; twice the deviation of the bubble shown in right view. (Courtesy of Kern Instruments, Inc.)

Figure 5-6 illustrates the *coincident* type of level bubble used on precise equipment. The bubble is centered by bringing the two ends together to form a smooth curve. A prism splits the bubble and makes the two ends visible simultaneously.

A noncoincident type of bubble can be centered with an accuracy of about one-tenth of its sensitivity. A coincident bubble, because of the opposite motions of the two ends and the magnification provided in an optical-reading instrument, can be centered with an accuracy of perhaps one-fortieth of its sensitivity. Thus the precision possible in centering, as well as the sensitivity of the bubble, must be considered in estimating the results to be obtained.

5-7. Telescopes. The telescope on an engineer's level, Fig. 5-7, is a metal tube containing four main parts: objective lens, negative lens, reticle, and eyepiece.

Objective lens. The objective lens is a compound lens, either doublet or triplet, securely mounted in the object end of the main tube with its optical axis reasonably concentric with the tube axis. Some light striking a lens is lost (approximately 4 to 5 per cent) by reflection and absorption, even though the rays are perpendicular to its surface. For a compound lens system the reduction of light is multiplied and may become critical. Loss by reflection is practically eliminated by means of a thin ($\frac{1}{4}$ wavelength of light) uniform coating evaporated onto the lens surfaces to be in contact with air. The coating material has an index of refraction less than that of glass. Loss of light by absorption usually is not serious unless the lenses are very thick.

Negative lens. The negative lens is mounted in a sliding tube so that its optical axis coincides with that of the objective lens. The purpose of the negative lens is to bring rays of light entering

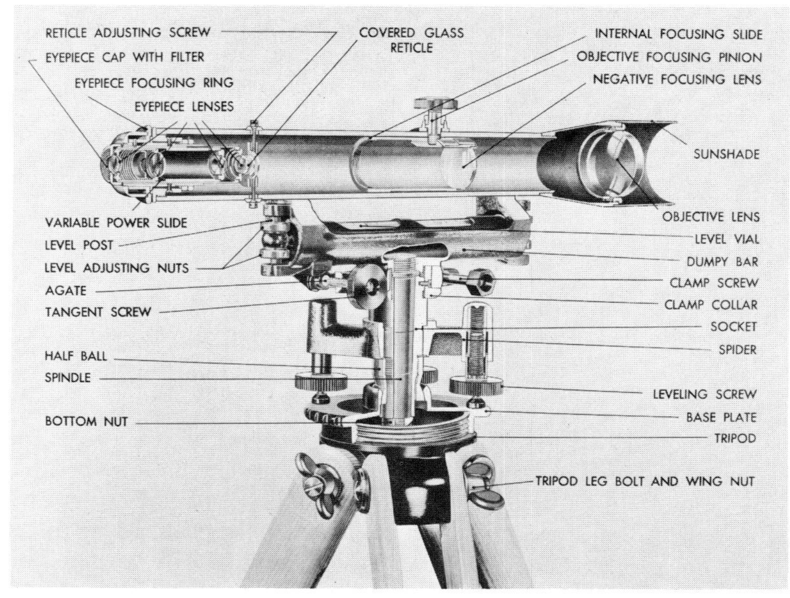

FIG. 5–7. Dumpy level.

through the objective lens to a focus at the plane of the reticle. It is important that the bearings of the slide and receive tubes be so fitted that there will be no deviation of either lens axis—objective or negative—during the focusing process from maximum to minimum distances. This combination of fixed objective lens and movable negative lens increases the effective focal length of the optical system by about 50 per cent. If an object infinitely far away is sighted (for example, a star), the distance from a lens to the image it forms is called its focal length, f. Practically, a sight distance of a few thousand feet reproduces this condition.

Reticle. A reticle is a pair of lines mounted near the eye end of the main tube and located at the point of principal focus of the objective optical system. The point of intersection of these lines, together with the optical center of the objective system, forms the directing axis of the telescope commonly called the line of collimation. The reticle is mounted by two pairs of opposing capstan screws placed at right angles to each other so that the

directing line of sight (line of collimation) can be adjusted at right angles to the trunnion axis of the telescope.

Cross hairs may be spider web, or filaments of platinum or glass stretched across an annular ring (doughnut). In many newer instruments, a thin glass plate with lines ruled, etched, and filaments of dark metal deposited in them serves as the reticle. Additional lines parallel to and equidistant from the cross lines are added when desired. If a glass reticle is used, the additional lines are shortened so as not to confuse them with the primary cross lines. The reticle is mounted to place the lines in a horizontal–vertical position.

Eyepiece. The eyepiece is a microscope with a magnification of about 35 diameters used to view the object image focused by the objective lens system at the plane of the reticle. It may consist of two lenses (giving an inverted image to the eye) or of four lenses (giving an erected image). The former gives a slightly better optical acuity but requires more training in its use.

The eyepiece is furnished with a focusing movement in order to accommodate the difference in vision of different observers.

Focusing. The process of focusing is the most important function to be performed in the use of a telescope. Telescopes today are generally internal focusing, but some older ones in service are external focusing. Dust and wear affect the slide and may disturb the optical axis of the external-focusing type of telescope. The internal-focusing telescope is more dust-resistant.

The fundamental principle of lenses is given by the formula

$$\frac{1}{f_1} + \frac{1}{f_2} = \frac{1}{f} \qquad (5\text{–}4)$$

where f_1 = distance from lens to image at plane of reticle;
f_2 = distance from lens to object, as on Plate A–9 in Appendix A;
f = focal length of lens.

The focal length is a constant for any particular set of lenses. Therefore, as the distance f_1 changes, f_2 must also change.

Since the reticle remains fixed in the telescope tube, the distance between it and the eyepiece must be adjusted to suit the eye of an individual observer. This is done by bringing the cross hairs to a clear focus, that is, making them appear as black as possible when sighting at the sky or a distant, light-colored object. Once this has been accomplished, the adjustment need not be changed for the same observer, regardless of the length of sight, unless his eyes tire from long observation or from high magnification of the telescope.

After the eyepiece has been adjusted, objects are brought to sharp focus at the plane of the cross hairs by moving the objective lens. If the cross hairs appear to travel over the object sighted when the eye is shifted slightly in any direction, parallax exists. Either the objective lens or the eyepiece must be adjusted to eliminate this effect if accurate work is to be done.

The "visible" rule *near-far, far-near* may be helpful to beginners using an old external-focusing telescope. If the object is near the observer, the objective lens is run far out. When the object is far away, the objective lens is brought nearer to the observer's eye.

The level vial is attached to the telescope tube in such a manner that (when in adjustment) the axis of the level bubble is exactly parallel to the line of sight through the center of the telescope. Centering the level bubble therefore makes the line of sight horizontal.

5-8. Optics.[1] A brief discussion of the optics of surveying instruments and optical tooling equipment (see Chapter 25) is desirable before leaving the subject of telescopes.

The purpose of a telescope is to create for the observer a picture that shows the position of the cross hairs on the target with the greatest possible clarity and precision. This end is attained by skillful design and perfection in manufacture to secure the combination of optical qualities best suited to a particular application. Optical factors include resolving power, magnification, definition, eye distance, size of pupil, and field of view.

Specifications alone can seldom indicate the true qualities of one telescope as compared with those of another. The most important test for a telescope, and in fact the only true test, is a simultaneous comparison with another telescope, under the same conditions.

[1] This section on optics includes some material from *Optical Alignment Equipment,* published by Keuffel and Esser Company and reprinted here by permission.

Some of the important optical terms will be explained.

Resolving power. The ability of a lens to show detail is termed resolving power. It is measured by the smallest angular distance, expressed in seconds of arc, between two points that are just far enough apart to be distinguished as separate objects, rather than as a single blurred object.

The maximum resolving power that theoretically can be attained with a telescope when the optical parts are perfectly designed, and exactly placed, depends entirely on the diameter of that part of the objective lens actually used (the effective aperture). The resolving power of an objective lens is independent of magnification. It can be computed by the formula

$$R = \frac{5.5''}{D} \qquad (5\text{-}5)$$

where R is the angle that can be resolved, in seconds, and D is the diameter of the effective aperture of the lens, in inches. For example, the objective lens of the jig transit shown in Fig. 25-7 has an effective aperture 1.18 in. in diameter. The resolving power is therefore 4.6 sec.

The accepted theoretical standard for the resolving power of the human eye is 60 sec, although a value between 80 and 90 sec would probably be more practical. Hence the resolving power of the telescope has to be brought at least to this limit by magnification. If the angular distance resolved by the telescope is 4.67 sec, this resolving power must be magnified 13 times to obtain 60 sec. Since the eyesight of different observers varies, more magnification is invariably used.

Magnification. The value of magnification (power) is the ratio of the apparent size of an object viewed through a telescope to its size as seen by the unaided eye from the same distance. Magnification varies slightly when the focus of the telescope is changed. Therefore it is affected somewhat by the distance to the object. For an optical-alignment telescope, the greatest magnification occurs at infinite focus, and this is the value used to describe the property.

Although telescopic magnification must be greater than $60/R$, there is a limiting point beyond which it is impossible to increase magnification without sacrificing definition. This point is reached when the magnification becomes greater than two or three times $60/R$. For larger values the quality of the image seen is impaired, and the accuracy with which the line of sight can be made to coincide with a target is reduced.

Certain disadvantages result from the use of too high a magnification, even when the objective lens is large enough to give the necessary resolution. With high magnification the field of view is reduced, and any heat waves, turbulence, or vibration cause the image of an object to move over the cross hairs too fast for accurate observation. The magnifying power of the telescopes on modern engineer's levels ranges from 26 to 41 diameters, and averages perhaps 32 diameters. The levels shown in Figs. 5–7, 5–9, 5–10, and 5–11 have variable-power eyepieces.

Definition. Definition is a term used to define the over-all results produced by a telescope. Better definition permits objects to be

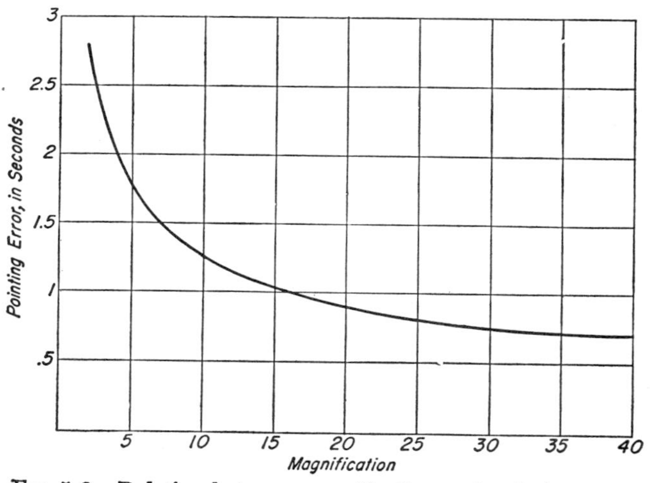

FIG. 5–8. Relation between magnification and pointing error. (Courtesy of Keuffel and Esser Company.)

seen more clearly through the telescope. It depends upon a number of optical features and is the quality that gives the greatest pointing accuracy.

Since definition is a relative term, it can be determined best by comparing the appearance of the same object when viewed through the telescope to be tested and a telescope with which the observer is familiar.

Pointing accuracy. The exactness with which the line of sight can be directed toward a target, or a target can be placed on the line of sight, is called pointing accuracy. It depends upon magnification, definition, the arrangement of the cross hairs, and the design of the target or scale sighted. The general relationship between magnification and pointing accuracy for telescopes having the same definition is shown in Fig. 5–8. Other tests show that the standard 1-min transit can be pointed within 2 to 5 sec of arc consistently in the field.

5–9. Level bar and supports. The telescope of a level rests upon vertical supports at each end of a horizontal bar called the level bar. The level bar in turn is centered on a vertical spindle which is accurately machined and rests in a conical socket of the leveling head. The spindle insures that the level bar will revolve in a horizontal plane when the instrument is properly adjusted.

The vertical supports for the telescope may be wyes which are adjustable in height and have clamps to hold the telescope tube, or they may be rigid and integral parts of the level bar as in the dumpy level. The dumpy and wye levels, shown in Figs. 5–7 and 5–9, respectively, differ primarily in the method of supporting the telescope.

5–10. Leveling head. The conical socket into which the vertical axis of the level bar fits is supported by four large thumbscrews called leveling screws. These rest upon a plate which is screwed on top of the tripod. The four leveling screws are in two pairs, those of each pair being directly opposite each other. The level vial is placed alternately over each pair of leveling screws, which are adjusted by turning them until the bubble remains in the center of the vial for a complete revolution. The line of sight will then generate a horizontal plane.

Most precise equipment has three rather than four leveling screws. The three-screw arrangement is faster and not subject to the rocking which takes place in the four-screw type when two opposite screws are turned up or down slightly more than the other two. The disadvantage of the three-screw type is that a slight difference in elevation of the line of sight results if all three screws are turned up or all are turned down. Manipulation of the four-screw head does not change the elevation of the telescope. Also, after the threads become worn on a three-screw leveling head, there is some loss of rigidity which must be eliminated by replacing the screws. Tightening one screw of each pair in the four-screw arrangement results in a clamping action and produces a stable setup.

5–11. Wye level. The wye level, Fig. 5–9, has a telescope tube A resting in supports B, called *wyes* because of their shape. Curved clips C, hinged at one end and pinned at the other, fasten the telescope in place. If the clips are raised, the telescope can be rotated in the wyes, or removed and turned end for end as part of the adjustment procedure. The sunshade X should always be used when observing, even in cloudy weather. A cap covers the objective lens when the instrument is not being used.

FIG. 5–9. Wye level. (Courtesy of W. & L. E. Gurley.)

The level tube D is attached to the telescope but can be adjusted in the horizontal and vertical planes by the capstan-headed screws Z. The wyes are supported by the *level bar* E which can be adjusted vertically by means of the capstan nuts F.

The *spindle* (visible in Fig. 5–7) is constructed at right angles to the level bar and rotates in a socket G (Fig. 5–9) in the leveling head H. A collar L turning on the socket can be secured by the *clamp screw* M to hold the telescope in any vertical plane. *Tan-*

gent screw N, also called the *slow-motion screw,* is operative only when the clamp screw is tightened. The tangent screw rotates the telescope through a small angle for precise settings.

Four *leveling screws O,* resting in cups *P* upon the bed or *foot plate Q,* control the leveling head. A ball-and-socket joint at the lower end of the spindle provides a flexible connection to the bed plate and a means of rotation as the leveling screws are raised or lowered. Threads on the inside of the bed plate fasten the head of the instrument to the tripod.

The wye level is simpler to adjust than the dumpy level because the telescope can be lifted from the wyes and turned end for end. This feature permits one man to make all adjustments by himself. Although more adjustments are required for the wye level, they are thus easier to make. The advantage is lost if the collars on the telescope, or the bearings of the wyes on which they rest, become worn. The instrument then must be adjusted in the same way as the dumpy level.

5–12. Dumpy level. In the dumpy level, Fig. 5–7, the telescope is rigidly attached to, and parallel with, the level bar. The level vial is set in the level bar and thereby protected somewhat. The level vial always remains in the same vertical plane as the telescope, but screws at each end permit vertical adjustment or replacement of the vial. Other construction details are similar to those of the wye level.

The advantages of the dumpy level over the wye level are (a) simpler construction with fewer movable parts, (b) fewer adjustments to be made, and (c) probably longer life for the adjustments. A disadvantage is that one of the adjustments requires a second man (rodman) and is more time-consuming. This difficulty can be eliminated if two points of known elevation are established several hundred feet apart and fixed targets set up on them.

The dumpy level is used almost exclusively in precise work and has replaced the older wye type in most fields.

5–13. Tripods. Several types of tripods are available. The legs may be fixed or adjustable in length, and solid or split. All types are shod with metallic conical points, and hinged at the top where they connect to a metal head. The adjustable-leg tripod is advantageous for setups in rough terrain or in a shop, but the type with a fixed-length leg may be slightly more rigid. The split-leg model is lighter than the solid type but less rugged. A wide-framed tripod, first used on European instruments, is now available from American manufacturers. A sturdy tripod in good condition is necessary to obtain the best results from a fine instrument.

In the past, many different thread types were used on tripods. The standard now adopted by all American manufacturers is eight threads per inch on a cap $3\frac{1}{2}$ in. in diameter.

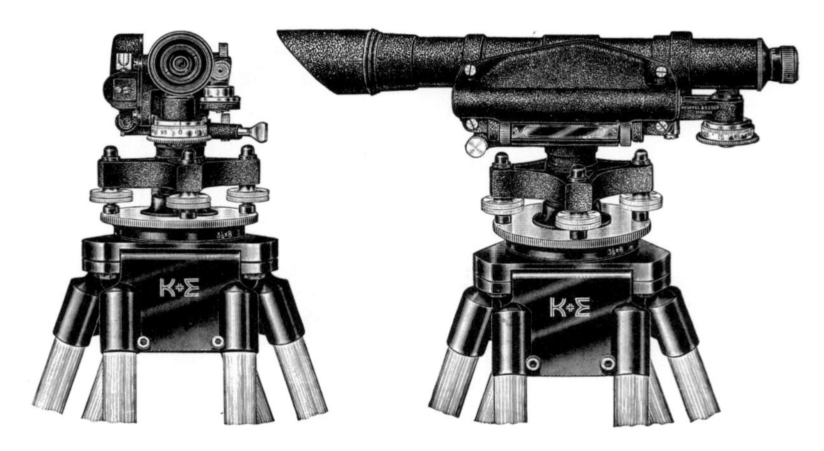

FIG. 5–10. Tilting level. (Courtesy of Keuffel and Esser Company.)

5–14. Tilting level. A tilting dumpy level, Fig. 5–10, is gaining favor for general use. Originally it was employed only on high-precision work. A bull's-eye (circular) spirit level is available for quick approximate leveling by means of the four leveling screws. Exact level is obtained by tilting or rotating the telescope slightly in a vertical plane about a fulcrum at the vertical axis of the instrument. A micrometer screw under the eyepiece controls this movement.

The tilting feature saves time and increases accuracy, since only one screw need be manipulated to keep the line of sight horizontal as the telescope is turned about a vertical axis. The telescope bubble is viewed through a system of prisms from the observer's normal position behind the eyepiece. The prism arrangement splits the bubble image into two parts. Centering the bubble is accomplished by making the images of the two ends coincide.

The instrument shown in Fig. 5–10 has 30x magnification; resolving power of 4 sec; minimum focusing distance of 6 ft; glass reticle; and sensitivity of level vial equal to 20 sec per 2 mm.

5–15. European instruments. European levels, one of which is shown in Fig. 5–11, are characterized by short telescopes, stream-

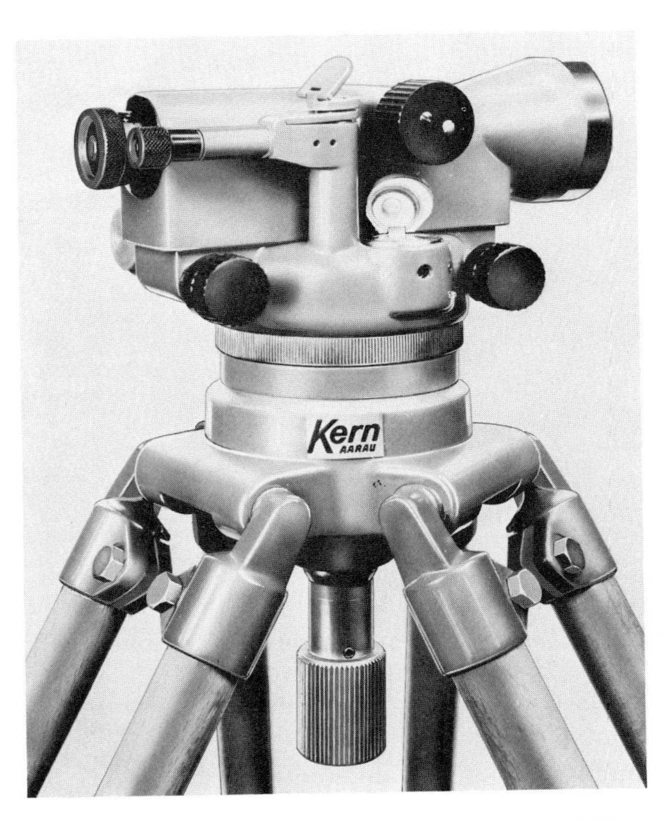

FIG. 5–11. GK23 European tilting level. (Courtesy of Kern Instruments, Inc.)

lined construction, small size, and light weight. Most such levels are of the dumpy type and include a tilting arrangement with a three-screw (or three-cam) leveling head. The GK23 has a telescope level vial with sensitivity of 18 sec per 2 mm and a centering precision of ±4 sec; telescope magnification of 30 power; stadia constant of 0; and weight of only 3.3 lb and 4.2 lb for the instrument and carrying case, respectively. A 2.44-in. horizontal glass circle, which can be read with its microscope by estimation to 1 min, is incorporated in the GK23C.

One type, Fig. 5–12, incorporates a self-leveling feature. If the instrument is leveled roughly, a compensator automatically levels

FIG. 5–12. Zeiss self-leveling level. (Courtesy of Keuffel and Esser Company.)

the line of sight and keeps it precisely level. The principles of operation of the compensator are shown schematically in Fig. 5–13.

Packaging is one of the outstanding innovations in European surveying equipment. Smoothly curved metal containers protect the instrument and make carrying convenient. European instruments compete in price with American equipment.

5–16. Hand level. The hand level, Fig. 5–14, is a hand-held instrument used on low-precision work and for checking purposes. It consists of a brass tube 5 or 6 in. long having a plain glass objective and a peep-sight eyepiece. A small level vial mounted above a slot in the tube is viewed through the eyepiece by means of a prism or a 45°-angle mirror. A horizontal hair extends across the tube.

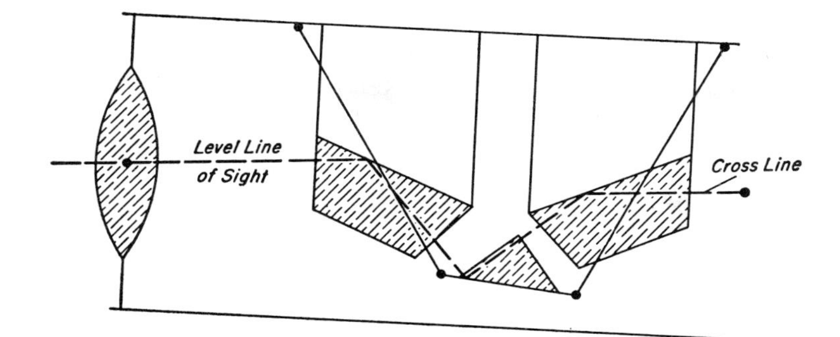

When Telescope Tilts Up, Compensator Swings Backward

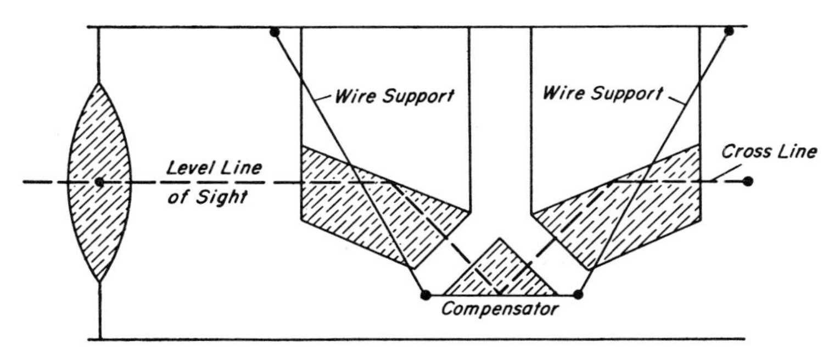

Telescope Horizontal

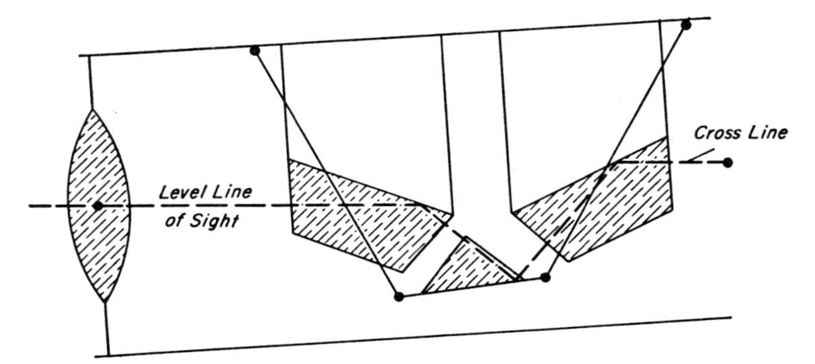

When Telescope Tilts Down Compensator Swings Forward

FIG. 5–13. Compensator of self-leveling level. (Courtesy of Keuffel and Esser Company.)

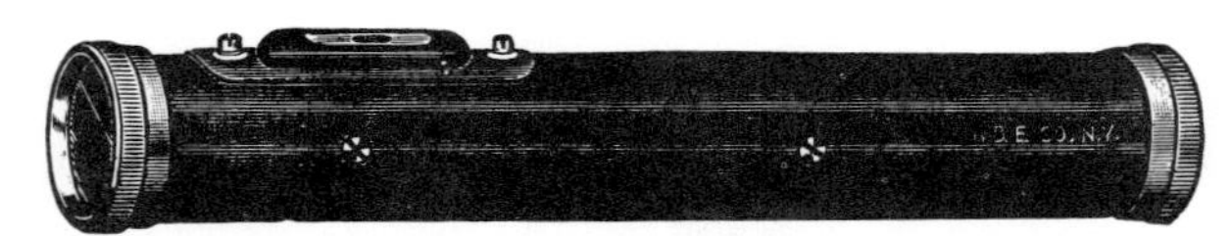

FIG. 5–14. Hand level. (Courtesy of Keuffel and Esser Company.)

The prism or mirror occupies only one half of the inside of the tube, the other half being open to provide a clear sight through the objective. Thus the rod sighted and the reflected image of the bubble are visible beside each other, with the cross hair superimposed.

The instrument is held in one hand and leveled by raising or lowering the objective end until the cross hair bisects the bubble. The tube can be steadied by making a tripod with the thumb on a cheekbone, the first finger on the forehead, and the eyepiece against the brow. Holding the level against a staff, or better still, resting it in a Y-shaped stick, increases the accuracy.

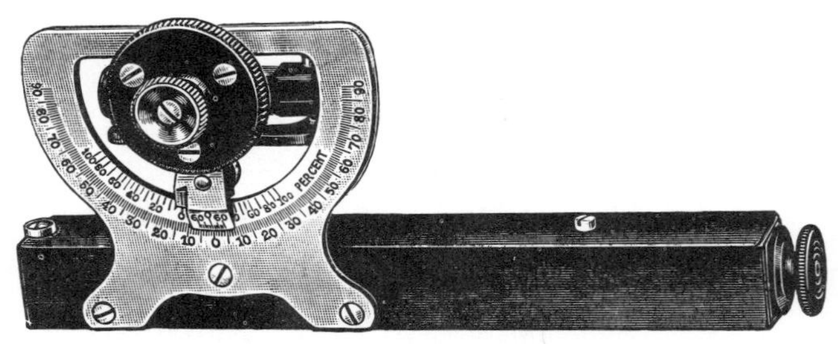

FIG 5–15. Abney hand level and clinometer. (Courtesy of Keuffel and Esser Company.)

Stadia hairs reading 1 : 10 may be included. Magnification of $1\frac{1}{2}$ diameters is usually provided for observing the bubble and the cross hair, but the rod is seen through plain glass. The length of sight possible is limited, therefore, to the distance at which a rod can be read with natural vision.

The *Abney hand level and clinometer*, shown in Fig. 5–15, has limited application in measuring vertical angles and slopes, and for direct leveling. It has an arc graduated in degrees up to 90°, a vernier reading to 10 min, and several scales for slopes ranging from a ratio of 1 to 1 to a ratio of 1 to 10.

5–17. Leveling rods. Rods used in most leveling work are made of wood and are marked with graduations in feet and decimals or in meters and decimals. There are two main classes of rods:

a. Self-reading rods which can be read by the levelman as he sights through the telescope and notes the apparent intersection of the horizontal cross hair with the rod.

b. Target rods which contain a movable target that is set by the rodman at the position indicated by signals from the levelman.

A wide choice of patterns, colors, and graduations on single-piece, two-section, and three-section leveling rods is available. The varied types, usually named for cities or states, include the Philadelphia, New York, Boston, Troy, Chicago, Florida, and California rods.

Rods for general leveling, and for special purposes such as slope staking, can be made by fastening a flexible ribbon of treated fabric to a wooden strip. Such strips, graduated in diverse ways, can be purchased from manufacturers.

A self-reading rod consisting of a wooden frame and an invar strip graduated in decimals of meters is used for precise work, to eliminate the effects of humidity and temperature changes. The invar strip is attached only at the ends and is free to slide in grooves on each side of the wooden frame.

The Philadelphia rod is a combination self-reading and target rod, and is the most popular type for all but precise leveling. The 7-ft by 13-ft model will be described in detail. Other lengths are also made, the 6.5-ft by 12-ft rod being common.

5–18. Philadelphia rod. The Philadelphia rod shown in Fig. 5–16 consists of two sliding sections graduated in hundredths of a foot and joined by brass sleeves *a* and *b*. The rear section can be locked in position by a clamp screw *c* to provide any length from a *short rod* for readings of 7 ft or less, up to a *long rod*, or *high rod*, of 13 ft which must be fully extended when used. The graduations on the front faces of the two sections read continuously from zero at the base to 13 ft at the top for the high-rod setting.

The rod graduations are accurately-painted alternate black and white spaces 0.01 ft wide. The 0.1- and 0.05-ft graduations are emphasized by spurs extending the black markings. Tenths are designated by black numbers, and foot marks by red numbers. The

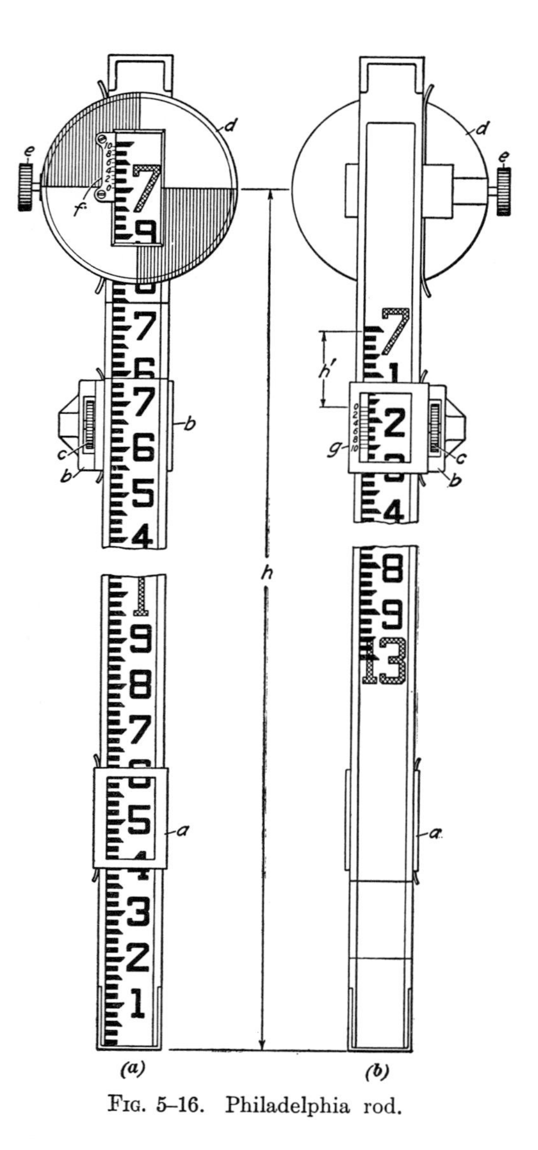

FIG. 5–16. Philadelphia rod.

figures straddle the proper graduation. A Philadelphia rod can be read accurately through the engineer's level at distances up to 250 ft.

On long sights, or when readings to the nearest 0.001 ft are desired, a target d is used. Circular, oval, and angular targets are made. All are approximately 5 in. high and painted red and white in alternate quadrants. A clamp e and a vernier scale f are part of the target. For readings of less than 7 ft, the target is set at the proper elevation in accordance with directions given by the instrumentman. When the rod is extended, the target is clamped at 7.000 ft and the rear section raised to bring the target to the correct height. Divisions on the back of the rod are marked from 7 to 13 ft in a downward direction. As the rod is extended, a fixed vernier scale g attached to sleeve b shows the target height.

Leveling rods are made of carefully selected, kiln-dried, well-seasoned hardwood, and are graduated in accordance with rigid specifications. They should not be used as seats, for pole vaulting, or left leaning against a tree or building, or laid on any surface with the graduated face down. Hands should be kept off the painted markings, particularly in the 3- to 5-ft section where a worn face will make the rod unfit for use. Letting the rod down "on the run" batters both sections and may change the vernier reading to less than 7.000 ft, for example to 6.998 ft. If this happens, the target must be set to the same reading, as 6.998 ft, for high rods.

5–19. Verniers. The fractional part of the smallest division of a scale or rod can be measured without interpolation by means of an auxiliary scale called a vernier. Figure 5–17 shows the simple type of direct vernier used on leveling rods.

As illustrated in Fig. 5–17a, the vernier has n divisions in a space covered by $(n - 1)$ of the smallest divisions on the scale. Then

$$(n - 1)d = nv \tag{5–6}$$

where d = the length of a scale division;
v = the length of a vernier division.

This is the fundamental basis for all vernier construction. For most leveling-rod verniers, $n = 10$, $d = 0.01$ ft, and $v = 0.09/10 = 0.009$ ft.

In Fig. 5–17a the reading is 0.300. If the vernier is moved so that its first graduation from zero coincides with the first graduation of the scale beyond 0.300, as in Fig. 5–17b, the vernier index has moved a distance equal to

$$d - v = 0.010 - 0.009 = 0.001 \text{ ft}$$

The reading is therefore 0.301 ft.

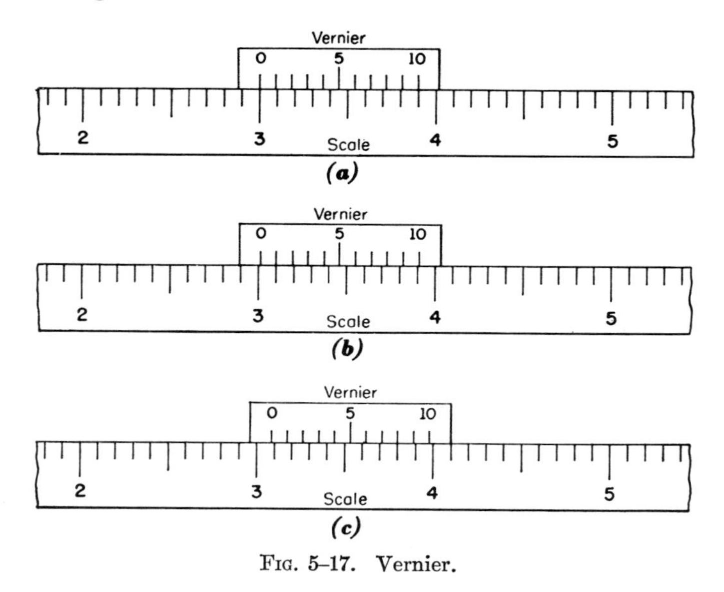

FIG. 5–17. Vernier.

If the vernier is moved so that the second graduation on the vernier is coincident with the graduation representing 0.32 on the scale, the movement from the position in Fig. 5–17a has been $2(d - v) = 0.002$ ft. Thus the fractional part of a scale division from the preceding scale graduation to the vernier index is read by determining the number of the vernier line which is coincident with *any* scale graduation. The reading for the decimal part of a foot in Fig. 5–17c is 0.308 ft.

From formula 5–6,

$$nv = (n - 1)d$$

or

$$nv = nd - d$$

and

$$(d - v) = \frac{d}{n} \qquad \textbf{(5–7)}$$

When a vernier is used, the smallest reading obtainable without interpolating is termed the *least count* of the vernier. For any vernier, the least count is given by the expression $(d - v)$ or d/n.

It is found most easily by the following relationship:

$$\text{Least count} = \frac{\text{value of the smallest division on the scale}}{\text{number of divisions on the vernier}}$$

An observer cannot be certain he is reading a scale and vernier correctly until he has determined the least count.

In selecting the vernier line which is coincident with a scale division, an observer must assume a position directly behind the lines, or over them, to avoid parallax. The second graduation on each side of the apparently coincident lines should be checked to see that a symmetrical pattern is formed about them. In Fig. 5–17c, vernier graduations 6 and 10 fall inside (toward division 8) the scale lines by equal distances, therefore 8 is the correct reading.

5–20. Setting up the level. The safest way to carry an engineer's level in a car is to keep it in its box. The box closes properly only when the instrument is set perfectly in the padded supports.

A level should be removed from its box by lifting on the level bar or base, *not* by grasping the telescope. The head must be screwed snugly on the tripod. If the head is too loose, the instrument is unstable; if too tight, it may freeze on the tripod. A grain of sand, roughness in the threads, or a change in temperature makes the head stick. Spreading the tripod legs until the head of the instrument almost touches the ground often helps to release the head in case it freezes.

The legs of a tripod must be tightened correctly. If each leg falls slowly of its own weight after being placed in a horizontal position, it is properly adjusted. Clamping the legs too tightly strains the plate and screws. If the legs are loose, the setup is wobbly.

Unlike the transit, the level is not set up over any particular point. It is inexcusable, therefore, to have the base plate badly out of level before using the leveling screws. Moving one leg radially, or circumferentially, will level the instrument. On side-hill setups, placing one leg on the uphill side and two on the downhill slope eases the problem. On very steep slopes some instrumentmen prefer two legs uphill and one down for a stable setup. The most convenient height of setup is one which enables the observer to sight through the telescope without stooping or stretching.

In leveling the four-screw head, the telescope is turned until it is over two opposite screws. The bubble is approximately centered by using the thumb and first finger of each hand to adjust the opposite screws. The procedure is repeated with the telescope over the other two leveling screws. Time is wasted by centering the bubble exactly on the first try, since it will be thrown off during the cross-leveling. Working with each pair of screws about three times should complete the job.

For a three-screw head, the telescope is aligned over one screw and is thus made perpendicular to the line through the other two. The telescope is leveled alternately by one screw and by the other two, and need not be rotated during the process.

The leveling screws are turned in opposite directions at the same speed by both hands, unless the intention is to tighten or loosen the leveling head. A simple rule is that the bubble follows the left thumb. This is illustrated in Fig. 5–18.

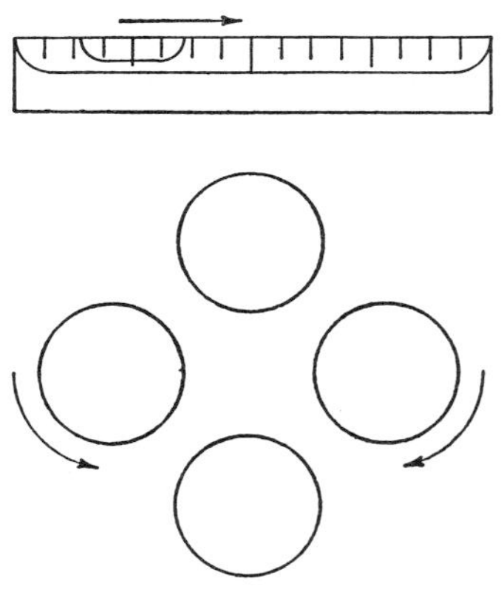

FIG. 5–18. Use of leveling screws.

If one hand turns faster than the other, the screws loosen and the head rocks on two screws, or the screws bind. Final precise adjustment may be made with one hand only. Leveling screws should be snug, not wrench-tight, to save time and avoid damage to the threads and base plate. A good instrumentman senses the proper setting of the leveling screws to permit ready movement without jamming the threads. The instrument should be leveled on the base plate before it is placed in the box.

5–21. Holding the rod. The duties of a rodman are relatively simple. Like a chainman, however, the rodman can nullify the best efforts of the observer if he fails to follow a few simple rules.

The leveling rod must be plumb to give a correct reading. In Fig. 5–19, point A is below the line of sight by a distance equal to AB. If the rod is tilted to position AD, an erroneous reading AE is obtained. It can be seen that the smallest reading possible, AB, is the correct one and this reading is secured only when the rod is plumb.

Formula 4–4 of Sec. 4–12 can be applied to determine approximately the error in reading caused by the rod not being plumb. For example, if $AB = 10$ ft and $EB = 6$ in. in Fig. 5–19, the error is

$$c = \frac{d^2}{2L} = \frac{0.5^2}{2 \times 10} = 0.012 \text{ ft}$$

Errors of this magnitude are serious, whether results are being carried to hundredths or thousandths, and make careful plumbing necessary for high rod readings.

On still days the rod can be plumbed by letting it balance of its own weight while lightly supported by the finger tips. The instrumentman makes certain the rod is plumbed in the lateral direction by checking its coincidence with the vertical cross hair and signaling for any adjustments necessary. The rodman can save time by squinting along the side of the rod to line it up with a telephone pole, a tree, or the side of a building. Plumbing on the line toward the instrument is more difficult, but holding the rod against the toes, stomach, and nose will bring it close to the plumb position.

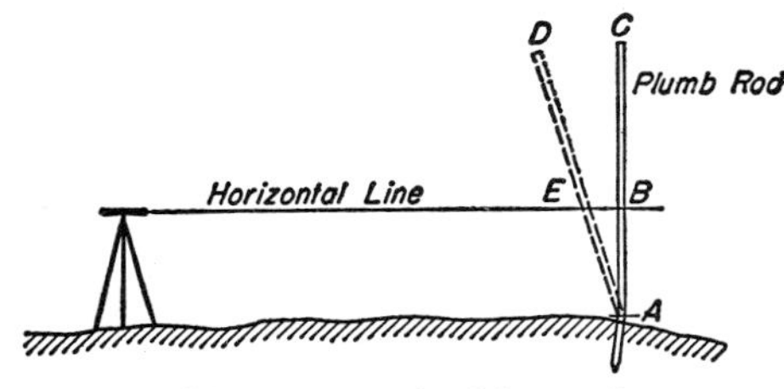

Fig. 5–19. Plumbing rod.

A rod level of the type shown in Fig. 5–20 insures fast and correct rod plumbing. Its L shape is designed to fit the rear and side faces of the rod while the bull's-eye bubble is centered to plumb the rod in both directions. The simplicity and reasonable cost of a rod level should encourage greater use of this accessory.

Fig. 5–20. Rod level. (Courtesy of Keuffel and Esser Company.)

5–22. Theory of leveling. Leveling is the process of determining the elevations or differences in elevation of points. Figure 5–21a illustrates the procedure in its basic form.

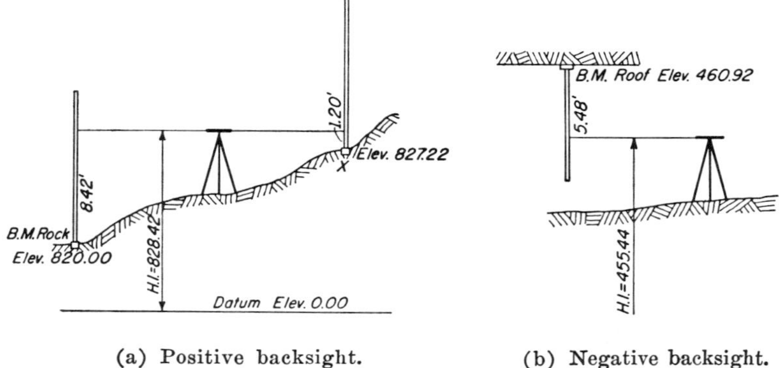

(a) Positive backsight. (b) Negative backsight.

Fig. 5–21. Theory of leveling.

A level is set up approximately halfway between B.M. Rock and point X. Assume that the elevation of B.M. Rock is known to be 820.00 ft. After leveling the instrument, a plus sight taken on a rod held on the B.M. gives a reading of 8.42 ft. A *plus sight* (+S), also termed *backsight* (B.S.), is the reading on a rod held on a point of known or assumed elevation. This reading is used in computing the height of instrument. The direction of the sight—whether forward, backward, or sideways—is not important. The term plus sight is preferable to backsight, but both are used.

Adding the plus sight 8.42 ft to the elevation of B.M. Rock, 820.00, gives the *height of instrument* (H.I.) as 828.42 ft. Height of instrument is defined as the vertical distance from the datum to the line of sight of the instrument.

Turning the telescope to bring into view the rod held on point X, a *minus sight* (−S), also called *foresight* (F.S.), of 1.20 ft is obtained. A minus sight is defined as the reading on a rod held at a point whose elevation is to be determined. The term minus sight is preferable to foresight, but both are used.

Subtracting the minus sight 1.20 ft from the H.I., 828.42, gives the elevation of point X as 827.22 ft.

All leveling theory and applications can thus be expressed by two equations which are repeated over and over:

$$\text{Elev.} + \text{B.S.} = \text{H.I.} \tag{5-7}$$

$$\text{H.I.} - \text{F.S.} = \text{Elev.} \tag{5-8}$$

Note that if a backsight is taken on a bench mark located in the roof of a tunnel, as in Fig. 5–21b, or in the ceiling of a room, with the instrument at a lower elevation, the rod would be held upside down, and the backsight must be subtracted to obtain the height of instrument. For similar conditions, a foresight would be added.

5–23. Types of leveling. The following types of leveling are performed in surveying: (a) differential leveling, (b) reciprocal leveling, (c) profile leveling, (d) barometric leveling, (e) trigonometric leveling, and (f) borrow-pit or cross-section leveling.

Each type has special applications or advantages under given conditions.

5–24. Differential leveling. Differential leveling is the process of determining the difference in elevation of two points. When the points are far apart, it may be necessary to set up the instrument several times. Figure 5–22 illustrates the procedure followed in differential leveling. Field notes for the work are shown in Plate A–3, Appendix A.

The points on which the rod is held to carry the line from one setup to the next are called *turning points* (T.P.). A turning point is defined as a solid point on which both a plus and a minus sight are taken on a line of direct levels. Horizontal distances for the plus and minus sights should be made approximately equal by pacing, by stadia measurements, by counting lengths of rails if working along a track, or by some other easy method. The effects of refraction, curvature of the earth, and lack of adjustment of the instrument are thereby eliminated by means of the principle of reversion (see page 327). On slopes, a zigzag path may be taken to utilize the longer rod length available on the downhill sights.

Bench marks are described the first time used, and are thereafter referred to by noting the page number on which detailed. The description should give the general location first and include enough particulars to enable a person unfamiliar with the area to find the mark readily. Bench marks are usually named for some prominent object which is nearby to aid in describing their location, one word

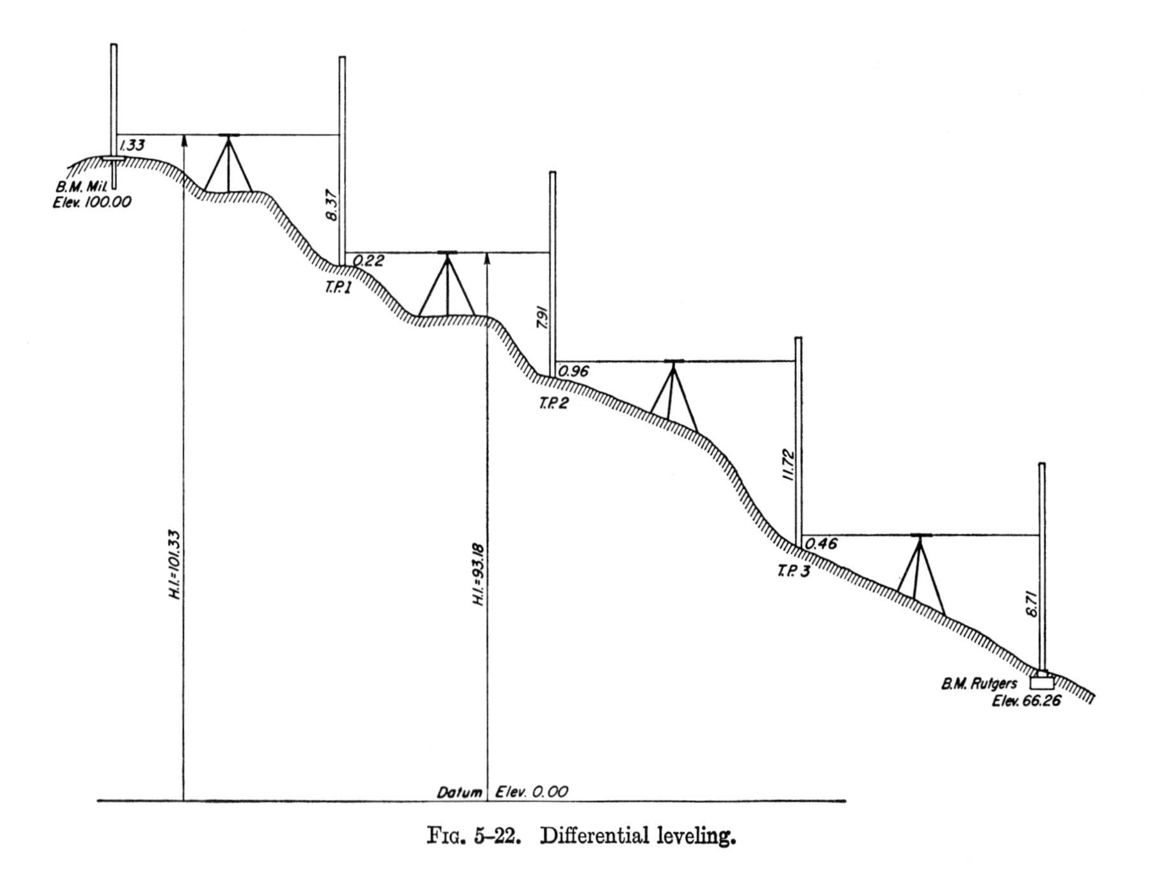

FIG. 5-22. Differential leveling.

being preferable. Examples are B.M. River, B.M. Tower, B.M. Corner, and B.M. Bridge. On extensive surveys, bench marks are given numbers. These have the advantage of giving relative positions along a line, but the numbers are more subject to mistakes in recording. Turning points are numbered consecutively along a line of levels but need not be described since they are merely a means to an end and will seldom, if ever, have to be relocated.

Before a party leaves the field, all possible note checks must be made to detect any mistakes in arithmetic. The algebraic sum of the plus and minus sights applied to the first elevation should give the last elevation. This computation checks the values of all H.I.'s and T.P.'s unless compensating errors have been made. When carried out for each left-hand page of tabulations, it is termed the "page check."

Important work is checked by leveling forward and backward between end points. The difference between the *rod sum* (algebraic total of plus and minus sights) on the run out and the rod sum on the run back is called the *loop closure.* Specifications, or the purpose of the survey, fix the permissible loop closure (see Sec. 5–34). If the permissible closure is exceeded, one or more additional runs must be made.

The difference in elevation between the end points is considered to be the average of the rod sums on the runs out and back. Where a number of interlocking "loops" are used in a network of levels, an approximate "loop adjustment" method or a least-squares adjustment can be employed to distribute the closure. True elevations can be obtained by starting from a bench mark whose elevation above mean sea level is known. If this is not possible, an assumed elevation may be used, and later all values can be converted to true elevations by application of a constant, as in Plate A–3.

Double-rodded lines of levels are sometimes used on important work. Plus and minus sights are taken on two rods from each setup of the instrument and carried in separate noteform columns. Special precautions are required to prevent errors in recording.

Flying levels may be run at the end of the work day, to check the results of an extended line run in one direction only. Longer sights and fewer setups are used, the purpose being to detect any large mistakes.

Three-wire leveling, formerly used mainly in precise work, now is employed on projects requiring only ordinary precision. Readings of the upper, middle, and lower cross hairs are averaged to obtain a better value. A check on the work is obtained by noting the difference between the middle and upper hairs, and the difference between the middle and lower hairs. If these differences fail to agree within one or two of the smallest units being read, the readings are repeated. The intercept between the upper and lower hairs gives the sight distance for checking the lengths of plus and minus sights.

5–25. Reciprocal leveling. Topographic features such as rivers, lakes, and canyons make it difficult or impossible to keep plus and minus sights short and equal. Reciprocal leveling is employed at such locations.

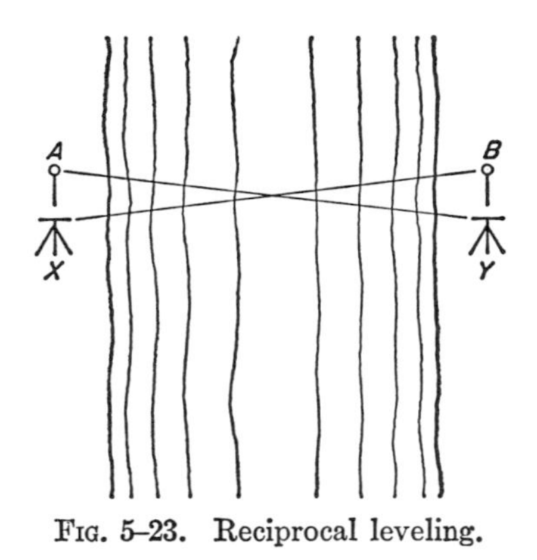

Fig. 5–23. Reciprocal leveling.

As shown in Fig. 5–23, the level is set up on one side of a stream at *X,* and rod readings are taken on points *A* and *B.* Since the sight *XB* is very long, several readings are secured for averaging. This is done by reading, turning the leveling screws to throw the instrument out of level, and then releveling and reading again. The process is repeated two, three, four, or more times. The instrument is now moved to *Y* and the same procedure followed.

The differences in elevation between *A* and *B* determined with the instrument at *X* and *Y* may differ because of curvature and refraction, and personal and instrumental errors. The average of the two differences in elevation is accepted as the correct value, if their precision is satisfactory. This procedure, which is a method of reversion, is used in making adjustments of levels and transits. Plate A–4 in Appendix A is a sample set of field notes for reciprocal leveling.

5–26. Profile leveling. On a route survey for a highway or a pipeline, elevations are required at every 100-ft station, at angle points (points marking changes in direction), at breaks in the slope of the ground surface, and at critical points such as roads, bridges, and culverts. These elevations when plotted show a *profile*—a vertical section of the surface of the ground along any fixed line. For most engineering projects the profile is taken along the center line, which is staked out in 100-ft stations or, if necessary, in 50-ft or 25-ft stations.

Profile leveling, like differential leveling, requires the establishment of turning points on which both plus and minus sights are taken. In addition, any number of *intermediate sights* (minus sights) may be obtained on points along the line from each setup of the instrument, as shown in Fig. 5–24. Plate A–5 is a sample set of level notes covering the profile shown in Fig. 5–24.

As shown in the notes, a plus sight is taken upon a bench mark, and intermediate sights are read on the stations, at breaks in the ground surface, and at critical points, until the limit of accurate sighting distance is reached. A turning point is selected, the instrument moved ahead, and the process repeated. The level itself is usually set up *off* the center line so that sights of more uniform length can be procured. Bench marks out of the way of future construction are established along the route on a long line.

It is evident that when the "page check" is made on arithmetic computations, only the minus sights taken on turning points can be used. For this reason, and to isolate the points to be plotted, a separate column is necessary for the intermediate sights.

Readings on paved surfaces, such as concrete roadways, curbs, and sidewalks, can be taken to 0.01 ft. Readings on the ground closer than 0.1 ft are not practical.

A modern development in highway surveys is the *Johnson elevation meter*. This device is towed behind a car or truck and produces a profile of the line traversed with 4th order accuracy. It can cover 100 miles of line per day at speeds up to 25 miles per hour.

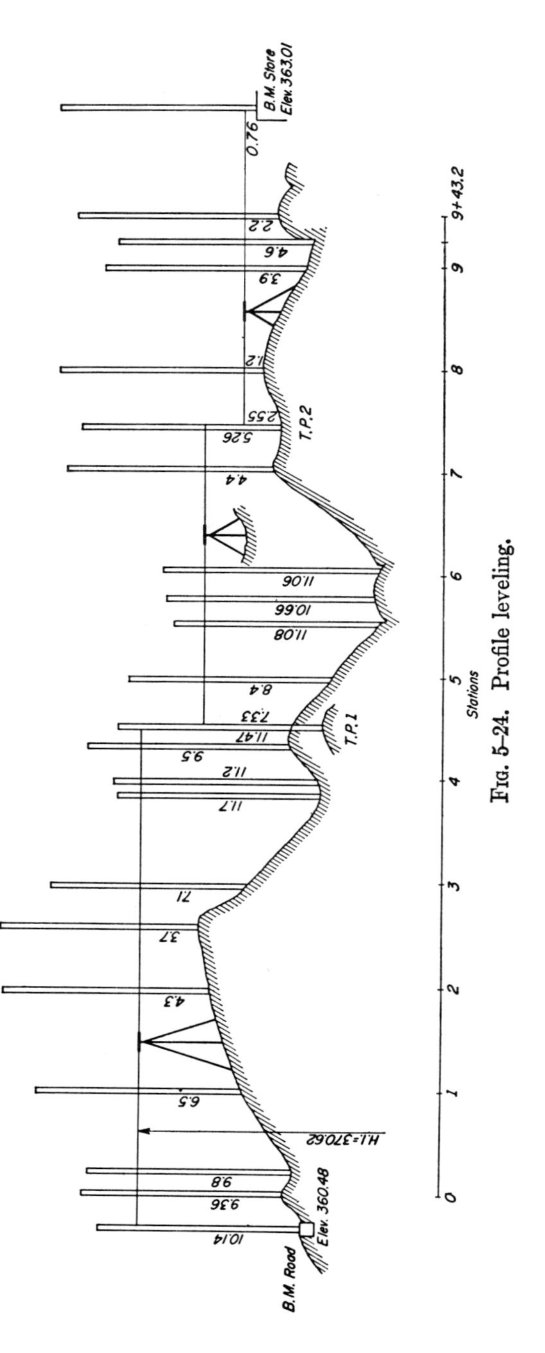

FIG. 5-24. Profile leveling.

5–27. Airborne profile recorder. Development of an airborne profile recorder was begun in the early 1940's to get a profile of the terrain beneath the path of an aircraft carrying radar equipment. The ground profile, with respect to an isobaric reference surface, was obtained by measuring the elapsed time required for radar signals to travel from the aircraft to the ground, be reflected, and return to the aircraft. By the late 1950's, the equipment was capable of vertical accuracies of 10 feet in flat terrain and 20 feet over mountainous areas.

Accuracy of airborne profile recording systems has now been improved through the use of LASER (a term coined from the descriptive phrase "Light Amplification by Stimulated Emission and Radiation"). The system operates similarly to electronic distance measuring devices in that the frequency of the LASER energy is varied, the output signal is modulated onto a carrier wave, and distance is determined by utilizing the phase shift principle.

The LASER system emits a continuous wave of extremely high intensity light which is generated by a neon-helium LASER. It is modulated onto a carrier and directed downward from the transporting aircraft in a narrow beam. The light is reflected back to the receiver which rejects the carrier and admits only the LASER light. The phase of the returning wave is compared with the phase of the outgoing wave so the distance from the aircraft to the ground is known to be an even number of half wavelengths plus the fraction indicated by the phase shift. Varying the frequency of the transmitted energy resolves the ambiguity of the unknown number of half wavelengths.

Flying heights are normally kept low during LASER profiling to get better accuracy. Altitudes of the aircraft above the terrain are usually recorded in digital form. The accuracy of LASER airborne profile recorders is quite phenominal, tests showing elevations to be correct to within one foot at flying heights of 1000 feet. Resolution of the system actually approximates 0.1 foot (a precision of 1/10,000) but the accuracy of the isobaric reference datum limits the terrain profile correctness to the higher figure.

5–28. Drawing and use of the profile. Profiles are usually plotted on a special paper called *Plate A profile paper,* which is ruled as shown in Fig. 5–25. Vertical lines are spaced $\frac{1}{4}$ in. apart, with every tenth line heavier. Horizontal lines are $\frac{1}{20}$ in. apart, with every fifth line heavier and every fiftieth line still heavier.

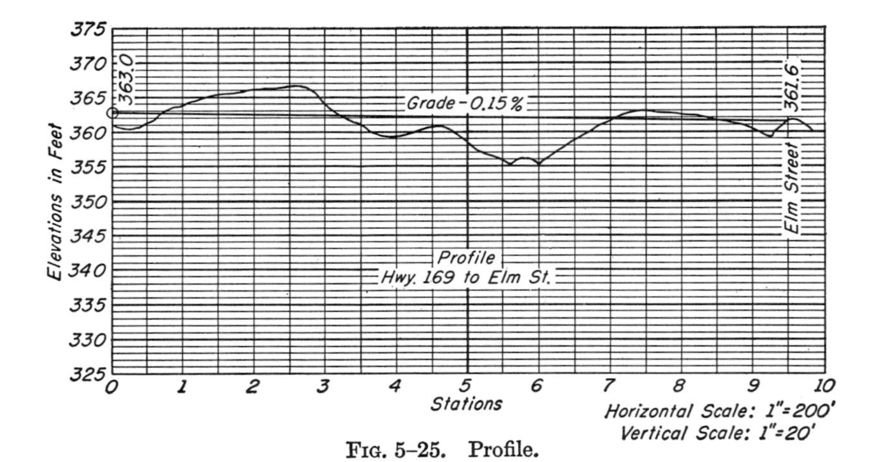

FIG. 5–25. Profile.

The vertical scale of a profile is generally exaggerated with respect to the horizontal scale in order to make differences of elevation more pronounced. A ratio of 10 to 1 is frequently used. Thus for a horizontal scale of 1 in. = 100 ft, the vertical scale would be 1 in. = 10 ft. The heavy lines of Plate A paper make blocks $2\frac{1}{2}$ in. by $2\frac{1}{2}$ in. and are best suited to a scale of 1 in. = 40 ft (or 400 ft) horizontally, and 1 in. = 4 ft (or 40 ft) vertically. The scale actually used should be plainly marked.

Curves, or in some cases straight lines, are drawn to connect the plotted elevations.

The plotted profile is used for many purposes, such as (a) determination of the depth of cut and fill on a proposed highway or airport, (b) study of grade-crossing problems, (c) investigation and selection of the most economical grade, location, and depth for sewers, pipelines, tunnels, irrigation ditches, and other projects.

Rate of grade (or *gradient*, or just *grade*) is the rise or fall in feet per 100 feet. Thus a grade of 2.5 per cent means a 2.5-ft difference in elevation per 100 ft horizontally. Ascending grades are plus, descending grades are minus. A grade line selected to give somewhat equal cuts and fills is shown in Fig. 5–25. The process of staking out grades is described in Chapter 24.

The term "grade" is also employed to denote the elevation of the finished surface of an engineering project.

5–29. Barometric leveling. The *barometer*, an instrument which measures air pressure, can be used to find relative elevations of points on the surface of the earth. Figure 5–26 shows a modern surveying *altimeter*. Calibration of the scale on different models is in multiples of 2 to 10 ft.

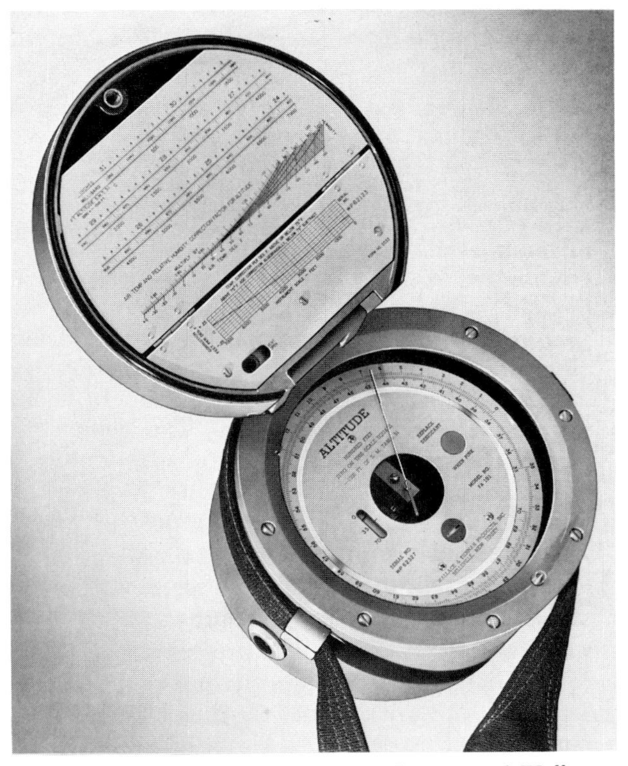

FIG. 5–26. Surveying altimeter. (Courtesy of Wallace and Tiernan.)

Air pressures are affected by conditions other than difference in elevation—for example, sudden changes in temperature and changing weather conditions due to storms. Also, there is during each day a normal variation in barometric pressure amounting to perhaps 100 ft difference in elevation and known as the diurnal range.

In barometric leveling, one or more control barometers remain on a bench mark (base) while the *roving* instrument is taken to points whose elevations are desired. The controls on the bases are read at stated intervals of time, perhaps every 10 min, and the elevations are recorded along with the temperature and time. Readings of the roving barometer are taken at critical points and adjusted later in accordance with changes observed at the control points. Methods of making field surveys by barometer have been developed in which one, two or three bases may be used. Other methods employ leapfrog or semi-leapfrog techniques.

The barometric method is particularly suited for work in rough country where a high order of accuracy is not necessary.

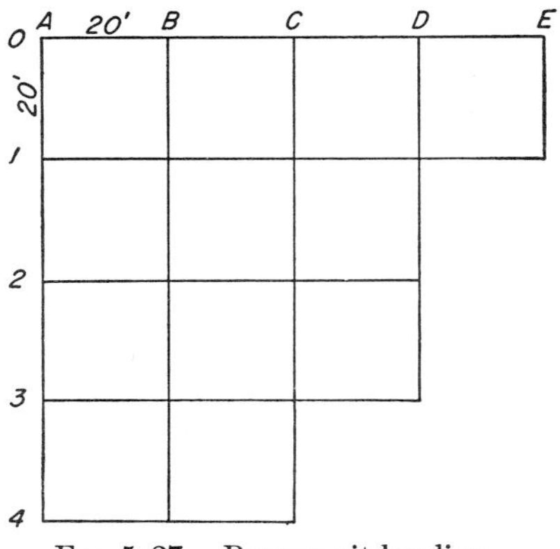

FIG. 5–27. Borrow-pit leveling.

5–30. Borrow-pit or cross-section leveling. The amount of earth, gravel, rock, or other material excavated or filled on a construction project may be determined by borrow-pit leveling. The quantities computed form the basis for payment to the contractor or materials supplier. The number of cubic yards of coal or other loose materials in stockpiles can be found in the same way.

As an example, assume the area shown in Fig. 5–27 is to be graded to an elevation of 358.0 for a building site. Notes for the field work are shown in Plate A–6.

The area to be covered is staked in squares of 10, 20, 50, 100, or more feet, the choice depending upon the size of the project and the accuracy desired. A transit and/or tape may be used for the layout. A bench mark of known or assumed elevation is established outside the area in a place not likely to be disturbed.

The level is set up at any convenient location from which a plus sight can be taken on the bench mark and minus sights are read on the corners of the squares. If the terrain is not too rough, it may be possible to select a point near the center of the area and take sights on all the corners from the same setup, as in this example.

Corners of the squares are designated by letters and numbers, such as *A–1*, *C–4*, and *D–2*. Since the site is to be graded to an elevation of 358.0, the amount of cut or fill at each corner can be obtained by subtracting 358.0 from its elevation. For each square, then, the average height of the four corners of the prism of cut or fill is determined and multiplied by the base area, 20 ft × 20 ft = 400 sq ft, to get the volume. The total volume is found by adding the individual values for each block and dividing by 27 to obtain a result in cubic yards.

As a simplification, the cut at each corner multiplied by the number of times it enters the volume computation can be shown in a separate column, a total secured, and divided by 4. The result multiplied by the base area of one block gives the volume. This procedure is shown in the sample noteform.

5–31. Use of the hand level. A hand level can be used for some types of work, such as differential leveling, when a high order of precision is not required. The instrumentman takes a plus and a minus sight while standing in one position, then moves ahead to repeat the process.

In hilly country it may be most convenient to locate contours without reading the leveling rod. A *contour* is a line that passes through points having the same elevation. The observer first measures his H.I.—the height of his eye above the ground. The H.I. above a datum is determined by taking a plus sight on a point of known elevation, as in differential leveling. The point at which a level line of sight strikes the ground will have the same elevation as the observer's H.I. After identifying this point on the ground, the instrumentman moves there and uses the elevation to compute a new H.I.

In leveling downhill, the observer finds by trial the point where he must stand to make a backsight strike the ground at the previously occupied position. The elevation of the trial point is then found by subtracting the observer's H.I., and the process is repeated. In locating 5-ft contours, adoption of an exact 5-ft H.I. by using a rod or forked stick is desirable to speed the work and simplify the arithmetic.

Cross-section leveling for a route survey is discussed in Chapter 24.

5–32. Size of field party. Ordinary differential leveling can be done efficiently by a two-man party. The instrumentman keeps notes if a self-reading rod is used; the rodman records if a target rod is employed. On precise leveling, an observer, an umbrellaman, a notekeeper, and two rodmen are required.

A self-reading rod is practically always used on borrow-pit and profile leveling, hence a party of two men is sufficient. Using a third man to keep notes relieves the observer of this task, however, and permits the party to move faster.

5–33. Signals. The distance between personnel, and noise from traffic or other sources, make it necessary to communicate by hand signals on many surveys. Some typical signals used in leveling will be listed in the following paragraphs of this section. Special signals to fit unusual requirements may be invented for a particular job. They should simulate as closely as possible the action to be taken. The instrumentman must remember that he has the advantage of telescopic magnification and give clear signals which cannot be misunderstood by a rodman using only natural vision. Equipping the rodman with a small telescope is helpful.

Plumb rod. If the rod is to the right of a plumb position, the instrumentman extends his right arm full length upward and inclined to the vertical. This position is maintained until the rod is plumb.

Give or take a T.P. Either the instrumentman or the rodman may give this signal by holding one arm straight up and moving it in a horizontal circle.

High rod. Instrumentman extends both arms horizontally and sideways, palms up, and brings them together over his head.

Raise for red. Instrumentman holds one arm straight forward, palm up, and raises the arm slowly to a position about 45° above the horizontal.

Raise target. Instrumentman raises an extended arm above his shoulder, holding it high if considerable movement is required. The arm is moved toward the horizontal position as the target approaches the desired setting.

Lower target. Same as "raise target," but the extended arm is held below the shoulder and moved up.

Clamp target. Instrumentman waves one hand in a vertical circle with his arm in a horizontal position.

T.P. or B.M. Rodman holds the rod horizontally above his head, then places it on the T.P. or B.M. Used in profile leveling to differentiate between intermediate sights and T.P.'s or B.M.'s for the benefit of the instrumentman and notekeeper.

All right. Arms are extended sideways, palms forward, and waved up and down several times. Used by any member of the party in all types of surveying.

Signals for numbers. One of the systems used is shown in Fig. 5–28.

Fig. 5–28. Signals for numbers.

5–34. Precision. Precision in leveling, as in taping, is determined by repeating the measurement or tying in to control points. The elevation of a bench mark may be obtained by leveling over two different routes, or by a closed circuit of levels returning to the point of beginning. If an accurately established bench mark is available at or near the end of the line run, a check can be made upon it.

Closures are compared with permissible values on the basis of either the number of setups or the distance covered in miles. The type of formula generally used to compute the allowable closure is

$$C = k\sqrt{M} \tag{5-9}$$

where C = allowable closure, in feet;
k = a constant;
M = distance, in miles.

The United States Coast and Geodetic Survey specifies constants of 0.017, 0.035, and 0.050 for its three classes of leveling now designated as first, second, and third order, respectively. As an example, if differential levels are run from an established B.M. A to a new B.M. B $\frac{3}{4}$ mile away, and back, with elevation differences of 10.025 ft and 10.047 ft, respectively, the closure is 0.022 ft. Then $k = C/\sqrt{M} = 0.022/\sqrt{\frac{3}{4} + \frac{3}{4}} = 0.018$ ft, and the leveling is of second-order accuracy. For leveling (fourth-order) a coefficient of 0.10 might be used.

If sights of 300 ft are taken, thereby spacing instrument setups at 600 ft, approximately nine setups per mile would be used. For ordinary work the allowable error of closure might then well be

$$E_C = 0.03\sqrt{N} \tag{5-10}$$

where E_C is the allowable closure, in feet, and N is the number of times the instrument is set up. Other values of k may be specified to correspond to the precision required and the average length of sight which can be taken on a very steep hillside. On first-order leveling the length of sight is varied during the day to conform to atmospheric conditions.

5–35. Adjustment of simple level circuits. Since permissible closures for level circuits are based upon lengths of lines or numbers of setups, it is logical to adjust elevations on these bases. Elevation differences and lengths of lines are shown for a circuit in Fig. 5–29. Adjustment of this level circuit will be demonstrated on the basis of the lengths of the lines. The closure found by algebraic summation of the elevation differences, is + 0.24 ft. Adding the lengths of the lines yields a total circuit length of 3.0 miles. Elevation adjustments are then (0.24 ft/3.0) times the corresponding

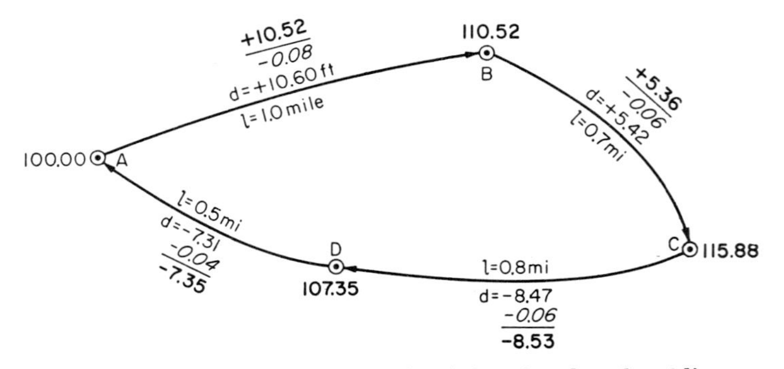

FIG. 5–29. Adjustment of level circuit based on lengths of lines.

lengths of lines, giving —0.08, —0.06, —0.06, and —0.04 ft as shown in italics. The adjusted elevation differences are used for finding the final elevations (shown in bold-face type) for the bench marks, which are based upon elevation of B.M. A = 100.00.

Level circuits with different lengths and routes are sometimes run from scattered reference points to obtain the elevation of a given bench mark. The most probable value of the bench-mark elevation can then be computed from the weighted mean of the observations, the weights varying inversely with the lengths of lines.

Adjustment of an interlocking network of level circuits is best accomplished by the least-squares method or one of the several approximate methods available. An interlocking two-loop level circuit and the corresponding field notes are shown in Fig. 5–30.

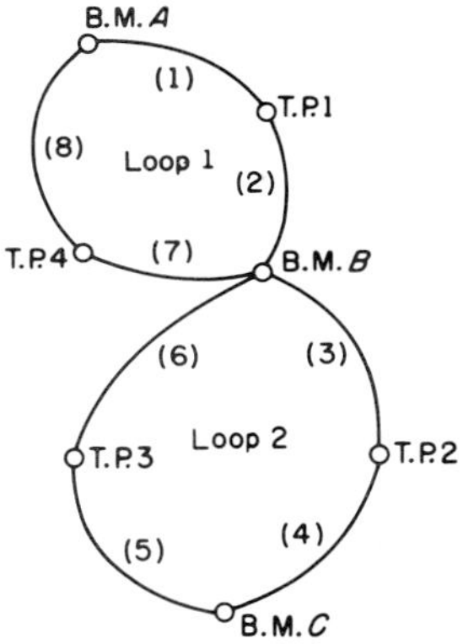

STA	+ Sight	H. I.	− Sight	Elev.
B.M. *A*	2.51	102.51		100.00
T.P. 1	2.97	104.44	1.04	46 101.~~47~~
B.M. *B*	7.53	108.15	3.82	60 100.~~62~~
T.P. 2	6.30	113.44	1.01	10 ~~12~~ 107.~~14~~
B.M. *C*	4.75	112.91	5.28	10 ~~12~~ 108.~~16~~
T.P. 3	0.96	107.39	6.48	35 ~~37~~ 106.~~43~~
B.M. *B**	4.32	105.02	6.69	60 ~~62~~ 100.~~70~~
T.P. 4	6.89	109.69	2.22	69 ~~72~~ 102.~~80~~
B.M. *A**			9.57	00 ~~04~~ 100.~~12~~

Fig. 5–30. Interlocking two-loop level circuit and corresponding field notes.

Adjustment by an approximate "loop circuit" method on the basis of the number of instrument set-ups will be described.

On Fig. 5–30, the instrument set-up number is given in parentheses. In the field notes, the second determination of the elevation of points (loop closures) is marked with an asterisk. The outer circuit, or loop 2, is first adjusted. From the notes it is seen that the elevation of BMB was first determined as 100.62 and after running levels around loop 2 and returning, the elevation of B.M.*B** was found to be 100.70 giving a closure error of + 0.08 ft for loop 2. This error of closure is distributed based on the number of instrument set-ups in the loop, as —0.08/4 = —0.02 ft per set-up. Corrections of —0.02 ft per set-up result in values of —0.02, —0.04, —0.06 and —0.08 ft respectively for TP-2, B.M.*C*, TP-3 and BM*B**. When applied to loop 2, the initial corrected elevations for these points are shown directly above the crossed out original unadjusted elevations in the notes.

Since the elevation of B.M.*B** has been corrected by —0.08 ft to 100.60, and set-ups (7) and (8) depend upon B.M.*B** as a reference base, the elevations of TP-4 and B.M.*A** which follow must also be corrected by —0.08 ft. As a result of the adjustment of loop 2, the adjusted elevation of B.M.*A*.* is 100.12 – 0.08 = 100.04. The final elevation of B.M.*A*.* must be 100.00, therefore a + 0.04 ft error of closure exists in loop 1 which is also distributed on the basis of the number of instrument set-ups in loop 1, as —0.04/4 = —0.01 ft per set-up. TP-1 and B.M.*B* are corrected by —0.01 and —0.02 ft respectively. Instrument set-ups (3) through (6) depend upon B.M.*B* as a reference elevation, and as B.M.*B* has just been corrected by —0.02 ft, the elevations of TP-2, B.M.*C*, TP-3 and B.M.*B** are also adjusted by —0.02 ft. Finally, continuing around loop 1, corrections of —0.03 and —0.04 ft are applied to TP-4 and B.M.*A** respectively. This completes the "loop circuit" adjustment and the final adjusted elevations for all points are shown above the crossed out original and initial corrected elevations.

5–36. Sources of error in leveling. All leveling measurements are subject to three sources of error: (1) instrumental, (2) natural, and (3) personal.

Instrumental Errors

Instrument not in adjustment. The most important adjustment of the level makes the line of sight parallel to the axis of the level vial so that a horizontal plane, rather than a conical surface, is generated as the telescope is revolved. Serious errors in rod readings result if the instrument is not so adjusted, unless the horizontal lengths of plus and minus sights are kept equal to utilize the principle of reversion (see page 327). The error is likely to be systematic, particularly in going up or down a steep hill where all plus sights are longer or shorter than all minus sights unless care is taken to run a zigzag line.

Rod not correct length. Incorrect lengths of the divisions on a rod cause errors similar to those resulting from incorrect markings on a tape. Uniform wearing of the shoe at the bottom of the rod makes H.I. values incorrect, but the effect is cancelled when included in both plus and minus sights. Rod graduations should be checked by comparing them with those on a standardized tape

Tripod legs loose. Tripod leg bolts which are too loose or too tight allow movement or strain which affect the instrument head. Loose metal tripod shoes cause unstable setups.

NATURAL ERRORS

Curvature of the earth. As noted in Sec. 5–3, a level surface deflects from a horizontal plane at the rate of $0.667M^2$, or about 8 in. in 1 mile. The effect of curvature of the earth is to increase the rod reading. Equalizing the lengths of plus and minus sights cancels the errors due to this cause.

Refraction. Light rays coming from an object to the telescope are bent, making the line of sight a curve concave to the surface of the earth and thereby decreasing the rod reading. Balancing the lengths of plus and minus sights usually eliminates the errors due to refraction, which are negligible in ordinary leveling. Large and sudden changes in atmospheric refraction may be important in precise work. Errors due to refraction tend to be accidental over a long period of time but could be systematic on one day's run.

Temperature variations. Heat causes leveling rods to expand, but the effect is not important in ordinary leveling.

If a level vial is heated, the liquid expands and the bubble shortens. This does not produce an error (although it may be inconvenient), unless one end of the tube is warmed more than the other and the bubble therefore moves toward the heated end. Other parts of the instrument warp because of uneven heating, and this distortion affects the adjustments. Shading the level by means of a cover when carrying it, and by an umbrella when it is set up, will reduce or eliminate heat effects. These precautions are followed in precise leveling.

Air boiling or heat waves, near the ground surface or adjacent to heated objects, make the rod appear to wave, and prevent accurate sighting. Raising the line of sight by high-tripod setups, avoiding sights which pass close to heat sources, such as buildings and stacks, and using the lower magnification of a variable-power eyepiece, reduce the effect.

Wind. Strong wind causes the instrument to vibrate and makes the rod unsteady. Precise leveling is not attempted on windy days.

Settlement of the instrument. Settlement of the instrument after the plus sight has been taken makes the minus sight too small and therefore makes the recorded elevation of the next point too great. The error is cumulative in a series of setups on soft material. Unusual care is required in setting up the level on spongy ground, blacktop, or ice, and readings must be taken in quick order, perhaps using two rods, and without walking around the instrument. Alternating the system of taking plus and minus sights helps somewhat.

Settlement of the rod. This condition causes an error similar to that resulting from settlement of the instrument. It can be avoided by selecting firm, solid turning points or, if none are available, by using a steel *turning pin.*

Personal Errors

Bubble not centered. Errors due to the bubble not being exactly centered at the time of sighting are the most important of any, particularly on long sights. If the bubble goes off between the plus and minus sights, *it must be recentered before the minus sight is taken.* Experienced levelmen develop the habit of checking the bubble before and after each sight, a procedure simplified by having a mirror-prism arrangement permitting simultaneous view of level vial and rod.

Parallax. Parallax caused by improper focusing of the objective and/or eyepiece lens results in incorrect rod readings. Careful focusing eliminates this condition.

Faulty rod readings. Incorrect rod readings result from parallax, poor weather conditions, long sights, improper target setting and rodding, and other causes including mistakes such as those due to careless interpolation and transposition of figures. Short sights selected to fit weather and instrument conditions reduce the number of reading errors. If a target is used, the rodman should read the rod for the plus sight and have the instrumentman check it independently as the rodman passes on the way to the next T.P. The observer stops to read the minus-sight setting as he moves ahead for the next setup.

Rod handling. Serious errors caused by improper plumbing of the rod are eliminated by using a rod level that is in adjustment. Banging the rod on a T.P. for the second (plus) sight may change the elevation of the point.

Target setting. The target may not be clamped at the exact place signaled by the instrumentman because of slippage. A check sight should always be taken after the target is clamped.

5-37. Mistakes. A few common mistakes in leveling are listed here for study.

Use of long rod. If the vernier reading on the back of a damaged rod is not exactly 6.500 or 7.000 for the short rod, the target must be set to read the same value before extending the rod.

Holding the rod in different places for the plus and minus sights on a T.P. The rodman can avoid such mistakes by using a well-defined point or by outlining the base of the rod with keel.

Reading a foot too high. This error occurs because the incorrect foot-mark is in sight near the cross line. For example, the observer may read 5.98 instead of 4.98. Noting the foot-marks above and below the vernier zero or cross line will usually prevent this mistake.

Waving the ordinary flat-bottom rod while holding it on a flat surface. This action produces an error because the rotation is about the rod edges instead of the center or the front face. Plumbing by using a rod level or by other means is preferable to waving which requires a mark with rounded top surface for good results.

Recording notes. Mistakes in recording, such as transposition of figures, entering values in the wrong column, and arithmetic mistakes, can be minimized by having the notekeeper mentally estimate the reading, by repeating the value called out by the observer, and by making the standard field-book checks on rod sums and elevations.

5-38. Reduction of errors and elimination of mistakes. Errors in leveling are reduced (but never eliminated) by careful adjustment and manipulation of both instrument and rod, and by establishing standard field procedures and routines. The following procedures eliminate most large errors or quickly disclose mistakes: (a) checking the bubble before and after each reading; (b) using a rod level; (c) keeping the horizontal lengths of plus and minus

sights equal; (d) running lines forward and backward; and (e) making the usual field-book arithmetic checks.

PROBLEMS

5–1. Compute and tabulate the combined effect of curvature and refraction on level sights of 100, 200, 300, 400, 500, and 1000 ft.

What error in elevation results from the combined effects of curvature and refraction on a downhill line of differential levels having successive B.S. and F.S. distances in feet as shown in problems 5–2 through 5–4?

5–2. 100, 200; 150, 250; 125, 150; 210, 250.
5–3. 160, 180; 160, 200; 220, 240; 90, 150.
5–4. 80, 120; 140, 180; 120, 160; 80, 150.

5–5. On a large lake without waves, how far from shore will a boat with a 20-ft mast disappear from view of a man lying on the beach at the water's edge?

5–6. Similar to problem 5–5 except the man is standing at the water's edge and his height of eye is 5.0 ft.

5–7. It is required to obtain the elevation of the top of a television tower located on the roof of a building. Since direct measurement is not feasible, the following field data were obtained: A line *AB*, 437.2 ft long, was staked out and the horizontal angles to the tower were measured to obtain $A = 46°18'$ and $B = 32°45'$. At point B a backsight of 5.27 ft was taken on a bench mark of elevation 625.83 ft, and the vertical angle to the top of the tower was found to be 56°52′. Calculate the elevation of the top of the tower.

5–8. A reading of 5.53 ft is taken on a rod 200 ft from a level with the bubble centered. The bubble is then moved 4 divisions off center, and a rod reading of 5.69 is obtained. If each division on the vial is 0.1 in., what is (a) the radius of curvature of the vial in feet, and (b) the angle in seconds subtended at the center by one vial division?

5–9. Similar to problem 5–8 except that each vial division is 0.01 ft.

5–10. Similar to problem 5–8 except that each vial division is 2 mm.

5–11. The bubble in a level vial on an engineer's level is drawn off 1½ divisions by heating of the forward (objective-lens) end of the tube. After releveling, if a plus sight of 8.16 ft is obtained on a shot of 200 ft, and the sensitivity of the vial is 20 sec per 2-mm division, what is the correct rod reading?

5–12. Similar to problem 5–11 except for a 30-sec per 0.01 ft level vial.

What is the sensitivity of a level vial for the conditions of problems 5–13 through 5–15?

5–13. Radius of curvature of 80 ft and 2-mm graduations.

5–14. Radius of curvature of 15 ft and ⅛-inch divisions.

5–15. Radius of curvature of 120 ft and 0.01-ft graduations.

5–16. What relationship should exist between the sensitivity of the level vial and the resolution-magnification of the telescope on an engineer's level?

5–17. A levelman fails to check the bubble and it is off 3 divisions on a sight of 200 ft. What error results for a 40-sec bubble? (Use the approximate relationships given in Sec. 8–1 to solve without tables or slide rule.)

5–18. Similar to problem 5–17, but assume 1½ division off, a 150-ft sight, and a 15-sec bubble.

5–19. An instrumentman using an unfamiliar engineer's level has been told that it has a 20-sec vial, but he does not know whether the divisions on the tube are 0.1 in., 0.01 ft, or 2 mm. What error in reading would result for each type if the bubble is off ½ division on a 240-ft sight?

5–20. Describe two ways of determining approximately the magnifying power of a telescope.

5–21. List in tabular form, for comparison, the advantages and disadvantages of the wye level, the dumpy level, and the European-type level.

5–22. A slope distance of 120 ft and a slope of 16 per cent are read with an Abney level. Compute the horizontal distance and the difference in elevation.

Sketch scales and verniers to fit the requirements given in problems 5–23 through 5–25.

5–23. Scale, ½-inch divisions; required reading to 1/32 inch. Scale, 1° graduations; required reading to 5 minutes.

5–24. Vernier has 12 divisions; required reading to 3 seconds. Scale, 4-minute spaces; least count if vernier has 48 divisions.

5–25. Barometer scale in inches and fiftieths; required reading to 0.002 inch. Scale in ½-pound units; required reading to 2 ounces.

5–26. Compute the distance a rod extended for a 13-ft reading must be out of plumb to introduce an error of 0.01 ft.

5–27. What error results on a 200-ft sight with an engineer's level if the rod reading is 9.00 but the top of the 12-ft rod is 4 in. out of plumb?

5–28. Complete the following set of "open-type" differential-leveling notes, and show the arithmetic check. What order of leveling does this work represent if the distance from B.M. 8 back to B.M. 8 along the circuit is 1000 ft?

Point	+S (B.S.)	H.I.	−S (F.S.)	Elev.
B.M. 8	11.46			126.22
T.P. 1	6.48		7.20	
B.M. 12			4.65	
B.M. 12	4.61			
T.P. 2	2.28		6.05	
T.P. 3	7.52		9.89	
B.M. 8			4.60	

5–29. Similar to problem 5–28, but assume that B.M. 12 is on the underside of a roof and above the H.I. and that the readings on it are obtained by holding the rod upside down on the B.M.

5–30. Complete the following set of "closed-style" differential-leveling notes, and show the arithmetic check. Note that readings marked with an asterisk (*) are "side shots" on critical points.

Point	+S (B.S.)	H.I.	−S (F.S.)	Elev.
B.M. 1	0.38			210.46
T.P. 1	4.27		12.10	
C.B.			*6.12	
T.P. 2	7.62		9.43	
T.P. 3	2.44		11.32	
Hyd.			*4.14	
T.P. 4	6.09		7.78	
B.M. 2			8.07	

5–31. Similar to problem 5–30, but assume that B.M. 1 is a mark on a building column above the line of sight and therefore the + S reading is −0.38 ft.

5–32. A differential-leveling loop started and closed on B.M. Post, the starting elevation of which was taken as 156.84 ft. The loop was run over fairly uniform terrain, and the B.S. and F.S. distances were kept approximately equal. Readings were taken in the following order: 2.34, 9.16; 2.05, 11.43; 5.21, 6.35; 10.34, 2.89; 12.87 and 2.94. Prepare a set of level notes on ruled paper and make the usual arithmetic check.

5–33. Complete the following differential-leveling notes for a double-rodded line. Show the standard page check, and determine the final elevation of B.M. 58.

Point	+S (B.S.)	H.I.	−S (F.S.)	Elev.
B.M. 57	8.487			835.124
T.P. 1H	3.820		7.875	
T.P. 1L	4.089		8.125	
T.P. 2H	3.924		8.306	
T.P. 2L	4.642		9.021	
T.P. 3H	3.875		10.430	
T.P. 3L	4.489		11.056	
B.M. 58			9.176	

5–34. For a level set midway between points *A* and *B*, the rod readings are 7.82 ft on *A* and 5.16 on *B*. The instrument is then moved to within a few feet of *A*, and readings of 6.95 ft on *A* and 4.21 ft on *B* are recorded. What is the true difference in elevation between *A* and *B*? With the instrument still near *A*, what rod reading is required on *B* to put the instrument in adjustment?

5–35. A level set up near point *A* reads 4.65 ft on *A* and 6.18 ft on point *B*.

The instrument is then moved near point B where readings of 5.98 ft on B and 4.51 ft on A are obtained. What is the true difference in elevation between points A and B? For the same setup near B, what should be the reading on a rod at A to put the instrument in adjustment?

5–36. The line of sight of a dumpy level is found by the peg adjustment test (similar to problems 5–34 and 5–35, and described in Chapter 16) to be inclined downward 0.009 ft per 100 ft of distance. What is the allowable difference between the B.S. and F.S. distances from each set up, if successive elevations are to be correct within 0.002 ft?

5–37. Complete the following set of profile-leveling notes, and show the arithmetic check.

Sta.	+S (B.S.)	H.I.	−S		Elev.
			F.S.	Int. F.S.	
B.M. *A*	1.24				126.40
T.P. 1	1.11		10.39		
1+00				6.4	
2+00				8.2	
3+00				7.6	
4+00				10.3	
T.P. 2	2.37		11.68		
4+40				1.1	
5+00				3.4	
6+00				5.8	
T.P. 3	3.64		7.29		
B.M. *B*			5.91		

5–38. Complete the following set of profile-leveling notes, and show the arithmetic check.

Sta.	+S (B.S.)	H.I.	−S (F.S.)	Elevations	
				B.M. & T.P.	Ground
B.M. *X*	3.519			412.635	
0+00			2.0		
1+00			7.3		
2+00			11.1		
3+00			10.4		
T.P. 1	5.466		6.872		
3+60			7.8		
4+00			3.9		
4+50			2.68		
5+00			1.4		
5+65			0.81		
BM *Y*			2.556		

5–39. Reciprocal leveling across a river gives the following readings in feet

from the first setup near *A*: on *A*, 3.256; on *B*, 6.814, 6.813, 6.815. For the setup near *B* the readings are: on *B*, 8.360; on *A*, 4.798, 4.797, 4.799. If the elevation of *B* is 348.212 ft, what is the elevation of *A*? What is the error of closure?

5–40. Reciprocal leveling across a canyon between B.M.'s X and Y gives the results shown. The correct elevation of X is 456.78 ft. What is the elevation of Y?
Inst. at X: $+S = 2.61$; $-S = 5.78, 5.79, 5.78$ ft.
Inst. at Y: $+S = 6.20$; $-S = 3.02, 3.02, 3.01$ ft.

5–41. A rectangular lot 150 ft N-S by 200 ft E-W is to be graded to an elevation of 215.0 ft. After division into 50-ft squares, rod readings are taken at the corners successively from east to west along E-W lines with an H.I. of 228.9. The readings in feet are as follows: 8.2, 7.6, 6.1, 5.0, 3.2; 7.5, 5.1, 4.3, 3.6, 3.2; 6.4, 3.9, 2.7, 2.3, 1.8; 5.3, 4.4, 1.7, 2.8, 0.9. Compute the volume of material (in cubic yards) to be excavated.

5–42. Similar to problem 5–41, but assume that the H.I. is 214.2 and the area is to be filled to an elevation of 215.0 ft.

5–43. Leveling between B.M.'s *A, B, C, D,* and *A* gives differences of elevation in feet of -15.630, $+32.452$, $+38.213$, and -55.022, and distances in miles of 0.4, 0.5, 0.5, and 0.6, respectively. If the elevation of *A* is 653.214 ft, what are the adjusted elevations of B.M.'s *B, C,* and *D*? What order of leveling does this represent?

5–44. Differential leveling from B.M.'s A to X, B to X, and C to X gives results in feet of -30.25, $+10.18$, and $+26.20$, respectively. The distances in feet are: $AX = 3000$, $BX = 2000$, and $CX = 4000$. Accepted elevations of the B.M.'s in feet are: $A = 570.86$, $B = 530.47$, and $C = 514.36$. What is the adjusted elevation of B.M. X?

5–45. A leveling rod $1\frac{1}{2}$ in. square is held on the flat top of a granite step for a turning point. The line of sight of the level is just above the base of the rod. Instead of plumbing with a rod level, the rodman waves the rod to let the instrumentman get the smallest reading; hence the rod tips alternately on the front and rear edges of the base plate. If a minimum reading of 0.16 ft is obtained while the rod is tipped on the rear edge, determine the true height of the instrument above the turning point.

5–46. A precise level is used on a tower to sight across a 20-mile fiord to a target on a tower of equal height. If the line of sight must be kept 8 ft above the water surface, what is the required height of tower above the shore line?

5–47. Readings on a line of differential levels are being taken to the nearest 0.01 ft. What is the maximum length of sight for which curvature of the earth and refraction can be neglected?

5–48. What two factors determine the size of squares used in borrow-pit surveys?

5–49. What are the primary differences between "ordinary" and "precise" leveling?

5–50. On what grade does the hypotenuse of a 30°–60° triangle lie, assuming one side of the 30° angle is horizontal?

5–51. Give two advantages of three-wire leveling over ordinary leveling.

5–52. What errors in leveling are eliminated by keeping the lengths of the plus sights and minus sights equal?

5–53. Explain how errors due to lack of adjustment of the instrument can be practically eliminated in running a line of differential levels.

5–54. Why should a long base be used in determining the height of an inaccessible point by trigonometric leveling?

5–55. List the considerations that govern a rodman in selecting turning points and bench marks.

5–56. How can errors due to settlement of the instrument and rod be reduced?

5–57. What is the simplest practical way to lay out a "level" ship model or missile test channel to contain water and have a track consisting of two rails 4000 ft long above it?

5–58. List 10 common errors in leveling, indicate whether they are systematic or accidental, and state how each can be reduced or eliminated.

5–59. From your study of the theory of errors, taping, and leveling, explain why it is usually easier to do good leveling than good taping.

What is the permissible error of closure for the lines of levels given in problems 5–60 through 5–62?

5–60. A line of third-order levels, 18-mile long circuit.

5–61. A second-order level circuit, 30 miles long.

5–62. A first-order level circuit of 50 miles.

5–63. What is the order of leveling of a 12-mile loop with a closure of 0.11 ft?

5–64. Compute and adjust the following notes for an interlocking two-loop level circuit using the approximate method discussed in Sec. 5–35.

Sta.	+ Sight	H.I.	− Sight	Elev.
B.M. *X*	3.87			735.20
T.P. 1	8.24		4.11	
Hyd	9.80		3.07	
B.M. *Y*	0.75		5.01	
T.P. 2	7.96		2.32	
T.P. 3	10.18		5.22	
B.M. *Y*	4.72		11.41	
T.P. 4	5.99		12.04	
T.P. 5	2.62		4.81	
B.M. *X*			6.16	

5–65. A level party began at B.M. *A* (Elev. 875.18) and ran to B.M. *B* (Elev. 987.50) but got an elevation of 987.43. Using elevation 987.43 for B.M. *B* as the starting value, return levels to B.M. *A* found its elevation to be 875.18. Rerunning the entire level circuit from B.M. *A* to B.M. *B* and back to B.M. *A* in the same manner produced identical results. What is the most probable explanation for this situation assuming B.S. and F.S. distances were balanced, and that 875.18 and 987.50 are correct elevations for B.M. *A* and B.M. *B* respectively?

chapter **6**

Angles, Bearings, and Azimuths

6–1. General. The location of points and the orientation of lines frequently depend upon the measurement of angles and directions. In surveying, directions are given by bearings and azimuths.

6–2. Units of angle measurement. An angle is the difference in direction of two intersecting lines. A purely arbitrary unit is used to define the value of an angle. The standard units in the United States are the degree, minute, and second (the sexagesimal system). Seconds are divided decimally. Radians may be used in computations, in fact are employed extensively in high-speed electronic computers.

6–3. Measurement of angles. Angles are measured directly in the field by using some device such as a compass, transit, theodolite, or sextant, or constructed without measurement on a plane table sheet. The compass, transit, and theodolite will be discussed in succeeding chapters.

An angle may be measured indirectly by the tape method, which has already been discussed, and by computing it from the relation to known quantities in a triangle or other simple geometric figure, as in trilateration.

Note that *there are three basic requirements in the determination of an angle.* As shown in Fig. 6–1, they are *(1) the base or starting line, (2) the direction of turning, and (3) the angular distance.* The method of computing bearings shown in illustration 6–1 is based on these three elements.

6–4. Measurement of direction. The direction of a line is its angle from an established line of reference, called a *meridian.* The compass and transit are used in plane surveying to measure directions of lines.

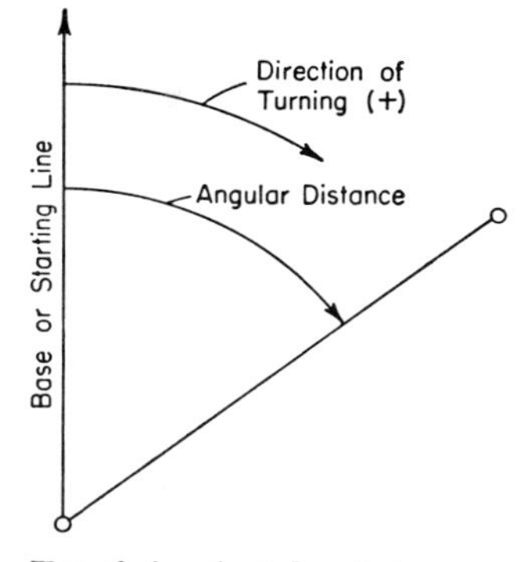

FIG. 6–1. Angular distance.

The reference line generally adopted is either the *true* (geographic) *meridian* or the *magnetic meridian*. If neither of these can be determined readily, any actual or assumed line can be selected and its relation to the true meridian ascertained later. The disadvantage of an assumed meridian is the difficulty, or perhaps impossibility, of reestablishing it if the original points are lost.

The true meridian for any one place upon the surface of the earth is the great circle, projected on the earth's surface, which passes through the observer's position and the north and south *geographic poles*. If a survey is based upon a plane coordinate system, such as a state plane coordinate system, a *grid meridian* is used as the reference meridian. Grid north is the direction of true north for a selected meridian. Thus grid north is parallel to the selected central meridian over the entire extent of area covered by the plane coordinate system.

The direction of a magnetic meridian is defined by a freely suspended, magnetized needle which is influenced by the magnetic poles only. A *magnetic pole* is the center of convergence of the magnetic meridians; also a place at which the magnetic dip is 90°.

6–5. Bearings. Bearings represent one system of designating directions. The bearing of a line is the acute horizontal angle between the meridian and the line. The angle is measured from either the north or south, toward the east or west, as may be necessary to give a reading less than 90°. The proper quadrant is shown by the letter N or S, preceding the angle, and the letter E or W, following it. An example is N 80° E.

True bearings are measured from the local geographic meridian;

magnetic bearings, from the local magnetic meridian. Magnetic bearings are obtained in the field by means of a magnetic needle in a compass box. These bearings may be used along with measured angles to obtain *computed bearings.*

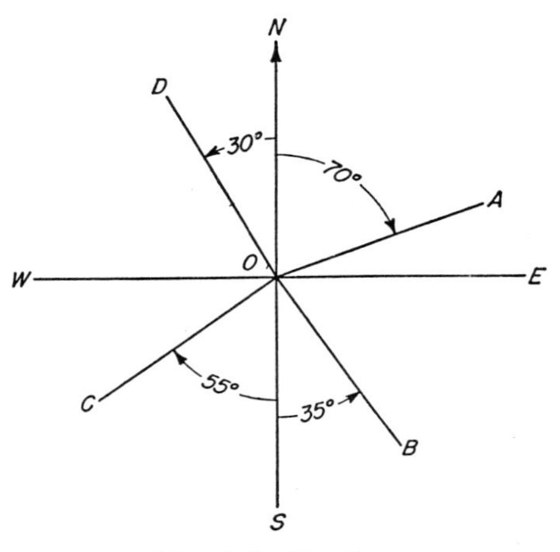

FIG. 6–2. Bearings.

In Fig. 6–2 all bearings in quadrant *NOE* are measured clockwise from the meridian. Thus the bearing of line *OA* is N 70° E.

All bearings in quadrant *SOE* are measured counterclockwise from the meridian. The bearing of *OB* is S 35° E.

Similarly, the bearing of *OC* is S 55° W and that of *OD* is N 30° W.

In Fig. 6–3 assume that a compass is set up successively at points *A, B, C,* and *D* and that the bearings are read on lines *AB, BA, BC, CB, CD,* and *DC.* The bearings of *AB, BC,* and *CD* are called *forward bearings,* and the bearings of *BA, CB,* and *DC* are then *back*

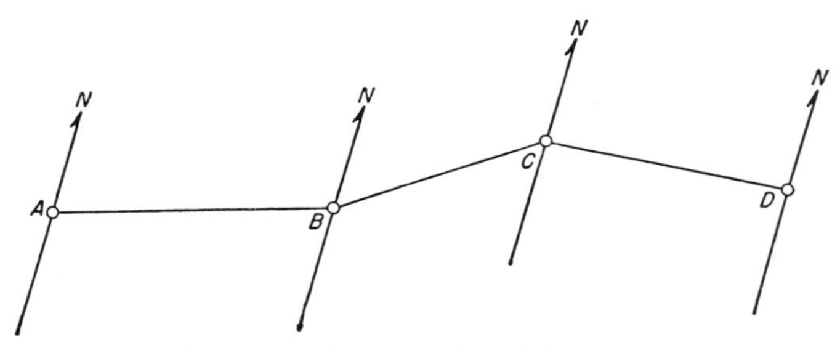

FIG. 6–3. Forward and back bearings.

bearings. Forward bearings have the same numerical value as back bearings but opposite letters. If the bearing of AB is N 72° E, the bearing of BA is S 72° W.

6–6. Azimuths. Azimuths are angles measured clockwise from any meridian. In plane surveying, azimuths are generally measured from north, but astronomers and the United States Coast and Geodetic Survey use south as the reference point.

As shown in Fig. 6–4, azimuths range from 0° to 360° and do not require letters to identify the quadrant. Thus the azimuth of OA is 70°; of OB, 145°; of OC, 235°; and of OD, 330°. It is necessary to state in the field notes, at the beginning of the work, whether azimuths are measured from north or south.

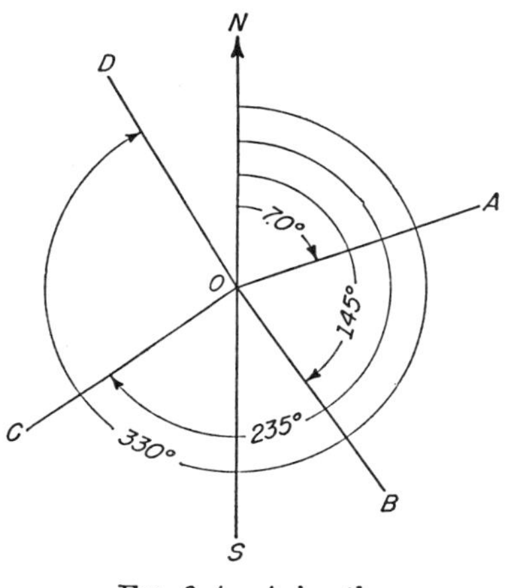

FIG. 6–4. Azimuths.

Azimuths may be *true, magnetic, grid* or *assumed,* depending upon the meridian used. They may also be *computed, forward, or back* azimuths. Forward azimuths are converted to back azimuths, and vice versa, by adding or subtracting 180°. For example, if the azimuth of OA is 70°, the azimuth of AO is 250°. If the azimuth of OC is 235°, the azimuth of CO is 235° − 180° = 55°.

Azimuths are read on the graduated circle of a transit after the instrument has been oriented properly. This can be done by sighting along a line with its known azimuth on the plates and then turning to the desired line. Azimuths (*Directions*) can be used advantageously in some data adjustment computations.

6–7. Comparison of bearings and azimuths. Because bearings and azimuths are encountered in so many surveying operations, the comparative summary of their properties given in Table 6–1 may be helpful.

6–8. Calculation of bearings. The calculation of bearings from azimuths consists of finding the quadrant in which the azimuth falls and then converting as shown in Table 6–1.

TABLE 6-1

COMPARISON OF BEARINGS AND AZIMUTHS

Bearings	Azimuths
Vary from 0° to 90°	Vary from 0° to 360°
Require two letters and a numerical value	Require only a numerical value
May be true, magnetic, calculated, forward, or back	Same
Measured clockwise and counterclockwise	Measured clockwise only
Measured from north and south	Measured from north only in any one survey, or from south only

Directions for lines in the four quadrants (azimuths from north):

N 54° E	54°	
S 68° E	112°	(180° − 68°)
S 51° W	231°	(180° + 51°)
N 15° W	345°	(360° − 15°)

Many types of surveys require the calculation of bearings for a *traverse*. A traverse is a series of distances and angles, or distances and bearings, or distances and azimuths, connecting successive instrument points. The boundary lines of a piece of property form a closed traverse. The survey of a highway from one city to another usually

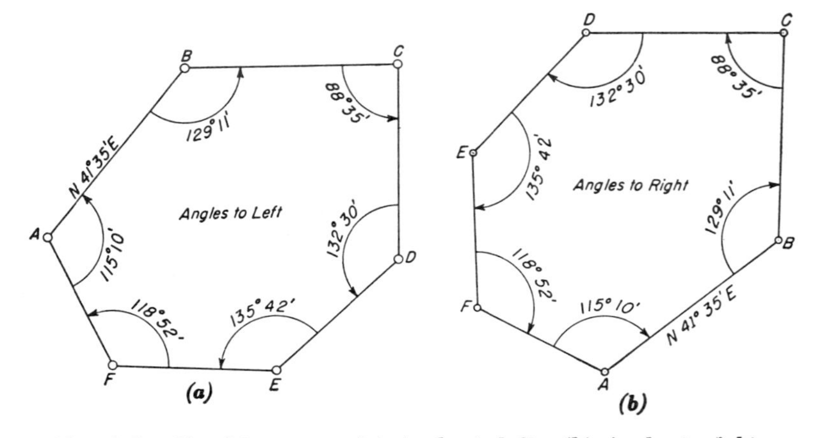

FIG. 6-5. Closed traverses. (a) Angles to left. (b) Angles to right.

is an open traverse, but if possible it should be closed by tying in on points of known coordinates near the starting and finishing points.

Figure 6–5 shows two closed traverses, in each of which the bearing of one line, *AB*, and the angles between the lines are known. Note that the polygons are "right" and "left," i.e., similar in shape but turned over like the right hand and left hand. Obviously, confusing angles measured to the left from a reference

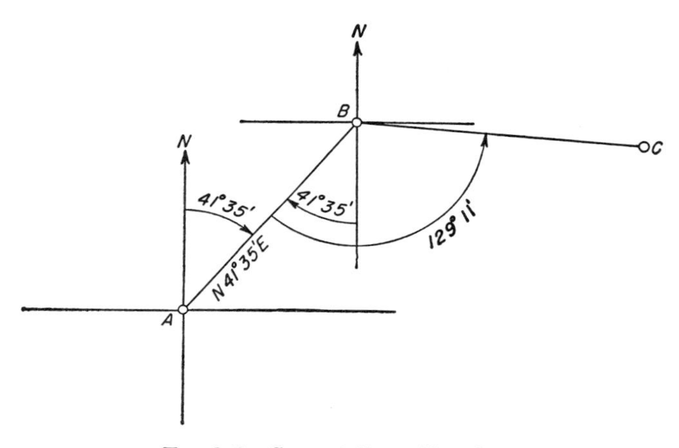

FIG. 6–6. Computation of bearings.

line with angles measured to the right will produce an incorrect location of a traverse.

Computation of the bearing of a single line is simplified by drawing a sketch similar to Fig. 6–6 and showing all the data. Assume that the bearing of a line *AB* is N 41° 35′ E, and the angle at *B* turned to the left (counterclockwise) is 129° 11′. Then the numerical value of the bearing of *BC* is 41°35′ − 129°11′ = −87°36′. By inspection of a sketch, the bearing of *BC* is S 87°36′ E.

If the bearings of a traverse having more than a few sides are to be computed, a tabular form of the type shown in illustration 6–1 is preferable. In this shorthand arrangement, northeast bearings, southwest bearings, and other angles measured clockwise (to the right) are called plus; southeast bearings, northwest bearings, and other angles measured counterclockwise (to the left) are considered minus. Bearing letters are repeated and shown reversed in parentheses, to call attention to the fact that *the angles are turned from the back bearings.* For example the bearing of *AB* is N 41° 35′ E, but at *B* the angle 129° 11′ is measured from the line *BA*, whose

Illustration 6–1

COMPUTATION OF BEARINGS

Angles to the Left					Angles to the Right				
AB	N	41° 35′ E	+	(SW)	*AB*	N	41° 35′ E	+	(SW)
		129° 11′	−				129° 11′	+	
BC	S	87° 36′ E	−	(NW)		S	170° 46′	+	
		88° 35′	−		*BC*	N	9° 14′ W	−	(SE)
	N	176° 11′	−				88° 35′	+	
CD	S	3° 49′ W	+	(NE)	*CD*	S	79° 21′ W	+	(NE)
		132° 30′	−				132° 30′	+	
	N	128° 41′	−			N	211° 51′	+	
DE	S	51° 19′ W	+	(NE)	*DE*	S	31° 51′ W	+	(NE)
		135° 42′	−				135° 42′	+	
EF	N	84° 23′ W	−	(SE)		N	167° 33′	+	
		118° 52′	−		*EF*	S	12° 27′ E	−	(NW)
	S	203° 15′	−				118° 52′	+	
FA	N	23° 15′ W	−	(SE)		N	106° 25′	+	
		115° 10′	−		*FA*	S	73° 35′ E	−	(NW)
	S	138° 25′	−				115° 10′	+	
AB	N	41° 35′ E	✓		*AB*	N	41° 35′ E	✓	

bearing is S 41° 35′ W. Since *BA* is 41° 35′ from south in a plus direction (Fig. 6–6) and angle *ABC* is turned 129° 11′ in a minus direction from *BA*, then the ***angular distance*** representing the algebraic summation of these two angular values is —87° 36′ measured from the south. Such an angular distance locates a line by giving the reference meridian, direction, and numerical value—in this case, from the south, in a minus direction (counterclockwise), 87° 36′.

Since the angular distance is less than 90°, the line *BC* must fall in the southeast quadrant, and the bearing is therefore S 87° 36′ E. When the algebraic summation gives an angular distance greater than 90°, the quadrant in which the line falls is determined and the bearing is found from its known relation to the north or south direction. Several examples of angular distances over 90° are shown in illustration 6–1.

The shorthand method is partially self-checking. If the back-bearing letters shown in parentheses are forgotten, the bearings of *BC*, *DE*, and *FA* would be wrong but those of *CD*, *EF*, and *AB* would be correct. Note that if deflection angles are used instead of direct angles, the base line for turning the deflection angle is the traverse line extended so its bearing letters remain unchanged. The bearing of the starting course *must* by recomputed as a check, using the last angle. A discrepancy shows an arithmetical error was made, or the angles were not adjusted properly prior to computing bearings.

Traverse angles should be adjusted to the proper geometric total before bearings are computed. If the angles total exactly $(n - 2)180°$, where n is the number of sides in the traverse, the original and computed bearings of *AB* should be the same. If the traverse angles fail to close by say 2 min and are not adjusted prior to the computation of bearings, the original and computed bearings of *AB* should differ by the same 2 min.

Angles are normally measured to the right (clockwise) in surveying, but their direction should be recorded in the field notes to remove any doubt. The first column in illustration 6–1 shows a computation for angles to the left. The second column gives the bearings for angles measured to the right.

In computing bearings for the courses of a traverse run by *deflection angles* (angles measured to the right or left from the prolongation of the preceding line, as explained in Sec. 8–16) the form shown in illustration 6–1 can be used; however, the reversed-bearing letters in parentheses are not needed. Since the angular value is turned from the line extended, the base or starting line retains its same direction.

6–9. Sources of error. Some of the sources of error in working with angles, bearings, and azimuths are:

a. Using an assumed reference line which is difficult to reproduce.
b. Forgetting to adjust the traverse angles before computing bearings.
c. Orienting an instrument by resighting on magnetic north.

6–10. Mistakes. A few of the mistakes made in work involving bearings and azimuths are:

a. Confusing magnetic and true bearings.
b. Mixing bearings and azimuths.
c. Failing to change the bearing letters when applying the direct angle at the forward end of a line.
d. Using the angle at the opposite end of a line in computing bearings, i.e., using angle *A* when starting with line *AB*.
e. Not using the last angle to recompute the starting bearing as a check, for example angle *A* in traverse *ABCDEA*.
f. Subtracting from 360°00′ as though it were 359°100′ instead of 359°60′, or using 90° instead of 180° in bearing computations.

PROBLEMS

6–1. Convert an angular reading of 114°54′ to radians; a reading of 127.42 grads to degrees, minutes, and seconds; and a reading of 3600 mils (clockwise from north) to a bearing in degrees, minutes, and seconds.

In problems 6–2 through 6–4 convert the azimuths (from south) to bearings.

6–2. 55°10′, 127°18′, 216°42′, 334°58′.
6–3. 38°32′, 162°07′, 191°21′, 343°44′.
6–4. 80°00′, 109°14′, 187°57′, 301°40′.

Convert the bearings given in problems 6–5 through 6–7 to azimuths (from north) and compute the acute angle between each pair of successive bearings.

6–5. N 15°17′E, S 24°43′E, S 56°01′W, N 78°35′W.
6–6. N 29°11′E, S 88°45′E, S 90°00′W, N 5°04′W.
6–7. N66°22′E, S 90°00′E, S 73°46′W, N 19°59′W.
6–8. What is the difference between the true bearing and the magnetic bearing of a line *AB*?

What is the azimuth (from South) of *CD* given the data in problems 6–9 through 6–12?

6–9. Azimuth *AB* = 120°14′, clockwise angles *ABC* = 162°38′, *BCD* =75°25′.
6–10. Azimuth *AB* = 186°34′, clockwise angles *ABC* = 131°00′, *BCD* = 219°42′.
6–11. Bearing *AB* = S 42°22′E, angles to left *ABC* = 147°40′, *BCD* = 188°06′.
6–12. Bearing *AB* = S 4°13′W, angles to left *ABC* = 36°52′, *BCD* = 315°28′.

In problems 6–13 through 6–16 given the bearing of *DE* = S 54°15′E, and direct angles *DEF* and *EFG*, compute the bearing of *FG*.

6–13. Angle *DEF* = 116°32′ right, *EFG* = 209°05′ right.
6–14. Angle *DEF* = 94°56′ right, *EFG* = 87°38′ left.
6–15. Angle *DEF* = 150°44′ left, *EFG* = 75°11′ right.
6–16. Angle *DEF* = 183°19′ left, *EFG* = 109°24′ left.

One side *AB* of a 5-sided field runs due north. Compute and tabulate the bearings and azimuths (from north) of each side given the interior angles listed in problems 6–17 through 6–20, all to right (R), or to left (L).

6–17. (R) $A = 60°00'$, $B = 134°17'$, $C = 128°42'$, $D = 79°25'$, $E = 137°36'$.
6–18. (R) $A = 140°15'$, $B = 109°30'$, $C = 85°00'$, $D = 50°45'$, $E = 154°30'$.
6–19. (L) $A = 167°51'$, $B = 43°22'$, $C = 98°34'$, $D = 135°08'$, $E = 95°05'$.
6–20. (L) $A = 120°30'$, $B = 84°15'$, $C = 108°45'$, $D = 75°00'$, $E = 151°30'$.

In problems 6–21 through 6–24 compute and tabulate the bearings of a regular hexagon given the starting bearing of side *AB* and direction of angle measurement.

6–21. Bearing of *AB* = N 32°25′E, angles to the right.
6–22. Bearing of *AB* = S 46°15′E, angles to the left.
6–23. Bearing of *AB* = S 55°30′W, angles to the left.
6–24. Bearing of *AB* = N 28°40′W, angles to the right.

6–25. Similar to problem 6–21 but for an octogon.
6–26. Similar to problem 6–22 but for an octogon.
6–27. Deflection angles (see Sec. 8–16) for an open traverse transmission line are listed. Sketch the traverse and compute and tabulate the bearings. Bearing from 0 + 00 to 12 + 10.4 is S 65°08′ W.

Station	Defl. Angle	Bearing
0+00		
12+10.4	10°16′ R	
27+48.9	14°28′ R	
38+81.6	22°02′ L	
65+75.0		

6–28. Similar to problem 6–27, except bearing of the first line = N15°14′W.
6–29. Sketch and complete the following notes for a closed traverse.

Sta.	Int. Angle	Defl. Angle	Forward Bearing	Back Bearing	Azim. (from S)
A			*AB* = S 78°22′ W		
B			*BC* = S 68°50′ E		
C		125°00′ R			
D		91°15′ L			
E	24°40′ L				
A					

6–30. Sketch and complete the following notes for a closed traverse.

Line	Bearing	Azim. (from N)	Defl. Angle	Angle to Right	Interior Angle
AB	N 25°15′ E				
BC	S 52°24′ E				
CD	due S				
DE	S 81°37′ W				
EA	N 44°46′ W				

6–31. An angle *AOB* is measured at different times, using different instruments and procedures. The results, which are assigned certain weights, are as follows: 38°10′34″, wt 3; 38°10′36″, wt 4; 38°10′37″, wt 1. What is the most probable value of the angle, the standard error of the mean, the probable error of the mean, and the 70 per cent error?

6–32. Similar to problem 6–31, but include an additional measurement of the angle which is 38°10′35″, wt 2.

6–33. The true (geodetic) azimuth of a long line *AB* is 56°37′42″. The true azimuth of *BA* is 236°37′45″. Explain the discrepancy.

6–34. Why have azimuths been measured from the south by astronomers and the United States Coast and Geodetic Survey?

6–35. At what location would all bearings be north?

6–36. List and outline three methods of determining the azimuth of a line.

Compute the bearings for a closed traverse *ABCDEFGHIJA* with interior angles (measured to the right) for an initial bearing as listed in problems 6–37 through 6–40. Similarly, compute azimuths (from north) for the initial azimuth recorded in problems 6–41 through 6–44.

A = 182°17′; B = 118°52′; C = 176°35′; D = 93°24′; E = 157°03′;
F = 144°39′; G = 161°48′; H = 125°00′; I = 97°15′; J = 183°07′.

6–37. Bearing *AB* = N 35°10′E.
6–38. Bearing *DE* = S 47°52′E.
6–39. Bearing *FG* = S 16°24′W.
6–40. Bearing *IJ* = N 68°31′W.
6–41. Azimuth *AB* = 125°10′.
6–42. Azimuth *DE* = 47°52′.
6–43. Azimuth *FG* = 334°26′.
6–44. Azimuth *IJ* = 208°08′.

chapter **7**

The Compass

7–1. General. The compass has been used by navigators and others for many centuries to determine directions. Prior to the invention of the transit and the sextant, the compass furnished surveyors with the only practical means of measuring directions and angles.

The surveyor's compass, like the Gunter's chain, has now become little more than a museum piece. Nevertheless, an understanding of the compass and its vagaries is necessary to check work already done. Also, the compass is still used on rough engineering surveys and is a valuable tool for geologists, foresters, and others.

The engineer's transit is equipped with a compass. In fact, the design of American transits is based upon the requirements of a long compass needle over the center of the instrument, and an erecting telescope. The small size of European transits and theodolites is obtained by means of an inverting telescope and omission of the compass (the compass is available for mounting as an accessory).

7–2. Theory of the compass. A compass consists of a magnetized steel needle mounted on a pivot at the center of a graduated circle. The needle points toward *magnetic north.*

The north and south magnetic poles are located approximately 1000 miles and 1560 miles, respectively, from the true geographic poles. The magnetic lines of force of the earth, which align the needle, pull or dip one end of the needle below a horizontal position. The angle of *dip* varies from 0° at the equator to 90° at the magnetic poles.

To balance the effect of dip in the northern hemisphere, the south end of the needle is weighted with a very small coil of wire. The position of the coil may be adjusted to conform to the latitude in which the compass is used. Weights on transit compasses are set for an average latitude of 40° N and usually do not have to be changed for any location in the United States.

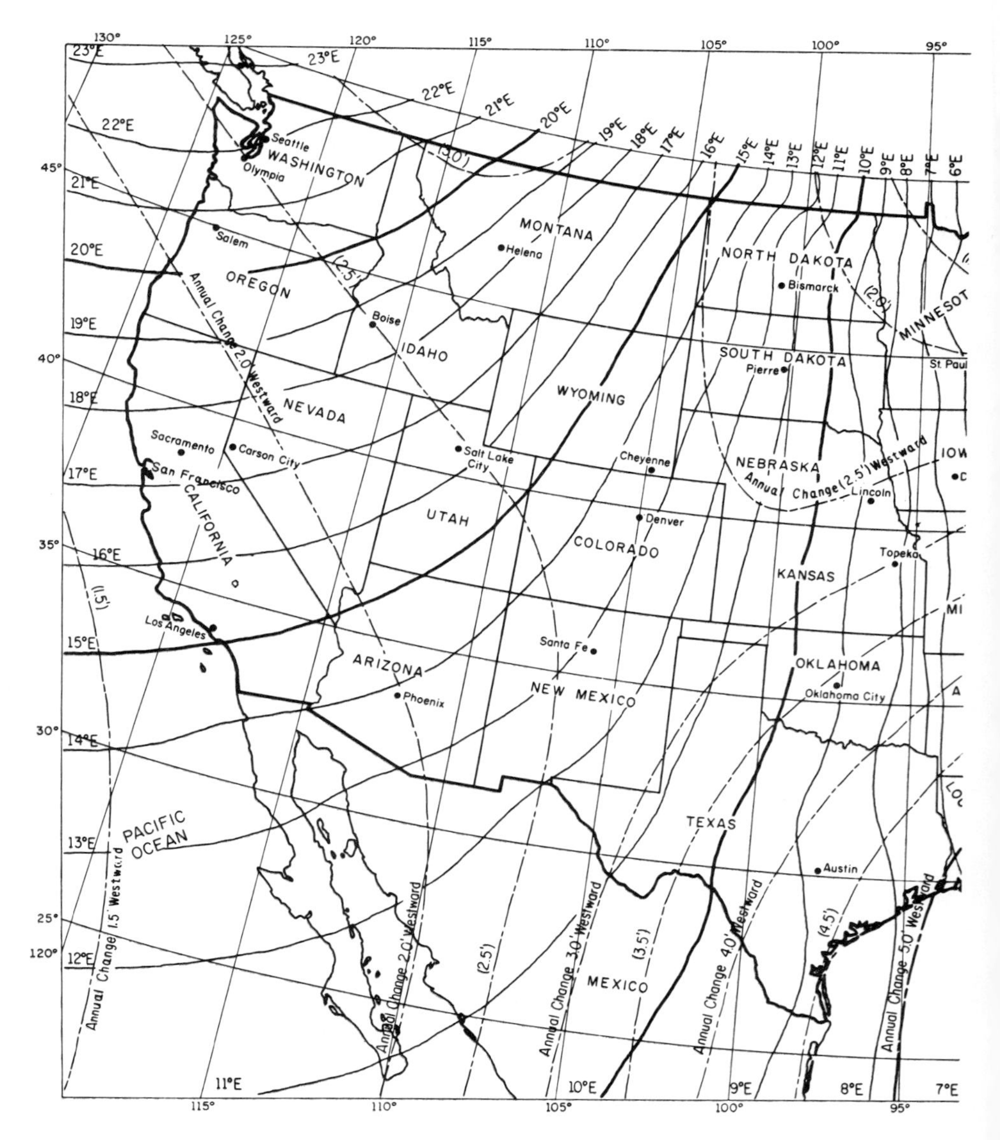

FIG. 7–1. Distribution of magnetic declination in the United States

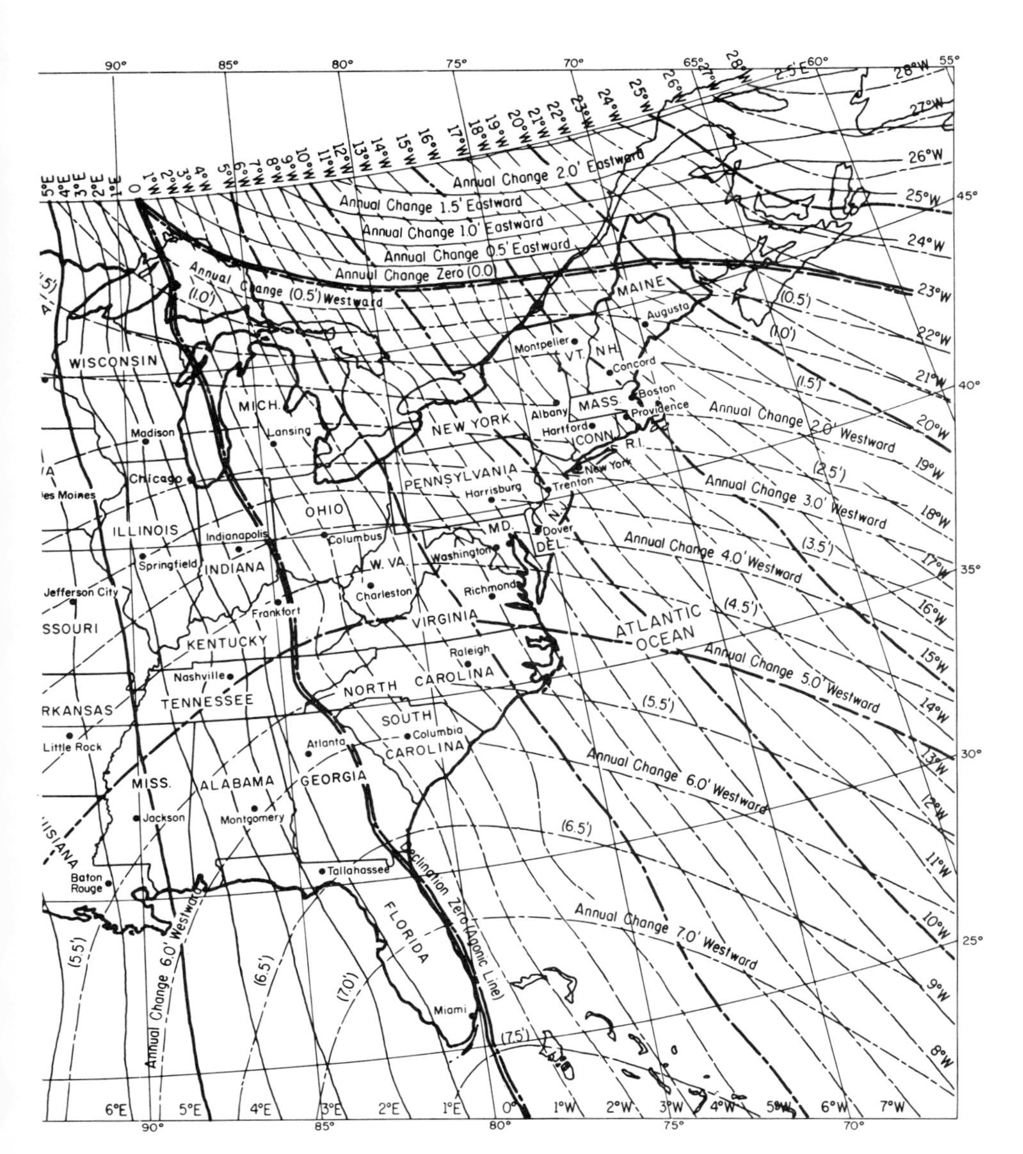

for 1965.0. (Courtesy of United States Coast and Geodetic Survey.)

As the compass box is turned, the needle continues to point toward magnetic north and gives a reading which is dependent upon the position of the graduated circle.

7–3. Magnetic declination. *Declination* is the horizontal angle between the magnetic meridian and the true or geographic meridian. Navigators call this angle the *variation* of the compass.

An east declination is obtained if the magnetic meridian is east of true north, and a west declination if it is west of true north. The value of the declination at any particular location can be obtained (if there is no local attraction) by establishing a true meridian from astronomical observations and then reading the compass while sighting along the true meridian.

A line drawn through points having the same declination is called an *isogonic line.* The line made up of points having a zero declination is termed the *agonic line.* On the agonic line the magnetic needle defines true north as well as magnetic north.

Figure 7–1 is an isogonic chart covering the United States, for the year 1965.0. The agonic line cuts diagonally across the country through Michigan, Indiana, Ohio, Kentucky, Tennessee, North Carolina, Georgia and Florida. Points to the west of the agonic line have an east declination, and points to the east of the line have a west declination. As a memory aid, the needle might be thought of as pointing toward the agonic line.

The annual change in declination shown on larger and more detailed charts aids in estimating the declination for a few years before and after the chart date. Secular change for longer intervals should be computed from available tables which extend back to the earliest times likely to be significant in such problems. The best way to determine the declination at a given location on any date is to make an observation. If this is not possible, the approximate declination can be obtained from the United States Coast and Geodetic Survey.

7–4. Variations in magnetic declination. Variations are periodic or irregular changes in the declination of the magnetic needle. These include secular, daily, annual, and irregular changes.

Secular variation. This is the most important of the changes because of its magnitude. Unfortunately, no general law or mathematical formula has been found to permit prediction of the secular change, and its past behavior can be described only by means of detailed tables and charts derived from observations. Records which have been kept at London for four centuries show a range in magnetic declination from 11° E in 1580, to 24° W in 1820, and back to 8° W in 1960. Secular variation changed the magnetic declination at Baltimore, Maryland, from 5° 11′ W in 1640 to 5° 41′ W in 1700, 0° 35′ W in 1800, 5° 19′ W in 1900, 7° 25′ W in 1950, and 7° 43′ W in 1960.

In retracing old property lines run by compass or based upon the magnetic meridian, it is necessary to allow for the difference in magnetic declination at the time of the original survey and at the present date. The difference, due mostly to secular variation, is generally ascribed merely to variation.

Daily variation. The daily variation of the declination of the magnetic needle causes it to swing through an arc averaging approximately 8 min for the United States. The needle reaches its extreme easterly position about 8 A.M., and its most westerly reading about 1:30 P.M. The mean declination occurs about 10:30 A.M. and 8 P.M. These hours and the amount of the daily swing vary with the latitude and the season of the year, but complete neglect of the daily variation is well within the range of error expected in compass readings. The term *diurnal variation* is often used in place of "daily variation."

Annual variation. This periodic swing amounts to less than 1 min of arc and can be neglected. It must not be confused with the annual change (the amount of the secular-variation change in one year), shown on some isogonic charts.

Irregular variations. Unpredictable magnetic disturbances and storms may cause irregular variations of a degree or more.

7–5. Local attraction. The magnetic field is affected by metallic objects and direct-current electricity, both of which cause a local attraction. If the source of the artificial disturbances is fixed, all bearings from a given station will be in error by the same amount. Angles calculated from the bearings taken at the station will be correct, however.

Local attraction is present if the forward and back bearings of a line differ by more than the normal observational errors. Consider the following compass bearings read on a series of lines:

AB	N 24° 15′ W	*CD*	N 60° 00′ E
BA	S 24° 10′ E	*DC*	S 61° 15′ W
BC	N 76° 40′ W	*DE*	N 88° 35′ E
CB	S 76° 40′ E	*ED*	S 87° 25′ W

The forward bearing *AB* and the back bearing *BA* agree reasonably well, indicating that local attraction does not exist at *A* or *B*. The same is true for point *C*. However, the bearings at *D* differ from the corresponding bearings taken at *C* and *E* by roughly 1° 15′. Local attraction therefore exists at point *D* and deflects the compass needle 1° 15′ to the west of north.

It is evident that to detect local attraction, successive stations on a compass traverse should be occupied, and forward and back bearings read, even though the directions of all lines could be determined by setting up the instrument only on alternate stations.

7–6. The surveyor's compass. The surveyor's compass is shown in Figs. 7–2 and 7–3. George Washington, and thousands of sur-

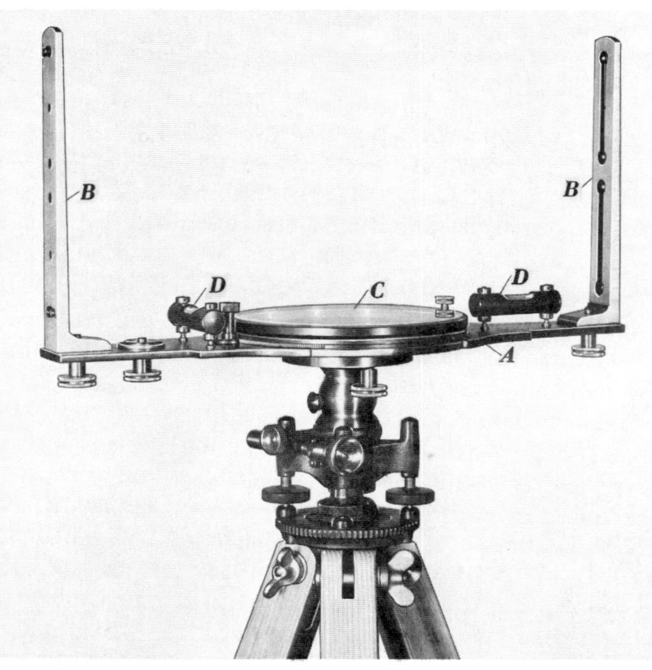

FIG. 7–2. Surveyor's compass. (Courtesy of W. & L. E. Gurley.)

veyors who followed him, used this type of instrument to run land lines—land lines which still determine property holdings and therefore must be retraced. The circle is graduated in degrees or half-degrees, but can be read to perhaps 5 or 10 minutes by estimation.

The instrument consists of a metal base plate *A*, Fig. 7–2, with two vertical sight vanes *B* placed at the ends and a round compass box *C* at the center. Two small level vials *D* are mounted on the plate. The sight vanes are strips of metal with vertical slits to define the line of sight.

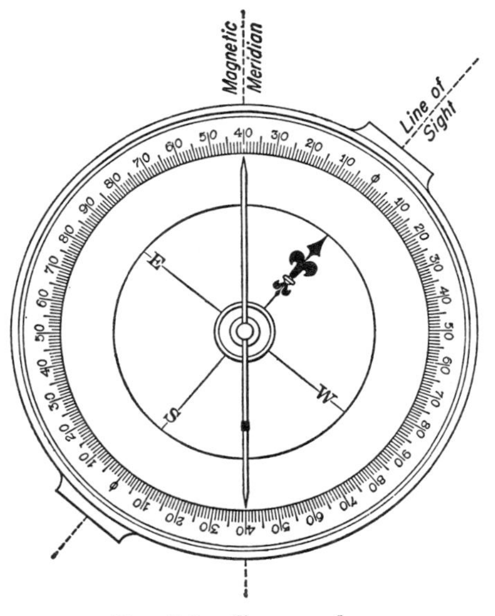

Fig. 7–3. Compass box.

The compass box, Fig. 7–3, has a conical point at its center to support the needle, and a glass cover to protect it. A circular scale at the outer rim of the box is graduated in degrees or half-degrees. The zero marks are at the north and south points and in line with the sight-vane slits. Graduations are numbered in multiples of 10°, clockwise and counterclockwise from 0° at the north and south to 90° at the east and west. As the sight vanes and compass box are revolved, the needle shows the bearing of the line observed.

The letters E and W on the compass box are reversed from their normal positions, to give direct readings. Thus in Fig. 7–3 the bearing of the line through the vanes is N 40° E.

Accuracy of a compass depends upon the sensitivity of the needle. A sensitive needle is one which is readily attracted toward a small piece of iron held nearby, but which settles in the original position each time the stimulus is removed. Sensitivity itself results from the needle having (a) proper shape and balance, (b) strong magnetism, (c) a sharp conical point, and (d) a smooth cup that bears on the pivot. Tapping the glass cover releases a needle which does not swing freely. Touching the cover with a moistened finger removes static electricity which may affect the needle.

Remagnetizing the needle is relatively easy, but resharpening the pivot is difficult. To retain the conical shape of the pivot and prevent blunting to a spherical or flat form, which produces sluggishness, the needle should be lifted from the pivot when not in use. A lever arm is provided to raise the needle off the pivot and press it against the glass cover when not needed.

Early compasses were supported by a single leg called a Jacob staff. A ball-and-socket joint and a clamp were used to level the instrument and set the plate in a horizontal position. Modern compasses are mounted on a base with a four-screw leveling head, as shown in Fig. 7–2.

The compass box of a transit is similar in construction to the surveyor's compass. The zero marks at the north and south points are on a line parallel with and beneath the telescope. A special adjustment on some instruments swings the graduated circle through an arc to compensate for a given declination, thus permitting true bearings to be read from the needle.

7–7. The forester's and geologist's compass. Figure 7–4 shows a type of compass employed by geologists and the United States Forest Service. It can be used as a hand-held instrument or can be supported on a staff or tripod.

The compass is made of aluminum and has brass sights. A declination adjustment is provided for the raised (upper) compass ring. The beveled (lower) ring is used to turn right angles, or to measure vertical angles by placing an edge of the base on a level surface.

FIG. 7–4. Compass. (Courtesy of Keuffel and Esser Company.)

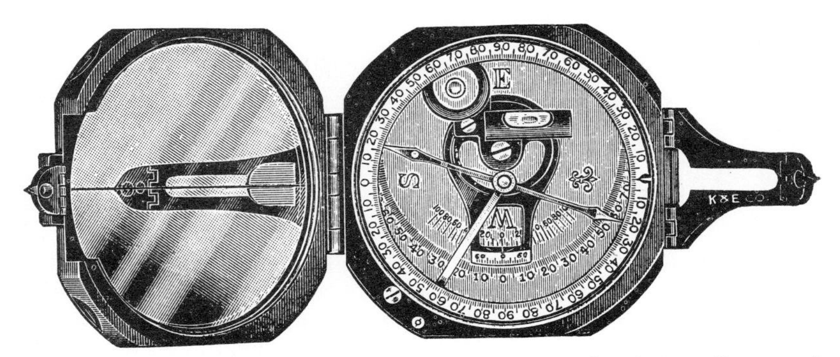

FIG. 7–5. Brunton pocket transit. (Courtesy of Keuffel and Esser Company.)

7–8. Brunton compass. Figure 7–5 shows a Brunton pocket transit, which combines the main features of a sighting compass, a prismatic compass, a hand level, and a clinometer. It is an accurate and convenient device for topographic and preliminary surveys of all kinds. It can be used as a hand-held instrument, or mounted on a Jacob staff or tripod.

The Brunton compass consists of a brass case hinged on two sides. The cover at the left has a fine mirror and a center line on the inside face. The hinged sighting point at the extreme left, and the sighting vane at the extreme right, are folded outward when the instrument is in use. The bearing of a line is determined from the compass-needle reading while the point sighted is reflected through the sight vane on the mirror.

The declination adjustment is made by revolving the raised compass ring. The clinometer (vertical-angle) arc inside the compass ring is graduated to degrees and can be read to the nearest 5 min by a vernier on the clinometer arm. Another arc gives grade percentages for both elevation and depression. The compass is held vertically, instead of horizontally, to read vertical angles or grade percentages. The instrument measures $2\frac{3}{4}$ in. by $2\frac{3}{4}$ in. by 1 in. and weighs approximately 8 oz.

The Brunton compass is widely used by geologists on field surveys.

7–9. Typical problems. Typical problems in compass surveys require the conversion of true bearings to magnetic bearings, magnetic bearings to true bearings, and magnetic bearings to magnetic bearings for the declinations existing at different dates.

As an example, assume that the magnetic bearing of a property line was recorded as S 43° 30′ E in 1862. The magnetic declination at the place of the survey was 3° 15′ W. The true bearing is needed for inclusion in the subdivision plan of the property.

A sketch similar to Fig. 7–6 makes the relationships clear and should be used by beginners to avoid errors. True north is designated by a full-headed long arrow, and magnetic north by a half-headed shorter arrow. The true bearing is seen to be S 46°45′ E. Using different color pencils to show the directions of true north, magnetic north, and lines on the ground helps clarify the sketch.

As another example, assume that the magnetic bearing of a line *AB* read in 1878 was N 26° 15′ E; the declination at that time

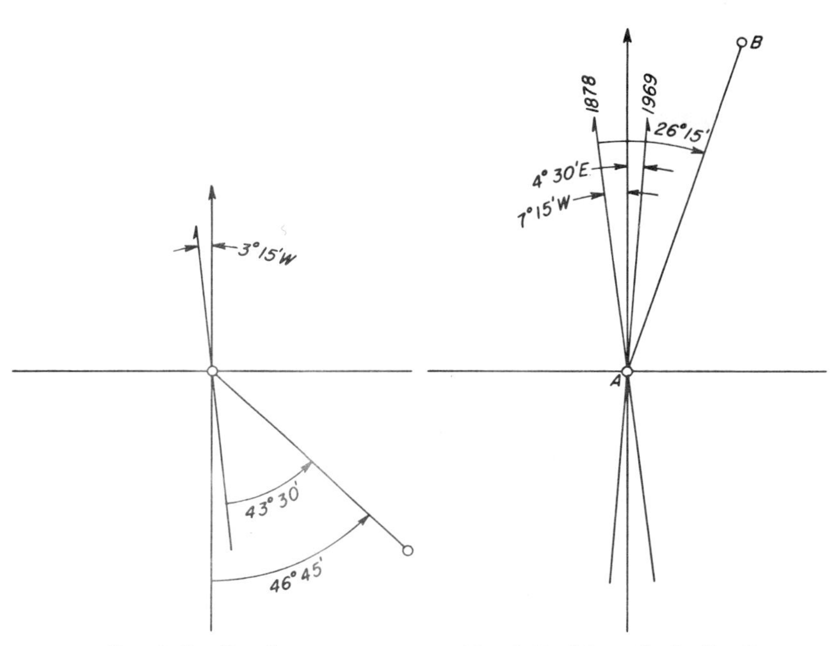

FIG. 7–6. Bearings.

FIG. 7–7. Magnetic declinations.

and place was 7° 15′ W; and in 1969 the declination is 4° 30′ E. The magnetic bearing in 1969 is needed. The declination angles are shown in Fig. 7–7. The magnetic bearing of the line *AB* in 1969 is equal to the bearing at the earlier date minus the sum of the declination angles, or N 14° 30′ E.

7–10. Sources of error in compass work. Some of the sources of error in using the compass are:

a. Compass badly out of level.
b. Pivot, needle, or sight vanes bent.
c. Magnetism of needle weak.
d. Magnetic variations.
e. Local attraction.
f. Chaining pins, metal range poles, axes, loose-leaf field books, or a penknife placed near the compass.

7–11. Mistakes. Some fairly common mistakes in compass work are:

a. Reading the wrong end of the needle.

b. Declination set off on wrong side of north.
c. Declination set off when reading magnetic bearings.
d. Parallax (reading while looking from the side of the needle instead of along it).
e. Failing to check the forward and back bearings when possible.
f. Failing to make a sketch showing known and desired items.

PROBLEMS

7–1. At the present rate of change of declination shown in Fig. 7–1, approximately how fast (in miles per year) is the agonic line moving in the United States, and in which general direction is it moving?

7–2. What is the total difference in magnetic declination between northeast Maine and northwest Washington?

7–3. Determine from Fig. 7–1 the approximate declination in 1965 at Seattle, Boston, Los Angeles, and Miami. What will be their approximate declinations in 1975?

7–4. Assuming that the rate of annual change of declination shown in Fig. 7–1 has been constant (although it has not been), compute the approximate declination for Chicago in 1895. Compare your result with the value shown on the isogonic chart for 1895 in one of the surveying books of that era in the college library.

7–5. The magnetic declination at a given place is 5°10′ W. What is the magnetic bearing of true north? Of true south? Of true east?

7–6. The magnetic bearing of a line of an old survey is recorded as N 12° E. It now is N 3° W. What has been the change in declination angle and the direction of the change?

7–7. At a certain location the declination was 3° 20′ W in 1750, 1° 38′ W in 1800, 0° 40′ E in 1850, 1° 00′ W in 1900, 4° 00′ W in 1950, and 4° 30′ W in 1960. If a line has a true bearing of N 3° 40′ W in 1970, what was its magnetic bearing in the other years listed?

7–8. Similar to problem 7–7 except that the magnetic bearing in 1970 is N 2°15′ E.

In problems 7–9 through 7–12 convert the following magnetic bearings to true bearings.

7–9. S 4°15′ W — Declination 5°30′ E.
7–10. N 1°30′ W — Declination 3°45′ E.
7–11. N 77°45′ E — Declination 10°00′ W.
7–12. S 7°00′ E. — Declination 9°15′ W.

What magnetic bearing should be used to retrace a line *AB* for the conditions given in problems 7–13 through 7–16?

	1850 Magnetic Bearing	1850 Declination	Present Declination
7-13.	N 22°15′ E	7°18′ E	11°42′ E
7-14.	S 64°30′ E	4°56′ W	8°33′ E
7-15.	S 70°45′ W	2°14′ E	4°29′ W
7-16.	N 89°45′ W	1°58′ E	3°16′ W

In problems 7-17 through 7-20 what was the magnetic declination of line *BC* in 1848 based on the following data from an old survey record?

	1848 Mag. Bearing	Present Mag. Bearing	Present Mag. Decl.
7-17.	S 61°15′ W	S 50°30′ W	16°47′ E
7-18.	N 22°00′ E	N 24°15′ E	4°15′ E
7-19.	S 2°30′ E	S 1°15′ W	3°00′ W
7-20.	N 0°15′ W	N 2°30′ E	5°20′ W

7-21. Does local attraction at a point affect the size of an angle computed from magnetic bearings read at that point? Explain.

7-22. After it was determined that no local attraction existed at point *A*, the following bearings were observed: $AB = \text{N } 62\frac{1}{2}° \text{ W}$, $BA = \text{S } 65\frac{1}{4}° \text{ E}$, and $BC = \text{S } 69° \text{ W}$. What is the acute angle at *B* and the correct bearing of *BC*?

7-23. The observed bearing of a line is N 88°45′ E. The correct bearing of the line is S 89°15′ E. What are the value and direction of the local attraction?

7-24. Determine the angle *XYZ* and the correct bearing of line *YZ* from the following data: correct bearing of *XY* is N 87° E, observed bearing of *YX* is S 82° W, and observed bearing of *YZ* is S 50° E.

Problems 7-25 through 7-27 list forward and back bearings. The correct bearing of *AB* is N 37° E. Compute the local attraction at each point possible, and the correct bearings of *BC* and *CD*.

7-25. $AB = $ N 37° E; $BA = $ S 38° W, $BC = $ S 65° E; $CB = $ N 68° W, $CD = $ Due S.

7-26. $AB = $ N 37° E; $BA = $ S 34° W, $BC = $ N 18° W; $CB = $ S 13° E, $CD = $ S 22° W; $DC = $ N 21° E.

7-27. $AB = $ N 37° E; $BA = $ S 35° W, $BC = $ N 71° E; $CB = $ S 64° W, $CD = $ S 86° E; $DC = $ S 89° W.

7-28. The following bearings were observed by compass: $AB = $ N 77°30′ E; $BA = $ S 79°15′ W; $BC = $ S 65°00′ W; $CB = $ N 64°45′ E; $CD = $ N 48°30′ W; $DC = $ S 48°30′ E. Find the magnetic bearing of *AB*. Where is the local attraction? Which way is the needle deflected at each point and how much?

7-29. From what point of beginning is it possible for a person to travel due north along a meridian for 1 mile, due east along a parallel of latitude for 1 mile, and then due south along a meridian for 1 mile and be back at the starting point?

7–30. Independent readings of a surveyor's compass to check the local declitions were recorded: N 13°45′ E, N 13°55′ E, N 13°40′ E, N 13°45′ E, N 13°50′ E, and N 13°50′ E. What is the most probable value of the declination and of the 90 per cent error?

7–31. Describe how you would determine in the field the magnetic declination at a given location.

7–32. Where on the earth's surface is the direction of the earth's magnetic lines of force horizontal, and where vertical? Why must an ordinary compass needle be balanced before it is used?

7–33. Classify the kind of error resulting from local attraction at a transit station.

7–34. Can local attraction be determined by setting up a compass at a single station? Explain.

Compute the interior angles of a 5-sided closed traverse *ABCDEA* having the surveyor's compass bearings listed in problems 7–35 through 7–38. Explain the closure.

	AB	*BC*	*CD*	*DE*	*EA*
7–35.	N 75°15′E	S 32°30′E	S 47°15′W	N 81°00′W	N 16°45′W
7–36.	N 14°00′E	N 58°45′E	N 79°15′W	S 33°30′W	N 82°30′E
7–37.	S 64°30′E	S 12°45′E	S 85°00′W	N 29°45′W	N 43°15′E
7–38.	S 35°20′W	S 57°25′W	Due East	N 48°50′W	N 09°10′W

chapter **8**

The Transit

8–1. General. The instrument used by surveyors and engineers for accurate measurement of angles is the engineer's transit. Practically all transits now are equipped with a vertical arc, thus permitting measurement of vertical as well as horizontal angles.

The standard transit is designed to give readings to the nearest minute of arc, but finer graduations are available at slightly greater cost and permit readings to 30, 20, 15, or 10 sec. Theodolites reading to the nearest second of arc are used on precise triangulation. Some European theodolites read directly to the nearest 0.2 sec with an accuracy to 0.01 sec.

In making measurements, it is helpful to keep in mind the relationship between angles and distances. Commonly used field conversions, shown in Fig. 8–1, are as follows:

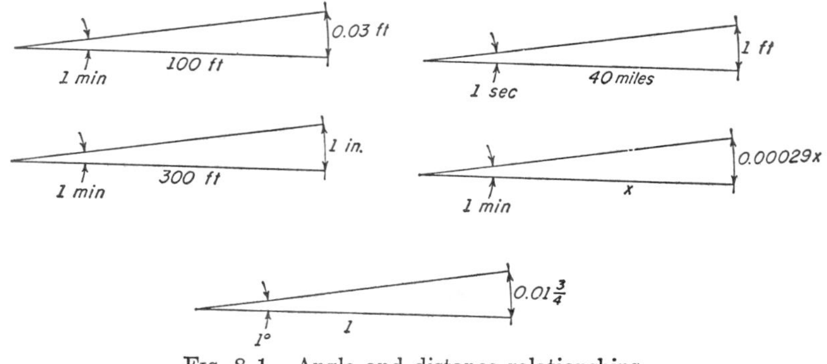

Fig. 8–1. Angle and distance relationships.

1 min of arc = 0.03 ft at 100 ft (approx)
1 min of arc = 1 in. at 300 ft (approx, actually 340 ft)
1 sec of arc = 1 ft at 40 miles (approx, useful in triangulation)

$$\sin 1\text{ min} = \tan 1\text{ min} = 0.00029$$

$$\sin 1^\circ = \tan 1^\circ = 0.01745 = 0.01\tfrac{3}{4}$$

A theodolite reading to the nearest 0.1 sec is theoretically capable of measuring the angle between two points 1 in. apart and 40 miles away.

8–2. The American transit. Transits are manufactured for general and special uses but all have three main parts: (1) upper plate, (2) lower plate, and (3) leveling head. These are shown in their relative positions in Fig. 8–2, and assembled in Fig. 8–3.

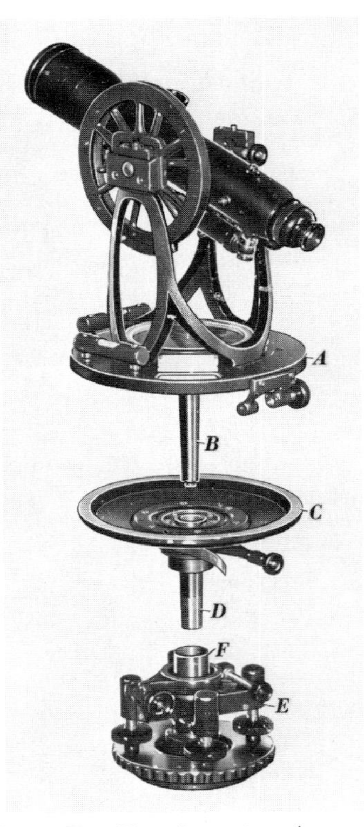

Fig. 8–2. Transit parts. *A*, upper plate; *B*, inner spindle; *C*, lower plate; *D*, outer spindle; *E*, leveling head; *F*, socket. (Courtesy of W. & L. E. Gurley.)

The various parts of a transit and its operation can best be learned by actually examining and handling an instrument. Once a transit has been taken apart and assembled, even though it be an old or damaged instrument retired from service, the precise machining and construction are certain to increase respect for this fine piece of equipment.

8–3. Upper plate. The upper plate, Figs. 8–2 and 8–3, is a hori-

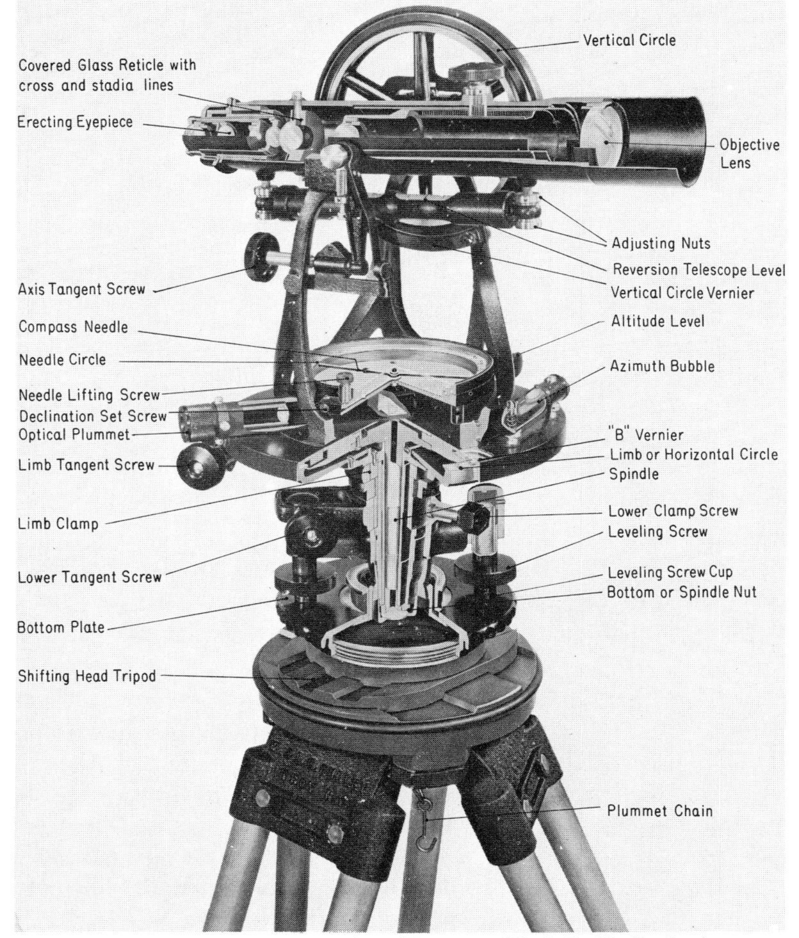

FIG. 8–3. Engineer's transit. (Courtesy W. & L. E. Gurley.)

zontal circular plate combined with a vertical *spindle* upon which it revolves about a true vertical axis. Attached to the plate are two level vials, one parallel with the telescope (altitude bubble) and the other at right angles to it (azimuth bubble) and two *verniers* set

180° apart. Provisions are made for adjusting the verniers and level vials.

Two vertical *standards,* of either the A or U type, are cast as an integral part of the upper plate, to support the horizontal *cross arms* of the telescope in bearings. The telescope revolves vertically about the center line through the arms. This line is called the *horizontal* (or *transverse*) *axis* of the telescope.

The telescope is similar to that of the engineer's level and contains an eyepiece, a reticle with one vertical and three horizontal lines, and an objective-lens system. The magnification range of transit telescopes is 18 to 28 diameters. A sensitive vial is attached to the telescope tube so the transit can be used in place of an engineer's level on work where the lower magnification and lesser sensitivity of the telescope vial are satisfactory.

When the level vial is below the telescope, the telescope is said to be in the normal, or direct, position. When the telescope is turned on its horizontal axis so that the level vial is above the telescope, the telescope is in its plunged, inverted, or reversed position. To permit use of the telescope for leveling in either the normal position or the inverted position, a reversion vial is desirable.

A *clamp screw* for the horizontal axis is tightened to hold the telescope level, or at any desired inclination to the horizontal. After the clamp screw is set, a limited range of vertical movement can be obtained by manipulating the horizontal-axis *tangent screw.*

A *vertical circle* or *arc* is supported by a cross arm and turns with the telescope as it is revolved. The arc normally is divided into half-degree spaces. Readings to the nearest minute are obtained from a vernier having thirty divisions. The vernier is mounted on one of the standards, with provisions for adjustment. If the vernier is set properly, it should read zero when the telescope bubble is centered. If it is out of adjustment, a constant error called the *index error* is found by reading the arc when the bubble is centered. The index error must be applied to all vertical angles, with appropriate sign, in order to obtain correct angles.

The upper plate also contains the *compass box,* and holds the *upper tangent screw.*

8-4. Lower plate. The lower plate, *C* in Fig. 8-2, is a horizontal circular plate graduated on its upper face. Its underside is attached to a vertical, hollow, tapered spindle *D* into which the upper plate fits precisely. The upper plate completely covers the lower plate, except for two openings where the verniers exactly meet the graduated circle.

The upper (limb) clamp screw, Fig. 8–3, fastens the upper and lower plates together. A small range of movement is possible after clamping, by using the upper (limb) tangent screw.

8–5. Leveling head. The leveling head consists of a bottom horizontal plate and a "spider," with four leveling screws between them. The leveling screws, set in cups to prevent scoring the bottom plate, are partly or completely enclosed for protection against dirt and injury. The bottom plate has a collar threaded to fit upon the tripod head.

The socket *F*, Fig. 8–2, of the leveling head includes a lower clamp screw, Fig. 8–3, to fasten the lower plate. The lower tangent screw is used to make precise settings after the lower clamp screw is tightened. The base of the socket is fitted into a ball-and-socket joint resting on the bottom plate of the leveling head, on which it slides horizontally. A plummet chain attached to the center of the spindle holds the plumb-bob string. An optical plummet, which is a telescope through the vertical center (spindle), is available on some transits. It will point vertically when the transit plate is level, but is viewed at right angles by means of a prism for ease of observation.

A recapitulation of the use of the various clamps and tangent screws may be helpful to the beginner (see also page 186). The clamp and tangent screw on one of the standards control the movement of the telescope in a vertical plane. The upper clamp fastens the upper and lower plates together. The upper tangent screw permits a small differential movement between them. The lower clamp fastens the lower plate to the socket. The lower tangent screw turns the plate through a small angle. If the upper and lower plates are clamped together, they will of course move freely as a unit until the lower clamp is tightened. Tangent screws are also called *slow-motion screws*.

Tripods for transits are used interchangeably for levels. Either fixed- or adjustable-leg tripods are suitable.

8–6. Scales. The horizontal limb of the lower plate may be divided in various ways but generally the horizontal circle is graduated into 30-min or 20-min spaces. For convenience in measuring angles to the right and to the left, the graduations are numbered from 0° to 360° clockwise, and from 0° to 360° counterclockwise. Figure 8–4 shows these arrangements. On newer transits, the numbers are slanted to show the direction in which the circle should be read. Different length lines mark the 10°, 5°, 1°, and other major graduations.

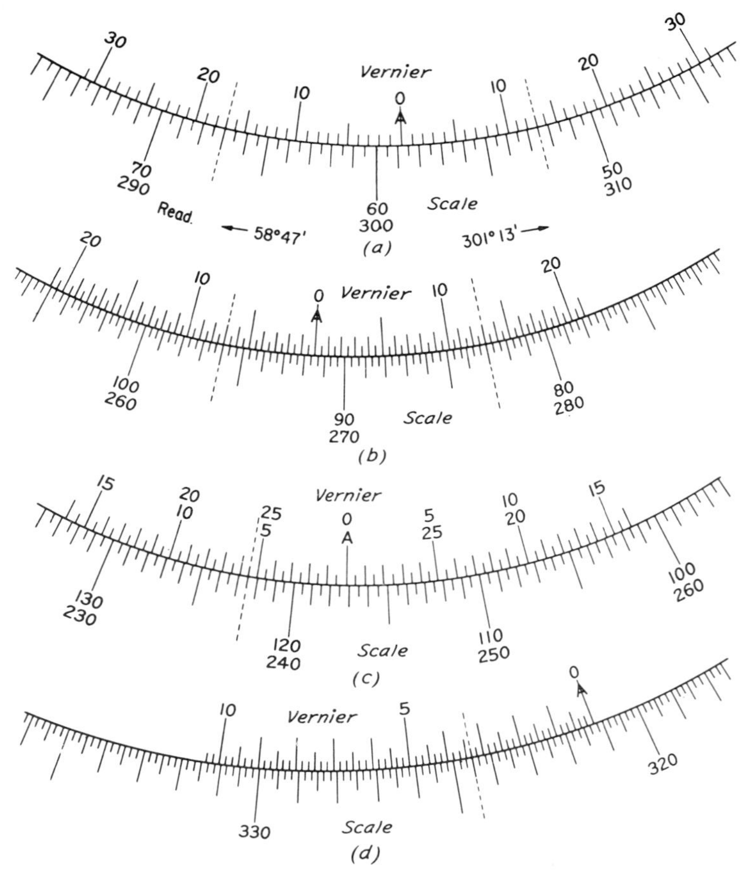

Fig. 8–4. Transit verniers.

The circles of more-precise instruments are graduated in divisions of 10 or 15 min. The outer set of numbers on some old transits runs from 0° to 180° and back to 0°. Obsolete instruments had the circle divided into quadrants like a compass box. Graduations from 0° to 360° facilitate the reading of azimuths and direct angles, and are therefore more advantageous than the quadrant system of numbering, which was used for bearings.

Transit circles are divided automatically by means of a precise dividing machine operating in an air-conditioned room. After each line is cut by a sharp tool, a precision gear moves the tool ahead for the next cut. Any small error in the gears is adjusted by a compensating cam. Under a microscope the division slashes look somewhat rough, but to the naked eye they are smooth. The marks are painted to make them stand out clearly. Graduations on transit scales are correct to within about 2 seconds of arc.

8–7. Verniers. The principle of the vernier was demonstrated in Sec. 5–19, and the least count given by the following relation:

$$\text{Least count} = \frac{\text{value of the smallest division on the scale}}{\text{number of divisions on the vernier}}$$

The combinations of scale graduations and vernier divisions generally used on transits are shown in Table 8–1.

TABLE 8–1

TRANSIT SCALES AND VERNIERS

Scale Graduations	Vernier Divisions	Least Count	Fig. No.
30 min	30	1 min	8–4a
20 min	40	30 sec	8–4b
30 min	60	30 sec	8–4c
15 min	45	20 sec	...
10 min	60	10 sec	8–4d

The three types of verniers shown in Fig. 8–4 are used on transits.

Direct, or *single, vernier* (Fig. 8–4d). This is read in only one direction and must therefore be set with the graduations ahead of the zero (index) mark in the direction to be turned.

Double vernier (Fig. 8–4a, b, and c). This vernier can be read either clockwise or counterclockwise, only one half being used at a time. Once the index mark has been set coincident with 0° 00′ on the circle, the observer is not limited to turning angles in one direction only.

Folded vernier (Fig. 8–4c). This type avoids the long vernier plate required by the normal double vernier. Its length is that of a direct vernier with half of the graduations placed on each side of the index mark. Except possibly for vertical arcs, the use of folded verniers is not justified by space or cost savings and is likely to cause reading errors.

8–8. Method of reading verniers. A vernier is read by finding the graduation on the vernier scale which coincides with *any* graduation on the circle. The vernier index shows the number of degrees (and sometimes also a fractional part of a degree) passed over on the scale. The coincident vernier graduation gives directly the additional fractional part of a degree. The second divisions on each side of the apparently coincident lines should be checked for symmetry of pattern.

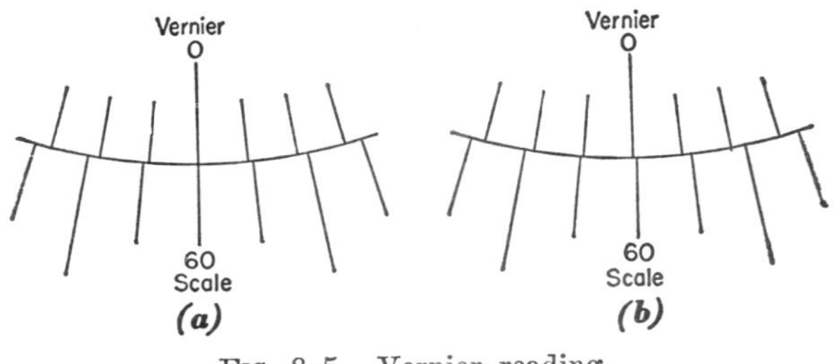

FIG. 8–5. Vernier reading.

In Fig. 8–5a, the zero mark on the vernier is set exactly opposite a graduation on the scale, since the distances between the second vernier and scale graduations on both sides of zero are equal. If two sets of lines appear to be almost coincident and a symmetrical pattern is formed, as in Fig. 8–5b by 0 and the first division on the left, a reading halfway between them can be interpolated.

Figure 8–4a shows two sets of numbers on the circle and a double vernier. The reading for the inner set is 58° 30′ + 17′ = 58° 47′. For the outer circle it is 301° 00′ + 13′ = 301° 13′. Note that *the vernier is always read in the same direction from zero as the num-*

bering of the circle, that is, on the side of the double vernier in the direction of the increasing angle.

The reading of the inner set of numbers from the double vernier in Fig. 8–4b is 91° 20′ + 07′ = 91° 27′; the reading of the outer set is 268° 20′ + 13′ = 268° 33′.

The folded vernier of Fig. 8–4c reads 117° 05′ 30″ on the inner set of numbers and 242°54′ 30″ on the outer set.

The direct vernier of Fig. 8–4d is the type used on a transit or a repeating theodolite and reads 321° 13′ 20″ for a clockwise angle.

An understanding of verniers is best obtained by practice in reading various types and by calculating and sketching the least count of different combinations of scale and vernier divisions. Typical mistakes in reading verniers include:

a. Not using a magnifying glass.
b. Reading in the wrong direction from zero, or on the wrong side of a double vernier.
c. Failing to determine the least count correctly.
d. Omitting 30 min when the index is beyond the half-degree mark.

8–9. Properties of an American-type engineer's transit. Transits are designed to have a proper balance between magnification and resolution of the telescope, least count of the vernier, and sensitivity of the plate and telescope bubbles. Thus the standard 1-min instrument has the following properties:

Magnification, 18 to 28 diameters.
Field of view, 1° to 1° 30′.
Resolution, 4 to 5 sec.
Minimum focus, about 5 to 7 ft.
Sensitivity of plate levels per 2-mm division, 70 to 100 sec.
Sensitivity of telescope vial per 2-mm division, 30 to 60 sec.
Weight of instrument head without tripod, 11 to 16½ lb.

Cross hairs usually include vertical and horizontal center hairs and two stadia hairs, as shown in Fig. 8–6b and Fig. 8–6c. Short stadia lines, used on glass reticles (Fig. 8–6c), avoid confusion between the center and stadia hairs.

A quarter hair, located halfway between the upper and middle hairs (Fig. 8–6d), is sometimes used to increase the range of stadia readings.

The X pattern (Fig. 8–6e) is incorporated in precise instruments

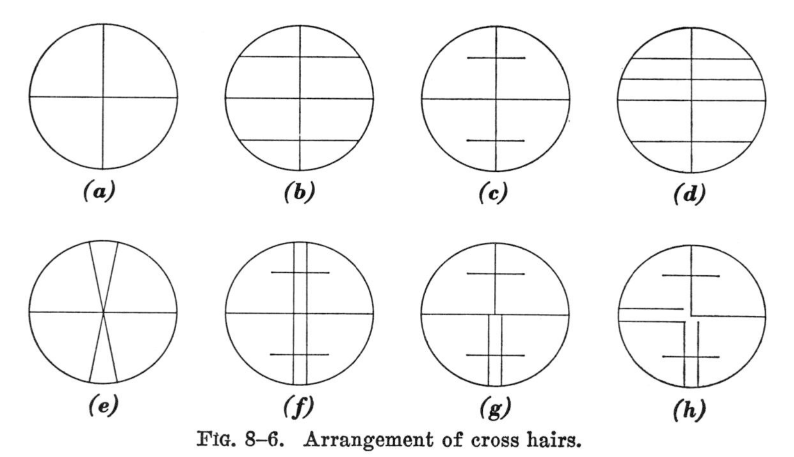

FIG. 8–6. Arrangement of cross hairs.

to prevent a rod or object seen at a long distance from being completely hidden behind the vertical hair. It also permits the observer to balance distances between the rod and hairs on both sides of the upper and lower sections to insure centering, a task the human eye does in a highly efficient manner. The arrangement shown in Fig. 8–6f, or one of the variations in Figs. 8–6g and 8–6h, likewise avoids covering the object sighted and aids in centering.

The tapered design of American transit spindles assures that despite wear, unless they are damaged by dirt or an accident, they will still seat and center properly.

The transit can be used to measure angles by repetition—that is, an angle may be repeated any number of times and the total angular distance added on the plates. A better value is thereby obtained for averaging, and errors are disclosed by comparing the values of the single and multiple readings.

8–10. Other American instruments. The *builder's transit-level* is a low-priced instrument for use on work requiring only short sights and moderate precision. The model shown in Fig 8–7 has a telescope with a magnification of 20 diameters and a resolution of 4.7 sec of arc, a telescope level vial with a sensitivity of 90 sec per 2-mm division, and a horizontal circle and vertical arc reading to 5 min. The cost is approximately one-third that of a standard engineer's transit.

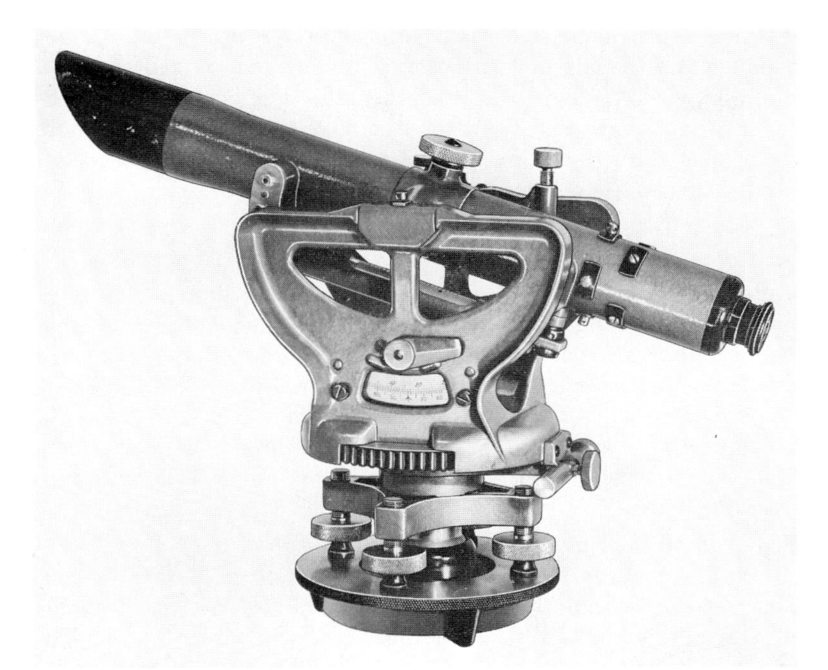

FIG. 8–7. Builder's transit-level. (Courtesy of Keuffel and Esser Company.)

Farm levels have a telescope with a magnification of 12 to 14 diameters, a compass, a telescope level vial with a sensitivity of 4 to 5 min, and a horizontal circle and vertical arc reading to 5 min.

Collimators are sighting devices used (a) in making adjustments of surveying instruments; (b) in establishing lines for fabrication of large objects in a shop; and (c) as optical plummets to center an instrument on a tower over a mark on the ground, or to set a point directly beneath the center of the instrument.

Theodolites are built by American manufacturers in limited quantity. A theodolite is a precision surveying instrument for measuring angles. There are two general classes of theodolites: repeating theodolites and direction theodolites (direction-instrument theodolites).

Repeating theodolites are precise transits which can be read to 10 sec or less.

A direction theodolite is a nonrepeating type of instrument which has no lower motion. Directions rather than angles are read. After a sight has been taken on a point, the direction of the line to the point is read on the circle. An observation on the next mark gives a new direction. The angle between the lines can be found by subtracting the first direction from the second. In some work, such as triangulation, directions are employed in the computations and adjustments in place of angles. Two or more sets of angles may be taken with the telescope in normal and reversed positions for averaging.

A modern theodolite is compact, simple in design, and light in weight, yet rugged enough to withstand severe usage. The telescope, circles, clamps, micrometers, microscope, and leveling screws are fully enclosed and virtually dust- and moisture-proof. An optical-micrometer system is used instead of a vernier. The telescope is short, inverting, internal-focusing, and equipped with a glass reticle. The horizontal and vertical circles are relatively small and are made of glass.

European theodolites are described in more detail in Sec. 8–24.

8–11. Setting up and handling the transit. A transit is taken from its box by holding the leveling head, the underside of the lower plate, or the standards (*not* by lifting the telescope), and is screwed securely on the tripod. A plumb-bob string is hung on the hook at the bottom of the spindle, using a slipknot to permit raising or lowering the bob without retying the string, and avoiding knots in the string.

A transit carried indoors should be balanced in a horizontal position under one arm, with the instrument-head forward. The same method is suitable in areas covered with brush. In open terrain the instrument may be balanced on a shoulder. The transit telescope should be clamped lightly in a position perpendicular to the plates. The plate clamps should be set lightly to prevent swinging, while still permitting ready movement if the instrument is bumped.

The wing nuts on the tripod must be tight, to prevent slippage and rotation of the head. They are correctly adjusted if each tripod leg falls slowly of its own weight when placed in a horizontal position. If the wing nuts are overly tight, or if pressure is applied to the legs crosswise instead of lengthwise to fix them in the ground, the tripod is in a strained position. The result may be an unnoticed movement of the instrument head after observations have begun. Tripod legs should be well-spread, to furnish stability and to place the telescope at a height convenient for the observer. Tripod shoes must be tight. Proper field procedures can eliminate most instrument inadjustments but there is no way to take care of a poor tripod with dried-out wooden legs except to discard or repair it.

The plumb bob must be brought directly over a definite point, such as a tack in a wooden stake, and the plates leveled. The tripod legs can be moved in, out, or sideways, to approximately level the plates before the leveling screws are used. Shifting the legs effects the position of the plumb bob and makes setting up a transit more difficult than for a level.

Two methods are used to bring the plumb bob within about $\frac{1}{4}$ in. of the proper point. In the first method, the transit is set over the mark and one or more legs are moved to bring the plumb bob into position. One leg may be moved circumferentially, to level the plates without greatly disturbing the plummet. Beginners sometimes have difficulty with this method because at the start the center of the transit is too far off the point, or the plates are badly out of level. Several movements of the tripod legs may then fail to both level the plates and center the plumb bob while maintaining a convenient height of instrument. If an adjustable-leg tripod is used, one or two legs may be lengthened or shortened to bring the bob directly over the point.

In the second method, which is particularly suited to level or uniformly sloping ground, the transit is set up near the point and the plates are leveled by moving the tripod legs as necessary. Then, with one tripod leg held in the left hand, another held under the left armpit, and the third held in the right hand, the transit is lifted and placed over the mark. A slight shifting of one leg should bring the plumb bob within $\frac{1}{4}$ in. of the proper position and leave the plates practically level.

The plummet is centered exactly by loosening the four leveling screws and sliding them on the bottom plate by means of the ball-and-socket shifting-head device, which permits a limited movement. To assure mobility in any direction, the shifting head should be approximately centered on the bottom plate before setting up the instrument.

The transit is accurately leveled by means of the four leveling screws somewhat in the same manner as described for the engineer's level. However, each level vial on the upper plate is set over a pair of opposite leveling screws, and it is not necessary to change the position of the telescope. If the plumb bob is still over the mark after leveling, the instrument is ready for use. But if the plates were badly out of level, or the leveling screws were not uniformly set, the plummet moves off the mark. The leveling screws must be loosened and shifted again, and the transit releveled. It is evident that time can be saved by starting with the plates reasonably level to eliminate excessive manipulation and possible binding.

8–12. Operation of the transit. Horizontal angles are measured with a transit by operating the upper clamp, lower clamp, and tangent screws. The telescope clamp and tangent screw are utilized to bring the object sighted to the center of the field of view.

Beginners may find it helpful to remember the following rules covering the use of the upper and lower clamps:

a. The lower clamp is used for backsighting only.
b. The upper clamp is used for setting the plates to zero, for setting the plates to a given angle, and for foresighting.

Expressed in a different way, it can be said that the lower clamp and tangent screw are used to bring the line of sight along the reference line from which an angle is to be measured. The upper clamp and tangent screw are used to set zero degrees on the plates before sighting along the reference line, and to obtain a differential movement between the plates when foresighting. The step-by-step procedure for measuring a *direct* (interior) angle ABC in Fig. 8–8 will be outlined to illustrate the use of the upper and lower motions:

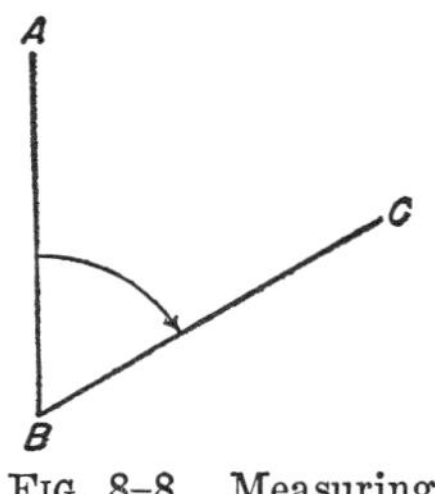

FIG. 8–8. Measuring an angle.

1) Set up the instrument over point B and level the plates. Loosen both motions. Estimate the size of the angle as a check on the value to be obtained.

2) Set the plates to read approximately zero by holding the upper plate while turning the lower by tangential pressure on its underside. Tighten the upper clamp (snug but not wrench-tight). The upper and lower plates are now locked together.

3) Bring the vernier zero exactly opposite the plate zero by means of the upper tangent screw. Use a positive (clockwise) motion. If

the zero is run beyond the point, back off and always finish with a positive motion. This prevents *backlash* (release of the spring tension, which can change the plate position).

4) Sight point A through the telescope. Set the vertical cross hair on, or almost on, the center line of the range pole or other object marking A, by turning the instrument with both hands on the plate edge or on the standards (*not* on the telescope).

5) Tighten the lower clamp. The plates, which were clamped together before, are now also fastened to the socket.

6) Set the cross hair exactly on the mark by means of the lower tangent screw, finishing with a positive motion. Both motions are now clamped, the plates read zero, and the telescope is pointing to A. The transit is therefore *oriented*, since the line of sight is in a known direction with the proper angle (zero degrees) on the plates. Read the compass bearing for line BA.

7) Loosen the upper clamp and turn the plate until the vertical hair is on, or almost on, point C. The lower plate containing the graduated circle is still clamped to the socket, and the zero graduation continues to point toward A. Tighten the upper clamp.

8) Set the vertical hair exactly on the mark by means of the upper tangent screw.

9) Read the angle on the plates, using the vernier ahead of the zero mark (in the same clockwise direction as the angle was turned). Read the compass bearing for line BC. Check the angle by comparing the measured value with the angle computed from the bearings.

8–13. Measuring direct angles by repetition. If an angle is to be measured by repetition (turned two, three, four, or six times), the method just described is followed for the first reading. Then, with the reading for the first angle left on the plates, a backsight is taken on A, as before, by using the lower clamp and tangent screw to retain the angle setting. The transit is now oriented in the starting position, but the single angle is set on the plates, instead of zero degrees.

The upper clamp is loosened and point C sighted again. Next, the upper clamp is tightened and the cross hair brought exactly on the mark with the upper tangent screw. The sum of the first two turnings of the angle is now on the plates. This process can be continued for the number of repetitions desired. The transit should be releveled if necessary after turning the angle, but *the leveling screws must not be used between the backsight and the foresight.* If an even number of repetitions is taken, half of them should be obtained with the telescope normal, and half with the telescope plunged, to eliminate by reversion the effects of some of the possible inadjustments of the instrument described in Chapter 16.

The total angle accumulated on the plates, divided by the number of repetitions, gives the average value. The total angle may be greater than 360°, making it requisite to add a multiple of 360° to the reading before dividing. It is always desirable, therefore, to determine the approximate (single) angle after the first foresight.

It might be assumed that turning an angle ten, fifty, or a hundred times would give an increasingly better answer, but this assumption would not be true. Experience shows that with a 1-min transit having the usual properties, an average observer can point the instrument (align the vertical wire) within about 2 to 5 sec of arc.

A 1-min vernier can be read to within 30 sec. An angle on the plates of (say) 42° 11′ 29″ would theoretically be called 42° 11′ by an experienced observer using a magnifying glass. If the angle on the plates is 42° 11′ 31″, presumably a reading to the nearest minute of 42° 12′ would be obtained. In either case, the recorded value would be within 30 sec of the correct angle.

If the transit is in adjustment, is level and exactly centered, and is being used by an experienced observer under suitable conditions, there are only two sources of error in measuring an angle—pointing the telescope and reading the plates. For a 5-sec average pointing error, and a maximum discrepancy of 30 sec in reading a 1-min vernier, the number of repetitions needed to strike a balance is six; for setting to zero, reading, and a 3-sec pointing error, it is ten.

Neglecting small plate graduation errors, using a 1-minute transit the maximum accidental error in measuring an angle by repetition, 2 direct and 2 reversed (2D, 2R) pointings, can be computed by noting that there are 2 readings—initial and final—and 8 pointings. Then the maximum accidental error $= \frac{1}{4}\sqrt{(30'')^2 + 8(3'')^2 + (30'')^2}$ $= \frac{1}{4}\ (43.3) = 10.8$ seconds. If measured 4 times independently and the results averaged, the maximum accidental error would be $(1/\sqrt{4})\ 85.3 = 42.6$ seconds showing the advantage of the repetition method.

Direct angles, measured singly or by repetition, are commonly used in boundary surveys, hydrographic work, and building construction. *Angles to the right* are direct angles measured clockwise from a backsight on the previous traverse point.

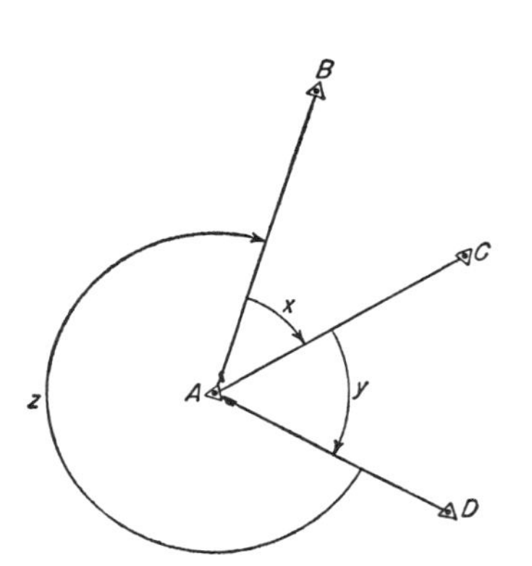

FIG. 8–9. Closing the horizon.

8–14. Closing the horizon. Closing the horizon is the process of measuring the angles around a point to obtain a check on their sum, which should equal 360°. For example, if in Fig. 8–9 only angles x and y are needed, it is desirable to also turn angle z to close the horizon at A. The method provides an easy way for the beginner to test his readings and pointings. Figure 8–10 shows the left page of notes covering the measurement of the angles in Fig. 8–9. The plate readings are changed slightly for each backsight to provide practice in reading the instrument, and for checking purposes.

The difference between 360° and the sum of angles x, y, and z is termed the *horizon closure*. Permissible values of this closure will determine whether the work must be repeated.

Plate A–7 shows a sample set of notes illustrating the measurement of angles by repetition to close the horizon. In this particular arrangement, the A vernier is set to read zero only at the beginning of the work and thereafter the final reading for each angle—for example, 253° 13′ 00″—becomes the initial reading for the next angle. Both a *vernier closure* (difference between the initial and final vernier readings) and a horizon closure are obtained in this rigorous procedure.

8–15. Laying off an angle. To lay off an angle BAC equal to 25° 30′ with the transit at point A (Fig. 8–11) the plates are set to read zero degrees and point B is sighted by using the lower motion. The upper clamp is loosened, the telescope is turned until the vernier reads 25° 30′, and the plates are again clamped. The line of sight establishes AC at the proper angle with AB.

To lay off an angle BAC equal to 25° 30′ 40″ by repetition with a 1-min transit, an angle BAC' of 25° 30′ is laid out as previously described and point C' marked.

CLOSING THE HORIZON					
Point Sighted	Plate Reading	Angle	Mag Bearing	Angle Comp. from Bear.	
	⊼ at point A				
B	3°26′		N22°15′E		
C	45°38′	42°12′		42°15′	
C	47°08′ ~~42°08′~~		N64°30′E		
D	107°04′	59°56′		60°00′	
D	110°35′		S55°30′E		
B	8°29′	257°54′		257°45′	
		360°02′			
	Closure	0°02′			

FIG. 8–10. Closing the horizon.

The angle BAC' is then measured by repetition as many times as the desired precision requires. The difference between the angle BAC' and 25° 30′ 40″ can be laid off by measuring the distance AC' and locating C by the following relation: Distance $C'C = AC' \tan C'AC$. The angle BAC can then be turned by repetition as a check.

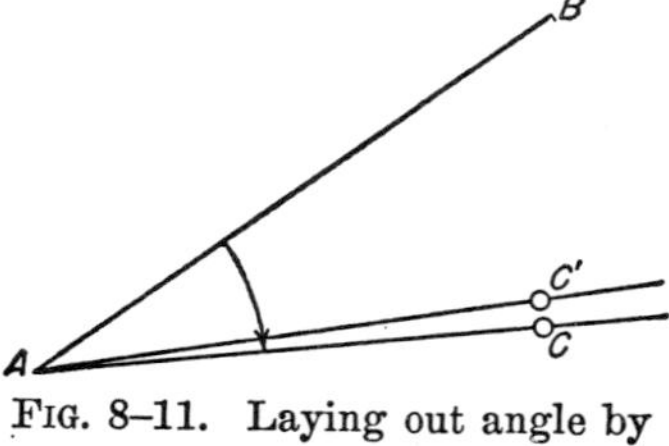

FIG. 8–11. Laying out angle by repetition.

In Fig. 8–11, if the angle BAC' is found by repetition to be 25° 30′ 20″, then $C'AC = 20''$. If distance AC' is 300 ft, then $C'C = 300 \tan 20'' = 0.029$ ft.

8–16. Deflection angles. A deflection angle is a horizontal angle measured from the prolongation of the preceding line, right or left, to the following line. In Fig 8–12a, the deflection angle at *F* is 12° 15′ to the right, written as 12° 15′ R. At *G* the deflection angle is 16° 20′ L.

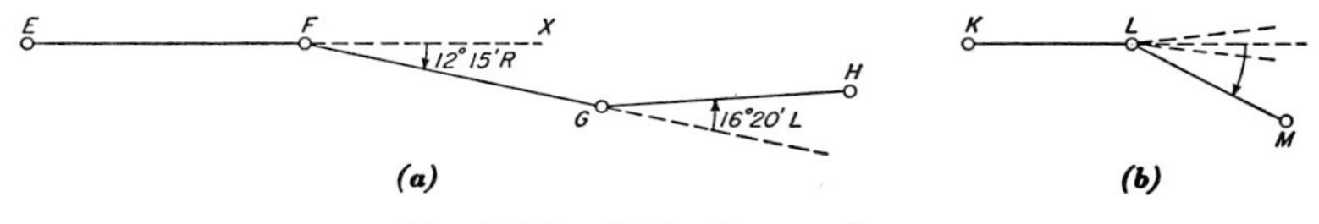

FIG. 8–12. Deflection angles.

A straight line between terminal points is theoretically the most economical route to build and maintain for highways, railroads, pipelines, canals, and transmission lines. Practically, obstacles and conditions of terrain require bends in the route, but deviations from a straight line are kept as small as possible. The use of deflection angles is therefore appropriate for easier visualization, sketching, and computation. Note that in the shorthand method of computing bearings from direct angles in illustration 6–1, the letters for back bearings shown in parentheses are not needed for deflection angles, since these angles are turned from the extended line.

The deflection angle at *F* (Fig. 8–12a) is measured by setting the plates to zero and backsighting on point *E* with the telescope plunged (level vial above the telescope). The telescope is then plunged (transited) by turning it 180° about its horizontal axis, which puts the level vial beneath the telescope. The line of sight is now on *EF* extended, and directed toward *X*. The upper clamp is loosened, point *G* sighted, the upper clamp tightened, and the vertical wire brought exactly on the mark by means of the upper tangent screw. The A vernier will be under the eyepiece end of the telescope, and so the observer can read the deflection angle without moving around the transit.

Deflection angles are subject to serious error if the transit is not in adjustment, as will be shown in Chapter 16. The deflection angle may be too large or too small, the nature of the error depending upon whether the line of sight after plunging is to the right or to the left of the true prolongation (Fig. 8–12b).

To eliminate the error caused by defects in adjustment of the transit, angles are usually doubled or quadrupled by the following procedure: The first backsight is taken with the telescope in the normal position. After plunging, the angle is measured and left on the plates. A second backsight is taken with the telescope plunged, the telescope is again transited back to the normal position for the foresight, and the angle is remeasured. Dividing the total angle by 2 gives an average angle from which the adjustment errors have been eliminated by cancellation. In outline fashion the method is as follows:

Backsight with telescope normal. Plunge and measure angle.
Backsight with telescope plunged. Plunge again and measure angle.
Read total angle and divide by 2 for an average.

8–17. Azimuths. Since azimuths are measured from a reference line, the direction of the initial line must be determined from a previous survey; from the magnetic needle; by a solar or star observation; or must be assumed. Suppose that in Fig. 8–13 the azimuth of line *AB* connecting two triangulation stations is known to be 132° 17′ from true north. To find the azimuth of any other line from *A*, such as *AC*, first set 132° 17′ on the scale numbered in a clockwise direction and backsight on point *B*. The transit is now oriented, since the line of sight is in a known direction with the appropriate angle on the plates. Loosen the upper motion and turn the telescope either clockwise or counterclockwise to *C*, but read the clockwise circle. In this case the reading would be, say, 83° 38′.

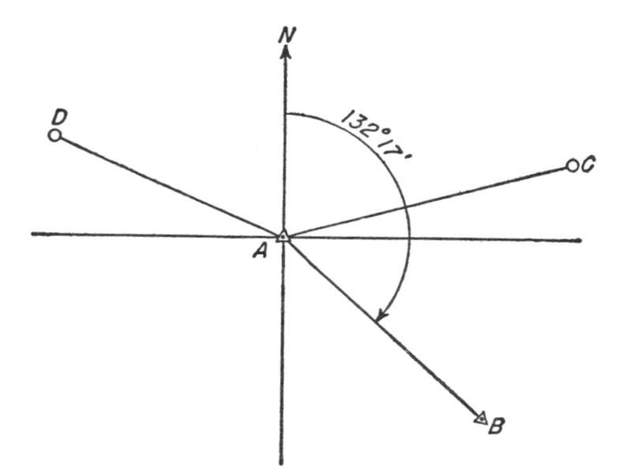

FIG. 8–13. Orientation by azimuths.

Note that after the lower clamp and tangent screw are used to backsight on point *B*, they are not disturbed regardless of the number of angles read from point A. When the plates read zero, the telescope is pointing true north. As a check, the compass needle can be lowered and read. If the telescope is pointing north, the needle should read the declination of the place. Figure 8–14 shows the left page of a sample set of notes for a traverse run by azimuths.

AZIMUTH-TAPE TRAVERSE

Point Occupied	Point Sighted	Distance	Azimuth	Mag. Bear.	
A	Mag. N		0°00'	Due N	
	B	126.24	23°32'	N23°30'E	
B	A		203°32'	S23°30'W	
	C	82.50	93°51'	S86°15'E	
C	B		273°51'	N87°00'W	
	D	122.58	137°39'	S42°30'E	
D	C		317°39'	N43°00'W	
	A	216.35	264°46'	S84°45'W	
A	D		84°46'	N84°45'E	
	B		23°34'	N23°30'E	
		Closure	0°02'		

FIG. 8–14. Azimuth traverse (initial orientation on magnetic north).

In Fig. 8–13, if the transit is set up at point B instead of at A, the back azimuth of AB (312° 17′) is put on the plates and point A sighted. The upper plate is loosened and sights are taken on points whose azimuths from B are desired. Again, if the plates are turned to zero, the telescope points true north.

An alternative method of orienting the transit at point B is to leave the azimuth of AB (132° 17′) on the plates while backsighting on point A with the telescope plunged. The telescope is then transited to the normal position to bring the line of sight along line AB extended with its proper azimuth on the plates. This method eliminates resetting the plates, but introduces an error if the instrument does not plunge perfectly.

8–18. Sights and marks. Objects commonly used for sights on plane surveys include range poles, chaining pins, pencils, and plumb-bob string. For short sights, string is preferred to a range pole because the small diameter permits more-accurate centering. Small red-and-white targets of thin-gage metal or cardboard placed on the string extend the length of observation possible.

An error is introduced if the range pole sighted is not plumb. The observer must sight as low as possible on the pole if the mark itself is not visible, and the rodman is obliged to take special precautions in plumbing the rod, perhaps employing a rod level.

In layout work on construction, and in topographic mapping, *permanent* backsights and foresights may be established. These can be marks on structures such as walls, steeples, water tanks, and bridges or can be fixed, artificial targets. They provide definite points on which the instrumentman can check his orientation without the help of a rodman.

8–19. Prolonging a straight line. On route surveys, straight lines may be continued from one transit hub through several others. To prolong a straight line from a backsight, the vertical wire is aligned on the back point by means of the lower motion. The telescope is plunged, and a point, or points, set on line.

To eliminate the effects of defective adjustment of the transit, the same procedure used in making a number of adjustments, known as the *principle of reversion,* is employed. The sighting method applied, actually *double reversion,* is termed *double centering.* Figure 8–15 shows a common use of the principle of reversion in drawing a right angle with a defective triangle. Lines OX and OY are drawn with the triangle in "normal" and "reversed" positions. The angle XOY represents twice the angular error in the triangle at the 90° corner.

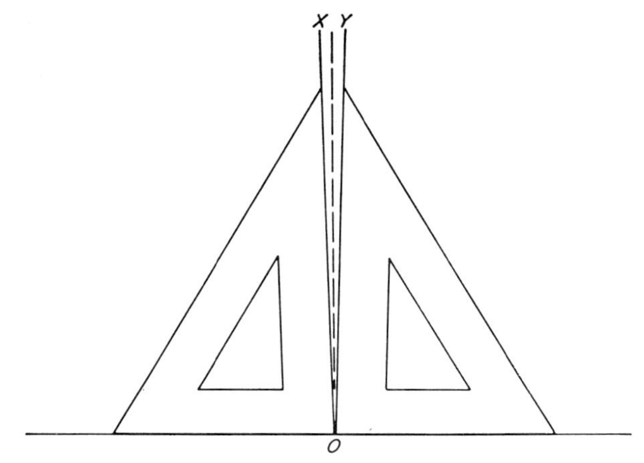

FIG. 8–15. Principle of reversion.

In double centering, after the first point *C′* in Fig. 8–16 has been located with the telescope plunged, the lower motion is released and a second backsight taken on point *A*, this time with the telescope still plunged. The telescope is transited again to its normal position and point *C″* is marked. The distance *C′C″* is bisected to get point *C*, which is on the line *AB* prolonged.

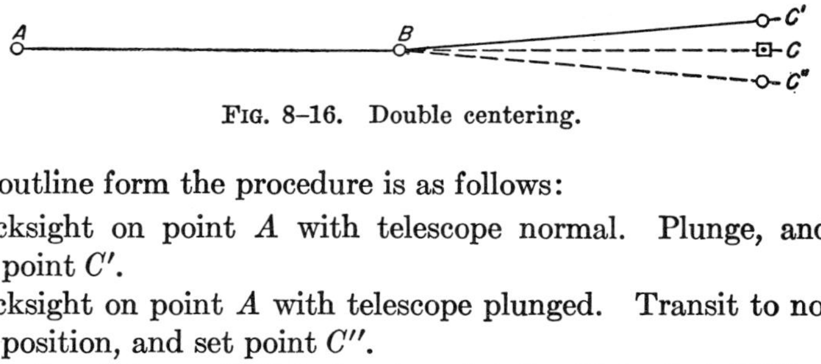

FIG. 8–16. Double centering.

In outline form the procedure is as follows:

Backsight on point *A* with telescope normal. Plunge, and set point *C′*.

Backsight on point *A* with telescope plunged. Transit to normal position, and set point *C″*.

Split the distance *C′C″* to locate point *C*. Note that *C′C″* represents twice the plunging error and, as described in Chapter 16, four times the adjustment error.

8–20. Prolonging a line past an obstacle. Buildings, trees, telephone poles, and other objects may block survey lines. Four of the various methods used for extending lines past an obstacle are: (a) the equilateral-triangle method; (b) the right-angle-offset method;

(c) the measured-offset method; and (d) the equal-angle method.

Equilateral-triangle method. At point *B*, Fig. 8–17a, a 120° angle is turned off from a backsight on *A*, and a distance *BC* of 80.00 ft (or any distance necessary, but preferably less than one tape length) is measured to locate point *C*. The transit is then moved to *C*; a backsight is taken on *B*; an angle of 60° 00′ is put on the plates; and a distance $CD = BC = 80.00$ ft is laid off to mark point *D*. The transit is moved to *D* and backsighted on *C*, and an angle of 120° 00′ is set on the plates. The line of sight *DE* is now along *AB* prolonged, if no errors have been made.

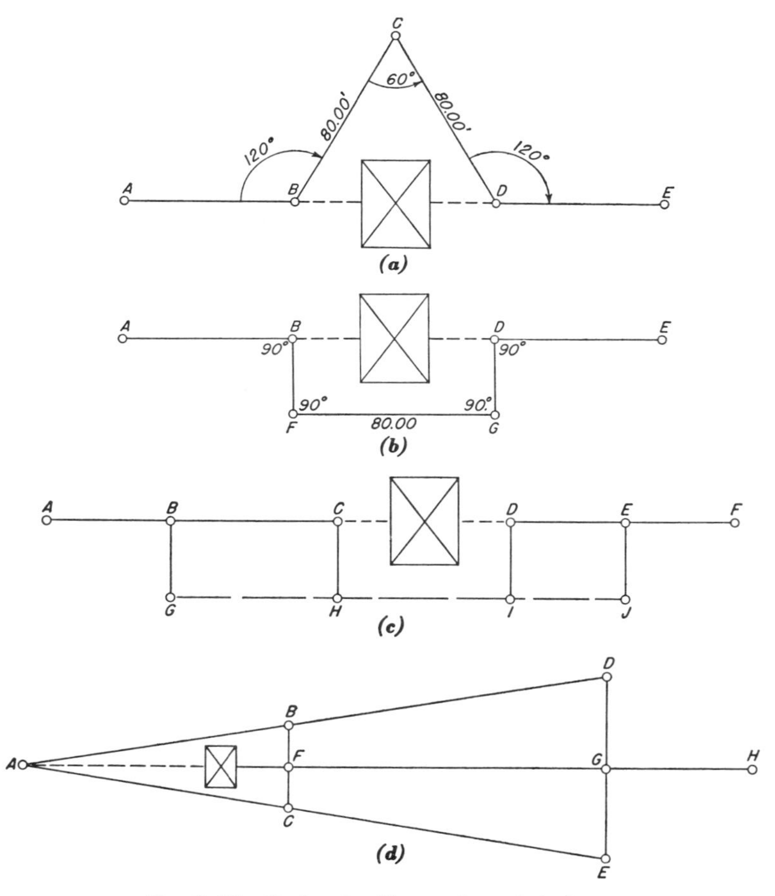

FIG. 8–17. Prolonging line past an obstacle.

Right-angle-offset method. The transit is set up at points *B, F, G,* and *D* (Fig. 8–17b) and 90° 00′ angles are turned off at each hub. Distances *FG* and *BF* (equal to *GD*) need only be large enough to clear the obstruction, but longer lengths permit more accurate sights.

The lengths shown in Fig. 8–17a and b permit students to check their taping and manipulation of the transit by combining the two methods.

Measured-offset method. To avoid the four 90° angles with short sights and consequently possible large errors, measured offsets obtained by swinging arcs with the tape can be used (Fig. 8–17c). A long base is established for check points on *GHIJ* if desired.

Equal-angle method. This method is an excellent one when field conditions are suitable. Equal angles just large enough to clear the obstacle are turned from the line at point *A*, and equal distances *AB* = *AC* and *AD* = *AE* are measured (Fig. 8–17d). The line through points *F* and *G* at the midpoints of *BC* and *DE*, respectively, provides an extension of line *AH* through the obstacle. Very little additional clearing of the route is necessary using this method to bypass a large tree on line in wooded or brushy areas.

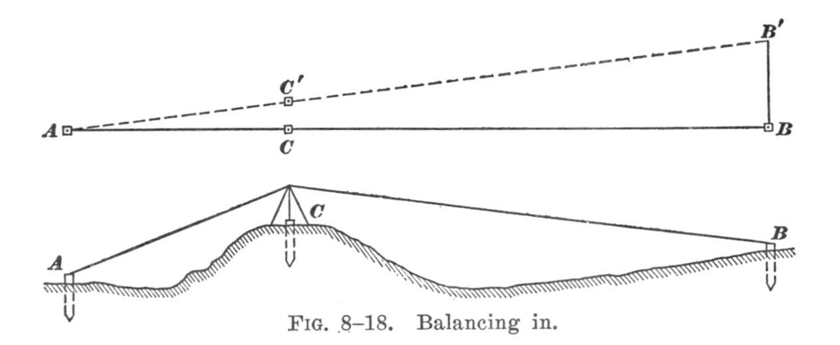

FIG. 8–18. Balancing in.

8–21. Balancing in. Occasionally it is necessary to set up the transit on a line between two points already established, for example points *A* and *B* (Fig. 8–18). This process is called "balancing in," or "wiggling in."

The location of a point *C* on line is estimated and the transit is set up over it. A sight is taken on point *A* from the trial point *C′*, and the telescope is plunged. If the line of sight does not pass through *B*, the transit is moved laterally a distance *CC′*, which is estimated from the proportion $CC' = BB' \times AC/AB$, and the process is repeated. Several trials may be required to locate point C exactly, or close enough for the purpose at hand. The shifting head of the transit is used to make the final small adjustment.

8–22. Random line. On many surveys it is necessary to run a random line from a point X to a nonvisible point Y, which is a known or indeterminate distance away. This problem arises repeatedly in surveys of the public lands.

On the basis of compass bearings or information from maps and other sources, a random line, such as XY' (Fig. 8–19) is run as close as possible by estimation to the true line XY. The distance XY', and the distance YY' by which the random line misses point Y, are measured and the angle YXY' is found from its calculated sine or tangent. The correct line is then run by turning off the computed angle $Y'XY$.

FIG. 8–19. Random line.

8–23. Measurement of a vertical angle. A vertical angle is the difference of direction between two intersecting lines measured in a vertical plane. As commonly used in surveying, it is the angle above or below a horizontal plane through the point of observation. Angles above the horizontal plane are called *plus angles*, or *angles of elevation*. Those below it are *minus angles*, or *angles of depression*. Vertical angles are measured in trigonometric leveling and in stadia work as an important part of the field procedure.

To measure a vertical angle, the transit is set up over a point and the plates are carefully leveled. The bubble in the telescope level vial should remain centered when the telescope is clamped in a horizontal position and rotated 360° about its vertical axis. If the vernier on the vertical arc does not read 0° 00′ when the bubble is centered, there is an index error which must be added to, or subtracted from, all readings. Confusion of signs is eliminated by placing in the field notes a statement such as "Index error is minus 2 min, to be subtracted from angles of depression and added to angles of elevation."

The horizontal cross line is set approximately on the point to which a vertical angle is being measured, and the telescope is clamped. Exact elevation or depression is obtained by using the telescope tangent screw. The vernier of the vertical circle is read and any index error applied, to obtain the true angle above or below the horizon. The observer must inform the recorder in advance whether he will call out the uncorrected reading of the angle or the adjusted value; the observed (uncorrected) quantity is preferable.

To eliminate the index error resulting from displacement of the vernier on the vertical arc, and from the lack of parallelism of the line of sight and the telescope level vial, an average of two readings may be taken. One is secured with the telescope normal, the second with the telescope inverted. This method requires the transit to be equipped with a complete vertical circle and reversible level vial.

Note that a transit can be used as a level, leveling by means of the telescope vial although it is not as sensitive as the vial on an engineer's level.

8–24. European theodolites. European theodolites differ from American transits in general appearance (they are compact, lightweight, and "streamlined") and in design by a number of features, the more important of which are as follows:

a) The *telescopes* are short and inverting, have reticles etched on glass, and are equipped with rifle sights for rough pointing.

b) The *horizontal and vertical circles* are made of glass. Graduation lines and numerals are etched on the surface of the circles. Although the lines are very thin (0.004 mm) and short (0.05 to 0.10 mm), they are more-sharply defined than can be achieved by scribing them on metal. Precisely graduated circles with small diameters can be obtained, and this is one reason for the compactness of the instruments. The circles are divided into the conventional sexagesimal degrees and fractions (360°), or into centesimal "grads" or "grades" (full circle divided into 400^g with each grade divided into 100^c or centesimal minutes, and each minute divided into 100^{cc} or centesimal seconds).

c) A *collimation level* or *index level,* sometimes of the coincidence type, is connected to the reading system of the vertical circle. This level provides a more accurate plane of reference for the measurement of vertical angles than the plate level.

d) The *reading system* for the circles consists basically of a microscope, the optics of which are situated inside the instrument. The reading eyepiece is generally adjacent to the telescope eyepiece or located on one of the standards. Some instruments have optical micrometers for fractional reading of circle intervals (micrometer scale visible through reading microscope); others are "direct-reading." The reading system of most European theodolites can be illuminated electrically for night or underground work.

e) The *vertical axis* is cylindrical, or a precision ball bearing, or a combination of both, and is generally made of steel.

f) The *leveling head* consists of three screws or cams.

g) The *bases* of some theodolites are designed to permit interchange of instrument and accessories (targets, subtense bar, etc.), without disturbing the centering over a survey point.

h) An *optical plummet,* built into the base or alidade of some theodolites, replaces the plumb bob and permits centering with great accuracy.

i) The *carrying cases* for theodolites are made of sheet steel or lightweight alloy or a combination of both. They are generally compact and watertight, and some can be locked.

j) *Distance-measuring devices* are sometimes permanent and integral parts of theodolites built specifically for tachymetry. Such instruments are generally self-reducing.

k) Various *accessories* increase the versatility of theodolites, adapting them for tachymetry and astronomical observations. The compass is an accessory, rather than an integral part, of the European theodolite.

l) *Tripods* are of the wide-frame type. Some are all-metallic and feature devices for preliminary leveling of the tripod head and for mechanical centering ("plumbing") to eliminate the need for a plumb bob or optical plummet.

European theodolites may be divided into two basic categories: *directional* (or triangulation) theodolites, and *repeating* (or double-center) theodolites.

Directional theodolites have a single vertical axis and therefore cannot be used to measure angles by the repetition method. They have a circle-orienting drive to set the horizontal circle to any desired position. On all European direction theodolites, each reading is the arithmetic mean of two opposed points of the circle graduation, so that eccentricity errors are automatically eliminated. This is equivelent to the average of the readings of the A and B verniers on a transit. Most direction theodolites are equipped with an optical micrometer which breaks down the circle graduations (20, 10, or 2 min) into fractional readings (10 sec, 1 sec, $\frac{1}{2}$ sec, or 1/10 sec of arc).

Repeating theodolites are equipped with a double vertical axis (similar to that of American transits, but usually cylindrical in shape), or with a repetition clamp. Only one side of either circle is read at one time, so that the readings are not automatically free from eccentricity errors. Such errors in the horizontal circle must be canceled by the repetition method. They cannot be eliminated

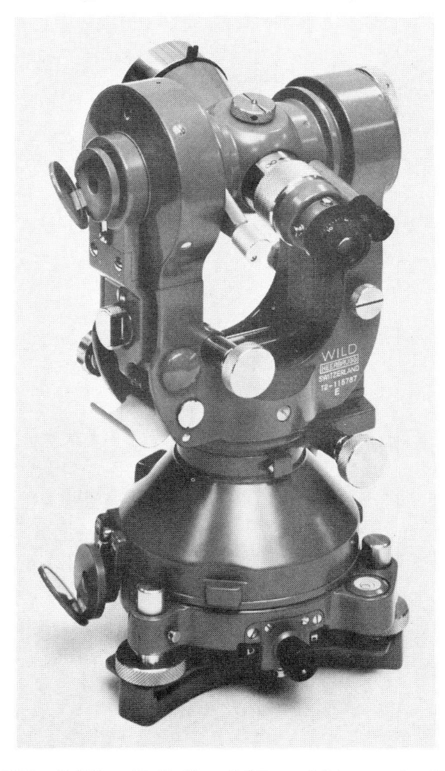

FIG. 8–20. T–2 theodolite which reads to one second. (Courtesy of Wild Heerbrugg Instruments, Inc.)

in the vertical circle. The circles are read by means of a reading microscope, either with an optical micrometer or by direct reading of a scale. Their least direct reading ranges from 1 min to 20 sec of arc.

Figure 8–20 shows the T–2 direction theodolite with which readings can be made directly to 1 sec of arc and can be interpolated to $\frac{1}{10}$ sec. Some practice is required to quickly center an instrument over a point with the optical plummet, but use of this device eliminates the error inherent in a swinging plumb bob on a windy day.

Figure 8–21 shows the arrangement of the horizontal and vertical circles, and the scale of the optical micrometer which is used with both

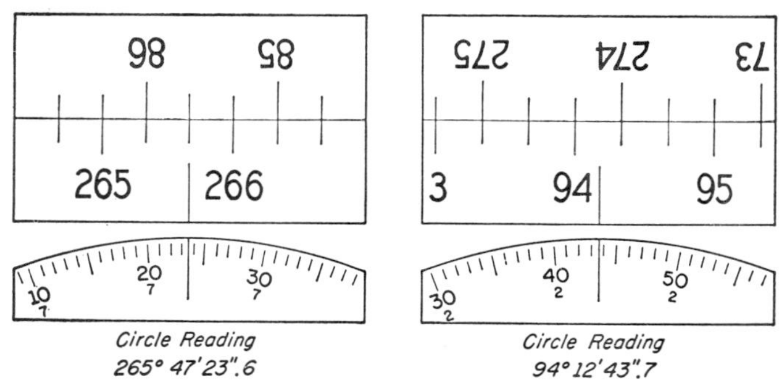

FIG. 8–21. Coincidence reading of horizontal circle.

circles. A knob is turned to shift the prism bringing the image to the eyepiece so that only one circle is visible at a time. The circles are divided into 20-min spaces and the whole-degree marks are numbered.

A reading is obtained by turning the micrometer head to make the division lines (which move in opposite directions) coincident. Since an average reading of the two sides of the circle is secured, each graduation is counted as 10 min to save a later division by 2. Movement of the micrometer is limited to 10 min. The set of numbers seen upside down is on the opposite side of the circle.

In Fig. 8–21 a direction of 265° 40′ is read by counting the number of divisions between 265° and its diametrically opposite graduation, 85°. The micrometer scale gives the additional minutes and seconds, in this example 7′ 23.6″. The vertical circle, read in the same manner, furnishes the zenith distance (the complement of the vertical angle) to avoid the need for signs.

If the horizontal circle is set to zero initially, an angle can be read directly. Usually, however, any value that happens to be on the circle when the telescope is sighted along the reference line is accepted as the first direction. An angle is then determined by subtracting this value from the direction read with the telescope pointed at the next target. To check an angle, or to average readings, the telescope is plunged and the sights are repeated.

Initial positions may be taken at different locations on the circle to reduce errors caused by eccentricity of the circle and possibly slightly imperfect graduations.

The double-circle theodolite shown in Fig. 8–22 incorporates the following important new features, among others:

a) The traditional foot-screws for leveling have been eliminated. Tripod heads are fitted with a ball-and-socket device to speed the rough leveling. A single turn of the three leveling knobs completes

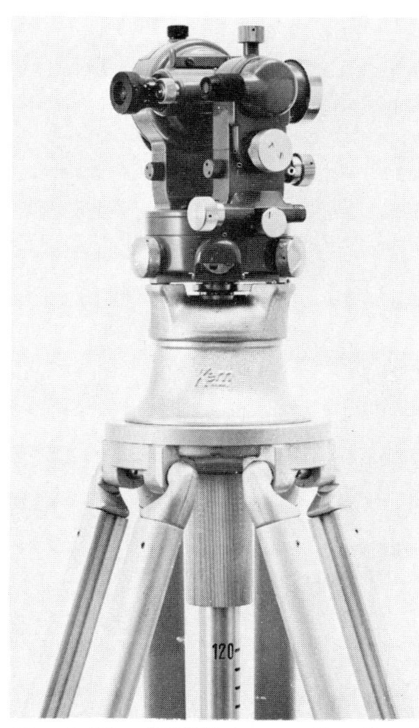

FIG. 8–22. DKMI ten-sec directional theodolite on centering tripod. (Courtesy Kern Instruments, Inc.)

the fine leveling process. These knobs are in fact eccentric cams with a horizontal axis. Their design permits a more compact arrangement and prevents lateral play and slippage which may occur because of worn threads in the standard three-screw leveling system.

b) The coincidence method of reading is replaced by a different system in which a double-line graduation can be seen in the field of view of the microscope (Fig. 8–23). One of these lines actually comes from a diametrically opposite point on a second concentric scale. Readings are taken by means of a micrometer pointer which is centered between the lines. This setting gives an average value for the two sides of the circle. The instrument shown is read directly to 10 sec and readings can be interpolated to 1 sec of arc.

c) A special ball-bearing axis is used to insure positioning of the upper part of the instrument within a few seconds. Obviously this is necessary in instruments reading to seconds.

d) A simplified method of clamping the theodolite on a tripod and in the carrying case is embodied.

The optical instruments described have many advantages. They also have disadvantages for some types of work, such as route surveys in which deflection angles are turned. On many surveys, readings to 1 min are sufficiently precise. A circle graduated in a counterclockwise direction is desirable in various kinds of surveying. Some American engineers brought up with older-type instruments consider the inverting telescope awkward in giving line and grade.

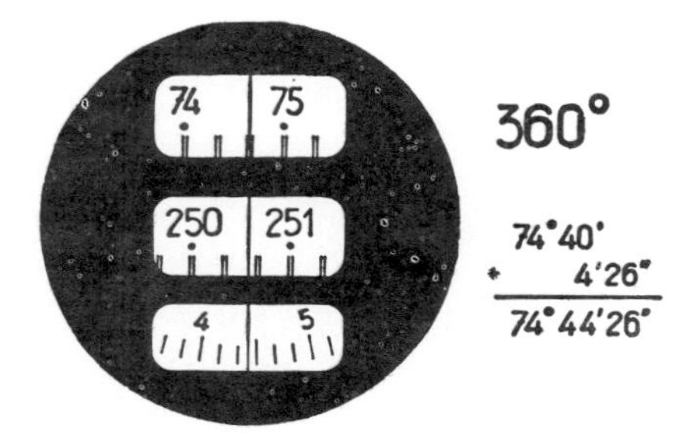

FIG. 8–23. Reading with a double-line graduation. (Courtesy of Kern Instruments, Inc.)

8–25. Sources of error in transit work. Errors in transit surveys result from instrumental, natural, and personal sources. Normally it is impossible to determine the exact value of an angle, and therefore the error in its measured value. Precise results can be obtained, however, by (a) following specified procedures in the field, (b) manipulating the instrument carefully, and (c) checking measurements. Probable values of accidental errors and the degree of precision secured can be calculated from the formulas given in Chapter 2.

Some Sources of Instrumental Errors

Improper adjustment of level vials, cross hairs, or standards. The adjustments described in Chapter 16 should be checked daily, or more often. Proper field procedures can be devised to counteract the effects of improper adjustments (eliminating parallax and sighting with the telescope in both the normal and plunged positions), but can be time-consuming and may make it necessary or desirable to adjust the instrument.

Damaged instruments. Transits should be repaired by experienced craftsmen; otherwise the equipment should be retired from use.

Eccentricity of centers or verniers. If the *A* and *B* vernier readings differ by exactly 180° for all positions, the circles are concentric and the verniers correctly set. If the readings disagree by a constant amount other than 180°, the verniers are offset and it is best to use only the *A* vernier. If the difference is not constant, eccentricity of the plates exists. Readings should be taken at several positions on the circle and the results of the *A* and *B* verniers averaged.

Some Sources of Natural Errors

Wind. Wind vibrates a transit and deflects the plumb bob. Shielding the instrument, or even suspending observations on precise work, may be necessary on windy days.

Temperature changes. Temperature differentials cause unequal expansion of various parts of the transit. A level bubble is drawn toward the heated end of the vial. This effect can be investigated by blowing on one end of the vial and noting the movement of the bubble, then releveling and checking the position of the cross hairs on a target. Temperature effects are eliminated by shielding the transit from sources of heat or cold.

Refraction. Unequal refraction bends the line of sight and may cause an apparent shimmering of the object sighted. It is desirable to keep the line of sight well above the ground and to avoid sights close to buildings, stacks, and even large individual bushes in generally open spaces. In some cases, observations may have to be postponed until atmospheric conditions have improved.

Settling of the tripod. The weight of the transit may cause it to settle in soft ground. If unfavorable terrain must be crossed, stakes can be driven to support the tripod legs. Work at a given station should be completed as quickly as possible. Stepping near one of the legs, or touching one while looking through the telescope, will demonstrate the effect of settlement upon the position of the bubble and cross hairs.

Some Personal Errors

Instrument not set up exactly over the point. The position of the plumb bob should be checked at intervals during the time a station is occupied, to be certain it is still centered.

Level bubbles not centered perfectly. The bubbles must be checked frequently, and must be releveled if necessary, before backsighting. Note that in leveling, or in measuring vertical angles, the level vial *under* the telescope is the critical one. For horizontal angles, the telescope can be elevated or depressed in the vertical plane without affecting measurements if the standards are properly adjusted. Hence the bubble *at right angles* to the telescope is the important one.

Vernier misinterpolated. The number of minutes on the scale passed over by the vernier index should be estimated to check a reading.

Improper use of clamps and tangent screws. An observer must form good operational habits and be able to identify the various clamps and tangent screws by their shape without looking at them. Final setting of tangent screws is always made with a positive motion to avoid backlash. Clamps should be tightened just once and not checked again and again to be certain they are secure.

Poor focusing. Correct focusing of the eyepiece on the cross hairs, and of the objective lens on the target, is necessary to prevent parallax. Objects sighted should be placed as near the center of the field of view as possible. Focusing affects pointing, an important source of error.

Overly careful sights. Checking and double-checking the position of the cross-hair setting on a target is wasteful of time and actually produces poorer results than one fast observation. The cross hair should be aligned quickly and the next operation begun promptly.

Unsteady tripod. The tripod leg bolts must be tight so that there is neither play nor strain (the legs can be tapped lightly to relieve any strain before the first sight is taken) and the shoes set solidly in the ground.

8–26. Mistakes. Some of the common mistakes to guard against are (a) sighting on, or setting up over, the wrong point; (b) calling out or recording an incorrect value; (c) reading the wrong circle; (d) using haphazard field procedures; and (e) turning the wrong tangent screw.

PROBLEMS

In problems 8–1 and 8–2 compute the natural sines and tangents of the angles listed using the conversion relationship for 1 min given in Sec. 8–1. Tabulate the answers beside exact values taken from tables. If the conversion relationship is used, what are the approximate limiting sizes of angles for acceptable errors of 1, 2, 2½, 5 and 10 per cent?

8–1. Angles of 0°05′, 0°10′, 0°15′, 0°30′, 1°00′, 2½°, and 5°00′.
8–2. Angles of 0°15′, 0°20′, 0°45′, 1°00′, 1½°, 2°00′, and 4°00′.

Problems 8–3 and 8–4 are similar to problems 8–1 and 8–2 but use the conversion for 1°.

8–3. Angles of 2½°, 5°, 7½°, 10°, and 15°.
8–4. Angles of 2°, 4°, 6°, 10°, and 20°.

Determine the angle subtended for the conditions given in problems 8–5 through 8–7.

8–5. Width of a 4-in. stake when observed by a transit 2 miles away.
8–6. A 3-in. pipe sighted by transit from 1 mile.
8–7. A ¼-in. diameter chaining pin seen by transit 150 ft distant.

In problems 8–8 through 8–11 what is the error in measured angle for the situations noted?

8–8. Sighting a range pole 4 in. off line on a 600-ft shot.
8–9. Setting a transit ¼ in. to the side of the tack on a 500-ft sight.
8–10. Lining in the edge (instead of center) of a 0.3-in. diameter pencil at 150 ft.
8–11. Sighting the edge (instead of center) of a 1-in. diameter range pole 200 ft away.

8–12. Brush obstructs the line of sight so only the top ½ ft of an 8-ft range pole can be seen on a 600-ft sight, and it is out of plumb by 0.3 ft perpendicular to the line. What angular error results?

In problems 8–13 through 8–16 sketch a vernier and find the number of its divisions or least count.

8–13. A transit circle divided into 2° spaces. Reading to 5 min required.

8–14. Transit plates with 5-min spaces. Reading to 20 sec required.

8–15. A protractor graduated to $\frac{1}{4}$° with 15 spaces on the vernier.

8–16. Theodolite circle marked to 10 min, and 40 vernier divisions.

8–17. What is the purpose of "selective magnification" telescopes?

8–18. What is the purpose of loosening wing nuts of a level or transit tripod before setting up?

8–19. Why are two plate levels used on a transit instead of just one?

8–20. Why are two or three verniers (or microscopes) read on a theodolite in accurate measurements of angles?

8–21. When and why are angles measured by repetition?

8–22. What are the primary sources of error in measuring angles with a transit assuming it is properly leveled and centered?

8–23. Outline a method of reading angles with a transit which will eliminate most instrumental errors caused by improper construction and poor adjustment of the transit.

8–24. An angle is to be measured precisely. (a) What error is eliminated when both the *A* and *B* verniers are read? (b) How is the error of graduation minimized? (c) What error is eliminated by measuring the angle the same number of times with the telescope normal and plunged?

In problems 8–25 and 8–26 compute the mean angle for the following field notes taken with the transit at point *B*. Under number of repetitions, C designates clockwise, and CC counterclockwise.

8–25.

Obj.	Tel.	No. of Reps.	Angle	Vernier *A*	Vernier *B*	Mean Angle
A	*D*	0	0°	00′ 00″	00′ 30″	
E	*D*	1 C	92°	22′ 30″	23′ 00″	
E	*D*	3 C	277°	08′ 00″	08′ 00″	
A	*R*	3 CC	0°	00′ 30″	00′ 30″	

8–26.

Obj.	Tel.	No. of Reps.	Angle	Vernier *A*	Vernier *B*	Mean Angle
A	*D*	0	164°	35′ 00″	35′ 20″	
F	*D*	1 C	331°	52′ 40″	52′ 40″	
F	*D*	6 C	88°	20′ 10″	20′ 20″	
A	*R*	6 CC	164°	34′ 40″	35′ 00″	

8–27. In Fig. 8–10, are the angles computed from bearings a reasonable check? Explain.

8–28. What are the advantages of closing the horizon around a point where only one angle is actually required?

8–29. Three angles at a point X were measured with a 10-sec theodolite to close the horizon. Based upon 12 sets of readings, the probable error of a single set of readings was found to be 3.2 sec. If the same procedure is used in measuring the angles of a triangle, what is the probable error of the triangle closure?

8–30. Similar to problem 8–29, but 4 angles, 8 sets of readings, and probable error of 4.0 sec.

8–31. An angle $ABC = 36°28'15''$ must be laid off with a 30-sec transit. After a 350-ft B.S. on point A, point C is marked 500 ft away with an angle of 36°28′ set on the plates. Angle ABC, measured by repetition 6 times, gives a reading of 218°48′30″. What offset at C will give the required angle?

8–32. Similar to problem 8–31, except the repeated angle is 218°28′.

8–33. How close to the true value should a horizontal angle be when read by repetition 6 times with a 15-sec. transit?

8–34. What is the correct sum of the deflection angles, 8 R and 2 L, in a closed traverse?

8–35. Why is an assumed reference line not desirable for azimuths?

8–36. List the advantages and disadvantages of using azimuths for control traverses, property surveys, and surveys for locating topography.

8–37. The line of sight of a transit is out of adjustment by 20 sec. (a) In prolonging a line by plunging the telescope between the backsight and foresight, but not double centering, what angular error is introduced? (b) What linear error results on a foresight of 800 ft?

A line RS is prolonged to point T by double centering. Two F.S. points T′ and T″ are set. What is the angular error introduced in a single plunging based upon the following data?

8–38. Length $ST = 522.60$ ft, T′T″ = 0.26 ft.
8–39. Length $ST = 387.15$ ft, T′T″ = 0.18 ft.
8–40. Length $ST = 876.54$ ft, T′T″ = 0.56 ft.

8–41. Because of obstacles, a random line is run from point A toward point F. What is the correct bearing of AF if AB = N48°E, 320 ft; BC = N 50°E, 180 ft; and the perpendicular offset from BC to F = 6.80 ft right?

8–42. Similar to problem 8–41, except AB = N 89°00′W, 10.0 ch.; BC = N 88°15′W, 10.4 ch., and the offset = 22 lk. left.

8–43. What is the index correction of a transit, and how is its exact value obtained? How is its effect eliminated?

8–44. The vertical transit angle to a point X is +2°40′, telescope normal, and + 2°46′ plunged. What is the index error? Find the correct vertical angle to a point Y if the telescope normal reading is −4°15′.

8–45. Similar to problem 8–44, except for readings of −6°57′ and −7°15′ on *X*, and −12°17′ on *Y*.
8–46. Similar to problem 8–44, except for readings of +0°04′ and +0°06′ on *X*, and −0°05′ on *Y*.
8–47. List the fundamental differences between a transit and a direction theodolite.

Observed directions with a theodolite, normal and plunged from *Y* to points *L, M* and *N* are listed in problems 8–48 and 8–49. Find the values of the 3 angles.
8–48. Normal: 36°27′18″, 102°55′41″, 266°07′24″. Plunged: 216°27′14″, 282° 55′43″, and 86°07′22″.
8–49. Normal: 117°52′05″, 198°34′37″, 301°44′28″. Plunged: 297°52′04″, 18° 34′40″, and 121°44′29″.

An instrumentman took a B.S. and a F.S. when the transit plumb bob was off the tack laterally. What maximum error in angle is introduced for the conditions noted in problems 8–50 through 8–52?
8–50. Plumb bob off $\frac{3}{8}$ in., 400-ft B.S., and 800-ft F.S.
8–51. Plumb bob off $\frac{1}{4}$ in., 500-ft B.S., and 200-ft F.S.
8–52. Plumb bob off $\frac{1}{2}$ in., 1000-ft B.S., and 300-ft F.S.

If pointings can be made with a probable error of ± 2 sec, and vernier readings and setting to zero to ± half the least count of the vernier, reading the *A*-vernier only what is the maximum accidental error in measuring an angle by repetition in the cases given in problems 8–53 through 8–55?
8–53. One-minute transit, 2D, 2R.
8–54. Thirty-second instrument, 4D, 4R.
8–55. Fifteen-second transit, 6D, 6R.

8–56. What is the required number of repetitions to measure a horizontal angle to an accuracy of 3 sec with a 20-sec transit? Assume reasonable pointing and reading errors.

What error in horizontal angles is consistent with the linear precisions listed in problems 8–57 through 8–59? Check by Table XI.
8–57. Linear precision: 1/400, 1/1200, 1/3000, 1/8000, and 1/20,000.
8–58. Linear precision: 1/300, 1/1000, 1/2500, 1/5000, and 1/10,000.
8–59. Linear precision: 1/500, 1/800, 1/1500, 1/4000, and 1/12,000.

8–60. Explain the difference between eccentricity and improper graduation of transit plates in their effect on measured angles.

chapter 9

Traversing

9–1. Definition and uses. The relative locations of points in a horizontal plane are determined by measurements from control points and control lines, collectively called *horizontal control.* In plane surveying, horizontal control usually consists of a traverse. A *traverse* is a series of consecutive lines whose lengths and directions are known.

An *open traverse,* Fig. 9–1, consists of a series of lines which are continuous but do not return to the starting point or close upon a point of equal or greater order of accuracy. This type is used in most route surveys.

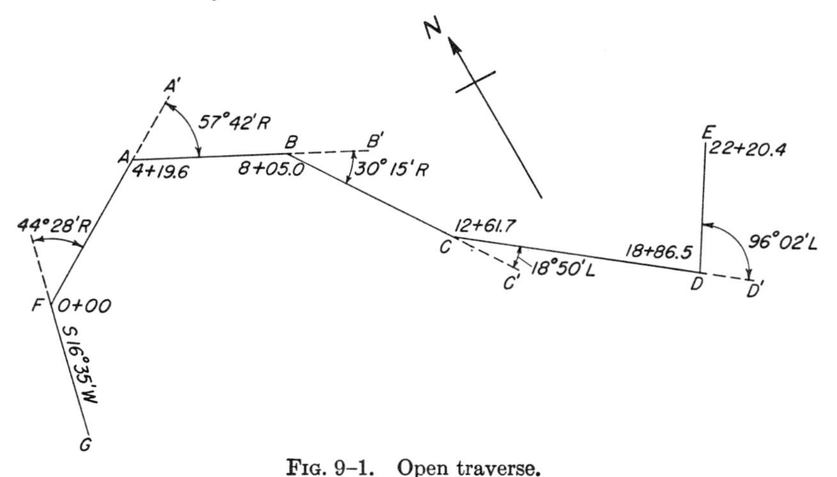

FIG. 9–1. Open traverse.

In a *closed traverse,* such as *ABCDEA,* Fig. 9–2, the lines return to the point of beginning or close upon a point of equal or greater order of accuracy. This type is run for property surveys, topographic maps, and construction projects.

A *hub* is set at each point, such as A, B, and C, Figs. 9–1 and 9–2, where a change of direction occurs. The hubs are termed *angle points* since an angle is measured at each one.

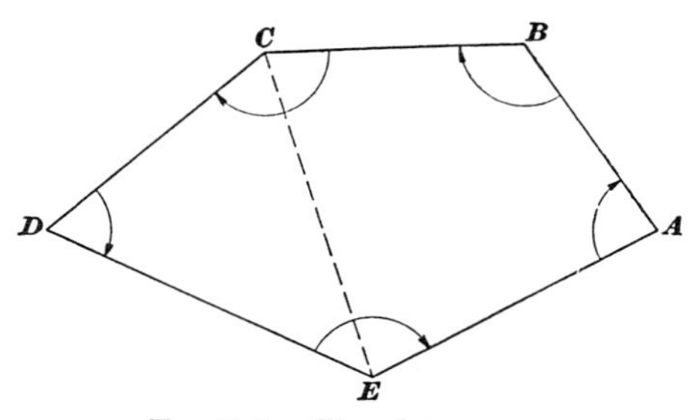

Fig. 9–2. Closed traverse.

Table 9–1 lists the permissible closures of traverses which form a closed loop, or which extend between adjusted positions of equal-order or higher-order control surveys.

TABLE 9–1

Permissible Closures of Traverses

Traverse	Maximum Permissible Closure	Maximum Probable Error of any Main-Scheme Angle
First order	1 part in 25,000	±1.5 sec
Second order	1/10,000 to 1/25,000	±3.0 sec
Third order	1/5000 to 1/10,000	±6.0 sec
Fourth order	> 1/5000	

9–2. Methods of running a traverse. Traverses are run by (a) compass bearings, (b) direct (interior) angles measured singly or by repetition, (c) deflection angles, (d) angles to the right, and (e) azimuths.

Distances are obtained by pacing, taping, stadia, subtense bar, electronic devices, or computation.

9–3. Traversing by compass bearings. The surveyor's compass was designed for use as a traversing instrument. Bearings are read directly on the compass as sights are taken along the lines (*courses*) of the traverse.

Calculated bearings, rather than observed bearings, are normally used in compass traverses run with a transit. The instrument is oriented at each hub by backsighting on the previous point with the back bearing set on the plates. The angle to the foresight is then read and applied to the back bearing to get the succeeding bearing. Some older transit circles were subdivided into quadrants to permit direct reading of bearings. Calculated bearings are valuable in retracing old surveys but are more important in office computations and mapping.

9–4. Traversing by direct angles. Direct angles, such as *ABC*, *BCD*, *CDE*, *DEA*, and *EAB*, Fig. 9–2, are used almost exclusively on property-survey traverses. They may be read either clockwise or counterclockwise, the survey party progressing around the traverse either to the right or to the left. It is good practice to measure all angles clockwise. Consistently following one method prevents errors in reading, recording, and plotting. The exterior angles should be measured to close the horizon.

9–5. Traversing by deflection angles. Route surveys are commonly run by deflection angles measured to the right or left from the lines extended, as indicated in Fig. 9–1. A deflection angle is not complete without a designation R or L, and of course it cannot exceed 180°. Each angle should be doubled or quadrupled to reduce instrument errors and an average value determined.

9–6. Angles to the right. Angles measured clockwise from a backsight on the previous line, Fig. 9–3, are called angles to the right,

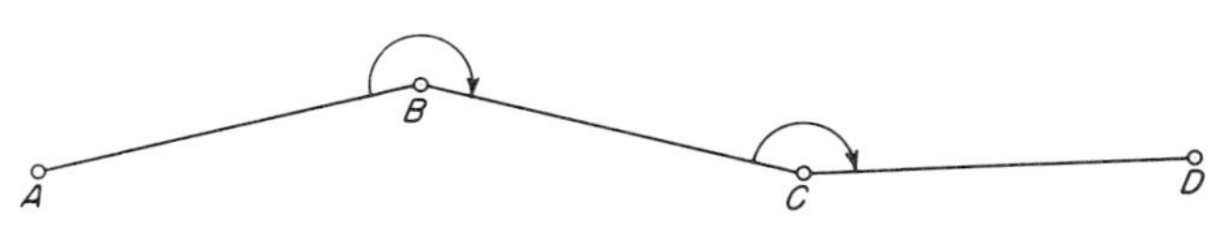

FIG. 9–3. Angles to the right.

or azimuths from the back line. The procedure used in the method is similar to running an azimuth traverse except that the backsight is taken with the plates set to zero instead of to the back azimuth. The angles can be checked (and improved) by doubling, or roughly tested by means of compass readings.

Principal use of the method is for open traverses, in locating topographic details, and on mine and tunnel surveys. Always turning angles in the clockwise direction eliminates mixups in recording and plotting, and is suited to the arrangement of plate graduations on all transits and theodolites, including direction instruments.

9–7. Traversing by azimuths. Topographic surveys are run by azimuths so that only one reference line, usually the true or magnetic north–south line, need be considered. In Fig. 9–4, azimuths are measured clockwise from the north end of the meridian through the angle points. The transit is oriented at each setup by sighting on the previous hub, one of the methods described in section 8–17 being used.

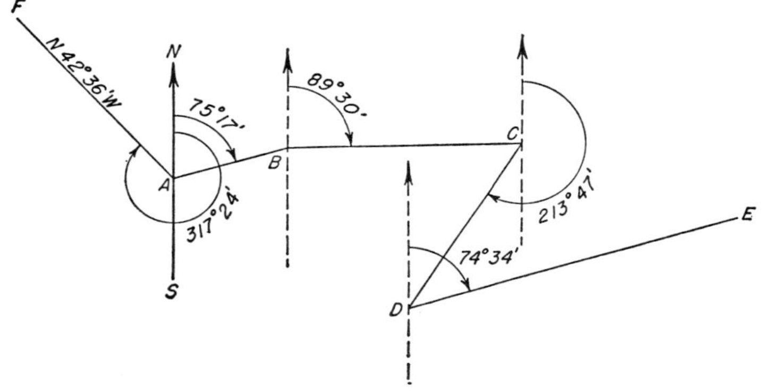

FIG. 9–4. Azimuth traverse.

9–8. Measurement of lengths. The length of each line is obtained by the simplest method capable of satisfying the required precision on a given project. For some geological and agricultural work, pacing is accurate enough. The precision specified for a traverse to locate boundaries is based upon the value of the land and the cost of the survey. On construction work, the allowable limits of closure depend upon the use and extent of the traverse, and the type of project. Bridge location, for example, demands a high degree of precision.

The 100-ft steel tape, read to hundredths on railroad and other surveys, and normally to tenths on highway work, is commonly used. It is good practice to measure lines in both directions as a check, and for open traverses, to start and close upon a point of known coordinates or location.

Single measurements by the tellurometer, geodimeter, and electrotape, one mile or longer, will give first- or second-order accuracy.

For sights of less than 500 ft, the subtense bar provides third-order results. Distances measured in both directions by stadia give low fourth-order control, good enough on some types of work such as low-precision topographic mapping.

In closed traverses each line is measured and recorded as a separate distance. On long open traverses for highways and railroads, distances are carried along continuously from the starting point. In Fig. 9–1, beginning with station 0 + 00 at point F, 100-ft stations are marked until hub *A* at station 4 + 19.6 is reached. Stations 5, 6, 7, 8, and 8 + 05.0 are set on the new course. The length of an open-traverse line is the difference between the stations of its ends.

9–9. Selection of traverse hubs. On property surveys, hubs are set at each corner if the actual boundary lines are not obstructed and can be occupied. If offset lines are necessary, a stake is set near each corner to simplify the measurements and computations. Long lines and rolling terrain may necessitate extra hubs to prolong the courses.

On route surveys, hubs are set at each angle point and at other locations at which they may be necessary to obtain topographic data or to extend the survey. Usually the center line is run before construction begins, and again after it is completed. An offset traverse may be necessary during the earth-moving and roadway-surfacing stages on a highway job.

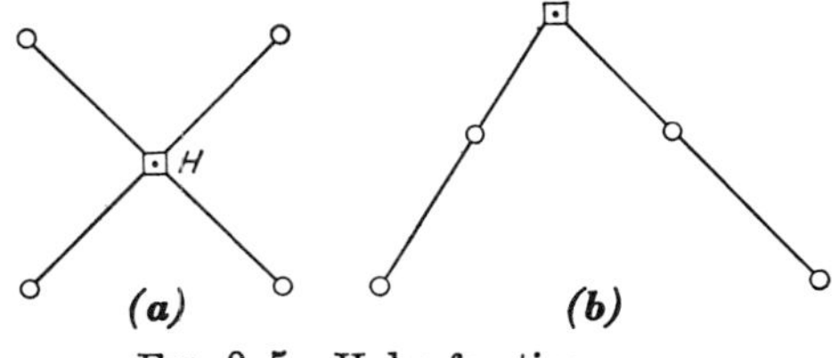

FIG. 9–5. Hubs for ties.

A traverse run for control of topographic mapping serves as a skeleton upon which are hung such details as roads, buildings, streams, and hills. Locations of hubs must be selected to permit complete coverage of the area to be mapped. *Spurs* consisting of one or more lines may branch off as open (*stub*) traverses to reach vantage points.

Traverse hubs, like bench marks, may be lost if not properly described and preserved. Ties are used to aid in finding a survey point, or to relocate one if it is destroyed. Figure 9–5a shows an arrangement of *straddle hubs* well suited to tying-in a point on a highway or elsewhere. The traverse hub H is found by intersecting strings stretched between diagonally opposite ties. Hubs in the position illustrated by Fig. 9–5b are sometimes used but are not so desirable for stringing.

Figure 9–6 and Plate A–1 in Appendix A illustrate typical traverse ties. Short lengths (less than 50 ft) permit use of the metallic tape,

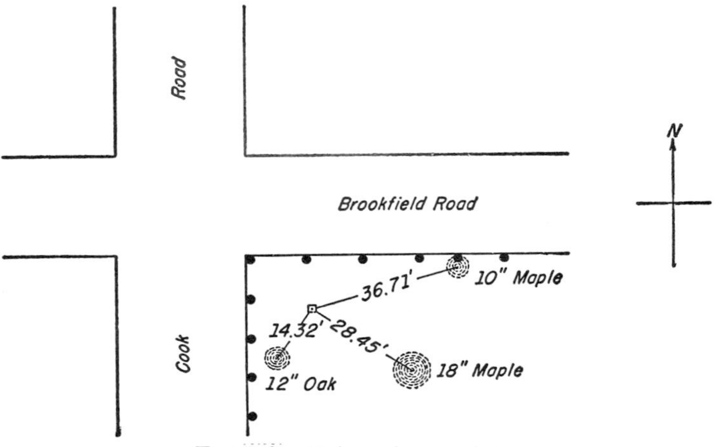

Fig. 9–6. Referencing a point.

but of course the distance to definite and unique points is a controlling factor. Two ties, preferably about at right angles to each other, are sufficient, but three allow for the possibility that one reference mark may be destroyed. Ties to trees can be measured in hundredths of a foot if nails aie driven into the sides of the trees.

Wooden and steel stakes, steel pins, pipes, and metal plates set in concrete are commonly used for hubs. Spikes, bottle caps, tops of tin cans ("shiners"), and chiseled marks are used on paved surfaces.

9–10. Organization of field party. The type of survey and terrain determine the size of party needed. One man pacing distances can run a compass traverse alone. A transitman-notekeeper with one rodman can lay out a stadia traverse. Three men—an instrumentman and two tapemen who also do the rodding—are enough for a transit-tape survey. Additional personnel for notekeeping and rodding always speed the work, but the increased production must be balanced against the greater cost of operating the party. A head chainman with energy and drive is more valuable than a fast instrumentman in keeping an engineering survey party moving.

On some surveys it is desirable to have a party chief who is free to move around and collect information on lines, hubs, reference marks, property-owners' names, and other items. The transitman or the head chainman may serve as party chief, but his range of movement as chief is then limited.

In brush or wooded country, one or two axemen may be needed to open lines.

9–11. Traverse notes. The importance of notekeeping was discussed in Chapter 3. Since a traverse is the end itself on a property survey, and the basis for all other data in mapping, a single error or omission in recording is one too many. All possible field and office checks must therefore be made. Examples of field notes for interior-angle and azimuth traverses are shown in Plate A–8 and Fig. 8–14, respectively.

9–12. Angle closure. The closure in angle for an interior-angle traverse is the difference between the sum of the measured angles and the geometrically correct total for the polygon. The sum of the interior angles of a closed polygon is equal to

$$(n - 2)180°$$

where n is the number of sides.

This formula is easily derived from known facts. The sum of the angles in a triangle is 180°; in a rectangle it is 360°; and in a pentagon it is 540°. Thus each side added to the three required for a triangle increases the sum of the angles by 180°.

Figure 9–2 shows a five-sided figure in which the sum of the measured interior angles equals 540° 02′, giving a closure of 2 min. The permissible closure is based upon the occurrence of accidental errors which may increase or decrease the measured angles. It can be computed by the formula

$$c = k\sqrt{n}$$

where n = the number of angles;
k = a fraction of the least count of the transit vernier used, in minutes or seconds.

For ordinary work, a reasonable value of k is $\frac{1}{2}$ to 1, and the permissible closure for a pentagon is 1 to 2 min.

The algebraic sum of the deflection angles in a closed traverse equals 360°, clockwise (right) deflections being considered plus and counterclockwise (left) deflections minus (lines not crossing). Reading the bearings aids in locating an angle turned, say, left, but mistakenly recorded to the right. A check is available on the bearings computed from deflection angles for an open traverse. The bearing of the last line is equal to the bearing of the first line plus the algebraic sum of the deflection angles. In a closed traverse, the bearing of the first line should be recomputed as a check by using the last angle after progressing around the traverse.

An azimuth traverse is checked by setting up on the starting point a second time after occupying the successive hubs around the traverse and orienting by back azimuths. The azimuth of the first side should be the same as the original value. Any difference is the closure. If the first point is not reoccupied, the interior angles computed from the azimuths will automatically check the proper geometric total even though one or more of the azimuths is incorrect.

In running a traverse by any method, it is advisable to record the magnetic bearings of all lines. Although these cannot be read closer than perhaps 15 min, they will disclose a serious mistake in an angle.

A *cutoff line,* such as CE in Fig. 9–2, run between two stations on a traverse, produces smaller closed figures to aid in checking and isolating errors.

Open traverses cannot be checked by a summation of angles. It is particularly important, therefore, to read the magnetic bearings of lines as a rough check. On long traverses, frequent observations on the sun or Polaris may be necessary to determine true bearings.

One way to check an open traverse is to run a separate series of lines with the same or a lesser degree of precision to close the traverse. Long sights and stadia distances may be used, for example, to get a rough check.

Another method for an open traverse is to obtain the coordinates of the starting and closing points by tying in to marks of known position, thereby making it a closed traverse, and comparing the computed difference in coordinates with the actual values. Statewide coordinate systems have been devised by the United States

Coast and Geodetic Survey for every state, and permanent monuments have been set for the use of all surveyors. Computation of coordinates for traverse courses is discussed in Chapter 10.

A graphical analysis to determine the location of a mistake can save a lot of field time.[1] For example, if the sum of the interior angles of a five-sided traverse gives a bad closure, say 10° 03′, it is likely that one mistake of 10° and several small errors of 1 min have been made. A method of locating the station at which the mistake occurred so that only one point need be reoccupied will be illustrated. The procedure shown for a five-sided traverse can be used for traverses having any number of sides.

In Fig. 9–7 the traverse has been plotted roughly by using the measured lengths and angles but has a linear closing error AA'. The perpendicular bisector of line AA' points to the angle in error—in this case, C. A correction applied to this angle will swing the traverse through an arc to eliminate the linear error AA'.

If, as in Fig. 9–8, a second angle point lies near the perpendicular bisector of AA', then the station at which the error has been made is the one for which the angular closure, when plotted at that station, is subtended by AA'. Stated differently, if the angle of closure is AEA', E is the error station; if ADA', D is the error station.

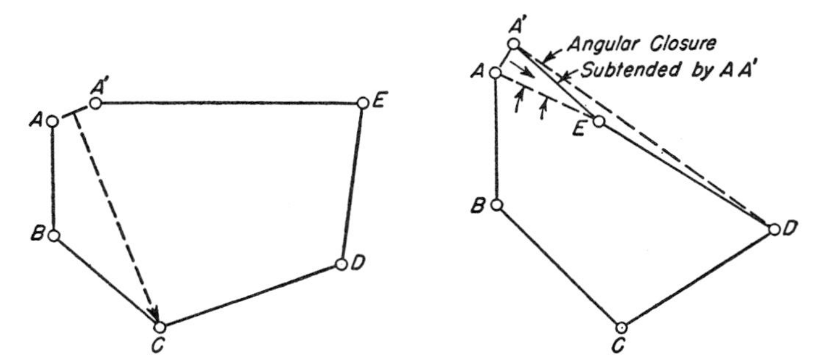

Fig. 9–7. Locating angular error. Fig. 9–8. Locating angular error.

9–12. Sources of error. Some of the sources of error in running a traverse include:

a. Errors in measurement of angles and distances.

[1] See Dana E. Low, "Finding Angle-reading Errors in Long Traverses," *Civil Engineering*, Vol. 24 (1954), p. 738.

b. Poor selection of hubs, resulting in bad sighting conditions due to (1) alternate sun and shadow, (2) visibility of only the top of the rod, (3) line of sight passing too close to the ground, (4) lines that are too long or too short, and (5) sighting into the sun.
c. Failing to double-center, or to double deflection angles.

9–13. Mistakes. Some of the mistakes to be guarded against in traversing include:

a. Occupying or sighting on the wrong hub.
b. Incorrect orientation.
c. Confusing angles to the right and left.

PROBLEMS

9–1. Compute and tabulate the bearings for the traverse of Fig. 9–1, and prepare a set of field notes for the data shown.

9–2. Prepare a set of typical field notes for the closed direct-angle traverse of Fig. 9–2. Assume that the bearing of *AB* is N 47°15′ W, scale the lengths of lines, and measure the angles with a protractor.

9–3. Similar to problem 9–2, but for a closed azimuth traverse.

9–4. Similar to problem 9–2, but for a closed deflection-angle traverse.

9–5. Prepare a set of typical field notes for the closed traverse of Fig. 6–5 (a) using azimuths. Scale the lengths of lines, and measure the angles with a protractor.

9–6. Similar to problem 9–5, but for a closed deflection-angle traverse using the data of Fig. 6–5(b).

9–7. If the bearings of all lines of a closed compass traverse *ABCDEFA* are given, how can the most-westerly hub be readily identified without drawing a sketch? The most northerly?

9–8. List four pertinent points to consider in selecting (a) hubs, (b) ties, and (c) stations.

9–9. Why is the method of placing reference hubs shown in Fig. 9–5(b) generally not as good as that of Fig. 9–5(a)?

9–10. What is the angular check on a closed interior angle traverse of 22 sides? On a closed deflection-angle traverse of 14 sides? On a closed azimuth traverse of 8 sides? On a closed azimuth traverse with a cutoff line?

9–11. What are the two primary reasons for referencing traverse hubs?

9–12. In running an azimuth traverse, if the plate bubbles on a transit run off center between the B.S. and F.S. readings, should they be recentered before taking the F.S. reading? Explain.

9–13. If the 70% error for each measurement of a traverse angle is 1 minute, what is the 70% error of closure in angle of a 4-sided traverse?

9–14. The true azimuth (from south) of a line *AB*, obtained from the coordinates of monuments at *A* and *B*, is 141°08′. After a transit at station

A has been oriented on this azimuth, point *C* is sighted and its azimuth is found to be 176°53′. An open deflection-angle traverse *ACDEFGHI* is then run, and the following angles are read: $C = 2°25'L$; $D = 8°46'L$; $E = 5°33'R$; $F = 1°16'R$; $G = 4°50'R$; and *H* (to *I*) $= 3°04'L$. The true bearing of *HI* determined by a Polaris observation is N05°39′W. Are the traverse angles acceptable for ordinary work?

9–15. Similar to problem 9–14, but assume that the azimuth of line *FG*, fixed by previous observations and use, is 173°34′ (from south).

9–16. Similar to problem 9–14, but assume that the azimuth of line *GH* is known to be 177°16′ (from south).

9–17. If hub *A* in problem 9–14 is at station 0 + 00 and hub *I* is at station 76 + 42.8, what precision of linear measurements will be consistent with that of the angles?

9–18. A transit-tape survey of a valuable farm is to be made in hilly country with a 30-sec transit and a 300-ft steel tape. The perimeter distance is about 5000 ft, and ten angles are to be turned. What procedures should be used to insure a minimum accuracy of 1/2000? Of 1/5000?

9–19. If the angles of a traverse are turned so that the 90 per cent error of any angle is 1 min, prove that the 90 per cent error of closure for a five-sided traverse is equal to 1 min $\sqrt{5}$.

The recorded bearings and lengths for a traverse are listed in problems 9–20 through 9–23. If the lengths are assumed to be correct, which bearing is wrong?

9–20. $AB =$ Due North, 200 ft; $BC =$ N 40°E, 150 ft; $CD =$ S 45°E, 200 ft; and $DA =$ S 56°22′W, 297.2 ft.

9–21. $AB =$ N 68°12′E, 340.2 ft; $BC =$ S 7°54′E, 261.7 ft; $CD =$ S 41°06′W, 413.5 ft; and $DA =$ N 4°48′W, 377.9 ft.

9–22. $AB =$ N 69°E, 437.0 ft; $BC =$ S 19°E, 236.1 ft; $CD =$ S 37°W, 243.9 ft; $DE =$ N 71°W, 324.2 ft; and $EA =$ N 19°W, 183.5 ft.

9–23. $AB =$ Due North, 995.5 ft; $BC =$ N 87°30′W, 597.5 ft; $CD =$ S 01°20′W, 548.0 ft; $DE =$ N 87°10′W, 335.0 ft; $EF =$ S 00°50′W, 564.0 ft; and $FA =$ N 75°35′E, 955.5 ft.

9–24. What problems arise in using a stub or spur traverse?

9–25. List a party organization suitable for "precise traversing" in hilly country using an invar tape, an engineer's level for determining tape elevations, and a transit for alignment.

chapter **10**

Traverse Computations

10–1. Purpose. Angles and directions for a traverse are readily investigated before leaving the field. Linear measurements, even though repeated, are a more likely source of error and must be checked by office computations to determine whether the traverse closes. The area inside the traverse can be computed by an extension of these calculations. Boundary lines of a piece of property are adjusted to form a closed figure for use in deed descriptions.

The first step in traverse computations is to balance (adjust) the angles to the proper geometric total. This is readily done since the total error is known, but not its exact distribution. Note that there are no extra observations beyond the required number of conditions [only one in this case, $\Sigma = (n - 2)180°$] to permit solid least squares analysis.

10–2. Balancing angles. Angles of a closed traverse can be adjusted to the correct geometric total by applying one of the three following types of correction:

1. Arbitrary corrections to one or more angles.
2. Larger corrections to angles where poor signals or observing conditions were present.
3. An average correction which is found by dividing the closure in angle by the number of angles.

These methods are demonstrated in illustration 10–1. For work of ordinary precision, it is reasonable to adopt corrections which are even multiples of the least count of the transit vernier or the smallest value obtained when measuring angles by repetition.

If method 1 is used, corrections of 30 sec are subtracted arbitrarily from any three angles to give the proper geometric total for a five-sided traverse. Selection of the angles at *A*, *C*, and *E* simply rounds off all values to the nearest minute.

If only the top of the rod was visible at C from the setup at B, method 2 might be used and the entire correction of 1 min 30 sec subtracted from the angle at B. Or 1 min might be subtracted from the angle at B and 30 sec subtracted from another angle suspected of being slightly off.

Illustration 10–1

ADJUSTMENT OF ANGLES

Point	Measured Angle	Method 1 Adjustment	Adjusted Angle by Method 1	Multiples of Avg. Corr.	Corr. Rounded to 30″	Successive Difference	Adjusted Angle by Method 3
A	100° 44′ 30″	30″	100° 44′	18″	30″	30″	100° 44′
B	101° 35′	0	101° 35′	36″	30″	0	101° 35′
C	89° 05′ 30″	30″	89° 05′	54″	60″	30″	89° 05′
D	17° 12′	0	17° 12′	72″	60″	0	17° 12′
E	231° 24′ 30″	30″	231° 24′	90″	90″	30″	231° 24′
Total	540° 01′ 30″	90″	540° 00′			90″	540° 00′

Method 3 consists of subtracting $1' \, 30''/5 = 18$ sec from each of the five angles. Since the angles were read in multiples of $\frac{1}{2}$ min, applying corrections of 18 sec gives a false impression of their precision. It is desirable, therefore, to establish a pattern of corrections, as shown on the right side of the table.

First a column consisting of multiples of the average correction of 18 sec is tabulated beside the angles. In the next column, each of these multiples is rounded off to the nearest 30 sec. Successive differences (adjustments) are then found by subtracting each value in the rounded-off column of corrections from the previous one. The adjusted angles obtained by using these adjustments must total to exactly the true geometric value. The adjustments fall into a pattern form and thus distort the shape of the traverse less than when all of the closure is put into one angle. This is particularly important in traverses of say twenty, fifty, or a hundred sides.

It should be noted that although the adjusted angles satisfy the geometric condition of a closed figure, they may be no nearer the true values than before adjustment. Unlike corrections for linear measurements, ***the adjustments applied to angles are independent of the size of an angle.***

10–3. Computation of bearings. Computation of bearings was discussed in Sec. 6–8. Angles adjusted to the proper geometric total must be used; otherwise the bearing of the first line will differ from its computed bearing, as found by applying the successive angles around a closed traverse, by the amount of the angle closure.

10–4. Latitudes and departures. Closure of a traverse is checked by computing the latitude and departure of each line (course). The latitude of a course is equal to the length of the course multiplied by the cosine of its bearing. Expressed in a different way, it is the length of the orthographic projection of a traverse course on a meridian. Latitude is also called *latitude difference* or *northing* or *southing*.

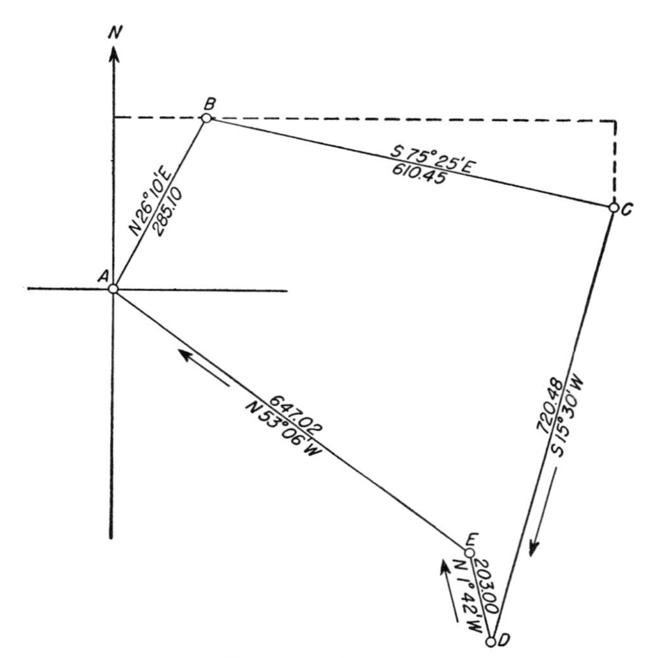

FIG. 10–1. Traverse.

The departure of a course is equal to the length of the course multiplied by the sine of its bearing. It is the length of the orthographic projection of a traverse course on a line perpendicular to the meridian. Sketching bearings in all four quadrants will show that the sine and cosine rules for latitudes and departures are independent of the bearing quadrant.

Departures and latitudes are merely the *X*- and *Y*-components used in algebra and mechanics.

North latitudes and east departures are considered plus; south latitudes and west departures are minus. The conditions of closure for a traverse are that (a) the algebraic sum of all latitudes must equal zero and (b) the algebraic sum of all departures must equal zero.

The computation of latitudes and departures, error of closure, and ratio of error will be illustrated by an example.

The interior angles of illustration 10–1 have been used to compute the bearings shown in Fig. 10–1. Note that all bearings are lettered on the outside of the traverse, and all lengths on the interior. This arrangement can be reversed, but a consistent policy should be followed. Arrows show the proper bearing directions.

Latitudes and departures are computed with the data and results usually inserted in a standard prepared form arranged as shown in illustration 10–2(a) for logarithms, in illustration 10-2(b) for desk calculator, and in illustration 10–4 for electronic computer. In practice the forms are printed with column headings and rulings to save time and to simplify checking.

Illustration 10–2(a)

COMPUTATION OF LATITUDES AND DEPARTURES USING LOGARITHMS

Course	Bearing Distance	Departure	Log Sine	Log Cosine	Latitude
AB	N 26° 10′ E		9.644423	9.953042	
	285.10		2.454997	2.454997	
		+125.72	2.099420	2.408039	+255.88
BC	S 75° 25′ E		9.985778	9.401035	
	610.45		2.785650	2.785650	
		+590.78	2.771428	2.186685	−153.70
CD	S 15° 30′ W		9.426899	9.983910	
	720.48		2.857622	2.857622	
		−192.54	2.284521	2.841532	−694.28
DE	N 1° 42′ W		8.472263	9.999809	
	203.00		2.307496	2.307496	
		− 6.02	0.779759	2.307305	+202.91
EA	N 53° 06′W		9.902919	9.778455	
	647.02		2.810918	2.810918	
		−517.41	2.713837	2.589373	+388.49
		Σ = + 0.53			Σ = − 0.70

Illustration 10–2(b)

COMPUTATION OF LATITUDES AND DEPARTURES USING A DESK CALCULATOR

Hub	Bearing	Length	Sine	Cosine	Departure	Latitude
A						
	N 26°10′ E	285.10	0.440984	0.897515	+125.72	+255.88
B						
	S 75°25′ E	610.45	0.967782	0.251788	+590.78	−153.70
C						
	S 15°30′ W	720.48	0.267238	0.963630	−192.54	−694.28
D						
	N 1°42′ W	203.00	0.029666	0.999560	− 6.02	+202.91
E						
	N 53°06′ W	647.02	0.799685	0.600420	−517.41	+388.48
A					Σ = + 0.53	= − 0.71

The computed latitudes and departures are then inserted in another prepared form similar to that shown in illustration 10–3. A combined table is preferred by some computers.

Summing the north and south latitudes and subtracting the smaller total from the larger gives the closure in latitude, 0.71 ft and the closure in departure is 0.53 ft. The *linear error of closure* is the hypotenuse of a small triangle with sides of 0.71 ft and 0.53 ft, and represents the distance from the starting point *A* to the computed point *A′* on the basis of the lengths and bearings used. In illustration 10–3 the error of closure is 0.89 ft. The term "error of closure" may be shortened to *closure*, but it should not be called the error since the actual error is never known.

The ratio of error for a traverse is expressed as a reciprocal. Its value is determined from the fraction having the error of closure as the numerator, and the perimeter of the traverse as the denominator. It is similar to the ratio of error in taping. In illustration 10–3 the ratio of error is 0.89/2466 = 1/2800. The denominator of the reciprocal is not carried beyond multiples of 100, or possibly 10.

The ratio of error is used as the measure of the precision and accuracy. On some surveys it must meet rigid specifications if the work is to be accepted and paid for. The following values of the ratio of error are reasonable for property surveys in the kind of area noted:

- Waste land. 1/500
- Ordinary farm land. . 1/1000
- Small community. . . . 1/2000
- Small city. 1/5000
- Metropolitan area. . 1/10,000

Illustration 10–3

BALANCING LATITUDES AND DEPARTURES BY THE COMPASS RULE

PT.	LENGTH	BEARING	LATITUDE North	LATITUDE South	DEPARTURE East	DEPARTURE West	BALANCED Lat.	BALANCED Dep.	TOTAL Lat.	TOTAL Dep.
A									0.00	0.00
			+0.08		*−0.06*					
	285.10	N 26° 10′ E	255.88		125.72		N 255.96	E 125.66		
B									N 255.96	E 125.66
				−0.18	*−0.13*					
	610.45	S 75° 25′ E		153.71	590.78		S 153.53	E 590.65		
C									N 102.43	E 716.31
				−0.21		*+0.15*				
	720.48	S 15° 30′ W		694.28		192.54	S 694.07	W 192.69		
D									S 591.64	E 523.62
			+0.06			*+0.05*				
	203.00	N 1° 42′ W	202.91			6.02	N 202.97	W 6.07		
E									S 388.67	E 517.55
			+0.18			*+0.14*				
	647.02	N 53° 06′ W	388.49			517.41	N 388.67	W 517.55		
A									0.00	0.00
	2466.05		847.28	847.99	716.50	715.97	0.00	0.00		
				847.28	715.97					
				0.71	0.53					

$$\text{Error of closure} = \sqrt{0.71^2 + 0.53^2} = 0.89 \text{ ft}$$

$$\text{Ratio of error} = \frac{0.89}{2466} = \frac{1}{2800}$$

A ratio of error of 1/4000 can be obtained with a 1-min transit and an ordinary 100-ft steel tape, although systematic errors such as tape too long or too short will enlarge or contract a traverse without changing its relative shape or ratio of error. To attain a *true* ratio of error of 1/10,000, a standardized tape must be used and corrections made for temperature, slope, and tension.

10–5. Methods for traverse adjustment. For any closed traverse the closure error must be distributed throughout the traverse to close the figure. This is true even though the closure is negligible in plotting the traverse at map scale. There are five basic methods for traverse adjustment: (a) the arbitrary method, (b) the transit rule, (c) the compass rule, (d) the Crandall method and (e) the method of least squares.

Arbitrary method. The arbitrary method of traverse adjustment does not conform to fixed rules or equations. Rather, the error of closure is distributed arbitrarily in accordance with the surveyor's analysis of the prevailing field conditions. For example, courses chained over rough terrain necessitating frequent plumbing and breaking chain will likely contain larger errors than courses on level ground, therefore they are arbitrarily given larger corrections. The total error of closure is thus distributed in this discretionary fashion to mathematically close the figure, i.e., make the algebraic sum of the latitudes and the algebraic sum of the departures equal zero. This method of traverse adjustment is simple to perform and provides a logical assignment of weights based upon the expected accuracy of individual measurements.

Transit rule. The transit rule theoretically is better for surveys where the angles are measured with greater accuracy than the distances, such as stadia surveys, but it seldom is employed in practice because different results are obtained for every possible meridian. Corrections are made by the following rules:

$$\frac{\text{Correction in latitude for } AB}{\text{Closure in latitude}} = \frac{\text{latitude of } AB}{\text{arithmetical sum of all latitudes}} \tag{10-1}$$

$$\frac{\text{Correction in departure for } AB}{\text{Closure in departure}} = \frac{\text{departure of } AB}{\text{arithmetical sum of all departures}} \tag{10-2}$$

Compass rule. The compass rule, suitable for surveys where the angles and distances are measured with equal precision, is the rule most commonly used in practice. It is appropriate for a transit-tape survey on which angles are measured to the nearest minute or half-minute and distances are taped to hundredths of a foot. Corrections are made by the following rules:

$$\frac{\text{Correction in latitude for } AB}{\text{Closure in latitude}} = \frac{\text{length of } AB}{\text{perimeter of traverse}} \tag{10-3}$$

$$\frac{\text{Correction in departure for } AB}{\text{Closure in departure}} = \frac{\text{length of } AB}{\text{perimeter of traverse}} \tag{10-4}$$

Application of both the transit and compass rules assumes that all lines were measured with equal care, and all angles taken with the same precision within themselves. Otherwise, suitable weights must be given to individual angles or distances. Small errors of closure can be apportioned by inspection.

The compass rule will be used to distribute the closures in latitude and departure for the traverse of Fig. 10–1. The equations were given in a form easy to remember. In applying them, however, it is simpler to use the following arrangement:

$$\text{Correction in latitude for } AB = \frac{\text{closure in latitude}}{\text{perimeter}} \times \text{length of } AB$$

The other corrections are likewise found by multiplying a constant—ratio of the closure in latitude (or departure) to the perimeter—by the successive course lengths. The computations can be done mentally if the total error is small, or by slide rule. Since the adjustments are made by an arbitrary rule, it is a waste of time to split hairs or carry values beyond the number of decimal places in the original measurements.

In illustration 10–3 the correction in latitude for AB is

$$\frac{0.71}{2466} \times 285 = 0.08 \text{ ft}$$

and that for BC is

$$\frac{1}{3500} \times 610 = 0.18 \text{ ft}$$

Each correction is generally lettered in different-colored ink or pencil above the latitude or departure to which it will be applied. On illustration 10–3, the corrections are shown in small italic numbers. In this example the adjustments are added to the north latitudes and subtracted from the south latitudes, to bring the totals of the north and south latitudes to the same intermediate value.

The corrections applied to the tabular values should produce a perfect closure. In rounding off, an excess or deficiency of 0.01 ft may result, but this is eliminated by revising one of the corrections.

Crandall method. In the Crandall method of traverse adjustment, the angular error of closure is first distributed in equal portions to all of the measured angles. The adjusted angles are then held fixed and all remaining corrections placed in the linear measurements through a weighted least squares procedure. The Crandall method is ideally suited for adjusting traverses where the linear measurements probably contain larger errors than the angular measurements, for example, stadia traverses.

Least squares method. The method of least squares, based upon the theory of probability, simultaneously adjusts the angular and linear measurements to make the sum of the squares of the residuals a minimum. The method is valid for any type of traverse survey regardless of the relative precision of angle and distance measurements since each measured quantity can be assigned a relative weight. It logically provides the best possible traverse adjustment method but has not been widely used because of the lengthy computations required. Invention of the electronic computer has now made long calculations routine and consequently the least squares method has been gaining popularity.

10–6. Traverse computation using electronic computers. Electroic computers have become commonplace in surveying computations and a modern textbook is incomplete without some discussion of them. Few modern surveying firms actually own computers but most do have access to computer facilities for solving lengthy and tedious calculations. The ordinary user of electronic computers does not need a thorough understanding of the internal operations of the machine. Familiarity with programming procedures for basic problems is important, however.

A computer program tells the machine in coded form precisely what it is to do. FORTRAN is one universal computer language, among others, which has been devised and used to program surveying and engineering problems. The simple FORTRAN program presented in illustration 10–4 performs the steps in illustration 10–3, including computation of latitudes and departures, their adjustment

Illustration 10–4

A FORTRAN PROGRAM FOR TRAVERSE COMPUTATION

```
C     TRAVERSE COMPUTATION                                                    001
C     COMPASS RULE, COORDINATE COMPUTATION AND AREA COMPUTATION               002
      DIMENSION DIST(25),DEG(25),AMIN(25),SEC(25),RAD(25),X(25),Y(25)         003
      DIMENSION XC(25),YC(25),XCOR(25),YCOR(25)                               004
      READ 10,N                                                               005
      DO 5 I=1,N                                                              006
      READ 15,DIST(I),DEG(I),AMIN(I),SEC(I)                                   007
      RAD(I)=(DEG(I)+AMIN(I)/60.+SEC(I)/3600.)/57.295828                      008
      X(I)=DIST(I)*SINF(RAD(I))                                               009
5     Y(I)=DIST(I)*COSF(RAD(I))                                               010
      DX=0.                                                                   011
      DY=0.                                                                   012
      PER=0.                                                                  013
      DO 20 I=1,N                                                             014
      DX=DX+X(I)                                                              015
      DY=DY+Y(I)                                                              016
20    PER=PER+DIST(I)                                                         017
      CLOS=SQRTF(DX**2+DY**2)                                                 018
      PREC=PER/CLOS                                                           019
      IPREC=PREC                                                              020
      PRINT 27                                                                021
      PRINT 28                                                                022
      XCOR(1)=0.                                                              023
      YCOR(1)=0.                                                              024
      DO 30 I=1,N                                                             025
      J=I+1                                                                   026
      XC(I)=X(I)-(DIST(I)*DX)/PER                                             027
      YC(I)=Y(I)-(DIST(I)*DY)/PER                                             028
      XCOR(J)=XCOR(I)+XC(I)                                                   029
      YCOR(J)=YCOR(I)+YC(I)                                                   030
      IF(J-N)30,30,35                                                         031
35    J=1                                                                     032
30    PRINT 40,I,J,YC(I),XC(I),YCOR(J),XCOR(J)                                033
      AC=0.                                                                   034
      DO 45 I=2,N                                                             035
      J=I+1                                                                   036
45    AC=AC+XCOR(I)*YCOR(J)-XCOR(J)*YCOR(I)                                   037
      AC=AC+XCOR(N+1)*YCOR(2)-XCOR(2)*YCOR(N+1)                               038
      ACRES=ABSF(AC)/(2.*43560.)                                              039
      PRINT 25,CLOS                                                           040
      PRINT 26,IPREC                                                          041
      PRINT 50,ACRES                                                          042
10    FORMAT(I5)                                                              043
15    FORMAT(F10.3,2F5.0,F10.2)                                               044
25    FORMAT(/4X,10HCLOSURE  =,F8.3,4H  FT/)                                  045
26    FORMAT(4X,19HPRECISION  =  1  IN,I7/)                                   046
27    FORMAT(20X,8HBALANCED)                                                  047
28    FORMAT(5X,6HCOURSE,6X,3HLAT,7X,3HDEP,6X,7HN COORD,3X,7HE COORD//)       048
40    FORMAT(3X,I3,2H -,I3,1X,4F10.3)                                         049
50    FORMAT(4X,6HAREA =,F10.3,6H ACRES)                                      050
      END                                                                     051
    5
 285.100  26.  10.      0.00
 610.450 104.  35.      0.00
 720.480 195.  30.      0.00
 203.000 358.  18.      0.00
 647.020 306.  54.      0.00
```

by the compass rule, and calculation of the linear error of closure and the ratio of error of the traverse. In addition, the program computes the traverse point coordinates, and the area within the traverse using the method of coordinates, discussed further in Secs. 10–8 and 11–6 respectively.

It is beyond the scope of this book to discuss in detail the FORTRAN language and computer programming procedures. Interested students are advised to secure a manual on the subject but a few brief comments will be made regarding the program of illustration 10–4. Note that each statement of the illustrated FORTRAN program contains a *card number* (on the right side) to serve as a reference number for that particular statement. Cards numbered 001 through 051 comprise the *source program* which is the means of conveying instructions to the computer for the particular problem in a language which the computer can understand. The source program is generally read into the computer on punched cards. Figure 10–2 illustrates the punched deck of cards for the

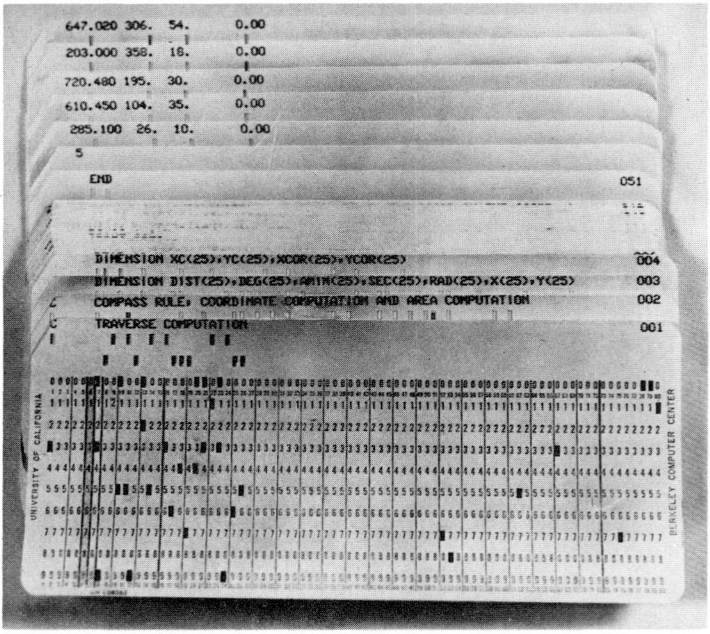

FIG. 10–2. Traverse computation deck.

program of illustration 10–4. It cannot be overemphasized that each card must be punched perfectly.

For the program of illustration 10–4, cards following number 051 are called *data cards* and are also read in the computer. They contain the *input* information required by the computer to perform the calculations, in this case the length and azimuth of each course of the traverse. The *output* (solution results) is printed by the computer in the format shown in illustration 10–5. The tabulated coordinates are based on assumed values at traverse point *A* of N = 0 and E = 0. Note that in the output, courses 1–2, 2–3, etc.,

Illustration 10–5

ELECTRONIC COMPUTER OUTPUT FOR ILLUSTRATION 10–3

COURSE	BALANCED LAT	BALANCED DEP	N COORD	E COORD
1 - 2	255.963	125.663	255.963	125.663
2 - 3	-153.528	590.651	102.435	716.314
3 - 4	-694.071	-192.693	-591.636	523.621
4 - 5	202.969	-6.067	-388.667	517.553
5 - 1	388.667	-517.553	0.	0.

CLOSURE = 0.884 FT

PRECISION = 1 IN 2788

AREA = 6.258 ACRES

correspond to courses *A–B, B–C,* etc., of illustration 10–3. This computer program is general and as presently dimensioned will accommodate any closed traverse containing up to 25 sides. The only input data required are the number of sides of the traverse to be computed and the length and azimuth of each course. Computation time was less than one second for the traverse of illustration 10–5.

Traverse computation is but one example of the application of computers to surveying problems; others include reduction of stadia notes, reduction of astronomical observations, circular and parabolic curve calculations, triangulation and trilateration calculation, adjustment computations, and earthwork determination.

10–7. Other methods of computing latitudes and departures. In the examples given, logarithms, natural functions of the bearing angles and a desk calculator, and the electronic computer were used to perform traverse computations. The limited number of significant figures which can be read on a slide rule makes it suitable only for calculations of corrections, for checking, and in stadia reduction as described in Chapter 12.

Traverse tables provide a rapid method of finding latitudes and departures in the field or office by using only addition (the arithmetic process in which the fewest mistakes are made). Sample pages of a traverse table are shown in Table I.

A traverse table is a compilation of the natural sines and cosines of angles multiplied by the integers 1 to 10, or perhaps by 1 to 100. The most complete tables give values for each minute of arc. Others, like Table I, show only the multiples of 15 min.

To find the latitude or departure of a line of given length and bearing, the answer for each digit is taken from the table and the decimal point is moved to the proper place. The quantities for all digits are added to obtain the result.

As an example, the latitude and departure for a course CD having a bearing of S 15° 30′ W and a length of 720.48 ft will be found by tabulation from a traverse table:

Length	*Latitude*	*Departure*	
720.	693.8	192.4	(from 72 with decimal point shifted)
0.48	0.46	0.13	
720.48	694.3	192.5	

Differences from the values shown in illustration 10–2 result from using a table carried only to hundredths, and then moving the decimal point. Since two of the values from the table are in tenths only, the answers cannot logically be carried out to hundredths.

10–8. Coordinates. In Fig. 10–1, if the coordinates of point A are 0, 0, then the departure and latitude of course AB are the X- and Y- coordinates of point B. The columns of total departures and latitudes in illustration 10–3 therefore give the coordinates of each hub with respect to the starting point A. In practice, one of the traverse hubs is arbitrarily taken as the origin of coordinates, or a tie line is run to some monument whose coordinates are known. To avoid negative values of X and Y, an origin may be assumed south and west of the traverse such that one of the hubs has the coordinates $X = 1000$, $Y = 1000$, or any other suitable values. In a closed traverse, assigning $Y = 0$ to the most southerly point, and $X = 0$ to the most westerly point, saves time.

In nationwide surveys spherical coordinates based on latitude and longitude must be used, but plane coordinates are satisfactory for ordinary surveys of limited extent.

Plane coordinates, and latitudes and departures, are employed in many ways. Examples are (a) map plotting, (b) location of inaccessible points, and (c) computation of areas, omitted measurements, and lengths and directions of property lines.

10–9. Lengths and bearings from latitudes and departures, or coordinates. The length and bearing of a line are readily obtained from the following relationships if the latitude and departure are known:

$$\tan \text{bearing} = \frac{\text{departure}}{\text{latitude}} \tag{10–5}$$

$$\text{Length of line} = \frac{\text{departure}}{\sin \text{bearing}} = \frac{\text{latitude}}{\cos \text{bearing}} \tag{10–6}$$

$$= \sqrt{(\text{departure})^2 + (\text{latitude})^2} \tag{10–7}$$

These formulas also can be applied to any line connecting two points whose coordinates are known. Thus the tangent of the bearing of AB equals the difference of the X-coordinates divided by the difference of the Y-coordinates:

$$\tan \text{bearing } AB = \frac{X_B - X_A}{Y_B - Y_A} \tag{10–8}$$

where X_A, Y_A and X_B, Y_B are the coordinates of A and B, respectively.

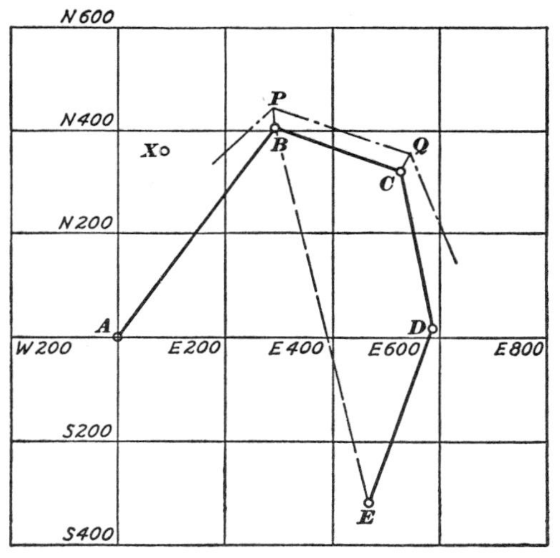

FIG. 10–3. Plot of traverse.

Likewise,

$$\text{Length } AB = \frac{X_B - X_A}{\sin \text{bearing } AB} = \frac{Y_B - Y_A}{\cos \text{bearing } AB} \quad (10\text{–}9)$$

$$= \sqrt{(X_B - X_A)^2 + (Y_B - Y_A)^2} \quad (10\text{–}10)$$

Figure 10–3 and illustration 10–6 will be used to explain these relationships. For example, suppose the length and bearing of a cutoff (or closing) line BE are required. The differences between the X- and Y-coordinates of hubs B and E are E 173.3 and S 719.2, respectively. Then

$$\tan \text{bearing} = \frac{173.3}{719.2} \text{ and bearing } BE = \text{S } 13° \, 33' \text{ E}$$

Illustration 10–6

COMPUTATIONS FOR A CLOSING LINE

Point	Length	Bearing	Latitude North	Latitude South	Departure East	Departure West	Coordinates Y	Coordinates X
A							0.0	0.0
	500.5	N 35° 30′ E	407.5		290.6			
B							N 407.5	E 290.6
	251.6	S 70° 10′ E		85.4	236.7			
C							N 322.1	E 527.3
	310.4	S 10° 50′ E		304.9	58.3			
D							N 17.2	E 585.6
	350.7	S 20° 18′ W		328.9		121.7		
E							S 311.7	E 463.9
						B	N 407.5	E 290.6
						Diff.	S 719.2	E 173.3

$$\text{Length } BE = \frac{719.2}{\cos 13° \, 33'} = \sqrt{173.3^2 + 719.2^2} = 739.8 \text{ ft}$$

10–10. Coordinate computations in boundary measurements. Computation of a bearing from the known coordinates of two points on a line is a common problem in boundary measurements. If the lengths and directions of lines from traverse points to the corners of a field are known, the coordinates of the corners can be determined and the lengths and bearings of all the sides can be calculated.

In Fig. 10–3, BC is a traverse line and PQ is the property line which cannot be run directly because of obstructions. The measured lengths and azimuths are: for BP, 42.5 ft and 354° 50′; for CQ, 34.6 ft and 26° 40′. From the latitudes and departures of these lines the coordinates of *P* and *Q* are found as follows:

	Y	*X*
B	N 407.5	E 290.6
BP	+ 42.3	− 3.8
P	N 449.8	E 286.8
C	N 322.1	E 527.3
CQ	+ 30.9	+ 15.5
Q	N 353.0	E 542.8

From the coordinates of *P* and *Q*, the length and bearing of the line *PQ* are found in the following manner:

	Y	*X*
Q	+353.0	+542.8
P	+449.8	+286.8
PQ	S 96.8	E 256.0

$$\tan \text{ bearing } PQ = \frac{256.0}{96.8} \text{ and } PQ = \text{S } 69° \, 17' \text{ E}$$

$$\text{Length } PQ = \frac{256.0}{\sin 69° \, 17'} = 273.7$$

By continuing this method around the field, the coordinates of all the corners and the lengths and bearings of all the lines can be determined.

10–11. Traverse orientation by coordinates. If the coordinates of one traverse hub, as *A* in Fig. 10–3, and a visible point *X*, are known, the direction of the line *AX* can be computed and used to orient the transit at *A*. In this way, azimuths and bearings of traverse lines are obtained without the necessity of making astronomical observations.

This procedure is followed in various cities which have control monuments and coordinate systems.

State and Federal mapping agencies will ultimately provide closely-spaced permanent monuments whose coordinates are based upon precise triangulation. Such marks will permit the accurate location of the corners of any piece of property, either by coordinates or by lengths and true bearings.

10–12. State plane coordinate systems. Plane rectangular coordinates are commonly used in surveys covering the limited areas of plane surveying. In 1933 the United States Coast and Geodetic

Survey established the North Carolina coordinate system, by means of which geodetic positions (latitude and longitude) of triangulation stations within the state could be transformed into plane rectangular coordinates (X and Y) on a single grid.

As additional points are set and their coordinates determined, they too become usable reference points. Finally, then, local surveys and the accurate restoration of obliterated or destroyed marks having known coordinates are simplified.

Some cities and counties have their own coordinate systems for use in locating street, sewer, property, and other lines. Because of their limited extent and the resultant discontinuity at city or county lines, such local systems are less desirable than a state-wide grid.

Military grids are used to pinpoint the locations of objects by coordinates, for fire-control and other purposes.

A more extended discussion of state plane coordinates is given in Appendix A.

10–13. Sources of error in traverse computations. Some of the sources of error in traverse computations include:

a. Failing to adjust the angles before computing bearings.
b. Improper adjustments of angles, latitudes, and departures.
c. Carrying out corrections beyond the number of decimal places in the original measurements.
d. Poor selection of coordinate axes.

10–14. Mistakes. Some of the more common mistakes in traverse computations are:

a. Applying angle adjustments in the wrong direction and failing to check the angle sum for proper geometric total.
b. Interchanging latitudes and departures, or their signs.
c. Confusing the signs of coordinates.

PROBLEMS

10–1. The sum of eight interior angles of a closed traverse, each read to the nearest minute, is 1079° 55′. Balance the angles by methods 1, 2, and 3 and state any assumptions made.

10–2. The closure of a 10-sided deflection-angle traverse *ABCDEFGHIJA* with 6 left deflections and 4 right deflection angles is +5 min. Balance the angles by methods 1, 2, and 3. State any assumptions made.

10–3. Balance the angles in the following azimuth (from north) traverse by any method. Compute and tabulate the bearings, assuming azimuth

AB is correct: *AB*, 147° 00′; *BC*, 22° 35′; *CD*, 52° 42′; *DE*, 156° 48′; *EF*, 89° 26′; *FA*, 273° 07′; *AB*, 163° 29′. Explain the closure.

10–4. Compute and tabulate for the following traverse (a) bearings, (b) latitudes and departures, (c) linear error of closure, and (d) accuracy. For what type of survey is the accuracy satisfactory?

Line	Interior Angle (Rt.)	Bearing	Length in Feet	Line	Interior Angle (Rt.)	Bearing	Length in Feet
AB	*A* = 125° 54′	East	542.0	*DE*	*D* = 121° 52′		1019.8
BC	*B* = 135° 00′		846.6	*EF*	*E* = 88° 59′		1118.0
CD	*C* = 114° 27′		845.4	*FA*	*F* = 133° 48′		606.8

10–5. In problem 10–4, if one side and/or angle is responsible for most of the error of closure, which is it most likely to be?

10–6. Use the data for the traverse of problem 10–4. (a) Adjust the latitudes and departures by using the compass rule. (b) If the coordinates of point *B* are 1000.0 N and 3000.0 E, determine the coordinates of all other points. (c) Compute the length and bearing of line *AE*.

For the traverses given in problems 10–7 through 10–9, compute and tabulate (a) the unbalanced latitudes and departures, (b) the latitudes and departures adjusted by both the compass and transit rules for comparison, (c) the linear error of closure, and (d) the precision (lengths are in feet).

10–7.	Course	*AB*	*BC*	*CD*	*DA*
	Bearing	N 07°30′E	N 86°15′E	S 14°00′W	N 69°30′W
	Length	320.0	491.2	515.7	435.5
10–8.	Bearing	N 30°E	S 60°E	S 75°W	N 60°W
	Length	100.0	300.0	141.1	200.0
10–9.	Bearing	S 73°47′E	N 19°58′E	N 52°22′W	S 20°00′W
	Length	295.4	778.0	308.6	891.2

10–10. Similar to problem 10–7, for the corrected traverse of problem 9–20.

10–11. Similar to problem 10–7, for the corrected traverse of problem 9–21.

10–12. Which of the traverse adjustment rules, compass or transit, swings the traverse around through a greater angle? Explain.

10–13. After adjustment of the latitudes and departures by the compass (or transit) rule, does a traverse actually close? Explain.

10–14. Compute the linear closure, the ratio of error, and the new bearings of the sides after the latitudes and departures are balanced by the compass rule in the following traverse.

Line	Length	Latitude	Departure
AB	357.65	N 326.41	W 146.18
BC	329.22	N 58.27	E 324.02
CA	423.60	S 384.52	W 177.70

Balance by the transit rule the latitudes and departures listed in problems 10–15 and 10–16. Compute the error of closure, ratio of error, and new bearings.

	Line	*AB*	*BC*	*CD*	*DA*
10–15.	Latitude	N 71.83	N 38.11	S 79.47	S 30.44
	Departure	W 7.70	E 162.93	E 18.35	W 173.34
10–16.	Latitude	N 146.35	N 315.76	S 138.27	S 323.64
	Departure	W 286.14	E 180.21	E 164.06	W 58.26

10–17. Using the traverse table, Table I in Appendix B, compute the ratio of error for the following compass traverse of an easement strip: *AB,* N 15°15′ W, 326.8 ft; *BC,* S 75°00′ W, 38.1 ft; *CD,* S 15°30′ E, 327.1 ft; *DA,* N 74°45′ E, 38.3 ft.

10–18. Similar to problem 10–17, but for the following traverse: *AB,* N 15° E, 600.0 ft; *BC,* S 75½° E, 7.2 ft; *CD,* S 15½° W, 600.2 ft; *DA,* N 75° W, 7.1 ft.

10–19. If the coordinates of Point *F* in Fig. 9–1 are E 2000.0 and N 2000.0, what are the coordinates of point *E*? What is the length and bearing of the line from *F* to *E*?

10–20. The following field notes are taken on a meander line along a stream: *AB,* N 89°30′ E, 364.1 ft; *BC,* S 89°15′ E, 317.6 ft; *CD,* N 88°45′ E, 442.5 ft. A point *E* is to be set on a continuation of line *AB* and 270.0 ft from point *D*. Calculate the bearing of *DE* and the angle turned from *DE* at point *E* which will prolong *AB*.

10–21. To speed up the field work in the survey of a closed boundary, two measurements which presented unusual difficulty in the field were omitted as shown in the field notes. Compute the missing data.

Line	Dist.	Bearing	Line	Dist.	Bearing
AB	518.9	Due West	*DE*	406.5	Due South
BC	409.0	....	*EA*	....	S 65° 35′ E
CD	481.8	N 52° 59′ E			

The azimuths of a closed traverse are: AB = 4°15′; BC = 44°32′; CD = 118°37′; DE = 208°26′; and EA = 271°04′. Which course is probably incorrect for the conditions given in problems 10–22 through 10–24?

10–22. Algebraic sum of latitudes = + 1.86 ft, of departures = + 0.09 ft.

10–23. Algebraic sum of latitudes = + 0.03 ft, of departures = −2.24 ft.

10–24. Algebraic sum of latitudes = +1.34 ft, of departures = +1.32 ft.

10–25. Determine the lengths and bearings of the sides of a lot whose corners have the following X- and Y- coordinates: *A* (0, 0); *B* (+80, +80); *C* (+120, −10); *D* (−10, −60).

10–26. Determine the azimuths and lengths of the lines of a closed traverse whose corners have the following X- and Y- coordinates: *A* (+0, +0); *B* (−20, +100); *C* (−150, +60); *D* (−120, +20); *E* (−140, −50).

10–27. Why should the ratio of error *not* be carried out to a reciprocal such as 1/3277.8?

10–28. Adjust the latitudes and departures in illustration 10–3 if the coordinates of points *D* and *E* are fixed by a previous survey of higher order.

10–29. What is the purpose of balancing the angles and latitudes and departures of a closed traverse? After balancing the latitudes and departures, will the angles close?

10–30. List 3 advantages to the private surveyor in using State Plane Coordinates.

10–31. For the traverse data of problem 10–4, compute (and balance) latitudes and departures, precision, and area using the computer program of Sec. 10–6.

10–32. Similar to problem 10–31 for the traverse of problem 10–7.

10–33. Similar to problem 10–31 for the traverse of problem 10–8.

10–34. Similar to problem 10–31 for the traverse of problem 10–9.

10–35. Similar to problem 10–31 for the traverse of problem 10–15.

10–36. Similar to problem 10–31 for the traverse of problem 10–16.

10–37. In searching for a record of the length and true bearing of a section of a town boundary which is straight between points *A* and *B*, the following notes of an old random traverse were found: (survey by compass and Gunter's chain, declination = 8°45′W)

Line	*A*-1	1-2	2-3	3-*B*	*B*-*A*
Mag. Bear.	North	N 30° E	East	S 4°30′W	
Dist. in Ch.	6.25	18.47	10.54	4.05	

Compute the bearing and length of *B*-*A*.

10–38. What uses are made of the ratio of error of a closed traverse?

10–39. In adjusting measured traverse angles, why are the adjustments not made in proportion to the size of angle?

10–40. Why are small angles (below 15°) undesirable in triangulation?

chapter **11**

Area

11–1. General. One of the reasons for making a property survey is to determine the area of land included within the boundaries. Most deeds give the area of a tract of land as well as a description of the property lines.

The unit of area for lots is the square foot, but for large tracts it is the acre. One acre = 43,560 sq ft = 10 sq ch (Gunter's). An acre lot, if square, would thus be 208.71 ft on a side.

11–2. Methods of measuring area. The methods of measuring area may be classified under two headings: (1) field measurements and (2) map measurements.

Field measurements may be carried out by means of one of the following methods:

a. Division of the area into triangles.
b. Offsets from straight lines.
c. Double meridian distances.
d. Coordinates.

Map measurements may be made by means of one of the following methods:

a. Division of the area into triangles.
b. Coordinate squares.
c. Use of a planimeter.

Each of these methods will be described and illustrated in the topics which follow.

11–3. Area by division into triangles. A field may be divided into simple geometric figures such as the triangles shown in Fig. 11–1. The area of a triangle whose sides are known may be computed by the formula

$$\text{Area} = \sqrt{s(s-a)(s-b)(s-c)} \qquad (11\text{–}1)$$

where a, b, and c are the sides of the triangle and

$$s = \tfrac{1}{2}(a + b + c)$$

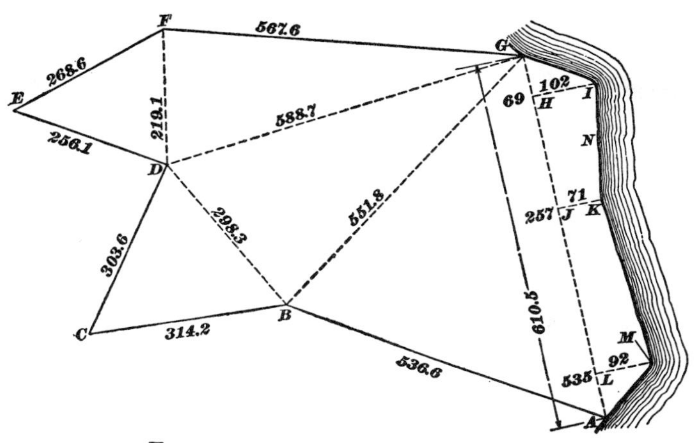

FIG. 11–1. Measurement by triangles.

Another formula for the area of a triangle is

$$\text{Area} = \tfrac{1}{2}ab \sin C \tag{11-2}$$

where C is the angle included between sides a and b.

The area of the field is the sum of the areas of all the triangles. Each side and division line must be measured. The triangle method was used more often prior to the invention of the transit for measuring angles.

11–4. Area by offsets from straight lines. Irregular fields can be reduced to a series of trapezoids by right-angle offsets from points at regular intervals along a measured line, as indicated in Fig. 11–2. This method is used also for cross sections and profiles.

The area is found by the formula

$$\text{Area} = b\left(\frac{h_0}{2} + h_1 + h_2 + \cdots + \frac{h_n}{2}\right) \tag{11-3}$$

where b is the length of a common interval between the offsets and $h_0, h_1, \ldots h_n$ are the offsets.

Offsets to an irregular boundary from a reference line AB are shown in Fig. 11–2. The enclosed area is

$$50\left(0 + 5.2 + 8.7 + 9.2 + 4.9 + 10.4 + 5.2 + 12.2 + \frac{2.8}{2}\right) = 2860 \text{ sq ft}$$

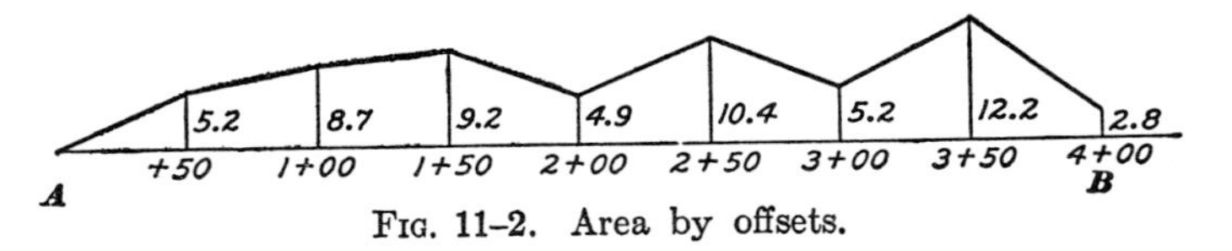

FIG. 11–2. Area by offsets.

For irregularly curved boundaries, Poncelet's rule and Francke's rule give somewhat better results more conveniently. For generally parabolic arcs, Simpson's one-third rule is applicable. Examples of the three rules are covered elsewhere.[1]

11–5. Area by double-meridian-distance method. It is convenient to compute the area of a closed figure by the double-meridian-distance method when the latitudes and departures of the boundary lines are known. The meridian distance of a traverse course is the right-angle distance from the center point of the course to the reference meridian. To ease the problem of signs, the reference meridian usually is passed through the most-westerly station of the traverse (one of the points where the bearings change from west to east).

In Fig. 11–3 the meridian distances of courses AB, BC, CD, DE, and EA are MM', PP', QQ', RR', and TT' respectively.

For the purpose of expressing PP' in terms of convenient distances, draw MF and BG perpendicular to PP'. Then

$$PP' = P'F + FG + GP = \text{meridian distance of } AB + \tfrac{1}{2} \text{ departure of } AB + \tfrac{1}{2} \text{ departure of } BC$$

Thus the meridian distance of any course of a traverse equals the meridian distance of the preceding course, plus one-half the departure of the preceding course, plus one-half the departure of the course itself. It is simpler to use full departures of courses. Therefore *double meridian distances* (D.M.D.'s), which are equal to twice the meridian distances, are employed. A single division by two is made at the end of the computation.

The D.M.D. for any traverse course is equal to the D.M.D. of the preceding course, plus the departure of the preceding course, plus the departure of the course itself. For the traverse in Fig. 11–3,

D.M.D. of AB = dep. of AB

D.M.D. of BC = D.M.D. of AB + dep. of AB + dep. of BC

Signs of the departures, east plus and west minus, must be considered. When the reference meridian is taken through the most-

[1] Tracy, John C. Surveying Theory and Practice.

westerly station of a closed traverse and calculations of the D.M.D.'s are started with a course through that station, the D.M.D. of the first course is its departure. A check on all computations is obtained if the D.M.D. of the last course after computing around the traverse is also equal to its departure but opposite in sign. If there is a difference, the departures were not correctly adjusted before starting, or an error has been made in the computations.

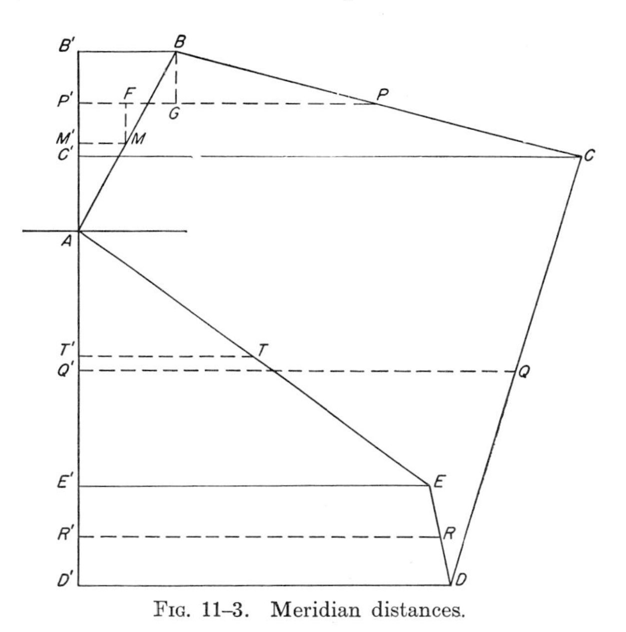

FIG. 11–3. Meridian distances.

The area enclosed by the traverse $ABCDEA$ in Fig. 11–3 is

$$B'BCC' + C'CDD' - (AB'B + DD'E'E + AEE')$$

The area of each of these figures equals the meridian distance of a course times its balanced latitude. The D.M.D. of a course multiplied by its latitude equals the double area. Summation of all the double areas gives twice the area inside the entire traverse.

Signs of the products of D.M.D.'s and latitudes must be considered. If the reference line is passed through the most-westerly station, all D.M.D.'s are positive. The products of D.M.D.'s and north latitudes are therefore plus, and the products of D.M.D.'s and south latitudes are minus.

Illustration 11–1

Computation of D.M.D.'s

Departure of AB =	+ 125.66	= D.M.D. of AB	
Departure of AB =	+ 125.66		
Departure of BC =	+ 590.65		
	+ 841.97	= D.M.D. of BC	
Departure of BC =	+ 590.65		
Departure of CD =	− 192.69		
	+1239.93	= D.M.D. of CD	
Departure of CD =	− 192.69		
Departure of DE =	− 6.07		
	+1041.17	= D.M.D. of DE	
Departure of DE =	− 6.07		
Departure of EA =	− 517.55		
	+ 517.55	= D.M.D. of EA	Check

To illustrate the procedure, the traverse of Fig. 11–3 having the departures in illustration 11–2 will be used. The calculations of D.M.D.'s are shown in illustration 11–1.

The computations for area are generally arranged as in illustration 11–2, although a combined form may be used. The smaller of the two double-area totals is subtracted from the larger, and the result divided by 2 to get the area in square feet. This area (272,610 sq ft), is divided by 43,560 sq ft to obtain the number of acres (6.258).

The fact that the total minus double area is larger than the plus double area signifies only that the D.M.D.'s were computed by going around the traverse in a clockwise direction. If the route had been from A through E, D, C, and B and back to A, the total plus double area would have been the greater. Areas carried out beyond the nearest square foot or 0.001 acre cannot be justified for a traverse on which distances were measured to the nearest 0.01 ft and angles were read to 1 min or $\frac{1}{2}$ min.

As a check, the area can be computed by *double parallel distances* (D.P.D.'s).

The D.P.D. for any traverse course is equal to the D.P.D. of the preceding course, plus the latitude of the preceding course, plus the latitude of the course itself.

Illustration 11–2

COMPUTATION OF AREA BY D.M.D.'S AND D.P.D.'S

Course	Balanced Latitude	Balanced Departure	D.M.D.	Double Areas +	Double Areas −	D.P.D.	Double Areas +	Double Areas −
AB	N 255.96	E 125.66	+ 125.66	32,164		+255.96	32,164	
BC	S 153.53	E 590.65	+ 841.97		129,268	+358.39	211,683	
CD	S 694.07	W 192.69	+1239.93		860,598	−489.21	94,266	
DE	N 202.97	W 6.07	+1041.17	211,326		−980.31	5,951	
EA	N 388.67	W 517.55	+ 517.55	201,156		−388.67	201,156	
Total	0.00	0.00		444,646	989,866		545,220	00
					444,646			
					2)545,220			

272,610 sq ft = 6.258 acres

The last three columns in illustration 11–2 show the computation of the area of the traverse in Fig. 11–3 by D.P.D.'s. Signs of the latitudes, north plus and south minus, must be used in calculating the D.P.D.'s.

All important engineering computations should be checked by using different methods, or by two persons who can employ the same method. As an example of good practice: An individual working alone in an office might use a calculating machine to compute latitudes and departures and check them by means of a traverse table. He might then calculate areas by D.M.D.'s and check his results by D.P.D.'s. Experienced surveyors and engineers have learned that a half-hour spent in checking computations may save a frustrating day of unsatisfactory closures in the field.

The area of a tract having a circular curve for one boundary, such as that in Fig. 11–4, can be found by dividing the figure into two parts: the polygon *ABCDEGFA* and the sector *EGF*. The radius $R = EG = FG$, and either the central angle *EGF* or the length *EF* must be known or computed, to permit calculation of the area of sector *EGF*. This area can then be added to the area of *ABCDEGFA*, which is found by the D.M.D. or coordinate method.

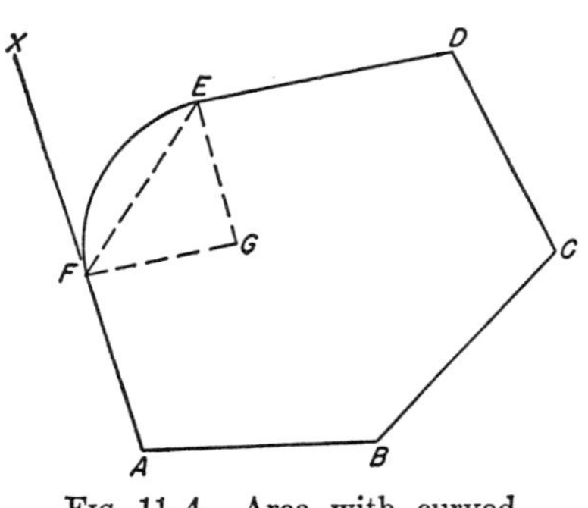

FIG. 11–4. Area with curved boundary.

As a variation, the length and direction of *EF* can be calculated and the area of segment *EF* added to that of *ABCDEFA*. Note that angle *XFE* equals one-half of angle *EGF*. The D.M.D. method is actually a special case of the coordinate method.

11–6. Area by coordinates. Determination of area from coordinates is a simple process for a closed traverse with known coordinates for each corner. The area is equal to one-half the sum of the products obtained by multiplying each *Y*-coordinate by the difference between the adjacent *X*-coordinates. The *X*-coordinates must always be taken in the same order around the traverse. The rule can also be stated in another form: The area is equal to one-half the sum of the products obtained by multiplying each *X*-coordinate by the difference between the adjacent Y-coordinates, taken in the same order around the figure. Using both rules provides a check on the answer.

For the traverse in Fig. 11–5, the rule is applied in the following relation:

$$\text{Area} = \tfrac{1}{2}[X_A(Y_E - Y_B) + X_B(Y_A - Y_C) + X_C(Y_B - Y_D) + X_D(Y_C - Y_E) + X_E(Y_D - Y_A)]$$

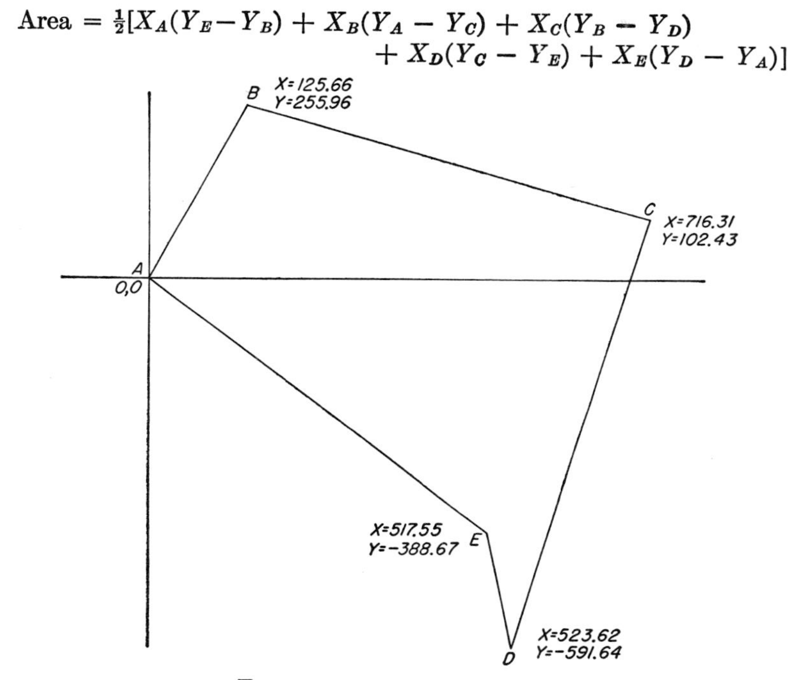

FIG. 11–5. Area by coordinates.

This formula is based on the summation of the areas of a series of trapezoids and can be derived readily or found in any book on coordinate geometry. The coordinates used are the total latitudes and total departures for the points. North and east coordinates are plus; south and west coordinates are minus. It is necessary to take account of these signs in using the formula. To avoid negative signs for coordinates, an origin may be selected to make all signs positive. Some surveyors let X equal zero for the most-westerly point, and let Y equal zero for the most-southerly station. The magnitudes of the coordinates and products are thereby reduced.

The total latitudes and total departures of illustration 11–2 will be used to demonstrate the computation of area by coodinates, with point

Illustration 11–3

COMPUTATION OF AREA BY COORDINATES

Point	Y	X	Difference of X's	Double Area +	Double Area −
A	0.00	0.00	517.55 − 125.66 = +391.89		0
B	+255.96	+125.66	0.00 − 716.31 = −716.31		183,347
C	+102.43	+716.31	125.66 − 523.62 = −397.96		40,763
D	−591.64	+523.62	716.31 − 517.55 = +198.76		117,594
E	−388.67	+517.55	523.62 − 0.00 = +523.62		203,515
A	0.00	0.00			
					2)545,219
					272,610 sq ft

			Plus	Minus
A	0.00	0.00		
B	+255.96	+125.66	0	0
C	+102.43	+716.31	+183,347	+ 12,871
D	−591.64	+523.62	+ 53,634	−423,798
E	−388.67	+517.55	−306,203	−203,515
A	0.00	0.00	0	0
			− 69,222	−614,442
			Subtract	− 69,222
				2)−545,220
				272,610 sq ft

A as the origin. The procedure and results are shown in illustration 11–3 at the top. The coordinate formula and computation as shown are more difficult to remember than the coordinate matrix solution shown in the second part of illustration 11–3 in which downward diagonal products are found directly. The method is readily adapted to electronic computers and is programmed in illustration 10–4.

11–7. Area from a map by triangles. A plotted traverse can be divided into triangles, the sides of each triangle scaled, and the areas of the triangles found by formula 11–1.

11–8. Area by coordinate squares. To find the area within a plotted traverse by coordinate squares, the map is marked off in squares of unit area. The number of complete unit squares included in the traverse is counted and the sum of the areas of the partial units is estimated. Areas of partial units can be computed by treating them as trapezoids, but generally this refinement is not necessary.

A simpler method results from the use of transparent paper marked in squares to some scale. The paper is placed over the traverse and the number of squares and partial units counted.

A third method consists of plotting the traverse on coordinate paper and determining the number of units in the manner just described.

11–9. Measurement of area by planimeter. The planimeter mechanically integrates area and records the answer on a drum and disk as a tracing point is moved over the outline of the figure to be measured.

Its major parts are a scale bar, graduated drum and disk, vernier, tracing point and guard, and anchor arm, weight, and point. The scale bar may be fixed, or adjustable as in Fig. 11–6. For the standard fixed-arm planimeter, one revolution of the disk (dial) represents 100 sq in. and one revolution of the drum (wheel) represents 10 sq in. The adjustable type can be set to read units of area directly for any particular map scale. The instrument touches the map at only three places—the anchor point, the drum, and the tracing-point guard.

As an example of the use of the planimeter, suppose the area within the traverse of Fig. 11–3 is to be measured. The anchor point beneath the weight is set in a position outside the traverse (if inside, a polar constant must be added), and the tracing point is brought over corner A. An initial reading, say 7231, is taken, the 7 coming from the disk, the 23 from the drum, and the 1 from the vernier. The tracing point is moved along the traverse lines from A to B, C, D, and E and back to A. The point may be guided by a triangle or a straightedge, but normally it is steered freehand. A final reading, perhaps 8756, is made. The difference between the initial and final readings, or 1525, represents the area if the bar was set exactly to the scale of the map. Since the bar setting may not be perfect, it is best to check the planimeter constant by running over the perimeter of a carefully-laid-out square 5 in. on a side, with diagonals of 7.07 in.

Assume that the difference between the initial and final readings for the 5-in. square is 1250. Then

$$5'' \times 5'' = 25 \text{ sq in.} = 1250 \text{ units}$$

or

$$1 \text{ unit} = \tfrac{25}{1250} = 0.02 \text{ sq in.}$$

and

$$1525 \text{ units} = 30.50 \text{ sq in.}$$

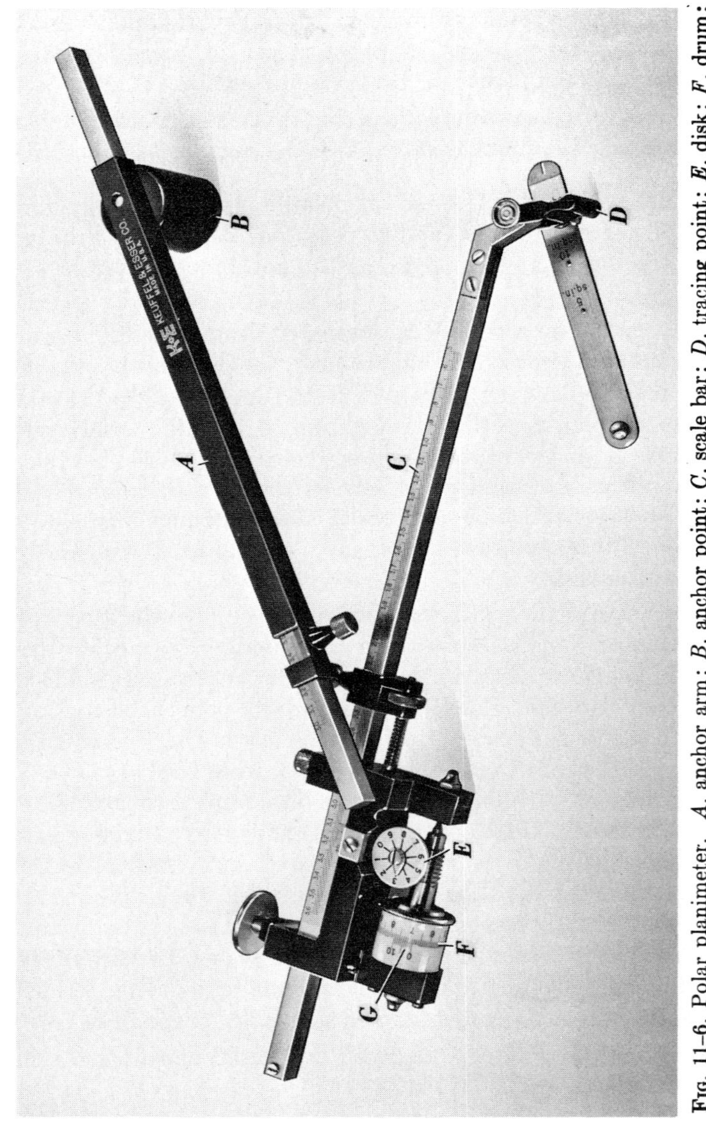

Fig. 11-6. Polar planimeter. *A*, anchor arm; *B*, anchor point; *C*, scale bar; *D*, tracing point; *E*, disk; *F*, drum; *G*, vernier. (Courtesy of Keuffel and Esser Company.)

For a map scale of 1 in. = 100 ft, 1 sq in. = 10,000 sq ft and the area measured is 305,000 sq ft.

As a check on the planimeter operation, the outline may be traced in the opposite direction. The initial and final readings at point *A* should agree within a limit of perhaps 2 to 5 units.

The precision obtained in using the planimeter depends upon the skill of the operator, the accuracy of the plotted map, the type of paper, and other factors. Results correct within $\frac{1}{2}$ to 1 per cent can be obtained by careful work.

The planimeter is most useful for irregular areas, such as that in Fig. 11–2, and has applications in many branches of engineering. Steam cutoff diagrams in mechanical engineering are readily investigated by planimetering. The planimeter is widely used in highway offices for determining areas of cross sections and is helpful in checking computed areas in property surveys.

11–10. Sources of error. Some of the sources of error in area computations include:

a. Poor selection of intervals and offsets to fit a given irregular boundary properly.
b. Failure to check an area computation by a different method.
c. Poor setting of the planimeter scale bar, and failure to check for the scale constant by tracing a known area.
d. Running off the edge of a map sheet with the planimeter drum.
e. Using different types of paper for the map and the planimeter-calibration sheet.
f. Poor selection of the origin, resulting in minus values for coordinates or D.M.D.'s.
g. Failure to draw a sketch to scale or general proportions for checking visually.
h. Adjustments of latitudes and departures not made in accordance with true conditions.
i. Using coordinate squares which are too large and which therefore make estimation of the partial blocks difficult.

11–11. Mistakes. In computing areas, common mistakes by students include:

a. Forgetting the division by 2 in any of the formulas or methods.
b. Confusing the signs of D.M.D.'s, coordinates, latitudes, departures, or areas.

PROBLEMS

11–1. Compute the area enclosed by the traverse *ABCDEFGHJLA* in Fig. 11–1.

11–2. Compute the area enclosed between the traverse line *GHJLA* and the shore line in Fig. 11–1 by the offsets method.

11–3. Compute the area between the traverse line from station 5 + 42 to station 7 + 00 and the creek in Fig. 13–3 by the offsets method.

11–4. Calculate the area between a lake and a straight line from which offsets are taken at irregular intervals (all values in chains) as follows:

Offset point	*A*	*B*	*C*	*D*	*E*	*F*	*G*
Distance from *A*	0	1.34	2.26	3.66	5.18	5.32	6.00
Offset	1.42	1.68	2.00	2.28	1.85	1.67	0.50

11–5. Compute the D.M.D.'s for the lines in problem 10–15.

11–6. List the D.M.D.'s and D.P.D.'s for the lines in problem 10–16.

11–7. For the data shown, where distances are given in 100-ft tape lengths, compute (a) the latitude and departure of the line *GA*, (b) the length of the line *GA*, and (c) the area within the traverse in acres both by D.M.D.'s and by coordinates.

Line	N Lat.	S Lat.	E Dep.	W Dep.
AB	6		6	
BC		8		7
CD		3	4	
DE		4		8
EF	8		2	
FG		0		5
GA				

11–8. Similar to problem 11–7 except that *AB* = N Lat. 8, E Dep. 8, and *DE* = S Lat. 5, W Dep. 9.

11–9. A traverse has courses with the tabulated latitudes and departures (in hundreds of feet). (a) Which is the most westerly station? (b) Which is the most southerly? (c) What are the length and bearing

Course	E Dep.	W Dep.	N Lat.	S Lat.
AB	5			4
BC		4		7
CD		3	2	
DE	2		5	
EF		3		4
FG		6	4	
GH	3			0
HA	6		4	

of the line connecting stations G and D? (d) Compute by D.M.D.'s the area enclosed, in acres. (e) Check by D.P.D.'s and by coordinates.

11–10. Similar to problem 11–9 except that AB = E Dep. 6, S Lat. 5, and FG = W Dep. 7, N Lat. 5.

11–11. Compute by D.M.D.'s the area in acres enclosed by the following traverse, placing the axis through the most westerly station. All distances are in engineer's chains.

Course	N Lat.	S Lat.	E Dep.	W Dep.
AB	6		8	
BC	5			3
CD		6		6
DE	4			4
EF		10	0	
FA	1		5	

11–12. Similar to problem 11–11 except that AB = N Lat. 7, E Dep. 9, and EF = S Lat 11, W Dep. 1.

11–13. Compute the area in problem 11–11 by using coordinates.

11–14. Compute the area in problem 11–12 by the matrix method described in Chapter 24.

11–15. For the following closed traverse (distances in Gunter's chains), (a) compute missing latitudes and departures, (b) plot the traverse by coordinates, and (c) calculate the area in acres by D.M.D.'s.

Line	Bearing	Dist.	Lat.	Dep.
AB	N 36° 54′ E	5	+ 4	+3
BC	Due East	5		
CD	Due South	10	−10	
DA	N 53° 06′ W	10		

Compute the area enclosed by the traverses of problems 11–16 and 11–17, selecting an axis so that there will be no negative D.M.D.'s.

11–16.

Course	*AB*	*BC*	*CD*	*DA*
Bearing	Due East	S 30°00′W	Due West	N 45°00′W
Length (ft)	256.0	150.0	50.0	Unknown

11–17.

Course	*AB*	*BC*	*CD*	*DA*
Bearing	N 60°00′W	Due South	N 45°00′E	Unknown
Length (ft)	400.0	240.0	141.4	Unknown

11–18. Calculate the area enclosed by the traverse of problem 10–4.

11–19. Determine the area within the traverse of problem 10–7.

11–20. Compute the area inside the traverse of problem 10–8.

11–21. Find the area surrounded by the traverse of problem 10–14.

11–22. Determine the area for the lot described in problem 10–15.

11–23. Calculate the area within the traverse of problem 10–16.

11–24. Compute the area of a piece of property bounded by a traverse and circular arc described as follows: *AB*, S 30°00′ W, 400.0 ft; *BC*, Due East, 400.0 ft; *CD*, N 45°00′ W, 200.0 ft; *DA*, a circular arc tangent to *CD* at point *D*.

11–25. Similar to problem 11–24, except that *CD* is 185.0 ft.

11–26. Divide the area of the lot in problem 11–24 into two equal parts by a line through point *B*. List in order the lengths and bearings of all sides for each part.

11–27. Partition the lot in problem 11–25 into two equal areas by means of a line parallel to *BC*. Tabulate in consecutive order clockwise the lengths and bearings of all sides.

11–28. Lot *ABCD* between two parallel street lines is 100.00 ft deep, has an 80-ft frontage (*AB*) on one street, and a 120-ft frontage (*CD*) on the other. Interior angles at *A* and *B* are equal, as are those at *C* and *D*. What distances *AE* and *BF* should be laid off by a surveyor in order to divide the lot into two equal areas by means of a line *EF* parallel to *AB*?

11–29. The area of a field plotted on a map to a scale of 1 in. = 400 ft is measured with a fixed-arm planimeter, for which the constant is 10, and the revolutions of the wheel are recorded as 3.268. What is the area of the field in square feet and in acres?

11–30. The tracing point of a planimeter with an initial setting of 1273 is run clockwise around a square 5 in. on a side, and the reading is 1398 when the tracing point is returned to the starting corner. The tracing point is then set on hub *A* of a plotted traverse with an initial reading of 1619 and is run around the traverse to *A* again, where the reading is 1994. If the map scale is 1 in. = 40 ft, what is the area enclosed by the traverse in square feet?

11–31. Using the data of problem 11–30 suppose that after a planimeter reading of 1398 is obtained, a check run in the counterclockwise direction around the square brings the reading back to 1270, instead of 1273, on the first corner. What per cent error does this result represent for the square? For the traverse? Is it satisfactory? What procedure is suggested?

11–32. Similar to problem 11–30 except that successive values are, respectively, 0020, 4-in. square, 0170, 0350, 0590, and scale 1 in. = 80 ft.

11–33. In problem 11–7, if the tape was assumed to be 100.00 ft long and after standardization it is found to be 100.08 ft long, what is the error in area?

11–34. If the Gunter's chain used in problem 11–15 was later discovered to have one missing link, what is the correct area within the traverse?

11–35. Write a computer program for calculating the area in problem 11–7 by the matrix method described in Chapter 24.

chapter 12

Stadia

12–1. General. The stadia method is a rapid and efficient way to measure distances accurately enough for trigonometric leveling, some traverses, and the location of topographic details. Furthermore, a two- or three-man party can replace the three- or four-man party required in transit-tape surveys.

The term *stadia* comes from the Greek word for a unit of length originally applied in measuring distances for athletic contests. The word denoted 600 Greek feet, or 606 ft 9 in. by present-day American standards.

The term "stadia" is now applied to the cross wires and rod used in making measurements, as well as to the stadia method itself. Stadia readings can be taken with modern transits, theodolites, and levels.

12–2. Measurement by stadia for horizontal sights. A transit equipped for stadia work has two additional horizontal cross wires spaced equidistant from the center of the telescope. The interval between stadia wires in most surveying instruments gives a vertical intercept of 1 ft on a rod held 100 ft away. Thus the distance to a rod decimally divided in feet, tenths, and hundredths can be read

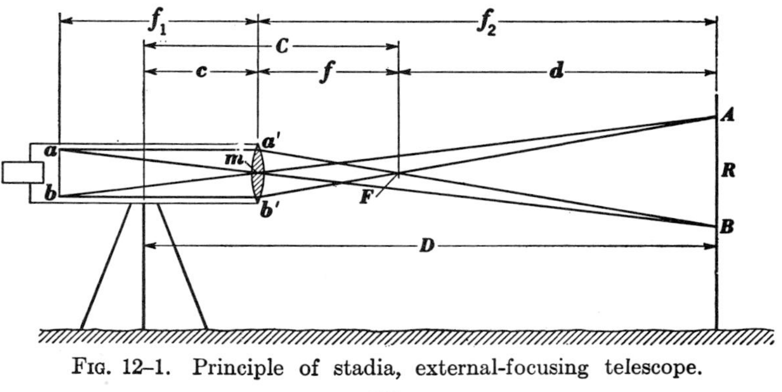

FIG. 12–1. Principle of stadia, external-focusing telescope.

directly to the nearest foot. This is sufficiently precise for locating topographic details (such as rivers, bridges, buildings, and roads) which are to be plotted on a map having a scale smaller than 1 in. = 200 ft.

The stadia method is based upon the principle that in similar triangles homologous sides are proportional. Thus in Fig. 12–1, which shows an external-focusing telescope, light rays from points A and B passing through the center of the lens form a pair of similar triangles AmB and amb. Here $AB = R$ is the rod intercept (*stadia interval*) and ab is the interval between the stadia wires.

Standard symbols used in stadia measurements, and their definitions, are as follows:

f = *focal length* of the lens (a constant for any particular compound objective lens). It can be determined by focusing upon a distant object and measuring the distance between the center (actually the *nodal point*) of the objective lens and the reticle.

f_1 = distance from the center (actually the nodal point) of the objective lens to the plane of the cross hairs when the telescope is focused on some definite point.

f_2 = distance from the center (actually the nodal point) of the objective lens to a definite point when the telescope is focused on that point. When f_2 is infinite, or very large, $f_1 = f$.

i = interval between the stadia wires (ab in Fig. 12–1).

f/i = *stadia interval factor*, usually 100.

c = distance from the center of the instrument (spindle) to the center of the objective lens. It varies as the objective lens moves in and out for different sight lengths but is generally considered to be a constant.

$C = c + f$. C is called the stadia constant although it varies slightly with c.

d = distance from the focal point in front of the telescope to the face of the rod.

$D = C + d$ = distance from the center of the instrument to the face of the rod.

Then from similar triangles,

$$\frac{d}{f} = \frac{R}{i} \quad \text{or} \quad d = R\frac{f}{i}$$

and

$$D = R\frac{f}{i} + C \tag{12–1}$$

Fixed stadia wires in engineer's transits, levels, and alidades are carefully spaced by instrument manufacturers to make the stadia interval factor f/i equal to 100. The stadia constant C ranges from about 0.75 to 1.25 ft for different telescopes but is usually assumed to be equal to 1 ft. The only variable on the right side of the equation, then, is R, the rod intercept between the stadia wires. In Fig. 12–1, if the intercept R is 4.27 ft, the distance from the instrument to the rod is $427 + 1 = 428$ ft.

The stadia interval factor should be determined the first time an instrument is used, although the exact value posted by the manufacturer on the inside of the box will not change unless the cross hairs, reticle, or lenses are replaced. Plate A–9 in Appendix A shows the notes for such a test.

The old-type external-focusing telescope has been described, since a simple drawing correctly shows the relationships. The objective lens of an internal-focusing telescope remains fixed in position while a movable negative focusing lens between the objective lens and the plane of the cross wires changes the direction of the light rays. As a result, the stadia constant is so small (perhaps a few tenths of a foot) that *it can generally be assumed equal to zero.*

Disappearing stadia hairs were used in some older instruments to prevent confusion with the center horizontal hair. Modern glass reticles with short stadia lines and a full-length center line accomplish the same result more efficiently.

12–3. Measurement by stadia for inclined sights. It is impractical to hold a rod at right angles to an inclined line of sight. The intercept on a plumbed rod is therefore read and corrected in order to get true horizontal and vertical distances.

In Fig. 12–2 the transit is set at M and the rod is held at O. With the middle cross hair set on point D to make DO equal to the height of instrument EM, the vertical angle (angle of inclination) is a. Note that in stadia work the height of instrument (*h.i.*) is defined as the height of the line of sight above the point occupied (*not* above the datum as in leveling).

Let S represent the slope distance ED; H, the horizontal distance $EG = MN$; and V, the vertical distance $DG = ON$.

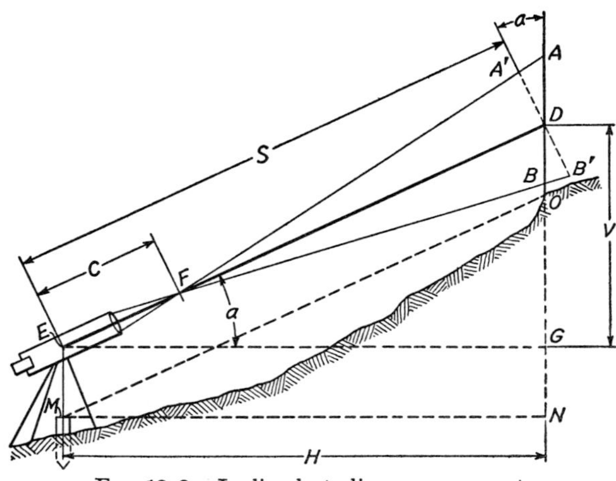

FIG. 12–2. Inclined stadia measurement.

Then $$H = S \cos a$$

and $$V = S \sin a$$

If the rod could be held normal to the line of sight at point O, a reading $A'B'$, or R', would be obtained, making

$$S = R' \frac{f}{i} + C$$

Since the rod is vertical, the reading is AB, or R. For the small angle at D on most practical sights, it is sufficiently accurate to consider that angle $AA'D$ is a right angle. Therefore

$$R' = R \cos a$$

and $$S = R \frac{f}{i} \cos a + C$$

or $$H = R \frac{f}{i} \cos^2 a + C \cos a \qquad (12\text{–}2)$$

For small angles, $C = 1$ ft approximately, and

$$H = R \frac{f}{i} \cos^2 a + 1 \qquad (12\text{–}3)$$

If $f/i = K$,

$$H = KR \cos^2 a + 1 \qquad (12\text{–}4)$$

To avoid multiplying R by $\cos^2 a$, which is a large decimal number, the formula for H is usually rewritten for use in computation as

$$H = KR - KR \sin^2 a + C \tag{12-5}$$

The vertical distance is found by the formula

$$V = S \sin a = \left(R \frac{f}{i} \cos a + C\right) \sin a$$

or

$$V = R \frac{f}{i} \sin a \cos a + C \sin a \tag{12-6}$$

For small angles, $\sin a$ is very small and the quantity $C \sin a$ can be neglected. Substituting for $\sin a \cos a$ its equal $\frac{1}{2} \sin 2a$, the formula becomes

$$V = KR(\tfrac{1}{2} \sin 2a) \tag{12-7}$$

In the final form generally used, K is taken as 100 and the formulas for reduction of inclined sights to horizontal and vertical distances are

$$H = 100\ R - 100\ R \sin^2 a + 1 \tag{12-8}$$

and

$$V = 100\ R(\tfrac{1}{2} \sin 2a) \tag{12-9}$$

Tables, diagrams, and special slide rules (see Fig. 12–3) enable the surveyor to obtain a quick solution to these formulas. Table II lists horizontal and vertical distances for a slope length of 100 ft and various vertical angles to expedite the "reduction" of notes.

An unfamiliar table should always be investigated by substituting values in it which will give known answers. For example, angles of 0°, 30°, and 45° can be used to check tabular results. Assuming a vertical angle of +30° 00′, a rod intercept of 1.00 ft, and a stadia constant C of 1 ft, the following results are secured: By Table II,

$$H = 75.00 \times 1.00 + 1 = 76 \text{ ft}$$

By formula 12–8,

$$H = 100 \times 1.00 - 100 \times 1.00 \times 0.50^2 + 1 = 76 \text{ ft}$$

An example of a typical computation required in practice will be given. Assume that in Fig. 12–2 the elevation of M is 268.2 ft, the h.i. is $EM = 5.6$ ft, the rod intercept is $AB = R = 5.28$ ft, the vertical angle a to point D at 5.6 ft on the rod is +4° 16′, and $C = 1$ ft.

From Table II, for an angle of +4° 16′ and a slope length of 100 ft,

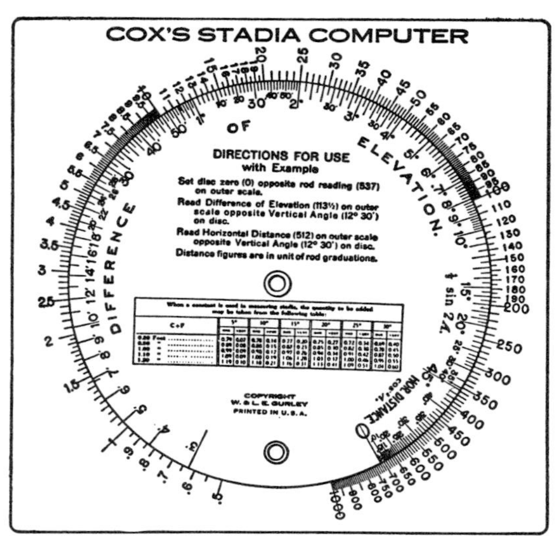

FIG. 12–3. Cox stadia computer.

the horizontal distance is 99.45 ft and the vertical distance is 7.42 ft. Then

$$H = (99.45 \times 5.28) + 1 = 525.1 + 1 = 526 \text{ ft}$$

and $$V = (7.42 \times 5.28) + 0.08 = 39.18 + 0.08 = 39.3 \text{ ft}$$

The elevation of point O, then, is

$$268.2 + 5.6 + 39.3 - 5.6 = 307.5 \text{ ft}$$

The advantage of sighting on the h.i. is evident. Since the rod reading and the h.i. are opposite in sign and cancel each other, they can be omitted from the elevation computation. If the h.i. cannot be sighted because of obstructions, setting the middle cross hair on the full foot mark just above or below the h.i. simplifies the arithmetic.

Determination of elevation differences by stadia can be compared with differential leveling. The h.i. corresponds to a plus sight, and the rod reading to a minus sight. On these is superimposed a vertical distance which may be either plus or minus, its sign depending upon the angle of inclination. On important sights to control points and hubs, instrumental errors should be reduced by good field procedures utilizing the principle of reversion, that is, reading vertical angles with the telescope in normal and plunged positions.

Direct rod readings (as in leveling), rather than vertical angles, are taken when possible to simplify the reduction of notes. Inspection of Table II shows that for vertical angles less than about 4°, the difference between the slope distance and the horizontal distance is negligible except on long sights (where the distance-reading error is also greater). Hence, inclining the telescope by several degrees for the stadia reading after taking a level "foresight" is permissible.

12–4. Stadia rods. Various types of markings are used on stadia rods but all have bold geometric figures designed for legibility at long distances. Stadia rods are graduated in feet and tenths. Hundredths must be interpolated. Different colors aid in distinguishing the numbers and the graduations.

One-piece, folding, and sectional rods having lengths of 10 or 12 ft are common. Longer rods increase the sight-distance limit but are heavy and awkward to handle. Often the lower foot or two of a 12-foot rod will be obscured by weeds or brush leaving perhaps only 10 feet visible so the maximum length of sight would be about 1000 feet. On longer sights, the half interval (intercept between the middle cross hair and upper or lower stadia hair) can be read. The full interval is obtained by doubling the half interval and using the standard equations for stadia reduction. With a quarter-hair between the middle cross hair and upper stadia hair, theoretically a distance of over 4000 feet can be read. On short sights the ordinary leveling rod is satisfactory.

12–5. Beaman arc. The Beaman arc, Fig. 12–4, is a device placed on some transits and alidades to facilitate stadia computations. It may be part of the vertical circle or may be a separate plate. The H and V scales of the arc are graduated in per cent. The V scale shows the difference of elevation per 100 ft of slope distance, whereas the H scale gives the correction per 100 ft to be subtracted from the stadia distance. Since V is proportional to $\frac{1}{2} \sin 2a$, and the correction for H depends upon $\sin^2 a$, spacing of the graduations decreases as the vertical angle increases. Therefore a vernier cannot be used, and an exact reading can be made only by setting the arc to read a whole number.

The indicator of the V scale is set to read 50 (or perhaps 30 on

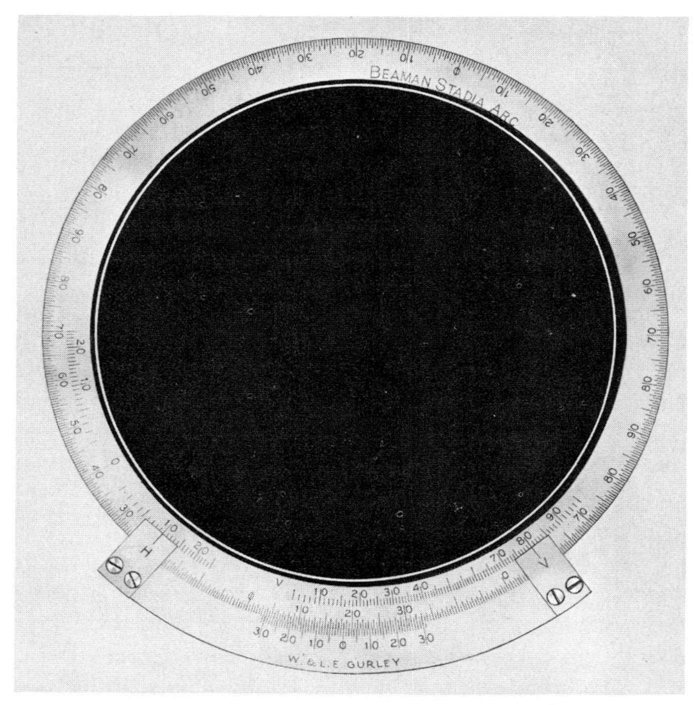

Fig. 12–4. Beaman arc (and vertical circle).

some instruments) when the telescope is horizontal. Minus values are thereby eliminated. Readings greater than 50 are obtained on sights above the horizon; readings less than 50, on sights below the horizon. In Fig. 12–4 the V scale reads 80, and the H scale reads 10. The arithmetic required in using the Beaman arc is simplified by setting the V scale on a whole number and letting the middle cross hair fall somewhere near the h.i. The H scale generally will not then read a whole number and the value must be interpolated, but this is unimportant since the arithmetic remains simple.

The elevation of a point B sighted with the transit set up over point A is found by the following formula:

$$\text{Elev. } B = \text{Elev. } A + \text{h.i.} + (\text{arc reading} - 50)\ (\text{rod intercept}) - \text{rod reading of center cross hair} \qquad (12\text{–}10)$$

Careful attention must be given to signs.

To illustrate the computation, assume that in Fig. 12–2 the V-scale reading is 56, the H-scale reading is 0.6, the rod intercept is 6.28 ft, the h.i. is 4.2 ft, the rod reading of the center wire is 7.3 ft, $C =$ 0 and the elevation of point M is 101.5 ft. Then

$$\text{Elev. } O = 101.5 + 4.2 + (56 - 50)(6.28) - 7.3 = 136.1 \text{ ft}$$

and $$H = (100)(6.28) - (0.6)(6.28) = 624 \text{ ft}$$

12–6. Instruments with movable stadia lines. Self-reducing European tachymeters, Fig. 12–5, and alidades have been developed in which *curved* stadia lines appear to move apart, or move closer

FIG. 12–5. Reduction tachymeter for use with vertical rod.

together, as the telescope is elevated or depressed. Actually, the lines are engraved on a glass plate which turns around a center (situated outside the telescope) as the telescope is transited.

In Fig. 12–6 the upper and lower lines (the two outer lines) are curved to correspond to the variation in the trigonometric function $\cos^2 a$ and are used for distance measurements. The two inner lines are employed to determine differences in elevation and are curved to represent the function $\sin a \cos a$. A vertical line, a center cross, and short stadia lines are marked on a second, but fixed, glass plate which is in focus simultaneously with the curved lines.

A constant stadia-interval factor of 100 is used for measurement of horizontal distances. A factor of 20, 50, or 100 is applied to measurements of elevation differences, its value depending upon the angle of slope. The factor to be used is indicated by short lines placed between the elevation curves.

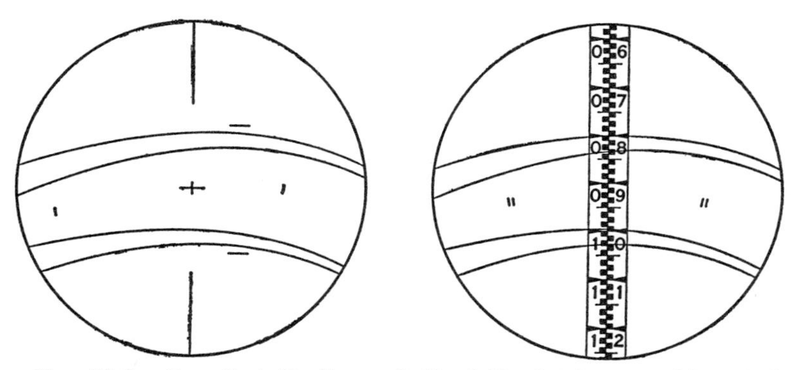

FIG. 12–6. Curved stadia lines. Left: field of telescope without rod. Right: field of telescope with rod (horizontal distance = 100 × 22.7 = 22.7 m; difference of height = 20 × 16.7 = 3.30 m). (Courtesy of Kern Instruments, Inc.)

12–7. Field notes. An example of stadia field notes is shown in Plate A–10. The transit is usually oriented by a meridional azimuth, and clockwise angles are taken to desired points.

In stadia work, many shots may be taken from one hub. *Orientation should therefore be checked by sighting on the control line after every ten or twenty topographic points and before leaving the hub.*

As previously stated, horizontal sights should be used in place of inclined sights whenever possible. Rod readings then become minus sights, as in profile leveling.

Columns in the notes are arranged in the order of readings—point sighted, rod intercept, azimuth, and vertical angle. A sketch and pertinent data are placed on the right-hand page as usual. The numbering of topographic points begins with 1 at the first hub and continues successively through all hubs, thus eliminating the possible duplication of numbers on any sketch. Reduction of the notes is done in the office, unless the information is needed immediately for control or is to be plotted on a plane-table sheet.

A new method showing possibilities under some circumstances is the use of a tape recorder for expediting the notekeeping.

12–8. Field procedure. Proper field procedures save time and reduce the number of mistakes in all surveying operations. The order of taking readings best suited for *stadia work involving vertical angles* is as follows:

a. Bisect the rod with the vertical wire.
b. With the middle wire approximately on the h.i., set the lower wire at a foot mark.
c. Read the upper wire, and subtract the reading of the lower wire from it to get the rod intercept. Record (or call out to the notekeeper) only the answer—the intercept.
d. Move the middle wire to the h.i. by using the tangent screw.
e. Release the rodman for movement to the next point by giving the proper signal.
f. Read and record the horizontal angle.
g. Read and record the vertical angle.

This procedure will enable an instrumentman to keep two or three rodmen busy in open terrain where points to be located are widely separated. The same order can be followed when the Beaman arc is used, but in step d the V scale is set to a whole number, and in step g the H- and V-scale readings are recorded.

12–9. Stadia traverses. In a transit-stadia traverse, distances, horizontal angles, and vertical angles are measured at each corner. Reduction of stadia notes as the survey progresses provides elevations to be carried from hub to hub. Average values of stadia distances and differences in elevation are obtained from a foresight and a backsight on each line. An elevation check should be secured by closing on the initial point, or on bench marks near both ends for an open traverse.

The horizontal angles should also be checked for closure. Any angular mis-closure should be adjusted, the latitudes and departures computed, and the precision of the traverse checked by the methods of Chapter 10.

12–10. Topography. The stadia method is most useful in locating numerous topographic details, both horizontally and vertically, by transit or plane table. In urban areas, angle and distance readings can be taken faster than a notekeeper is able to record the measurements and prepare a sketch.

The use of stadia in topographic work will be covered in more detail in the chapters on topography and the plane table.

12–11. Stadia leveling. The stadia method is adaptable to trigonometric leveling. The h.i. is determined by sighting on a point of known elevation, or by setting the instrument over such a point and measuring the height of the horizontal axis above it with a stadia rod. The elevation of any point can then be found by computation from the rod intercept and the vertical angle. If desirable, a leveling circuit can be run to establish and check the elevations of two or more points.

12–12. Precision. A ratio of error of 1/500 can be obtained for a transit-stadia traverse run with ordinary care. Short sights, a long traverse, and careful work may give ratios up to 1/1000. Errors in stadia work are usually the result of poor rod readings rather than incorrect angles. An error of 1 min in reading a vertical angle does not appreciably affect the horizontal distance. The same 1-min error produces a difference in elevation of less than 0.1 ft on a 300-ft sight for any vertical angle.

Figure 12–7 shows that if stadia distances are read to the nearest foot (the usual case), horizontal angles *to topographic points* need be read only to the nearest 5 or 6 min for comparable precision on 300-ft sights. A stadia distance to the nearest foot is assumed to be correct to within about $\frac{1}{2}$ ft. Allowing the same $\frac{1}{2}$-ft error laterally,

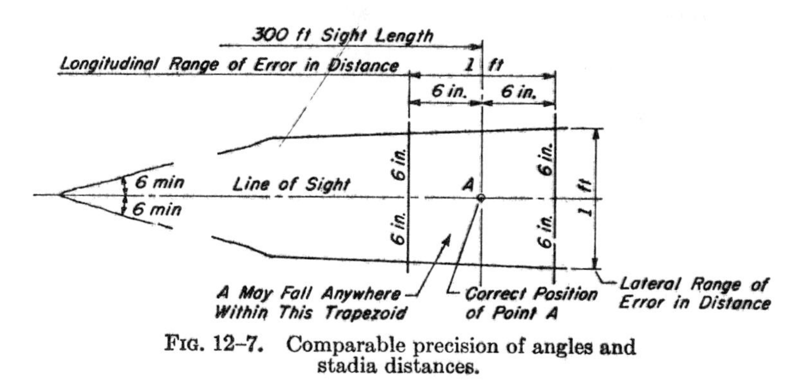

FIG. 12–7. Comparable precision of angles and stadia distances.

the direction can be off about 5 min (readily computed from sin 1 min = 0.00029). Angles therefore can be read without using the vernier, merely by estimating the position of the vernier index.

The accuracy of trigonometric leveling by stadia depends upon the lengths of sights and the size of the vertical angles required.

12–13. Sources of error in stadia work. The errors which occur in transit operations are inherent in stadia work too. Additional sources of error include the following:

Instrumental errors

a. Improper spacing of the stadia wires.
b. Index error.
c. Incorrect length of rod graduations.
d. Transit axis of sight not parallel with the axis of the telescope bubble.

Personal errors

a. Rod not held plumb.
b. Incorrect rod readings resulting from long sights.
c. Careless leveling for vertical-arc readings.

Most errors in stadia work can be eliminated by (a) careful manipulation of the instrument, (b) limiting the length of sights, (c) using a good rod and a rod level, and (d) averaging readings in the forward and backward directions. Line of sight error cannot be corrected by field procedures—the instrument must be adjusted.

12–14. Mistakes. Some typical mistakes in stadia work are the following:

a. Index error applied with wrong sign.
b. Confusion of plus and minus vertical angles.
c. Arithmetic mistakes in computing the rod intercept.
d. Use of incorrect stadia interval factor.

PROBLEMS

12–1. An external-focusing telescope has a focal length of 10 in. What error in spacing the stadia lines will raise the stadia interval factor from 100.00 to 101.50?

12–2. Determine the stadia interval factor for an external-focusing telescope having an average C of 1.10 ft. Stadia distances (constant not included) and corresponding taped distances (from instrument center), in feet, are as follows:

Taped	111.52	195.23	263.65	355.17	418.82
Stadia	111	196	264	357	420

12–3. Similar to problem 12–2, but the transit has an internal-focusing telescope with a negligible C and a focal length of 0.5 ft.

12–4. On stadia measurements, why is it advisable to keep the line of sight through the lower cross line several feet above all intervening ground?

Compute the distance from transit to rod for the sights in problems 12–5 through 12–7.

12–5. Horizontal sight, intercept = 3.44 ft, $K = 102.0$, $C = 1.0$ ft.

12–6. Horizontal sight, intercept = 4.81 ft, $K = 99.0$, $C = 0.9$ ft.

12–7. Horizontal sight, intercept = 2.65 ft, $K = 98.8$, $C = 0$.

Compute the horizontal distance and difference in elevation for the data given in problems 12–8 through 12–10.

12–8. Rod intercept = 1.92 ft, $a = +3°00'$, $K = 100.0$, $C = 1.0$ ft.

12–9. Rod intercept = 5.54 ft, $a = -4°10'$, $K = 99.0$, $C = 0$.

12–10. Rod intercept = 6.32 ft, $a = -10°15'$, $K = 101.4$, $C = 1.0$ ft.

What is the rod intercept for the conditions given in problems 12–11 through 12–13.

12–11. Horizontal distance = 100.0 ft, $K = 101.0$, $a = +9°00'$, $C = 0$.

12–12. Horizontal distance = 200.0 ft, $K = 98.0$, $a = -5°00'$, $C = 0$.

12–13. Horizontal distance = 150.0 ft, $K = 99.2$, $a = +6°30'$, $C = 1.1$ ft.

12–14. The specifications for a stadia traverse being run over the site of a proposed reservoir require an accuracy not lower than 1/400. What is the lower limit of vertical angles for which it is necessary to reduce inclined sights to horizontal distances?

12–15. Same as problem 12–14 except for an accuracy of 1/50.

12–16. Compute the ground elevations for the following stadia notes. The transit is at station A, elevation 622.4 ft, h.i. $= 5.1$ ft, and $C = 1$. All sights are on the h.i. except as noted.

Point	Inter-cept	Azimuth	Vert. Angle	Point	Inter-cept	Azimuth	Vert. Angle
B	1.93	169° 30′	− 1° 21′	16	0.44	270° 45′	− 4° 50′
12	0.42	225° 10′	− 2° 34′	17	0.56	296° 10′	− 6° 34′
13	0.56	226° 35′	0° 00′ on 12.35 ft	18	0.49	321° 55′	−12° 22′
14	0.92	241° 55′	+ 4° 10′	19	0.36	342° 35′	+ 1° 49′ on 6.5 ft
15	0.85	262° 00′	−10° 08′	20	0.65	349° 20′	+ 5° 26′

12–17. Same data as problem 12–16 but transit has a +0°03′ index error.

12–18. Stadia notes for a transit at E, elevation 436.8 ft, h.i. $= 5.0$ ft, and $C = 0$ are as follows. Compute the ground elevations for sights on the h.i. except as noted.

Point	Inter-cept	Azimuth	Vert. Angle	Point	Inter-cept	Azimuth	Vert. Angle
F	4.07	8° 06′	+5° 12′	57	2.25	6° 20′	−7° 48′
53	2.44	350° 55′	−6° 45′	58	2.37	13° 45′	−6° 04′ on 6.1 ft
54	2.18	350° 55′	−8° 37′	59	1.60	39° 30′	−4° 36′
55	2.64	350° 55′	−8° 29′	60	2.83	50° 50′	−3° 45′
56	3.16	350° 55′	−5° 56′	61	3.31	54° 00′	3.0 ft

12–19. Same data as problem 12–18 but transit has a −0° 04′ index error.

A sight is taken on B.M. Stone, elevation 628.7, with a transit at hub X. In problems 12–20 through 12–23 calculate the elevation of hub X.

12–20. Rod intercept = 4.36 ft, $a = +\,6°\,12'$, h.i. $= 5.4$ ft, and $C = 1.0$.

12–21. Rod intercept = 6.17 ft, $a = -\,4°\,18'$, h.i. $= 4.9$ ft, and $C = 0$.

12–22. Rod intercept = 7.02 ft, $a = -\,3°\,35'$, h.i. $= 5.2$ ft, and $C = 0.6$.

12–23. Rod intercept = 6.00 ft, $a = +1°\,30'$ to 7.7 ft on the rod, h.i. $= 4.8$ ft, $C = 0$. Use the approximate relationships of Sec. 8–1.

At hub C, elevation 520.0, a F.S. is taken on hub D, the transit is moved to D and a B.S. taken on C. In problems 12–24 through 12–26 determine the distance CD and the elevation of hub D. $K = 100.0$ and $C = 0$.

12–24. At C, $R = 2.99$, $a = +2°\,20'$ to 9.6 on the rod, h.i. $= 5.0$ ft.
At D, $R = 3.00$, $a = -2°\,40'$ to 1.4 on the rod, h.i. $= 4.8$ ft.

12–25. At C, $R = 4.52$, $a = -3° 15'$ to 7.2 on the rod, h.i. = 5.1 ft.
At D, $R = 4.50$, $a = +3° 18'$ to 3.5 on the rod, h.i. = 5.2 ft.

12–26. At C, $R = 2.67$, $a = +1° 52'$ to 7.6 on the rod, h.i. = 5.4 ft.
At D, $R = 2.67$, $a = -1° 46'$ to 3.3 on the rod, h.i. = 4.9 ft.

In problems 12–27 through 12–30 determine the horizontal distance AB and the elevation of point B for the Beaman-arc readings noted with the instrument at A, elevation 415.6 ft, and h.i. = 5.2.

12–27. Rod intercept = 4.28 ft, V-scale = 72 to 2.3 on rod and H-scale = 5.2.

12–28. Rod intercept = 5.66 ft, V-scale = 83 to 4.8 on rod and H-scale = 12.

12–29. Rod intercept = 7.80 ft, V-scale = 40 to 6.9 on rod and H-scale = 1.

12–30. Rod intercept = 2.82 ft, V-scale = 25 to 5.0 on rod and H-scale = $6\frac{1}{2}$.

For consistent precision of angles and distances in transit-stadia readings, how close must the angles be read under the conditions of problems 12–31 through 12–33?

12–31. Stadia constant = 1 ft, a sight of 500 ft.

12–32. Stadia constant = 0, a sight of 800 ft.

12–33. Stadia constant = 1 ft, a sight of 1,000 ft (estimated to nearest 5 ft).

12–34. In using a one-minute transit, at what distance does a 1-ft stadia constant become negligible?

In problems 12–35 and 12–36 compute the error of closure and adjust the elevations of the following azimuth-stadia traverse by distributing the closure in proportion to the differences in elevation between adjacent hubs. The elevation of A is 536.2 ft, and the stadia distances and vertical angles to the h.i. of 5.7 ft are the averages of foresights and backsights.

12–35.

Course	Intercept	Azimuth	Vert. Angle
AB	3.27	89° 16′	+4° 32′
BC	6.22	14° 28′	+3° 51′
CD	5.09	269° 10′	−5° 04′
DA	6.10	177° 14′	−2° 10′

12–36.

Course	Intercept	Azimuth	Vert. Angle
AB	3.62	82° 06′	−4° 10′
BC	3.85	349° 30′	+2° 46′
CD	2.11	263° 22′	+3° 38′
DA	4.14	191° 37′	0° 00′ on 11.9 ft

12–37. Same as problem 12–35 except elevation $A = 406.8$ ft and the closure is to be adjusted in proportion to course lengths.

12–38. Same as problem 12–36 except elevation of $A = 281.1$ ft and the closure is to be adjusted in proportion to course lengths.

12–39. Is it good practice to read the A and B verniers in stadia work? Explain.

12–40. Why is it more difficult to obtain good elevation closures than satisfactory distance closures in transit-stadia surveying?

12–41. What assumption is made in the stadia formula for inclined sights to simplify it?

12–42. What check should be made during the progress of work in taking many azimuths and distances from a hub in a topographic survey?

12–43. Upon what principle is the stadia method based?

12–44. Is the subtense bar method actually stadia in a horizontal plane? Explain.

12–45. What are the advantages of the Beaman stadia arc in reading distances and elevation differences?

12–46. What error in stadia distance results if instead of being properly plumbed the top of a 12–ft stadia rod is inclined 6 in. away from the transit when a rod intercept of 6.50 ft is obtained on a horizontal sight?

12–47. Why is it important in stadia work to have the line of sight of a transit parallel with the axis of the telescope bubble?

12–48. In running stadia traverses it is recommended that sights be taken both forward and backward on each course. Why?

chapter **13**

Topographic Surveys

13–1. General. Topographic surveys are made to determine the configuration (relief) of the surface of the earth and to locate the natural and artificial features thereon. A topographic map is a large-scale representation, by means of conventional symbols, of a portion of the earth's surface, showing the *culture, relief, hydrography,* and perhaps the *vegetation.* Products of man, such as roads, trails, buildings, bridges, canals, and boundary lines, are termed artificial features (culture). Names and legends on maps are included in this classification.

Topographic surveys are made and used by engineers to determine the most desirable and economical location of highways, railroads, bridges, buildings, canals, pipelines, transmission lines, and other facilities; by geologists to investigate mineral, oil, water, and other resources; by foresters in locating fire-control roads and towers; by architects in housing and landscape design; by agriculturists in soil conservation; and by geographers and scientists in numerous fields.

A *planimetric map,* or *line map,* shows the natural and/or cultural features in plan only. A *hypsometric map* presents relief by conventions such as *contours, hachures, shading,* and *tinting.* Several methods of locating topographic details in both horizontal and vertical position will be discussed. The preparation of maps from topographic surveys is covered in Chapter 14.

13–2. Control for topographic surveys. The first requirement of any topographic survey is good control.

Horizontal control is provided by two or more points on the ground accurately fixed in position horizontally by distance and direction. These points provide data to orient and check traverses, plane-table surveys, and terrestrial and aerial photographs. A system of horizontal-control points is established by triangulation, trilateration, or traversing.

Horizontal control for ordinary work is generally a traverse, although one line may suffice for a small area. Triangulation and trilateration are the most economical basic control for surveys extending over a state or the entire United States. Monuments of the state plane coordinate systems are excellent for all types of work but unfortunately there are not enough of them available in most areas.

Vertical control is provided by two or more bench marks which are in or near the area to be surveyed. A vertical-control net is established by lines of levels starting from, and closing on, bench marks. Elevations may be determined for the traverse hubs, with provision in some cases for marks set nearby and out of the way of construction. A lake surface is a continuous bench mark and should be utilized when possible. Even a gently flowing stream may serve as supplementary control. Barometric leveling is now employed to extend the vertical-control net in rugged terrain.

Topographic details are usually built upon a framework of traverse hubs whose positions and elevations have been established. Any errors in the hub positions or their elevations are reflected in the location of topography. It is advisable, therefore, to run, check, and adjust the traverse and level circuits before topography is taken, rather than to carry on both processes simultaneously. This is particularly true in plane-table work where an error in the elevation or position of an occupied station will displace the plotted locations of cultural features and contours.

The kind of control (triangulation, trilateration, or traverse), and the method selected to obtain topography, govern the speed, cost, and efficiency of a topographic survey.

13–3. Topographic details. Objects to be located in a survey may range from single points to meandering streams and complicated geological formations. The process of tying topographic details to the control net is called *detailing*.

13–4. Methods of locating points in the field. Seven methods used to locate a point P in the field are illustrated in Fig. 13–1. One line (distance) must be known in each of the first four methods. The positions of three points must be known or identifiable on a chart to apply the seventh method, which is called the three-point problem. Known or measured distances are shown in the figure as full lines. The quantities to be measured in the respective diagrams are:

1. Two distances.
2. Two angles.
3. One angle and the adjacent distance.

4. One angle and the opposite distance.
5. One distance and a right-angle offset.
6. The intersection of string lines from straddle hubs.
7. Two angles at the point to be located.

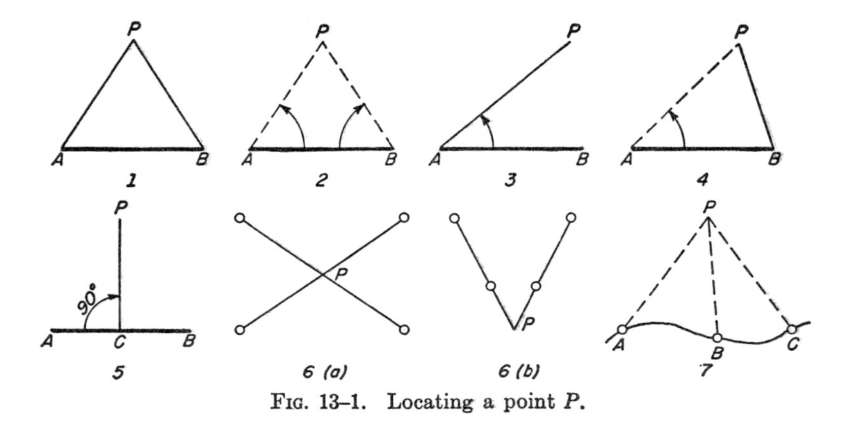

FIG. 13–1. Locating a point *P*.

Straddle hubs are generally used as references for important points which might be disturbed or lost, rather than to locate topography. An experienced party chief employs whichever method is appropriate in a given situation; he must consider both field and office (computation and map) requirements.

13–5. Location of lines. Most objects can be located by considering them to be composed of straight lines, each line being determined by two points. Irregular or curved lines may be assumed straight between points sufficiently close together. Thus detailing becomes a process of locating points. Two examples will illustrate the measurements made to locate straight and curved lines.

In Fig. 13–2, the house *abcdea* is to be tied to traverse line *AB*. The building is so shaped that only two main corners, such as *a* and *e*, need be found. Corner *a* could be located by any of the first five methods of Fig. 13–1 but an angle and a distance are used as shown. It is good practice to locate a third corner, if possible, to provide a check. All sides of the house are measured by tape and the lengths are recorded on a sketch. Location of the barn by measurements from the house is satisfactory if only a tape is available, but two angles and distances from the traverse hubs normally provide a

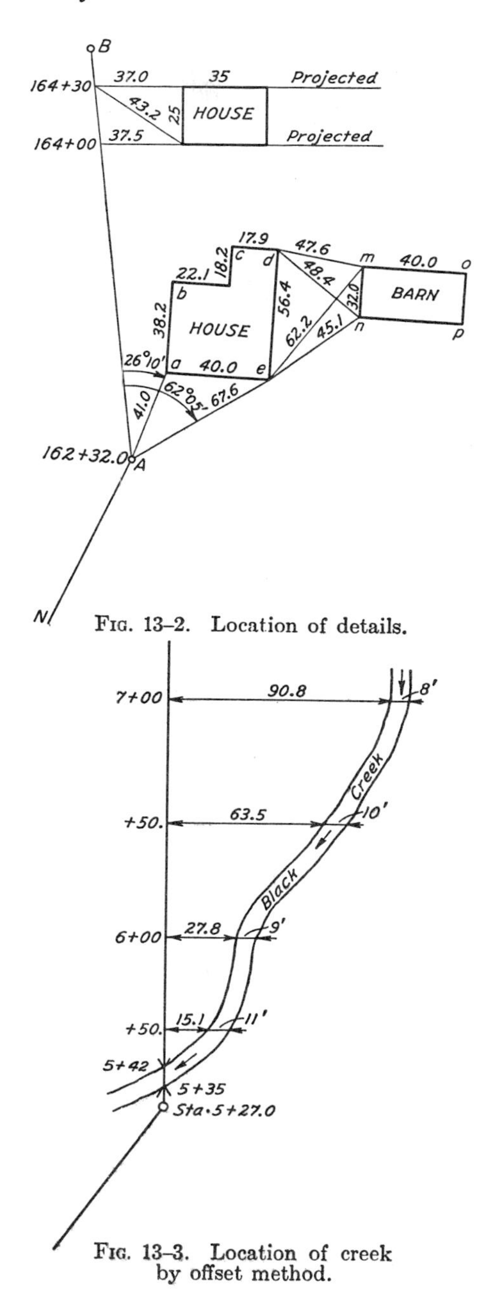

FIG. 13-2. Location of details.

FIG. 13-3. Location of creek by offset method.

quicker and independent result. Stations projected on the center-line traverse from the second house are practical in route surveys, but again angles and distances are a strong alternate.

Since a transit is used to measure the angle, the distances to *a* and *e* could be measured by tape or stadia. All the measurements may be shown on the sketch, but generally the angles and the distances from the traverse points are tabulated on the left page of the field book to avoid crowding the sketch. The topographic points are then identified by consecutive numbers, rather than letters, as on Plate A–10.

The location of a crooked stream by using the offset method is shown in Fig. 13–3. At intervals along the traverse line, offsets to the edges of the creek are measured. The offsets can be taken at regular intervals or spaced at distances which will permit the line to be considered straight between the successive offsets.

13–6. Location of lines from a single point. The single-point method may be used to locate lines of a closed figure, such as the boundaries of a field. A point *O* is chosen from which all corners can be seen, as in Fig. 13–4a. The direction to each corner is found by measuring all the central angles, or by azimuths from point *O*. The lengths of all radiating lines, such as *OA* and *OB*, are taped or read by stadia, and the sides of the field are computed by trigonometry since two sides and the included angle in each triangle are known. As a check, the coordinates of each corner can be calculated from the lengths and directions of the radiating lines and used to determine the boundary lengths.

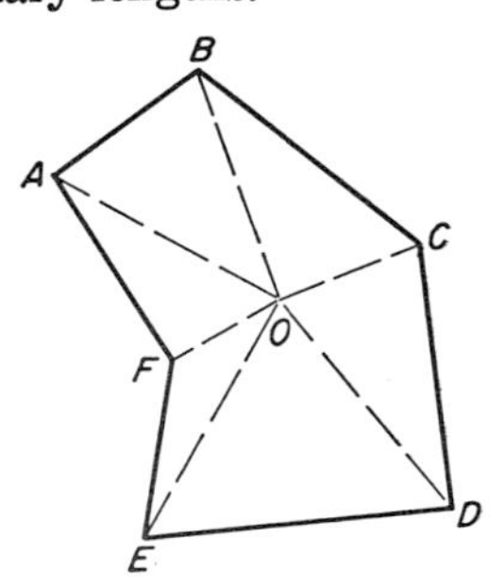

FIG. 13–4a. Location of lines.

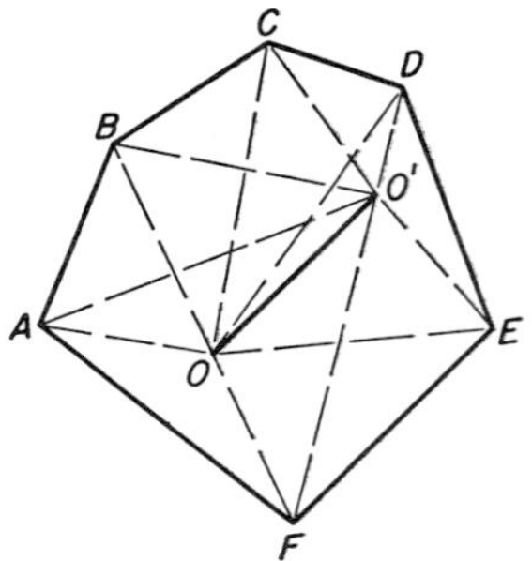

FIG. 13–4b. Location of lines.

A more-rigid solution of the boundary measurements can be obtained by using the method shown in Fig. 13–4b. A line *OO′* is chosen as the base and its length is carefully measured. Angles are taken from each end of the base to all corners, and the lengths of the radiating lines are taped or read by stadia as in the single-point method.

13-7. Contours. The best method of quantitatively representing hills, mountains, depressions, and ground-surface undulations on a two-dimensional sheet of paper is by contours. *A contour is a line connecting points of equal elevation.* Contours may be visible on the ground, as in the case of a lake shore-line, but usually the elevations of only a few points are found and the contours are sketched between these controls.

Contours are shown on maps as the traces of level surfaces of different elevations, as in Fig. 13–5. Thus level surfaces cutting a vertical cone form circular contours, and level surfaces intersect a sloping cone to produce elliptical contours. On uniformly sloping surfaces, such as those in highway cuts, contours are straight lines.

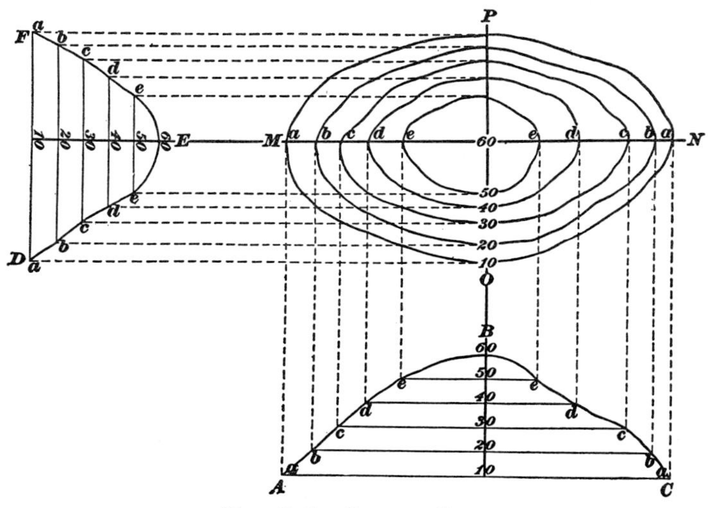

Fig. 13–5. Contour lines.

Most contours are irregular lines like the closed loops for the hill in Fig. 13–5. The vertical distance between the level surfaces forming the contours is called the *contour interval.* For topographic quadrangles at 1:24,000 scale, the United States Geological Survey uses one of the following contour intervals: 5, 10, 20, 40, or 80 ft. The contour interval selected depends on the scale of the map and the amount of relief in the area mapped; where extensive flat areas occur in mountainous regions, supplementary contours at one-half or one-fourth the basic interval are employed (and shown as dashed lines.)

Figure 13–6 is a contour map showing 25-ft contours. *Spot elevations* are usually given for critical points such as peaks, sags, streams, and highway crossings. Sketching ridge, valley and drainage lines (dashed) prior to drawing contours is desirable.

13–8. Characteristics of contours. Certain characteristics of contours are fundamental in their location and plotting.

a. Evenly spaced contours show a uniform slope.

b. The distance between contours indicates the steepness of the slope. Wide spacing denotes flat slopes; close spacing, steep slopes.

c. Contours which increase in elevation represent hills. Those which decrease in elevation portray valleys. A contour forming a closed loop around lower ground is termed a *depression contour*. Contour elevations are shown on the uphill side of the lines, or in breaks, to avoid confusion.

d. Irregular contours signify rough, rugged country. Smooth lines designate gradual slopes and changes.

e. Contour lines tend to parallel each other on uniform slopes.

f. Contours never meet except on a vertical surface such as a wall or cliff. They cannot cross except in the unusual case of a cave or an overhanging shelf. Knife-edge conditions are seldom found in natural formations so that infrequently, if ever, will a single contour of higher elevation exist between two contours of lower elevation, and vice versa.

g. Valleys are usually characterized by V-shaped contours, and ridges by U-shaped contours.

h. The *V*'s formed by contours crossing a stream point upstream.

i. The *U*'s made by contours crossing a ridge line point down the ridge. Contours cross ridge and valley lines at right angles.

j. Contours tend to parallel streams and have an M shape just above stream junctions.

k. Contours cross curbs and a crowned sloping street in typical U-shaped curves.

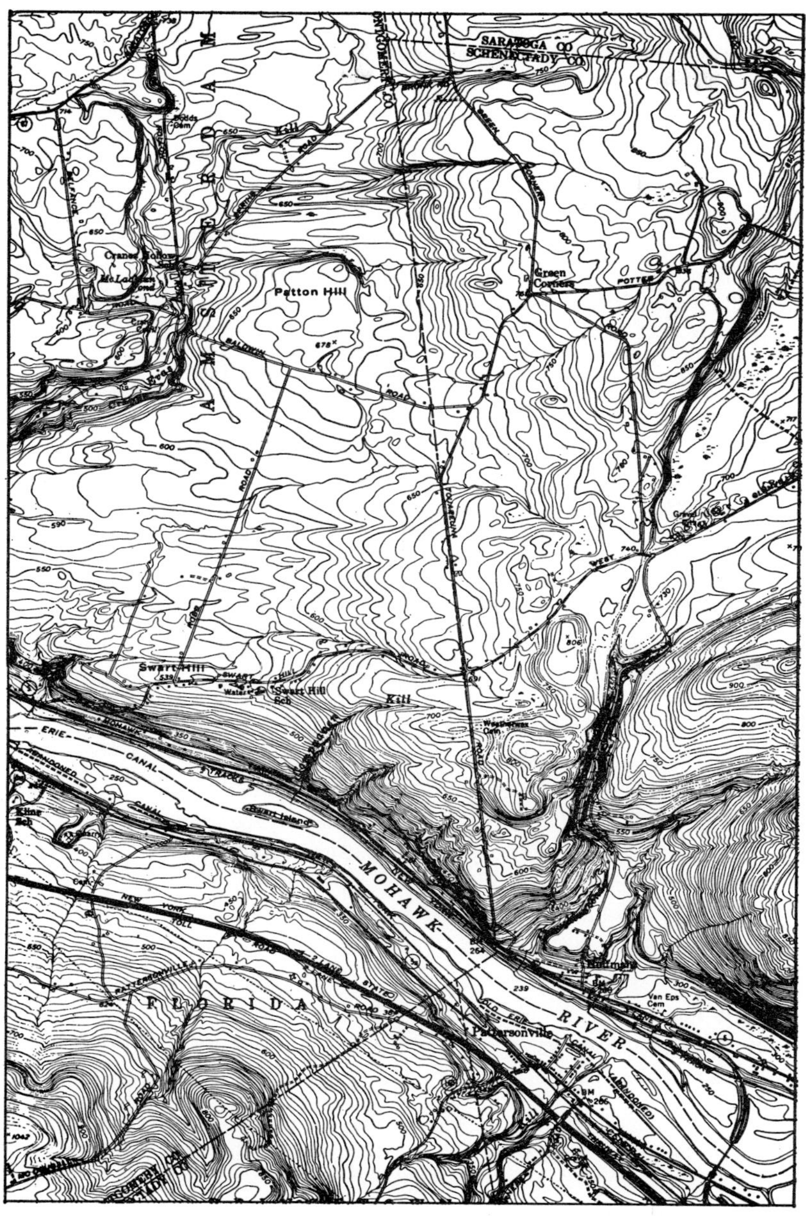

FIG. 13–6. Contour map. (Courtesy U. S. Geographical Survey.)

l. Contours are perpendicular to the direction of maximum slope.
m. Since the earth is a continuous surface, all contours must close upon themselves. This closure may occur within the area being mapped, but often happens outside the area and therefore will not appear on the map sheet.

Keeping these principles in mind will make it easy to visualize contours when looking at an area, and will prevent serious mistakes in sketching. Numerous points may be necessary to locate a contour in certain types of terrain. For example, in the unusual case of a level field which is at the contour elevation, exact location of a single contour would be time-consuming, or perhaps impossible.

13–9. Methods of obtaining topography. Location of topographic details is usually accomplished by one of the following methods:

a. Transit-tape.
b. Transit-stadia.
c. Plane table.
d. Coordinate squares (grid).
e. Offsets from a center line (cross sectioning, cross profile).
f. Photogrammetry.

A brief explanation of the use, advantages, and disadvantages of each will be given.

13–10. Transit-tape method. The transit-tape method is the most accurate, but it is also the slowest and is therefore too costly for ordinary work covering a large area. Angles are measured with a transit, and distances are taped from the traverse hubs to the required points. After corners of buildings, bridges, and other objects have been located, their lengths, widths, and projections may be taped and sketched in the field book.

Transit-tape surveys are made in cities where distances must be accurately known and recorded, even though a normal-scale map cannot show this precision. In topographic mapping the transit-tape method is usually reserved for traverse lines, and the details are taken by some other system.

13–11. Transit-stadia method. The transit-stadia method is rapid, and sufficiently accurate for most topographic surveys. Stadia distances, azimuths, and vertical angles are read for lines radiating from the transit to the required points.

Contours may be found by either the *direct method* (*trace-contour method*) or the *indirect method* (*controlling-point method*). In the direct method, the rod reading (foresight) which must be subtracted from the h.i. to give the contour elevation is determined. The rodman then selects trial points which he believes will give this minus sight, and is directed uphill or downhill by the instrumentman until the required rod reading is actually secured (within 0.1 to 0.5 ft, the allowable discrepancy depending upon the terrain and the required accuracy).

For example, in Fig. 13–7 if the transit is set up at point *A*, whose elevation is 674.3 ft, and if the h.i. is 4.9 ft, the elevation of the line of sight is 679.2 ft. If 5-ft contours are being located, a reading of 4.2 or 9.2 with the telescope level will place the rod on a contour point; for example in Fig. 13–7, the 9.2-ft rod reading means that the point *X* lies on the 670-ft contour. After one of these points has been located by trial, the distance and azimuth are read and the process is repeated. The work is speeded by using a piece of red cloth perhaps 0.2 ft wide which can be moved up and down on the stadia rod to mark the required reading and eliminate searching for a number on any length sight.

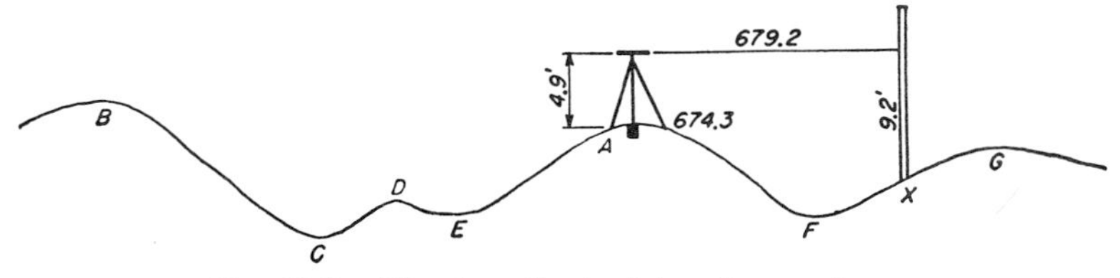

FIG. 13–7. Direct method of locating contours.

The maximum distance between contour points is determined by the terrain and the accuracy required. The tendency for beginners is to take more sights than necessary in ordinary terrain. Contours are sketched between the located points as a part of the drafting-room work, but they may also be drawn in the field book to clarify unusual conditions.

Unless the rod can be read by using the transit as a level, the direct method of locating contours is impractical. Too much time is wasted if the combination of a vertical angle and a stadia distance must be juggled to get a required difference in elevation.

For the indirect method, the stadia rod is set on critical points where there are changes in ground slope, such as *B*, *C*, *D*, *E*, *F*, and *G* in Fig. 13–7. Elevations are obtained with the telescope level whenever possible, to save time in reducing notes and to increase the accuracy. Points are selected at random, or along desired azimuth lines. The rodman moves clockwise to facilitate notekeeping and plotting. Contours are interpolated between the high and low points whose elevations are found.

The direct method is advantageous in gently-rolling country. The indirect method is better in rough, rugged terrain. Plate A–10 in Appendix A shows sample notes for a transit-stadia survey.

13–12. Plane-table method. In the plane-table method an alidade is sighted on a rod held at the point to be located, and the stadia distance and the vertical angle (or Beaman arc) are read. The direction of the line is drawn along the alidade ruler, thus eliminating the need for measuring or recording any horizontal angles. Vertical angles too are avoided, if possible, by using the alidade as a level.

The instrumentman sketches contours by either the direct or indirect method while looking at the area. Since the map is plotted in the field, coverage can be checked by observation. Plane-table usage is discussed in Chapter 15.

13–13. Coordinate squares. The method of coordinate squares (grid method) is better adapted to locating contours than culture, but can be used for both. The area to be surveyed is staked in squares 10, 20, 50, or 100 ft on a side, the size depending upon the terrain and the accuracy required. A transit may be used to lay out lines at right angles to each other, such as *AD* and *D3* in Fig. 13–8. Grid lengths are marked on these lines and the other corners are staked by intersections of taped lines. Corners are identified by the number and letter of the intersecting lines. If a transit is not available, all layout work can be done with a tape.

To obtain elevations of the corners, the engineer's level is set up in the middle of the area, or in a position from which level sights can be taken on each point. Contours are interpolated between the corner elevations (along the sides of the blocks but sometimes diagonally also) by estimation, or by calculated proportional distances. Except for plotting the contours, this is the same procedure as that used in the borrow-pit problem.

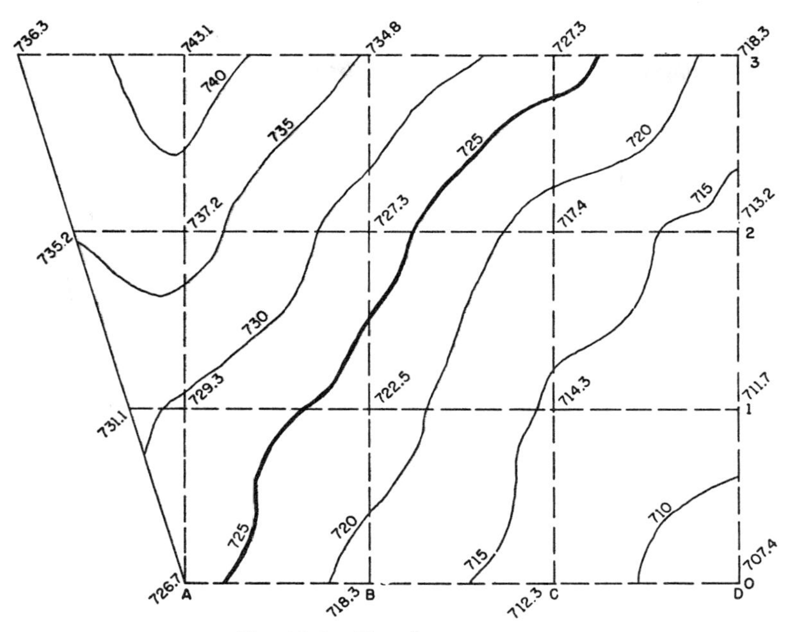

FIG. 13–8. Coordinate squares.

13–14. Offsets from the center line. After the center line for a route survey has been run by transit and tape, a profile is taken to get elevations at the regular stations and at critical points. Details such as fences and buildings are then located by right-angle offsets. The right angle can be measured by a *pentagonal prism* of the type shown in Fig. 13–9, or estimated by bringing the palms of the hands

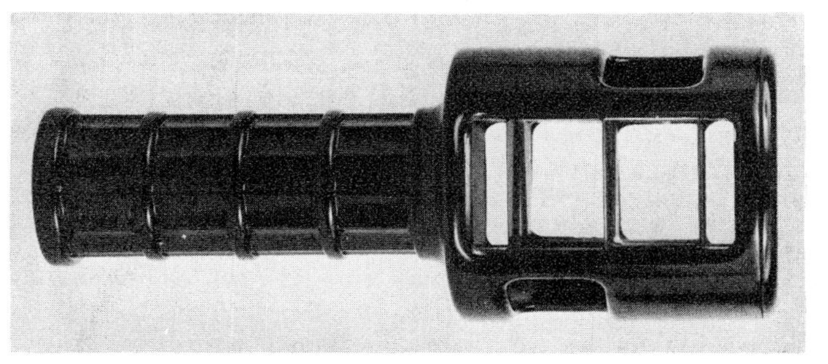

FIG. 13–9. Double pentagonal prism (on side). (Courtesy of Kern Instruments, Inc.)

together with arms outstretched in front of the body (eyes should be closed to prevent unconsciously pointing to a prominent object near the perpendicular line).

Cross-sectioning consists in taking a vertical section of the surface of the ground at right angles to the center line of a route survey. In effect, it is profiling normal to the center line. Rod readings are secured at all breaks in the ground surface and are recorded with their respective distances out. Plate A–11 in Appendix A is a sample set of cross-section notes.

Cross sections are usually plotted on special paper. This paper is most commonly ruled in 1-in. squares divided decimally with lighter lines 0.1 in. apart horizontally and vertically. The cross-section notes—when plotted along with the design template (base width and side slopes cut in a piece of plastic) for a proposed highway or canal—outline the areas of cuts and fills, which are readily measured by planimetering. Excavation quantities can be calculated from the areas of the cross sections and their known distances apart.

In recent years, electronic computers have generally eliminated the need to plot cross sections and planimeter the areas. Instead, the cross-section field notes and design-template values are read into the computer together with an appropriate program. The computer calculates and lists the cut and fill areas as well as the excavation or fill volumes. Cross sections with superimposed design templates can be plotted automatically by the electronic computer if desired.

On some surveys, contour points are located along with any decided changes in slope of the ground. For example, if 5-ft contours are being delineated, a typical set of notes on the right-hand page of a field book would be in the following form:

L				C.L.	R			
$\frac{85}{63}$	$\frac{80}{45}$	$\frac{75}{28}$	$\frac{70}{9}$	72.4	$\frac{75}{12}$	$\frac{80}{27}$	$\frac{85}{38}$	$\frac{90}{56}$

The numerator of a fraction represents the contour elevation, and the denominator is the distance out from the center line.

The contour points can be plotted on cross-section paper to obtain the equivalent of a set of cross-section notes, or the contours might be drawn on the plan view of a proposed highway.

Readings are taken with an engineer's level or with a hand level.

13–15. Photogrammetry. Plotting topography from measurements on aerial photographs is one phase of photogrammetry. Topographic maps of the United States Geological Survey are now prepared almost exclusively by this method which is particularly advantageous for large and rugged areas.

A separate chapter (Chapter 18) is devoted to the fundamental principles of photogrammetry.

13–16. Selection of field method. Selection of the field method to be employed on any topographic survey depends upon the following considerations:

a. Purpose of the survey.
b. Map use (accuracy required).
c. Map scale.
d. Contour interval.
e. Size and type of area involved.
f. Cost.
g. Equipment and time available.
h. Experience of the personnel.

Items a, b, c, d, and e are interdependent. The cost will be a minimum if the most suitable method is chosen for a project. On large-scale work, personnel cost rather than equipment investment will govern (except perhaps in photogrammetric offices). A private surveyor making a topographic survey of 50 or 100 acres, however, may be governed in his choice of method by the equipment he owns.

Special training is necessary before the average surveyor can do photogrammetric work. Likewise, relatively few men have had enough planetable experience to become efficient in its operation.

13–17. Specifications for topographic surveys. Specifications for topographic surveys and maps of large areas generally state the maximum errors permitted in horizontal positions and vertical elevations. Typical specification clauses might be as follows:

Horizontal accuracy: For maps to scales larger than 1:20,000, not more than 10 per cent of the points tested shall be in error by more than $\frac{1}{30}$ in. For maps to scales smaller than 1:20,000, the limit of error is $\frac{1}{50}$ in., which represents 40 ft. These limits of accuracy shall apply in all cases to positions of well-defined points only, such as monuments, bench marks, highway intersections, and corners of large buildings.

Vertical accuracy: Not more than 10 per cent of the elevations tested shall be in error by more than one-half the contour interval.

The accuracy of any map may be tested by comparing the positions of points whose locations or elevations are shown upon it with corresponding positions as determined by surveys of a higher accuracy.

Published maps meeting certain accuracy requirements may have noted in their legends that "This map complies with the national standard map-accuracy requirements."

The plotted horizontal positions of objects are checked by an independent traverse, triangulation, or trilateration run to points selected by the person or organization for whom the survey is made. A profile run from any point, in any arbitrary direction, is compared with one made from the plotted contours. Thus both field work and map drafting are checked.

13–18. Sources of error in topographic surveys. Sources of error in topographic surveys include the following:

a. Control not established, checked, and adjusted before topography is taken.
b. Control points too far apart.
c. Control points poorly selected for proper coverage of area.
d. Poor selection of points for contour delineation.

13–19. Mistakes. Typical mistakes in topographic surveys include the following:

a. Improper selection of contour interval.
b. Unsatisfactory equipment or field method for the particular survey and for the terrain conditions.
c. Insufficient horizontal and vertical control of suitable precision.
d. Too few contour points taken.
e. Omission of some topographic details.

PROBLEMS

13–1. List six examples of topographical details classified as "culture."

13–2. Tabulate, in order of descending precision, the methods and/or equipment for establishing horizontal control for (a) large-area topographic surveys and (b) small-area topographic surveys. Do the same for vertical control.

13–3. In a topographic survey, what advantages does a lake have as a bench mark?

13–4. Describe the best method for determining the water-surface elevation of a lake disturbed by light winds.

13–5. Sketch and name the seven field methods for locating points, and give an example of the best application of each.

13–6. Prepare a set of field notes to locate the buildings and walks of Fig. 14–7, from the traverse shown, by the most appropriate method or methods.

13–7. Contours with a 5-ft interval are spaced $\frac{3}{8}$ in. apart on a map having a scale of 1 in. = 200 ft. What is the average slope of the ground between the contours?

13–8. On a map to a scale of 1 in. = 400 ft, how far apart would 5-ft contours be on a uniform slope (grade) of 6 per cent?

13–9. List the relative advantages and disadvantages of each of the following methods of determining the locations and elevations of ground points: (a) grid; (b) controlling-point; (c) trace-contour; (d) cross-profile.

13–10. Give examples and sketch terrain conditions for which the best method of locating contours would be (a) the grid method, (b) the direct method, and (c) the indirect method.

13–11. Prepare a set of field notes for locating the topographic details shown in Fig. 13–2. Scale the distances and angles. Use different methods from those shown, including each of the first five methods of Sec. 13–4.

13–12. If the locations of contours for an area are to be based upon barometric readings, which method of location is best and why?

13–13. Draw the 2-ft contours for problem 5–41.

13–14. Draw the 1-ft contours for problem 5–42.

13–15. Draw the 1-ft contours for problem 12–16.

13–16. Draw the 1-ft contours for problem 12–18.

13–17. Draw 5-ft contours for problem 12–35.

13–18. Draw 5-ft contours for problem 12–36.

13–19. Explain in detail how to test the accuracy of the map of Fig. 13–6.

13–20. List and give an example of each of the pertinent factors to be considered in selecting the proper scale for a particular transit-stadia topographic-mapping project.

13–21. List the reasons for completing up-to-date topographic maps like that of Fig. 13–6 for the entire United States immediately.

13–22. In outline form, describe a method of making a topographic survey of a lot 100 ft by 120 ft having a maximum difference of elevation of about 10 ft. Assume that only a tape is available. Improvise your own equipment.

13–23. Sketch a contour crossing a 60-ft wide street which is on an 8 per cent grade, has a parabolic crown 1 ft high at the center line, and 6-in. deep curbs.

13–24. List an additional property of contours not given in Sec. 13–8.

13–25. Draw a profile of the terrain alongside and parallel with Ross Road from Swarthill to Baldwin Road for the contours in Fig. 13–6.

chapter 14

Mapping

14-1. General. There is an increasing use of maps by engineers and the general public. All engineering construction projects require maps or site plans. Industries searching for suitable locations for new plants must have maps showing terrain conditions. If maps are not available, the best location may be passed over or time lost while a survey is run.

The military services have always depended upon a steady flow of up-to-date maps. During World War II, the Army Map Service prepared more than 40,000 maps of all types covering approximately 400,000 square miles of the earth's surface; 500,000,000 copies of maps were printed. The Normandy Invasion alone required 70,000,000 copies of 3000 different maps. In the first four weeks of the Korean War, the Army Map Service and the Far East Command printed and distributed 10,000,000 copies of maps—more than printed during all of World War I—and by July 1953, over 70,000,000 copies were distributed to the United States, South Korean, and United Nations military services. While the requirements have greatly diminished, the flow of maps continues. Because of the size and dispersion of military forces, the weather, and terrain in Vietnam, the Army Map Service had dispensed more than 150,000,000 map copies by the end of 1967.

An engineer must know how to make maps, to be better able to interpret them. Maps for laymen must be clear and easily understood if they are to give maximum service.

Most maps are made for some definite reason; for example, to show topography, boundaries of property, precise location of traverse points, routes of highways or railroads, soil-erosion control areas, forested and reforestation areas, mineral lands, and other special features.

14-2. Mapping agencies. Maps are prepared by private surveyors, industries, cities, counties, states, and several agencies of the Federal government. Unfortunately, the results of some surveys have been squandered by failure to coordinate them with earlier or later work. Valuable maps are gathering dust, unknown and unavailable to prospective users. A start has been made to improve this deplorable situation by setting up depositories in some cities and states, where every obtainable map of the area will be filed for the use of interested persons. The Map Information Office of the Geological Survey provides information useful to surveyors, engineers, cartographers, and other technical map users, as well as the general public. This office, which is a central source of information on surveys and maps, provides comprehensive data on topographic maps, aerial photography, and geodetic control surveys, and also basic data needed for engineering and construction programs, such as irrigation projects, highways, railroad location, urban and rural development, transmission and pipe lines, airports, and the location of radio and television facilities.

Some of the agencies producing maps and charts on a national scale are the Geological Survey, Navy Oceanographic Office, and the Coast and Geodetic Survey.

The Geological Survey began publishing topographic maps in 1886 as an aid to scientific studies. Standard sheets cover 7½- or 15-min quadrangles and show the works of man in black, contours in brown, water features in blue, and woodland areas in green. The price of individual maps is 50 cents and over 6,000,000 a year are sold. About 75 per cent of the United States is covered. An index map giving the status of topographic mapping in the United States and its territories and possessions is available free of charge from the Geological Survey. Other index maps showing the status of aerial photography, and aerial mosaics in the United States are published by the same agency.

The Coast and Geodetic Survey provides basic position and elevation control for mapping, and nautical and aeronautical charts which are fundamental tools in maintaining the nation's air and sea transportation systems, as well as tide and current surveys, seismological and geomagnetic surveys.

The Army Map Service devotes a substantial measure of its peacetime efforts to the development of mapping techniques, the improvement of mapping specifications, and the accumulation, evaluation, and maintenance of existing maps. Its primary effort, however, is the compilation of new mapping on a world-wide basis to provide in peace what would be needed in war.

14-3. Map drafting. Map drafting may be divided into four parts: (a) plotting the traverse; (b) plotting the details; (c) drawing the topography and special data; and (d) finishing the map.

The discussion in this chapter will be confined to large-scale maps primarily. Map scales are generally classified as follows:

Large scale, 1 in. = 100 ft or less

Intermediate scale, 1 in. = 100 ft to 1000 ft

Small scale, 1 in. = 1000 ft or more

Although measurements to 0.01 or 0.001 ft can be made in the field, the accuracy of maps can only be as good as drafting procedures permit.

14-4. Plotting the traverse. Measurements of lengths and angles are used to plot a traverse.

Distances are plotted from the field data to a selected scale, such as 1 in. equals 10, 20, 40, 50, or 100 ft. An engineer's scale is used but is supplemented by steel scales and dividers for plotting control points accurately to perhaps $\frac{1}{50}$ or $\frac{1}{100}$ in.

The choice of a scale depends upon the purpose, size, and required precision of the finished map. The dimensions of a standard sheet, the type and number of topographic symbols, and the need for scaling distances from it are some of the considerations involved.

Map scales are given in three ways: (a) by a *ratio* or *representative fraction*, such as 1/2000; (b) by an *equivalence*, such as 1 in. = 200 ft; and (c) *graphically*. Two graphical scales at right angles to each other and in diagonally opposite corners of a map sheet permit accurate measurements to be made even though the paper changes dimensions.

A map drawn to any scale can be enlarged or reduced by means of a *pantograph*, by an opaque projector, or photographically.

Angles for a traverse are plotted by (a) coordinates, (b) tangents, (c) chords, (d) protractors, and (e) drafting machines.

a) *Coordinate method.* The most accurate method of plotting a traverse is by coordinates. If the latitudes and departures of traverse lines are available from area computations, the total X-coordinate and the total Y-coordinate for each angle point are readily determined and plotted from an origin through one corner of the traverse. For extensive surveys, the map sheet is first laid out accurately in a grid pattern with unit squares of convenient size such as 100, 400, 500, or 1000 ft. The unit squares are checked by measuring the diagonals.

Any errors in plotting the coordinate points are detected by comparing the scaled distance and bearing of each line with the length and direction measured in the field or computed.

Small circles, $\frac{1}{8}$ in. or less in diameter, are used to mark the traverse hubs. Lines should extend only to the circumference of these circles to preserve the exact center point. On most topographic maps the hub locations and traverse lines are omitted from the finished drawing. If shown, they may be drawn in red ink to make them less prominent on a print.

Bearings and lengths of traverse lines are always given unless the coordinates are tabulated. They are placed on the lines so that they can be read easily when the user looks at the sheet from the bottom or from the right-hand side. The bearings should be forward bearings and continuous around the traverse. When the bearing is read from left to right, but actually runs from right to left, an arrow is used to show the correct direction of bearing, as in Fig. 14–7.

b) *Tangent method.* To lay off an angle by the tangent method, a convenient distance is measured along the reference line to serve as a base. Thus, in Fig. 14–1, to plot a 12° 14′ deflection angle at

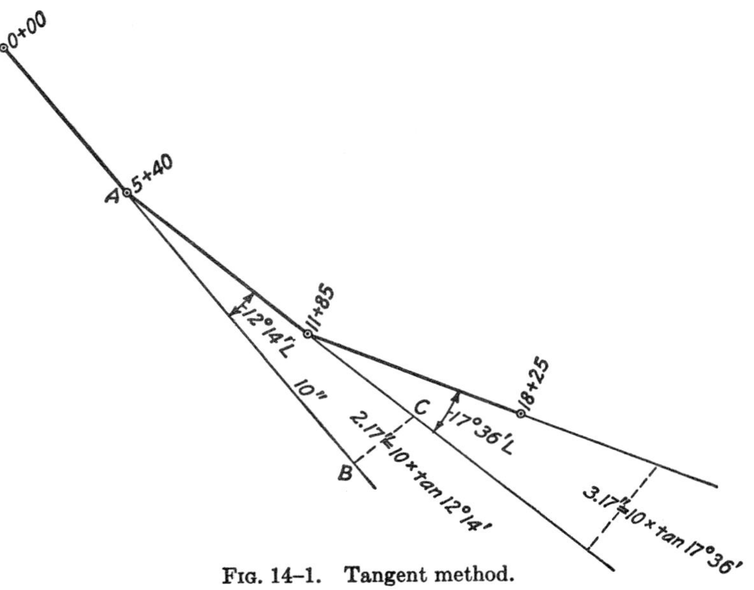

FIG. 14–1. Tangent method.

point A, a length AB equal to 10 in. is first marked on the prolongation of the back line. A perpendicular with a length equal to the distance AB times the natural tangent of 12° 14′ (equal to 2.17 in.) is erected at B to locate point C. The line connecting A and C makes the desired angle with AB. Any length of base can be used, but a distance of 10 or 100 units requires only the movement of the decimal place in the natural tangent taken from a set of tables.

The tangent method is employed extensively for plotting deflection angles. It is not so advantageous for large direct angles.

c) *Chord method.* To lay off an angle by the chord method, as indicated in Fig. 14–2 a convenient base length of 10 units is first marked on one side, BA, giving point D. With the vertex B as the center and a radius of 10 units, an arc is swung. Then with point D as

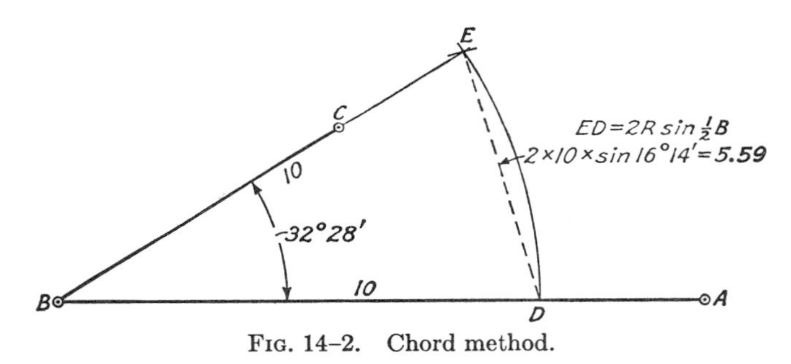

FIG. 14–2. Chord method.

the center, and a radius equal to the chord for the desired angle, another arc is swung. The intersection of the two arcs locates E. The line connecting points B and E forms the other side of the angle.

The chord for a desired angle may be found by multiplying twice the distance BD by the sine of half the angle. Thus in Fig. 14–2 the chord is $2 \times 10 \times \sin 16° 14'$. By making the base length 10, 50, or 100 units to some scale, the chord is readily calculated by using natural sines taken from a trigonometric table.

d) *Protractor method.* A protractor is a device made of paper, plastic, or metal, cut in a full circle or a semicircle, with angle graduations along the circumference. A fine point identifies the center of the circle. The protractor is centered at the vertex of the angle, with the zero line along one side, and the proper angle point is marked along the edge.

Protractors are available in sizes having radii from 2 to 8 in. A metal circle with a movable arm extending beyond the edge, as illustrated in Fig. 14–3, is often used. The arm *a* rotates about the center *b* and has a vernier *c* reading to minutes. A similar arrangement on drafting machines makes plotting fast and accurate.

Protractors are universally used for the plotting of details but are not suitable for high-precision work on traverses or control.

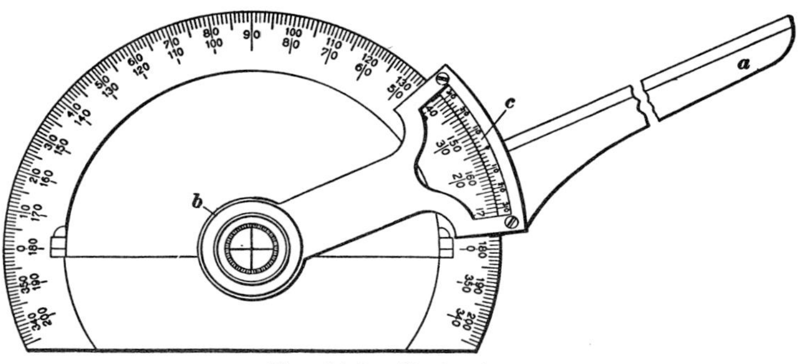

FIG. 14–3. Protractor.

14–5. Advantages and disadvantages of the different methods. The computations of coordinates for hubs in closed traverses are readily checked. Any large errors in plotting are found by scaling. Correction usually involves replotting only one point. For example, if the scaled lengths of lines *CD* and *DE* of a plotted traverse are not equal to their field measurements, point *D* is off. The independence of each point in the plotting procedure is a definite advantage.

The tangent method is accurate and probably the best way to lay out a single angle precisely. In plotting a traverse, however, an error in any angle or distance is carried through the remainder of the traverse. If a traverse fails to close, each angle and line must be checked. If the closure is due to drafting rather than the field work, it may be necessary to rotate each line, except the first, slightly and progressively.

The chord method has the disadvantage that errors in any traverse line are passed along to succeeding courses. Erection of a perpendicular is eliminated but determination of the chord lengths is somewhat more laborious than finding perpendicular offsets.

Layout of angles by protractor is the fastest but least accurate of the four methods.

14–6. Plotting details. Boundary corners and important points upon which construction work may depend are plotted by coordinates or the tangent method, but the protractor is used for most details. Orientation of the protractor zero line is by meridian for details located by azimuths or bearings, and by the backsight line if direct angles were measured. Angles are marked along the edge of the protractor, and distances are scaled from the vertex to plot the detail. To avoid obliterating the vertex, and to reduce the erasures, the engineer's scale can be laid along the line through the plotted point and the distance marked. An alternative method consists of drawing a short line through the point at the approximate distance by estimation, and then marking along it.

For accuracy, the distance to the detail should be less than the radius of the protractor unless the type with an extended arm is used.

14–7. Plotting contours. Points to be used in plotting contours are located in the same manner as those for details. Contours found by the direct method are sketched through the points. Interpolation between the plotted points is necessary for the indirect method.

Interpolation to find contour points between points of known elevation can be done in several ways:

a. Estimating.
b. Scaling the distance between points of known elevation and locating the contour points by proportion.
c. Using a rubber band graduated to some scale and stretching it to make convenient marks fall on the known elevations. Special devices known as *variable scales* are available which contain a graduated spring. The spring may be stretched to make convenient marks fall on the known elevations.
d. Using a triangle and scale, as indicated in Fig. 14–4. To interpolate for the 420-ft contour between point *A* at elevation 415.2 and point *B* at elevation 423.6, first the 152 mark on one of the engineer's scales is set opposite *A*. Then with one side of the triangle against the scale and the 90° corner at 236, the scale and the triangle are pivoted together around *A* until the perpendicular edge of the triangle passes through point *B*. The triangle is then slid to the 200 mark and a dash drawn to intersect the line from *A* to *B*. This is the interpolated contour point *P*.
e. Using a converging-line device, such as that in Fig. 14–5, which

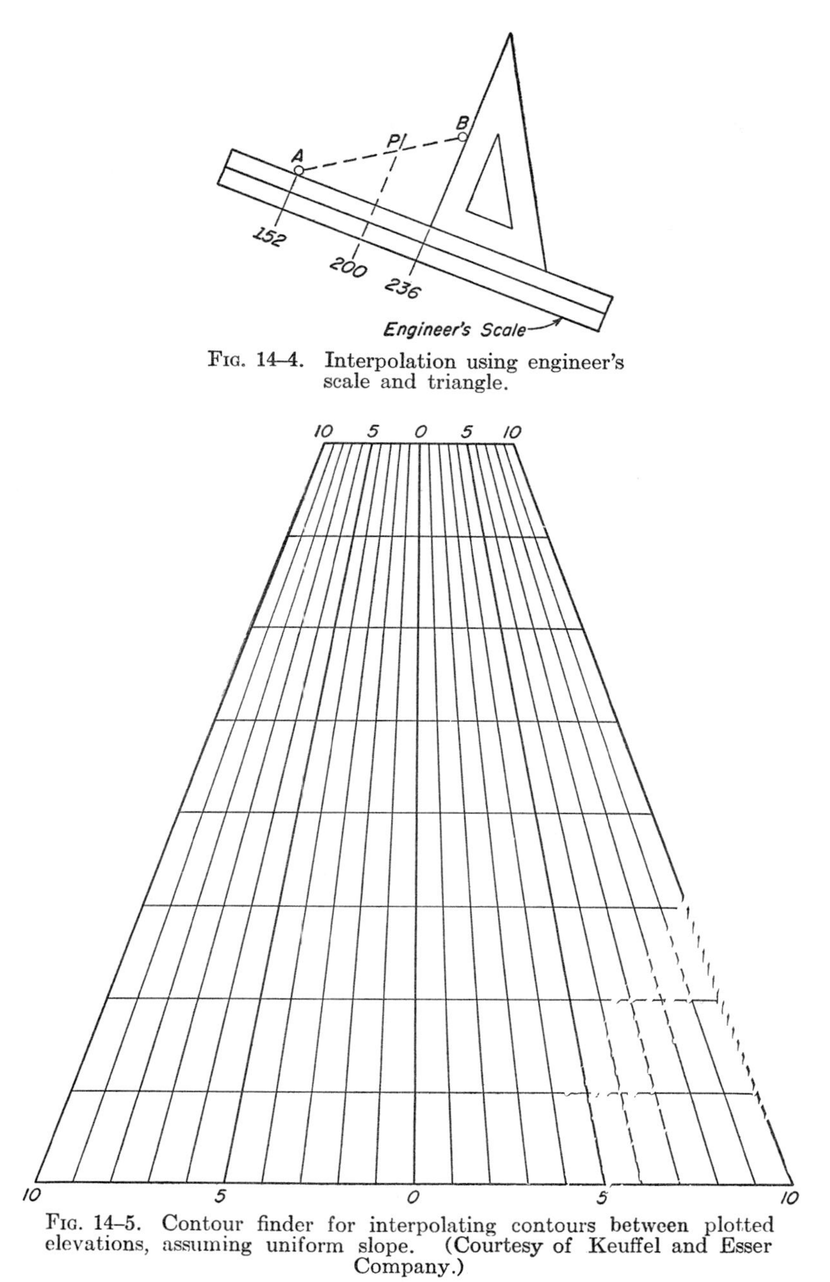

Fig. 14–4. Interpolation using engineer's scale and triangle.

Fig. 14–5. Contour finder for interpolating contours between plotted elevations, assuming uniform slope. (Courtesy of Keuffel and Esser Company.)

can be pivoted and adjusted to fit the difference in elevation between any two points. The procedure is illustrated by the following example: Assume two plotted points have elevations of 17.6 and 25.9. Draw a straight line between these points. Determine the difference between the given elevations to the nearest whole number. Since 25.9 − 17.6 = 8.3, use 8. Place the contour finder over the map so that its horizontal lines are parallel to the line drawn and so that 8 intervals nearly fill the space between the plotted points. Adjust the finder until the 17.6 point lies 0.6 into the first interval and the 25.9 point lies 0.9 beyond the eighth interval. With a needle point, punch through the contour finder at the contour points desired.

The standard color for inked contour lines is sepia. Every fifth or tenth line is heavier. The contour interval depends upon the terrain, the scale, and the purpose of the map.

Electronic computers can be used to plot contours automatically. The required input to the computer includes the control traverse data, topographic detail information, map scale, contour interval, and the appropriate contour program.

Hills, mountains, depressions, and ground-surface undulations can also be represented by hachures and other forms of hill-shading. Hachures are short lines drawn in the direction of the slope. They are heavy and closely spaced for steep slopes, and fine and widely separated for gentle slopes. They show the general shape of the ground rather than actual elevations and thus are not suitable for most engineering work.

14–8. Topographic symbols. Standard symbols are used to represent special topographic features, thereby making it possible to show many details on a single sheet. Figure 14–6 gives a few of the hundreds of symbols employed on topographic maps. Considerable practice is required to draw these symbols well at a suitable scale. Before placing symbols on a map, such things as buildings, roads, and boundary lines, are first plotted and inked. The symbols are then drawn, or cut from standard sheets having an adhesive on the back and pasted on the map. A fully detailed map with coloring and shading is a work of art.

14–9. Locating the traverse on a map sheet. The appearance of a finished map has considerable bearing on its acceptability and value.

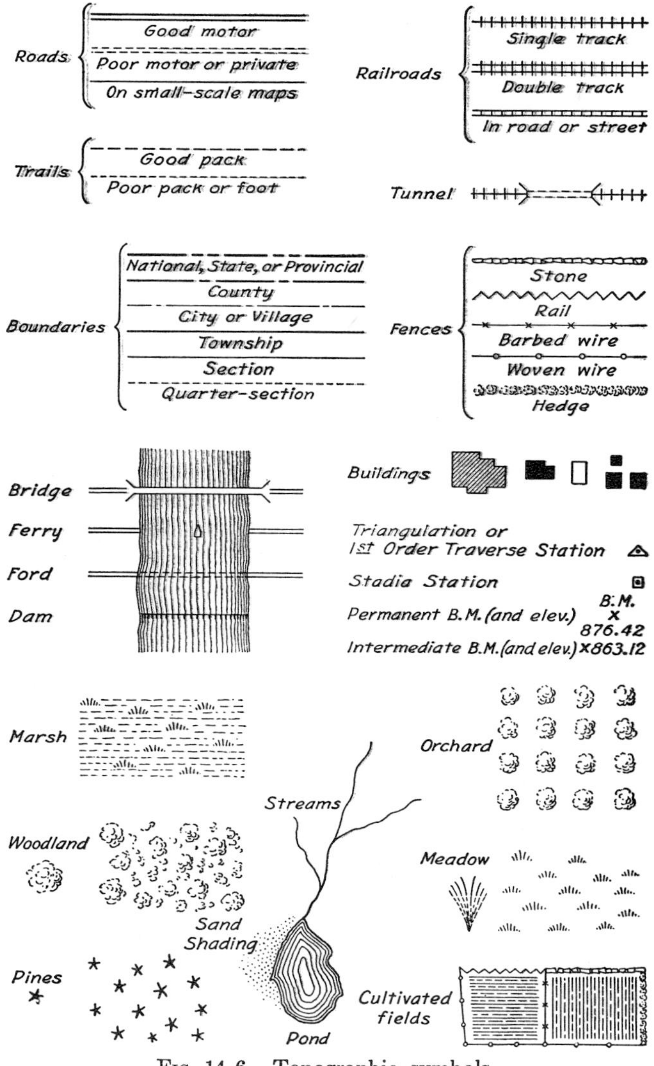

Fig. 14–6. Topographic symbols.

A map which is poorly arranged, carelessly lettered, and unfinished-looking does not inspire confidence in its accuracy. A border line somewhat heavier than all other lines improves the appearance of the sheet.

The first step in map arrangement is to determine the position of the traverse and topography which will properly balance the sheet. Figure 14–7 shows a traverse (no topography outside the traverse) suitably placed. Before any plotting is done, the proper scale for a sheet of given size must be determined.

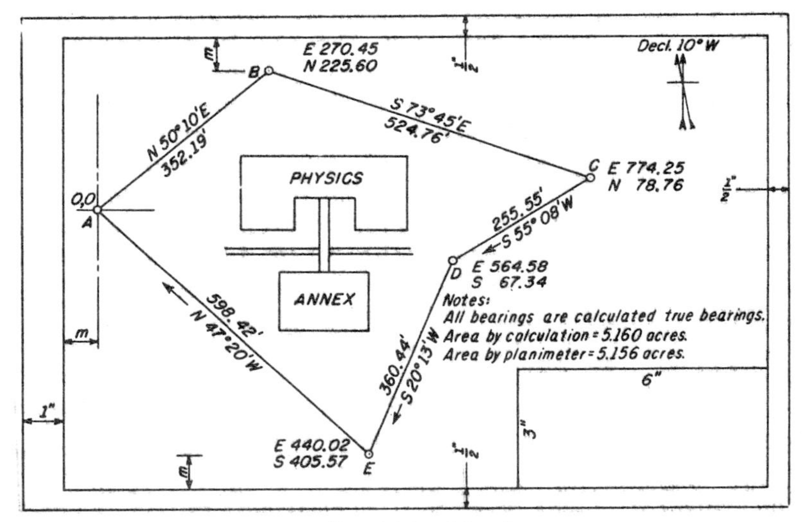

Fig. 14–7. Map layout.

Assume that an 18-in. by 24-in. sheet with a 1-in. border on the left (for possible binding), and a ½-in. border on the other three sides, is to be used. If the most-westerly station *A* has been selected as the origin of coordinates, then divide the total departure to the most-easterly point *C* by the number of inches available for plotting in the east-west direction. The maximum scale possible in Fig. 14–7 is 774.25 divided by 22.5, or 1 in. = 34 ft. The nearest standard scale which will fit is 1 in. = 40 ft.

This scale must be checked in the *Y*-direction by dividing the total *Y*-coordinate of 225.60 + 405.57 = 631.17 ft by 40 ft, giving 15.8 in. required in the north-south direction. Since 16½ in. is available, the scale of 1 in. = 40 ft is satisfactory, although 1 in. = 50 ft provides a better border margin.

In Fig. 14–7 the traverse is centered on the sheet in the Y-direction by making each distance m equal to $\frac{1}{2}[17 - (631.17/40)]$, or 0.61 in. The weights of the title, notes, and arrow compensate for the traverse being to the left of the center of the sheet.

If topography is to be plotted outside the traverse, the maximum north-south and east-west distances to topographic features must be added to the traverse coordinates before computations are made for the scale and centering distances.

On some types of maps it is permissible to skew the meridian with respect to the borders of the sheet in order to better accommodate a traverse relatively long in the NE–SW or NW–SE direction, or to make street lines parallel with the borders. When a traverse with topography is to be plotted by any means other than coordinates, it is advisable to first make a sketch showing controlling features. If this is done on tracing paper, orientation for the best fit and appearance is readily determined by rotating and shifting the tracing.

14–10. Meridian arrow. Every map must display a meridian arrow for orientation purposes. It should preferably be near the top of the sheet, although it may be shifted elsewhere for balance. An arrow cannot be so large, elaborate, or heavily blacked in that it becomes the focal point of the sheet, as was true on maps of fifty years ago.

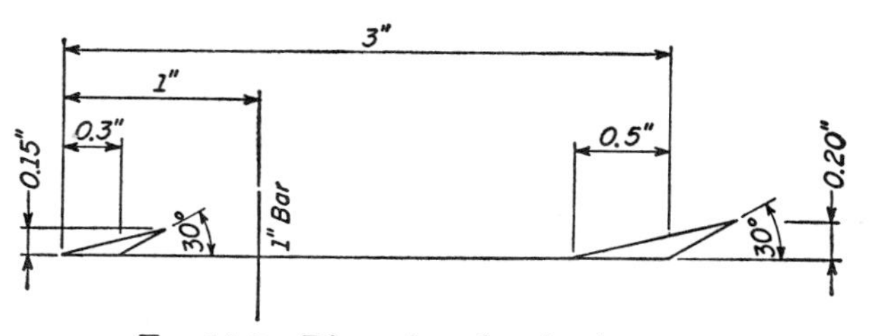

Fig. 14–8. Dimensions for simple arrow.

Either the magnetic or true meridian, or both, may be shown. A true-meridian arrow is identified by a full head and full feather, and a magnetic arrow by a half-head and half-feather. The half-head and half-feather are put on the side away from the true north arrow to avoid bumping it. Figure 14–8 shows dimensions of an arrow that is suitable for routine maps. In practice, an arrow is traced from a sheet of standards, or cut out for pasting on the map.

14–11. Title. The title may be placed wherever it will better balance the sheet, but it is always kept outside the property lines on a boundary-survey sheet. Usually the title occupies the lower right-hand corner, with any pertinent notes just above or to the left of it. A search for a particular map in a bound set or in a loose pile of drawings is facilitated if all titles are in the same location. Since sheets are filed flat, bound on the left border, or hung from the top, the lower right-hand corner is the most convenient position.

The title should state the type of map; the name of the property or project and its owner or user; the location or area; the date completed; the scale; and the name of the surveyor. Additional data may be required on special-purpose maps. Lettering should be simple in style rather than ornate, and should conform in size with the individual map sheet. Emphasis is placed on the most important parts of the title by increasing the height of letters and using upper-case letters.

Perfect symmetry of outline about a vertical center line is necessary, since the eye tends to exaggerate any defection. Also, an appearance of stability is obtained by having a full-width bottom line.

An example of a title and its arrangement for the 18-in. by 24-in. sheet of Fig. 14–7 is shown in Fig. 14–9. The letters in lines 1 and 2

University of Hawaii
Civil Engineering Department
TRANSIT-STADIA SURVEY OF ENGINEERING CAMPUS
Scale 1 inch = 20 feet Date 8 Nov. 1955
Survey by L. Gill, T. Hall, A. Aker Map by J. Jones

Fig. 14–9. Title arrangement.

could be $\frac{1}{4}$ in. high; those in line 3, $\frac{5}{16}$ in. high; and those in the last two lines, $\frac{1}{8}$ in. high.

No part of a map better portrays the artistic ability of a draftsman than a neat, well-arranged title.

14–12. Notes. Notes cover special features pertaining to the individual map, such as the following:

All bearings are true bearings [or calculated, or magnetic].
Area by calculation is *X* acres.
Area by planimeter is *Y* acres.
Legend [explanation of unusual symbols; for example, * represents cooling towers].

Notes must be in a prominent place where they are certain to be seen upon even a cursory examination of the map.

14–13. Paper. Drawing paper for surveying maps should be of excellent quality, take ink without spreading, stand erasures, and not deteriorate with age. Good brown detail paper and high-grade white drawing paper are commonly used. Cloth-backed white paper and aluminum-backed paper are desirable for important work. They are expensive but more resistant to dimensional changes and rough handling.

Many maps are drawn or traced on transparent paper or fine linen so that blueprints can be made from them. An original paper drawing may be finished in pencil for printing, but maps of a permanent nature should be inked.

Blueprinted maps have the disadvantage of changing dimensions because of variations in temperature and humidity, and the printing process itself. Surveys of city lots are normally drawn on small sheets and reproduced by the ozalid method, which does not entail wetting and drying the printing paper.

14–14. Sources of error in mapping. Sources of error in mapping include the following:

a. Not checking (scaling) distances when plotting by coordinates.
b. Plotting by protractor.
c. Using a soft pencil for plotting.
d. Variations in dimensions of map sheet.

14–15. Mistakes. Some of the more common mistakes in mapping include the following:

a. Improper orientation of topographic notes in field and office.
b. Using wrong edge of engineer's scale.
c. Making arrow too large, complex, or black.
d. Omitting the scale or necessary notes.
e. Failing to balance the sheet by making a preliminary sketch.

PROBLEMS

14–1. To keep plotting errors within 5 ft when using an engineer's scale, what is the smallest feasible map scale?

14–2. On a plane-table sheet having a scale of 1/3600, what is the smallest distance that can be plotted with an engineer's scale?

14–3. How would the maps of Figs. 13–2, 13–3, 13–6, 13–8, and 14–7 be classified on the basis of scale?

14–4. What over-all size change in an 18 in. by 24 in. blueprinted map sheet is possible for an increase of 50 deg F and 90 per cent humidity?

14–5. List side by side the various methods of plotting traverses, and show under each its relative advantages and disadvantages.

14–6. Similar to problem 14–5, except for the methods of plotting topographic details.

14–7. Plot 1-ft contours for the data in Plate A–6 of Appendix A.

14–8. Plot 5-ft contours for the data in Plate A–11.

14–9. Plot the topography in the notes of Plate A–10.

14–10. Sketch the conventional symbols for woodland, pines, orchards, ponds, and cleared land.

14–11. Sketch the conventional symbols for a cultivated field, a meadow, a stream with a bridge, a vineyard, and cactus.

14–12. Using an $8\frac{1}{2}$ by 11-in. quadrille sheet, plot in conformance with Sec. 14–9 the traverse of problem 10–9 to a suitable scale. Allow a $\frac{1}{2}$-in. border on the left side and a $\frac{1}{4}$-in. border on the other three sides.

14–13. Similar to problem 14–12, but use the traverse of problem 11–15.

14–14. Similar to problem 14–12, but assume that topography coverage extends about 250 ft to the north side of line *BC* only.

14–15. Similar to problem 14–13, but assume that topographic details are included for a distance out of about 200 ft perpendicular to line *CD* and for about 300 ft perpendicular to line *DA* only.

14–16. Plot the traverse of problem 10–21 by bearings and distances; then check the positions of the corners by coordinates.

14–17. Plot the traverse of problem 12–35 by azimuths and distances; then check the positions of *A, B, C,* and *D* by coordinates.

14–18. Plot the following azimuth traverse by coordinates to a scale of 1 in. = 40 ft: $AB = 320.4$, 340° 12′; $BC = 618.6$, 5° 38′; $CD = 654.3$, 96° 26′; $DE = 200.6$, 174° 34′; $EF = 447.5$, 182° 27′; $FA = 633.7$, 251° 53′.

14–19. In problem 14–18, azimuths and distances to a lake shore were taken from the traverse points as follows: *A*, 276° 00′, 352.2 ft; *A*, 300° 54′, 431.6 ft; *B*, 290° 26′, 216.3 ft.; *B*, 340° 12′, 269.1 ft; *C*, 342° 44′, 241.5 ft; *C*, 298° 06′, 168.8 ft; *C*, 5° 38′, 171.0 ft; *C*, 55° 30′, 190.3 ft; *D*, 354° 36′, 255.7 ft. Plot these points and join them with lines to represent a lake shore. Finish the map, giving proper symbols, lengths and bearings of the fence lines, an arrow, notes, and a title. Show the land along the shoreline changing from marsh to pines to orchard.

14–20. Why are traverse lines inked in red on a survey map? Should the hubs be in red or black? Why?

14–21. If it is necessary to keep plotting errors within 20 ft on a large map, using an engineer's scale, what is the smallest map scale suitable?

14–22. For a survey and student map similar to Fig. 14–7, how close would you expect the computed and planimetered area within the traverse to agree?

14–23. List the important items usually lettered in upper-case letters, and those in lower-case letters, (a) on a topographic map and (b) on a subdivision map of property bordering a river or lake. What is meant by a "stickup" on a map?

14–24. Why do the features shown on a topographic map of a large area differ from those on a map of a small area?

14–25. Is there any difference between a map and a chart? Between a map and an aerial photograph?

14–26. What statements should be in the title of a profile map? Of a property survey map?

14–27. Why are map titles usually placed in the lower right-hand corner of the sheet?

14–28. What is the first step in drawing a surveying map?

14–29. Describe two mapping conditions where it might be desirable to skew the meridian with respect to the border, and one situation where the arrow must be parallel with the border.

14–30. Assume that only part of a building can be shown on a topographic map because of the sheet size and scale being used. Why should the building lines which must be cut off *not* be run into the border line?

14–31. List 6 topographic features that would be classified as relief, and 6 which would be termed culture.

14–32. What would characterize a contour map which "lacks expression"?

14–33. What does T & S mean on a surveying map?

14–34. Which of the U.S. Government agencies produces most of the topographic maps of the United States?

14–35. Why are countours the best method of showing elevations on a map?

chapter **15**

The Plane Table

15-1. General. The plane table is one of the oldest types of surveying instruments. In its modern form a plane-table outfit consists of a tripod; drawing board; an alidade equipped with stadia wires; a stadia rod; and a tape. In the field a sheet of drawing paper or plastic is fastened to the board and a map is made by plotting directions and distances obtained by sighting with the alidade.

Mapping topographic features while they are in full view is advantageous in many types of surveys in civil and mining engineering, forestry, geology, agriculture, and military operations. Geologists use the Brunton compass and plane table almost exclusively in their surveys. A small board, a light tripod, and a peep-sight alidade, collectively known as a *traverse table,* are standard equipment in military mapping. The table is leveled and oriented by movement of the tripod legs.

Although the plane table is considered obsolete by some people, and relatively few private surveyors own or have used this equipment, the plane table is a valuable piece of surveying equipment.

15-2. Description of the plane table. The drawing board of a plane table (usually 24 in. × 31 in.) is carefully made in such a way that it resists warping and other damage from weathering. The upper side is smooth but has some means, such as brass screws, for attaching the paper to the board. At the center of the underside of the board is a socket with threads which fit a head fastened to the top of the tripod.

Two radically different tripod heads, the *Coast and Geodetic Survey type* and the *Johnson head,* are available for leveling and orienting the board. The Coast and Geodetic Survey type has four leveling screws, and a clamp and a tangent screw like those of a transit. Leveling and orientation are accomplished easily.

FIG. 15–1. Johnson-type head.

The Johnson head, Fig. 15–1, has a ball-and-socket arrangement to hold the board in position after leveling, and to prevent its turning in a horizontal direction. Clamp *A* regulates the motion of the table as it moves on the larger ball joint. It is tightened, after the board has been leveled by pressing or lifting on one side. The lower clamp *B* controls the movement of the board about the vertical axis and is clamped after orientation.

Keeping the table level is the most difficult part of plane-table work for beginners. Even light pressure on an edge of the board applies an effective turning moment on the relatively small supporting area of the leveling head.

An alidade, Fig. 15–2, consists of a telescope supported by a

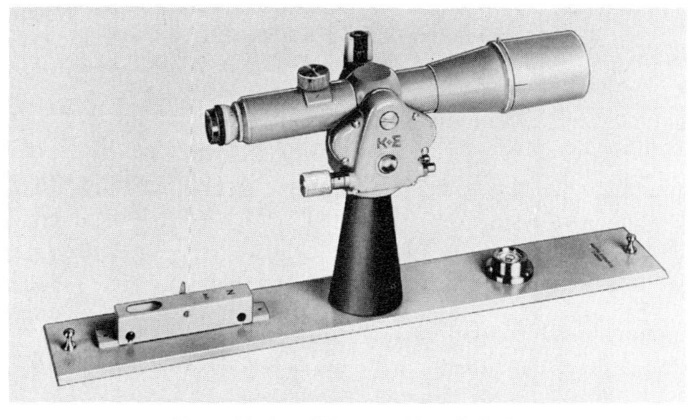

FIG. 15–2. Plane-table alidade.

pedestal rigidly attached to a base or blade that is up to 18 in. long on some instruments. The telescope is similar to that of a transit and is equipped with one vertical and three horizontal cross hairs. A sensitive level vial and bull's-eye level, a vertical arc and/or Beaman arc, a compass needle, and lifting knobs are provided. The telescope may be centered over the blade, or be offset to place the line of sight along the edge.

Accessories used with the plane table include a scale, triangles, a compass, a magnifying glass, a French curve, a slide rule, erasers,

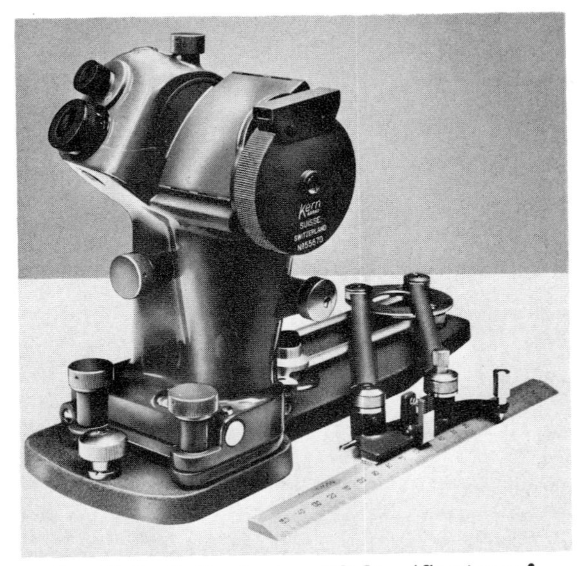

Fig. 15-3. Self-reducing alidade. (Courtesy of Kern Instruments, Inc.)

a protractor, a declinator, and Scotch drafting tape. A *declinator* is a brass plate on which a compass box and one or two level vials are mounted. Two edges of the plate are parallel with the north-south line of the compass circle. The declinator is used, in the absence of a compass on the blade, to determine bearings and to orient the alidade by placing an edge against the base.

The alidade of European design shown in Fig. 15-3 is a self-reducing stadia instrument and has a parallel-ruler plotting device.

15-3. Use of the plane table. The plane table is best suited for traversing and taking topography. It is seldom used to run boundary lines or route surveys, although some details may be added by plane table after transit-tape surveys have been made, or on an aerial photograph.

With a sheet of drawing paper firmly fastened on the board, the table is set up over any point such as *A*, Fig. 15–4, and leveled. The beveled edge of the alidade blade is then placed over the corresponding point *a* on the paper. This point may be selected in a

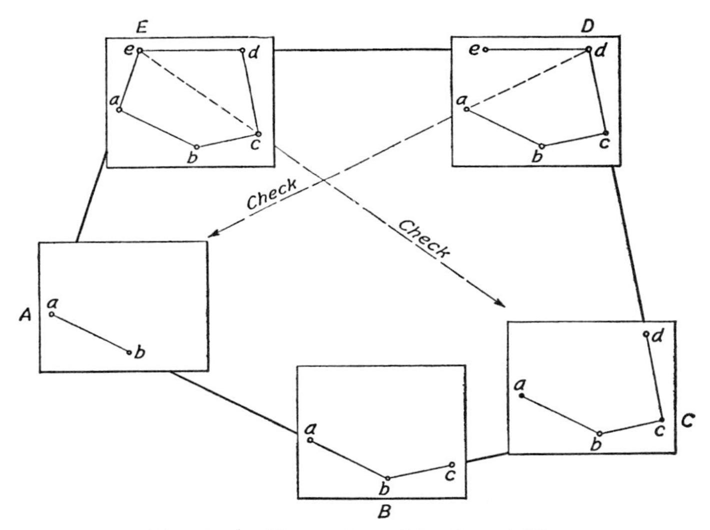

FIG. 15–4. Traversing with plane table.

convenient position on the board if not previously plotted. The direction to a point *B* is obtained by sighting through the telescope to align it and then drawing a line along the beveled edge. The stadia distance and vertical angle are read and reduced immediately, and *b* is located by scaling the horizontal distance on line *ab*. The difference in elevation computed from the stadia reading is applied to the elevation of *A* (assuming a reading on the h.i.) to obtain the elevation of *B*. This value is noted beside the plotted point for ready reference.

On precise work, horizontal control in the form of triangulation or traverse stations, and vertical control represented by bench marks scattered over the area to be mapped, must be plotted carefully on the sheet in advance of any field work. Stadia readings are reduced and plotted as soon as taken, since notes are not recorded. *Sketching and interpolation for contours should be kept abreast of the plotting while the ground location of all mapped points is still fresh in the observer's mind.*

Obviously the drawing paper must be resistant to moisture and temperature changes, because the same sheet may be used in the field for extended periods. A hard pencil, 6H or 8H, is necessary to avoid smudges. The board may be covered with a waterproof material to protect the drawing. A small "window" is left open over the working area. Plastic sheets which can be marked with a stylus are employed in modern practice.

15–4. Setting up and orienting the plane table. After the board is screwed on the tripod head and the paper is fastened, the tripod is set up so that the plotted point is over the corresponding hub on the ground. Generally this relationship is estimated by eye for small-scale maps. If greater accuracy is required, a plumb bob is used, or a pebble is dropped from the underside of the board to check the position. The table must be carefully leveled by means of the leveling screws (or by the ball-and-socket movement of the Johnson head).

For convenience in drafting and to avoid pressure on the table which will disturb its level, the board should be placed at a height about 1 in. lower than the observer's elbows.

The table may be oriented by (a) compass, (b) backsighting, or (c) resection.

a) To orient by compass, the alidade is laid in the direction deemed most desirable for the north-south line, and the board is turned until the compass on the blade, or the declinator against the ruler, reads north. The board is then clamped and a line is drawn along the blade edge for future reference.

If a line AB of known bearing, say S 28° 30′ E, has already been plotted, the alidade is placed along AB and the table turned until the compass reads S 28° 30′ E. The table is then clamped. Orientation by needle is not recommended if the backsighting or resection method can be employed readily.

b) The backsighting method is most commonly used in traversing, although the table may initially be oriented by compass to get north at the top or left side of the sheet. After a line AB has been drawn by sighting from a setup at A to point B, the table is moved to B, set up, and leveled. With the alidade along BA, the table is turned until point A is lined in, and the board is then clamped. The directions of other sights taken with the table in this position will automatically be referred to the same reference line or meridian as AB.

c) Resection will be discussed in Sec. 15–8.

After the table has been oriented by any of the three methods, prominent distant points should be sighted and short lines drawn near the edge of the sheet in their direction. At intervals the alidade can be laid on these lines and the occupied station point to recheck the orientation. By observing on the *permanent backsights* the services of a rodman are not required. The term *effective eccentricity* is used to describe the combined effects of both the rod and the plane table being off their corresponding ground points.

15–5. Traversing. To run a traverse, the table is set up over the initial point A of the survey, leveled, and clamped. Point a on the plane-table sheet, Fig. 15–4, is marked to represent this hub. With the edge of the ruler on a and the alidade sighted at B, line ab is drawn. The distance AB may be determined by tape or stadia and its length marked off according to the scale of the map to locate b. It is essential that this first course be plotted accurately because it serves as the base line for all other measurements.

The table can now be moved to hub B, set up, and leveled. It is oriented by placing the edge of the blade on line ba and turning the board on its vertical axis until the alidade sights point A. The distance BA is measured and the average of AB and BA is used in laying out ab. The next hub C is observed with the blade touching b, the distance BC is determined, and the length bc is plotted. In similar fashion, succeeding points can be occupied and the traverse lines plotted on a map. Whenever possible, check sights should be taken on previously occupied hubs. Small discrepancies are adjusted, but if a plotted point is missed by an appreciable distance, some or all of the measurements should be repeated.

Details can be located while the traverse is being run, or later. It is desirable to close and adjust the traverse before taking topography, inasmuch as all plotting done at an incorrectly located station will be offset. Two methods are used to obtain details—radiation and intersection.

15–6. Radiation method. With the table oriented at any station of the traverse, radiating lines may be drawn to points whose locations are desired, as indicated in Fig. 15–5. Generally, distances are measured by stadia, but a cloth tape is suitable for short sights. The radiating lines, or *rays*, are drawn along the edge of the blade and the distances are scaled.

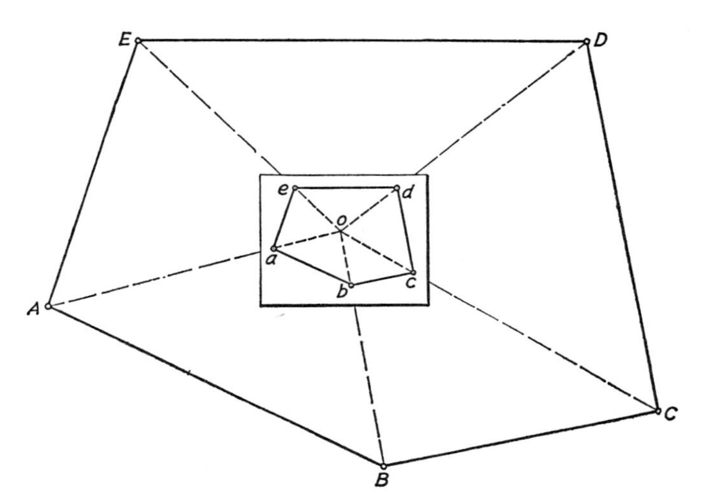

Fig. 15–5. Radiation with plane table.

15–7. Intersection, or graphical-triangulation, method. In the intersection method, shown in Fig. 15–6, rays of indefinite length are drawn toward the same point from at least two setups of the plane table. The intersection of the rays is the location of the point desired. One measured line serves as a base. No other distances are required.

The term graphical triangulation is sometimes applied to this method, which corresponds to triangulation by transit intersections. *The method is particularly important in locating inaccessible and distant points.* In actual field practice, both radiation and intersection may be used on the same setup. Most shots are taken by radiation.

The elevation of an inaccessible point located by intersection can be found by reading the vertical angle to the point (middle cross hair on the point), scaling the plotted map distance to it, and computing the elevation by a trigonometric formula.

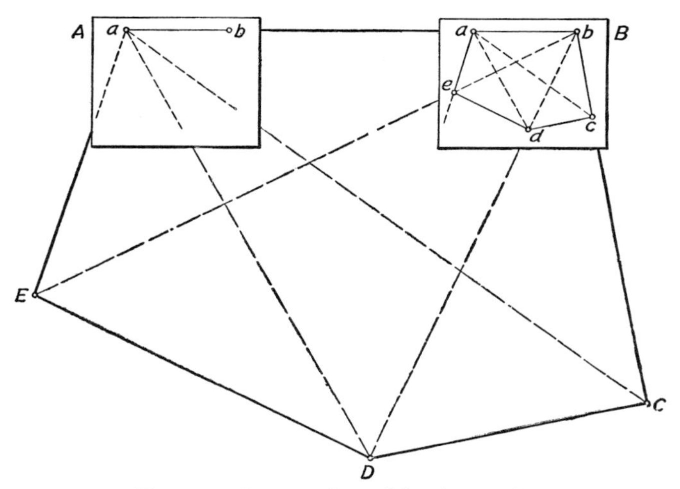

FIG. 15-6. Intersection with plane table.

15-8. Resection. Resection is a method of orientation employed when the table occupies a position not yet located on the map. Solutions for two field conditions will be described. In the first, called the *two-point problem,* the length of one line is known. In the second, the *three-point problem,* the locations of three fixed points are known.

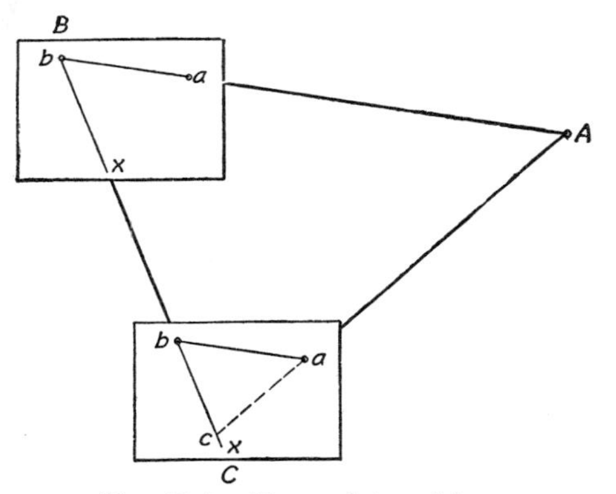

FIG. 15-7. Two-point problem.

15-9. The two-point problem. In Fig. 15-7 line *ab* represents the known length *AB* on the ground. With the table oriented at *B* and the alidade aligned on *C*, a line *bx* of indefinite length is drawn. The distance *bc* is not measured since the observer may wish to reserve judgment on the most advantageous position for his next instrument station, or he may wish the rodman to remain at point A. The table is moved to hub C (any point on line bx), oriented by backsighting along xb, and clamped. The alidade blade is set on a and pivoted around this point until the corresponding station A is sighted. A line drawn along the ruler from point a to intersect line bx gives the location c of the occupied point. Line ac is called a resection line.

15–10. The three-point problem. The three-point method of location has many varied uses. It permits the topographer to set up the plane table at any favorable position for taking topography and then to determine its location on a map by sighting three plotted points. These points might be church steeples, water towers, flagpoles, lone trees, radio masts, jutting cliff formations, and other prominent signals. Measurement of the three distances by electronic devices offers a new approach. The three-point method has long been employed in navigation to ascertain the position of a ship by observing on three shore features recognizable on a coastal chart.

The three-point problem is discussed frequently in technical literature. Trigonometric, mechanical, and graphical solutions have been devised. The tracing-paper, three-arm-protractor, and trial solutions will be described.

If the plane table is on a great circle through the three known points, its location is indeterminate. A strong solution results (a) when the table is well inside the great triangle formed by the three points, or (b) when the table is not near the great circle passing through the three points.

Tracing-paper method. A piece of transparent paper is fastened to the table as shown in Fig. 15–8. From any point p' on the paper, three radiating lines $p'a'$, $p'b'$, and $p'c'$ are drawn toward hubs A, B, and C, which can be observed through the alidade. The tracing paper is then moved until the three radiating lines pass through the corresponding points a, b, and c previously plotted on the map. The vertex of these lines, p'', is the correct location of the table. This point is marked and the board turned to make the lines radiate to the hubs A, B, and C on the ground. The board is then clamped.

Three-arm-protractor method. A similar solution can be obtained mechanically by means of a device called a three-arm protractor. This instrument has a center (somewhat like that of a drafting machine) around which two of the arms can be rotated. The desired

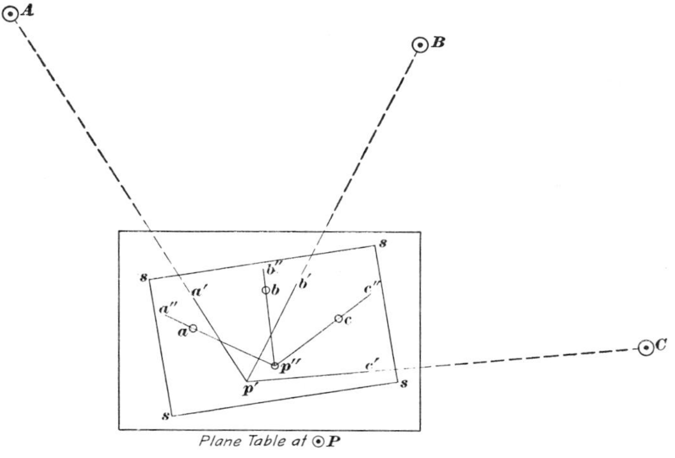

FIG. 15–8. Tracing-paper method.

angle between each rotating arm and the fixed arm is set by a vernier on a 360° graduated circle.

Angles $Ap'B$ and $Bp'C$ in Fig. 15–8 are read with a sextant or a transit. The protractor arms are clamped to these angles and then made to pass through the plotted points a, b, and c by trial. The three-arm protractor is not used on plane-table work but is a valuable tool in hydrographic surveying and coastal navigation.

Lehmann's method. If tracing paper is not available, a *trial solution* by Lehmann's method can be used. In this method the table is oriented by estimation, and resection lines are drawn through the plotted positions a, b, and c of the three reference points. The three lines will form a *triangle of error* if the orientation is not perfect. An experienced observer can reduce the triangle to a point by two or three trials.

Three rules which facilitate the procedure will be stated. In these rules the *point sought* is the point on the plane-table sheet representing the station occupied. The point sought may be inside the triangle of error, or in one of the six areas formed by extending the resection lines beyond the triangle of error, as shown in Fig. 15–9. The three rules successively eliminate the impossible locations of the point sought, and define its true position. The surveyor is assumed to be facing the signals, and the directions right and left are given accordingly.

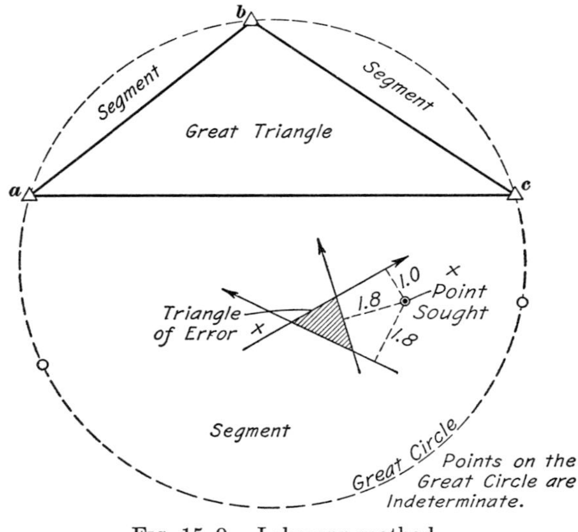

FIG. 15–9. Lehmann method.

RULE 1. The point sought is inside the triangle of error if the station occupied is within the *great triangle* (the triangle formed by the three signals), as shown in Fig. 15–10. This is the simplest case. The exact location of the point sought is determined by rule 3.

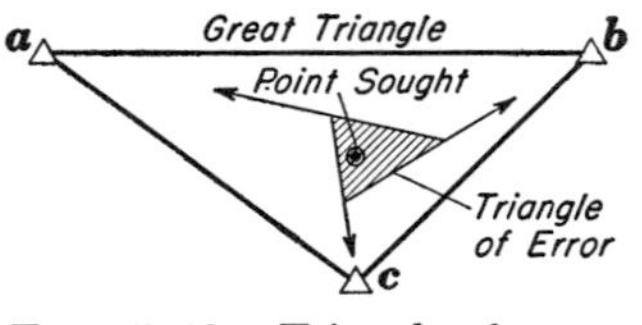

FIG. 15–10. Triangle of error.

RULE 2. The point sought is either to the right of all three resection lines drawn from the fixed points, or to the left of all three resection lines. In Figs. 15–9 and 15–11 the point sought must be in one of the two areas marked with an x. Elimination of one of these possible areas, and the exact location of the point sought, are determined by rule 3.

RULE 3. The point sought is always distant from each of the three resection lines in proportion to the distances from the respective signals to the plane-table station. As an example of the use of this rule, assume that in Fig. 15–9 the estimated distances (or possibly the distances read by stadia) from the plane table to the three signals A, B, and C are 1800 ft, 1800 ft, and 1000 ft respectively. The distances are therefore in the proportion of 1.8 to 1.8 to 1.0.

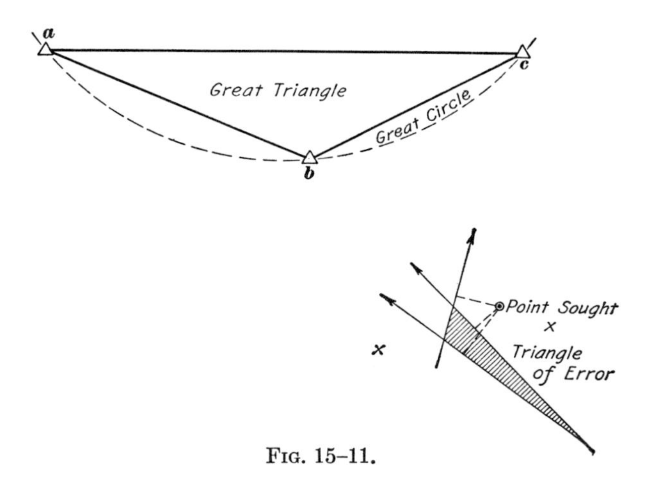

FIG. 15–11.

Using trial sketches, perpendiculars are erected on each of the three resection lines until the unique arrangement is obtained which will give perpendiculars having lengths in the required proportion. Their intersection is the point sought.

After the trial location, T, of the table has been found, the blade of the alidade is laid along point T and the plotted position of the most distant station, a in Fig. 15–9. The table is then reoriented by sighting on the signal at A, and the board is clamped. New resection lines are drawn through points b and c. If the position of the table found by trial is not perfect, a small triangle of error is formed again and the process may have to be repeated once more.

Although the three rules given are sufficient for all cases (actually only the last two are needed, since rule 1 is a special case), two additional rules are helpful:

a) When the point sought is outside the great circle, as in Fig. 15–11, it is always on the same side of the resection line from the most distant point as the intersection of the other two lines.

b) When the point that is sought falls within one of the three segments of the great circle, as in Fig. 15–9, the line drawn from the middle point lies between the point sought and the intersection of the other two lines.

15–11. Leveling. Elevations are obtained by using the alidade as a level, or by reading vertical angles and slope distances. After the alidade has been leveled carefully, a sight may be taken upon a rod set on a bench mark to obtain the H.I. Often the h.i. is simply found by measuring the height of the telescope above the point occupied, by standing the rod alongside the table. The H.I. being known, the rod readings are taken as in ordinary leveling to determine the elevations of points.

Trigonometric leveling is commonly used where considerable difference of elevation exists. This method requires the use of vertical angles and stadia distances, from which differences of elevation are obtained by calculation. A Beaman arc on the alidade facilitates computations, but a stadia-reduction chart, slide rule, or stadia table may be used.

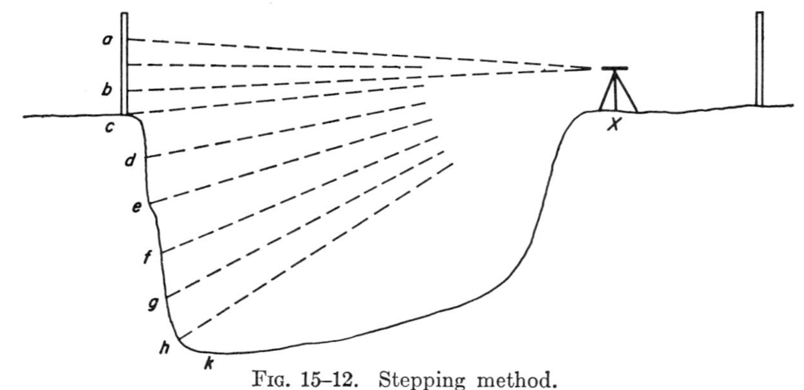

Fig. 15–12. Stepping method.

The *stepping method* of obtaining elevations is sufficiently accurate for some purposes, and in rough terrain it saves time and effort on the part of the rodman. In Fig. 15–12, assume that the elevation of point *k* at the bottom of a gorge is needed on a survey employing two rodmen, one on each side of the canyon. After the intercept on the rod between stadia hairs, *ab*, is found to be say 5.2 ft, the upper hair is set on the base of the rod at *c*. The position *d* where the lower cross hair strikes the ground is noted, and the upper cross hair is depressed to sight this point. The process is continued until point *h* is found, and the remaining drop to the lowest spot *k* is estimated at perhaps 2 ft. The difference in elevation between *c* and *k* is therefore 28 ft.

Another version of the stepping method utilizes the fact that the axis of sight through the upper (or lower) cross hair is on a $\frac{1}{2}$% grade from the axis through the middle hair.

Differences of elevation are generally computed in the field and plotted on the plane-table map. The topographer may compute (if he does not have a self-reducing tachymeter) as well as plot, or he may have one member of the party make the stadia reductions. Contours are drawn through points established by the direct method, or are interpolated if breaks in slope have been taken. Spot elevations are shown for grade crossings, for peaks and depressions which do not fall at contour levels, and for all other critical points.

The number of points taken need be only 50 to 60 per cent of the number on a comparable transit-stadia survey to locate contours with the same degree of accuracy. Furthermore, features shown on the map can be compared with the terrain as work progresses, and any discrepancies readily discovered.

One of the greatest difficulties arising in the use of the plane table is keeping it level. Pressing on one corner of the board is a practical way to center the bubble and complete a few sights with the Johnson head. This procedure eliminates the necessity of loosening the ball-and-socket joint, which operation usually disturbs the orientation.

15–12. Advantages and disadvantages of the plane table. Some of the advantages of the plane table over the transit-stadia method follow:

a. The map is made while looking at the area.
b. Irregular lines, such as stream banks and contours, can be sketched.
c. Notes are not taken.
d. Fewer points are required for the same precision in locating contours.
e. A map is produced in less over-all time (field plus office).

Some disadvantages of the plane table in comparison with the transit-stadia method are as follows:

a. More field time is necessary.
b. Bad weather may halt field work.

c. Control must be plotted in advance for precise work.
d. More time is required to become proficient with the plane table than with the transit.
e. The table is set lower than a transit normally would be, thus providing less sight clearance.
f. Distances must be scaled if lengths are to be taken from the map, and areas computed.
g. Many awkward items must be carried.
h. The plane table is unsuitable for wooded country.

15–13. Plane-table pointers. One of the troublesome problems in operating a plane table is to keep the alidade blade on the plotted position of the occupied point, such as *P* in Fig. 15–13. As the alidade is moved to sight a detail, the edge moves off point *P*. A solution sometimes tried is to use a pin at *P* and pivot around it, but a progressively larger hole is gouged in the paper with each sight.

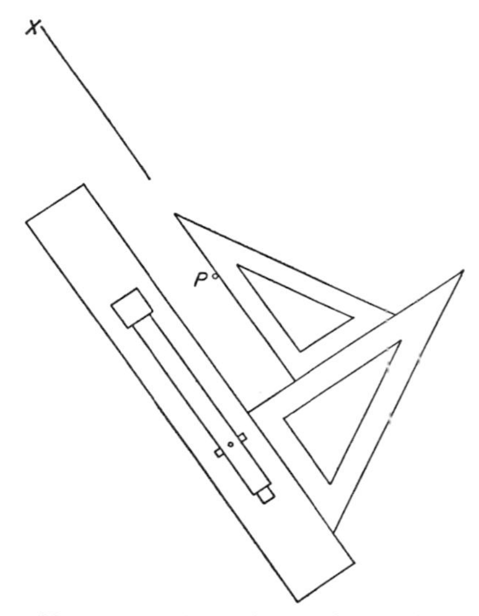

FIG. 15–13. Transfer of pivot point.

Some alidades, such as that in Fig. 15–3, have a parallel ruler attached to the blade, and this provides the best answer. The next best is to place the alidade within an inch or two of the point *P*, pivot as necessary for a sight, and then transfer the line through *P* by using two triangles as shown in Fig. 15–13. The small error produced by the eccentric sight is no greater than that resulting from not being exactly over the ground point *P*, or even that caused by the telescope axis not being over the edge of the blade.

Other pointers on the use of the plane table that may be helpful are as follows:

1. Use buff or green detail paper to lessen the glare.
2. Plot and ink the traverse in advance of the detailing, showing

lengths of traverse lines, coordinates of triangulation stations, and useful signals available.

3. Have at least one vertical control point for each three hubs of a traverse, and show all known elevations.
4. Have all of the accessories.
5. Cover the portion of the map not being used.
6. Set up the table slightly below elbow height.
7. Check the orientation on two or more lines if possible.
8. Check the distance and elevation difference in both directions when setting a new hub. Read vertical angles with the telescope in normal and plunged positions, if possible. Use the average value in plotting.
9. Read the distance first, then the vertical angle; or with a Beaman arc, read the H scale and then the V arc.
10. Lift the forward end of the alidade blade to pivot, instead of sliding the blade, to keep paper cleaner.
11. Clean the blade frequently to remove graphite.
12. Reduce stadia readings in the following preferred order: self-reducing tachymeter, Beaman arc, slide rule, charts, tables, formulas.
13. Check the location of hubs by resection and by *cutting in* (sighting and plotting) prominent objects.
14. Draw short lines at the estimated distance on the map to plot points. Do not start the lines at the hub occupied.
15. Keep in mind the comparable limits of accuracy for direction (angle) and distance.
16. Identify points by consecutive numbers or names as they are plotted. Number the contours.
17. Have the rodman make independent sketches on long shots, for later transference to the plane-table map.
18. Use walkie-talkie sets to enable the rodmen to describe topographic features when the observer cannot identify them because of distance and obstacles.
19. Use the same points to locate details and contours whenever possible.
20. Locate contours by the direct method in gently sloping and rolling terrain. Use the indirect method in rugged country and areas having uniform slopes.
21. Sketch contours after three points have been plotted. Points

on the map lose their value if they cannot be identified on the ground.

22. Show spot elevations for summits, sags, bridges, road crossings, and all other critical points.
23. Tie a piece of colored cloth on the stadia rod at the required rod reading to speed work in locating contours by the direct method.
24. Use cotton rope $\frac{3}{8}$ in. in diameter soaked in hot paraffin, with knots of different colors at 5- or 10-ft intervals, in making measurements in areas covered by brush and timber.
25. Employ the stepping method in areas where access is difficult.
26. Draw lines at 1-in. intervals on all four margins of the plane-table sheet at the time the first plotting is done. The lines provide a means of determining the expansion and contraction of the paper.
27. Utilize vertical aerial photographs for plane-table sheets. The planimetric details can be checked and contours added.
28. Use a 6H or harder pencil to avoid smudging.

15–14. Sources of error in plane-table work. Sources of error in plane-table operation include the following:

a. Table not level.
b. Orientation disturbed during the detailing.
c. Sights too long for accurate sketching.
d. Poor control.
e. Traversing and detailing simultaneously.
f. Too few points taken for good sketching.

15–15. Mistakes. Some typical mistakes made in plane-table work are as follows:

a. Detailing without proper control.
b. Table not level.
c. Orientation incorrect.

PROBLEMS

15–1. List side by side for comparison the respective advantages and disadvantages of the Johnson-type tripod head and the Coast and Geodetic Survey head.

15–2. Tabulate three methods of orienting a plane table, their advantages and disadvantages, and an example of the practical use of each method.

15–3. For what field conditions is the plane table preferred to transit-stadia in making a topographic survey?

15–4. Compare the radiation and intersection methods for general plane-table operations, and give an example of the most desirable application of each.

15–5. Locate the position in a classroom of a point *C* on your desk by the two-point method. Mark point *A* on the blackboard and point *B* in a convenient manner on another desk. Measure the horizontal distance *AB* and plot it to scale as *ab* on a sheet of paper fastened to a drawing board. Then locate point *c* on the sheet over point *C*. Sight over one edge of an engineer's scale to get the resection line.

15–6. Locate your position in the classroom by the three-point method. Mark three points, *A* in one corner of the room and *B* and *C* on the blackboards. Measure the horizontal distances from *A* to points *B* and *C*. Plot the points to scale on a sheet $8\frac{1}{2}$ in. by 11 in. Sight over an edge of an engineer's scale to obtain the resection lines. Change positions so that you are first inside the great triangle, then outside the great triangle but inside the great circle, and finally outside the great circle.

15–7. A plane table is set inside the triangle formed by three known points. Show how the position of the table is located by Lehmann's method.

15–8. Sketch the strongest and weakest 3-point location positions and comment on them.

15–9. Upon what factors does accurate plane-table orientation depend?

15–10. Upon what two things does the size of the triangle of error depend in the Lehmann solution of the three-point problem?

15–11. List side by side for comparison advantages and disadvantages of 3 methods of solving the three-point problem given in Sec. 15–10.

15–12. What are the three main methods of locating points with a plane table?

15–13. Given points *C* and *D* plotted on a plane-table sheet. Describe the procedure in locating points *E, F,* and *G* with a nontelescopic alidade by making only two setups and without measuring any distances or setting up the table at *D*.

15–14. List the pertinent factors to be considered in selecting the proper scale for any specific plane-table project.

15–15. On what maximum length of sight can the effects of curvature and refraction be neglected in determining elevations with a plane table?

15–16. How can the plane table be used in connection with topographic maps prepared photogrammetrically?

15–17. What is the purpose of a cover sheet (sheet mask) with a "window" on plane table work in the field?

15–18. Are spur traverses desirable with a plane table? Explain.

15–19. How is a plane table map checked for accuracy?

15–20. A plane table is oriented at hub *A* by compass only to start a survey and as a result the first line is off by 10 minutes from the assumed direction. If the line is extended 3 miles and a hub B set, what is its position error on the plane table sheet if a scale of 1 in. = 2000 ft is used?

15–21. On a traverse 2 miles long, an experienced plane-table man uses an alidade with a 16-power telescope, Beaman arc, 45-seconds per 2-mm striding level vial, and a Philadelphia rod, and balances sights limiting them to 500 ft. What maximum errors in distance and elevation are reasonable?

15–22. The alidade on a plane table is placed 1½ in. off the proper point in sighting at an object 100 ft away. The map scale is 1 in. = 40 ft. Is the error significant? Explain.

15–23. The telescope of an alidade is centered over a blade 3 in. wide. What error is introduced in plotting a shot 200 ft long if the map scale is 1 in. = 100 ft? If the shot is 400 ft and the map scale is 1 in. = 200 ft? If the sight is 600 ft and the map scale is 1 in. = 200 ft?

15–24. The stepping method is used to determine the elevation of a point k in a canyon (see Fig. 15–12) from an instrument set up over X, elevation 825.0 ft. With the telescope level, an intercept $ab = 3.58$ ft is obtained on a rod held at c, and the reading of the middle cross line on the rod is 5.2 ft. If six "steps" are taken from the bottom of the rod at c, and an estimated vertical distance of 1 ft remains to point k, what is the elevation of k? The H.I. is 5.0 ft.

15–25. Similar to problem 15–24, except that the rod intercept is 4.16 ft, 7 steps are taken, and the estimated vertical distance remaining to point k is 2 ft.

15–26. What is the effect upon orientation of a plane table if both table and rod are eccentric (a) in the same amount and direction, and (b) in opposite directions normal to the line of sight? For the backsight distances and effective eccentricities given in the following tabulation, fill in the respective errors in orientation.

Effective eccentricity (ft)	1.2	1.2	1.2	2.0	2.0	2.0
Backsight distance (ft)	100	300	500	100	300	500
Error in orientation, angular						
Error in orientation, ft/100 ft						

15–27. Similar to problem 15–26, but for effective eccentricities of 3.0 and 1.5 ft.

15–28. The total length of a closed plane-table traverse is 8400 ft. The traverse fails to close by 0.40 in. on the sheet. If the scale being used is 1 in. = 400 ft, what ratio of error has been obtained?

15–29. Similar to problem 15–28, except the closure is 0.24 in. and the scale 1 in. = 1000 ft.

15–30. A plane-table traverse is run for a geology project with vertical control consisting of B.M. Lafayette, elevation 746.52, near hub A, and B.M. Lehigh, elevation 771.89, adjacent to closing hub D. Determine the closure in elevation and adjust the traverse hub elevations. Readings are taken on the h.i. for vertical angles, and h.i. = 5.0 ft at *A* and 5.5 ft at *D*.

Hub	Pointed Sighted	Stadia Dist. in ft	Vert. Angle, or Rod Reading with Telescope Level
A	B.M. Lafayette		3.68 ft
	B	240	+3° 18′
B	*A*	240	−3° 18′
	C	468	+2° 30′
C	*B*	470	−2° 32′
	D	390	−1° 24′
D	*C*	389	+1° 25′
	B.M. Lehigh		3.16 ft

15–31. What are the main troubles which arise in using the plane table and how are they best corrected?

chapter 16

Adjustment of Instruments

16–1. General. Surveying instruments are designed and constructed to give correct horizontal and vertical measurements. A good instrument, properly handled, may stay in adjustment for months or longer, and last a lifetime. Nevertheless, equipment should be tested periodically, and adjusted to maintain its accuracy. A level, for example, should be checked each day it is used on important work.

It is not feasible to rigidly set the level vials and cross hairs in perfect position on most levels and transits, since replacement or adjustment of parts would then necessarily become a factory operation. Temperature changes, jarring, and undue tension on adjusting screws may cause an instrument to go out of adjustment.

Proper field procedures, such as double centering, keeping backsights and foresights equal, and releveling when the bubble goes off as the telescope is rotated about a vertical axis, permit accurate work to be done with an instrument out of adjustment. Frequently, however, a few minutes spent in making adjustments will reduce the time and effort required to operate an instrument efficiently. Furthermore, some errors may be introduced that can be eliminated only by adjusting the instrument.

Precise adjustment of transits, levels, and alidades in the field is made easier if the following conditions exist:

a. Terrain permitting solid setups, level sights of at least 200 ft for levels, 400 ft for transits, and measurement of a vertical angle of 45° or more.

b. Good atmospheric conditions—no heat waves, and no sight lines passing through alternate sun and shadow or directed into the sun.

c. Instrument in the shade or shielded from direct rays of the sun.

Three permanent points set approximately 300 ft apart on a straight line, on level or nearly level ground, and preferably at the same elevation, expedite adjustments. Organizations having a number of instruments in use, and surveyors working in one area over a long period of time, find it profitable to set such permanent points.

Standard methods and a prescribed order must be followed in making adjustments. Correct positioning of parts is attained by loosening or tightening the proper adjusting nuts and screws with special pins. Time is wasted if each adjustment is perfected on the first trial, since some adjustments affect others. The complete series of tests may have to be repeated several times if the instrument is badly off. A final check of all the adjustments should be made to insure that none has been disturbed. The simplest adjustment of all, the removal of parallax by careful focusing of the objective lens and eyepiece, must be kept in mind at all times.

Straight, strong adjusting pins that fit the capstans should be used, and the capstans should be handled with care to avoid damaging the soft metal. Adjustment screws have been properly set when an instrument is shipped from the factory. Tightening the capstan screws too much (or not enough) nullifies otherwise correct adjustment procedures and may leave the instrument in worse condition than it was before testing.

16–2. Tapes. Tapes are tested by comparing them with a standard. Many surveyors and surveying organizations have a standardized tape which is used only to check other tapes, or they have fixed marks set a known distance apart for the same purpose. The National Bureau of Standards will test any steel tape for a nominal fee and will furnish a certificate giving its correct characteristics for the tension and method of support desired.

Hard usage, kinks, and repairs change the length of a tape. Comparison with a standardized tape at frequent intervals permits even an old tape to be used with assurance.

16–3. Principle of reversion. Most adjustments of surveying instruments are checked by the method of reversion. The method consists of reversing the instrument in position to double the error and make it more apparent. Assume that in Fig. 16–1 the angular error between the correct and unadjusted lines is ε. This error is caused by the difference in length of a and b (for example, the heights of the two ends of a level vial or the telescope) as shown in position *1*. After the telescope is turned 180° in azimuth, the unadjusted line occupies position *2* because a and b have changed places.

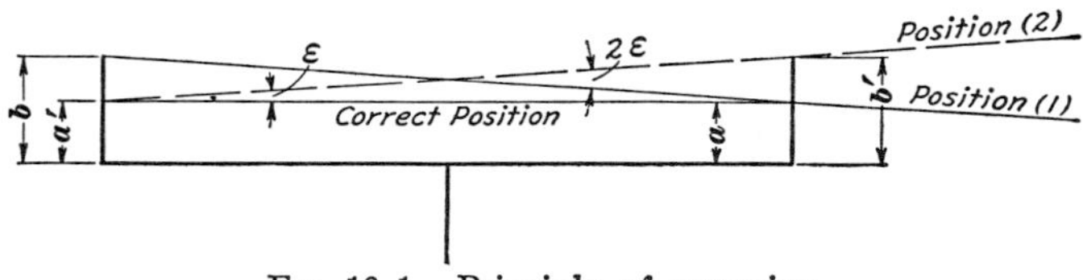

FIG. 16–1. Principle of reversion.

Since the angle between positions *1* and *2* is 2ϵ, it follows that single reversion doubles the error. The correct adjustment is secured by making a equal to b.

16–4. Adjustment of the wye level. The purpose of an engineer's level is to establish a horizontal plane of sight when the telescope is revolved about a vertical axis. The principal lines of a wye level (Fig. 16–2) which must be adjusted to maintain this condition are the following:

a. Axis of sight. The line from the optical center of the objective lens to the intersection of the cross hairs.
b. Axis of the collars. The line joining the centers of the collars.
c. Axis of the level bubble. The line tangent to the circular arc of the level vial at its midpoint.
d. Axis of the bottom of the wyes. The line tangent to the inner surfaces of both wyes at the supporting (lowest) points for the telescope.
e. Axis of the level bar. The lengthwise reference line or axis of the level bar which is perpendicular to the vertical axis.
f. Vertical axis. The line through the spindle about which rotation takes place.

For perfect adjustment, the axis of sight and the axis of the collars must coincide. Lines a, b, c, d, and e must be parallel to each other

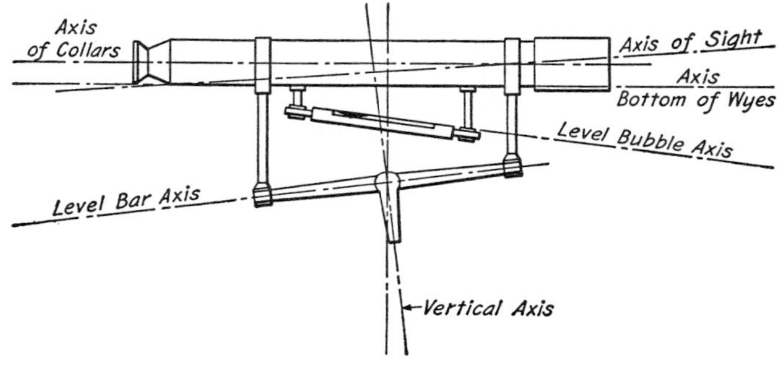

FIG. 16–2. Wye level.

and at right angles to the vertical axis. To provide this relationship in a wye level, it is necessary that the cross hairs, level vial, and wyes be adjustable. The three adjustments required will be described.

Adjustment of Cross Hairs

a) *Purpose.* To make the horizontal cross hair truly horizontal when the instrument is leveled.

Test. Set up the level and sight a sharply defined point with one end of the horizontal cross hair. Turn the telescope slowly on its vertical axis so that the cross hair moves across the point. If it does not remain on the point for the full length of the cross hair, the instrument is out of adjustment.

Correction. Loosen the four capstan screws holding the reticle. Rotate the reticle in the telescope tube until the horizontal hair is in a position where it will remain on the point as the telescope is turned. The screws should be carefully tightened in their final position.

b) *Purpose.* To make the axis of sight coincide with the axis of the collars.

Test. Level the instrument and sight the intersection of the cross hairs upon some sharply defined point. Clamp the vertical axis. Unfasten the wye clips. Revolve the telescope in the wyes and note whether the intersection of the cross hairs remains on this point. If a displacement occurs with the telescope inverted, an adjustment is necessary so mark a second point beside the first one.

Correction. Bring the intersection of the cross hairs halfway back to the original point by means of the capstan screws which hold the reticle in the telescope tube. Repeat the operation until the intersection remains on the original point during a complete revolution of the telescope.

Adjustment of Level Vial

Purpose. To make the axis of the level bubble parallel to the axis of sight.

Test. Level the instrument over two opposite leveling screws. Clamp the vertical axis. Open the wyes, lift the telescope, turn it end for end, and replace it in the wyes. The distance the bubble moves off center represents double the error.

Correction. Turn the capstan nut at one end of the level vial to move it up or down until the bubble is brought halfway back to the central position. Level the instrument and retest.

Adjustment of Wyes

Purpose. To make the axis of the level bubble perpendicular to the vertical axis. The axis of sight, the axis of the level vial, and the axis of the collars are in correct relation following the previous adjustments. The axis of the level bar and the vertical axis are fixed by the manufacturer and are not adjustable. To bring the axis of sight parallel to the level bar it is necessary to make the axis of the wyes parallel to the level bar. The wyes must be of equal height.

Test. Level the instrument with the telescope directly over two opposite leveling screws. Then revolve it 180° about its vertical axis. The distance the bubble has moved off center represents double the error.

Correction. Change the height of one wye by means of the adjusting screws on it to bring the bubble halfway back toward the centered position. Level the instrument with the leveling screws and repeat the test until the level bubble remains centered as the telescope is revolved about the vertical axis.

Peg Adjustment

The foregoing three adjustments are called indirect adjustments. It is good practice to test their accuracy by the peg adjustment described in section 16–5 for the dumpy level. The process is the same for both types of levels. However, the error is corrected by moving the level vial of the wye level, and by moving the cross hairs for the dumpy level. An alphabetical memory aid for this adjustment is: c-d (cross hair-dumpy), v-w (vial-wye). If the wyes or the collars are not properly turned or have become worn, the peg adjustment must be used instead of reversing the telescope in the wyes. The chief advantage of the wye level—ease of adjustment by removing and turning the telescope—is thereby lost.

16–5. Adjustment of the dumpy level. In the dumpy level, Fig. 16–3, the standards are not adjustable and the telescope tube cannot be revolved or reversed in them. Accordingly there are fewer adjustable parts. The principal lines, defined as for the wye level, are the following:

a. Axis of sight.
b. Axis of the level bubble.
c. Axis of the level bar.
d. Vertical axis.

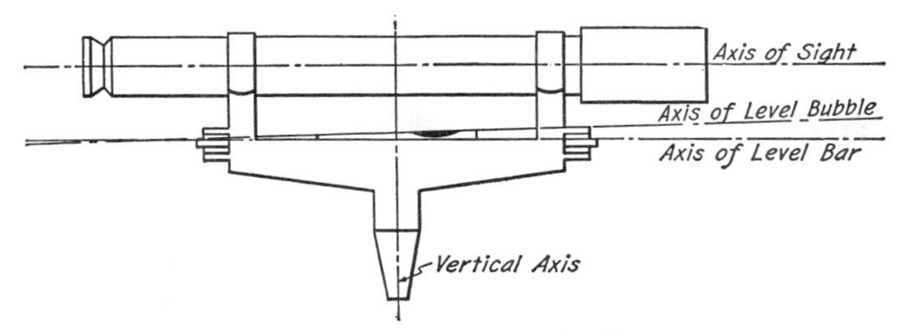

FIG. 16–3. Dumpy level.

For perfect adjustment, it is necessary that the axis of sight, the axis of the level bubble, and the axis of the level bar be parallel to each other and perpendicular to the vertical axis. There are two adjustable parts: the cross hairs and the level vial.

ADJUSTMENT OF LEVEL VIAL

Purpose. To make the axis of the level bubble perpendicular to the vertical axis.

Test. Set up the level, center the bubble, and revolve the telescope 180° about the vertical axis. The distance the bubble moves off the central position is double the error.

Correction. Turn the capstan nuts at one end of the level vial to move the bubble halfway back to the centered position. Level the instrument by means of the leveling screws. Repeat the test until the bubble remains centered during a complete revolution of the telescope.

PRELIMINARY ADJUSTMENT OF HORIZONTAL CROSS HAIR

Purpose. To make the horizontal cross hair truly horizontal when the instrument is leveled.

Test and Correction. Same as (a), first adjustment of cross hairs for the wye level.

PEG ADJUSTMENT

Purpose. To make the axis of sight perpendicular to the vertical axis and thus parallel to the axis of the level bubble. This adjustment is also called the *two-peg method* and the *direct adjustment.*

Test. Level the instrument over a point C halfway between two stakes A and B about 300 ft apart. See Fig. 16–4. Determine the difference in rod readings a_1 and b_1 on A and B, respectively. Since the distance to the two points is the same, the true difference in elevation is obtained even though the axis of sight is not exactly horizontal.

Then set up the instrument at D on line with the stakes and close to one of them—A in this case—and level. With the eyepiece only a few inches from the rod, the reading on A is taken by sighting through the objective-lens end of the telescope. Usually a pencil is centered in the small field of view. A rod reading b_2 is taken on B.

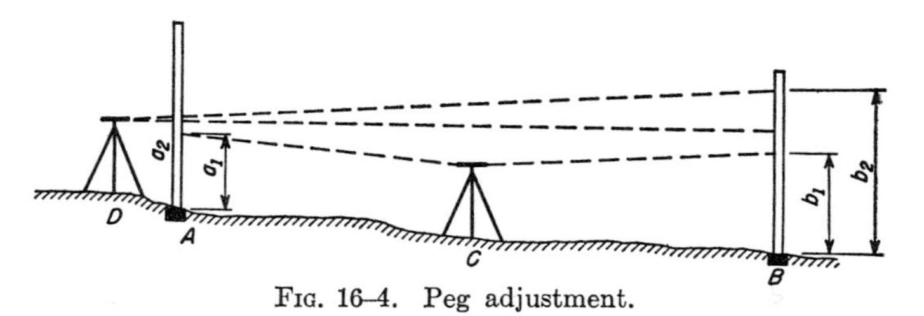

FIG. 16–4. Peg adjustment.

If the axis of sight is parallel to the axis of the level bubble (that is, horizontal), the rod reading b_2 should equal the rod reading at A plus the difference in elevation between A and B, or $(b_1 - a_1) + a_2$. The difference, if any, between the computed and actual readings is the error to be corrected by adjustment.

Correction. Loosen the top (or bottom) capstan screw holding the reticle, and tighten the bottom (or top) screw to move the horizontal hair up or down and give the required reading on the rod at B. Several trials may be necessary to get an exact setting.

An alternative method of testing the adjustment is by reciprocal leveling. A setup is made close to A and readings are taken on A and B. The level is moved to a position near B and similar sights are taken. The difference in elevation is computed and the reticle is shifted to give a reading on the distant rod equal to the reading on the near point plus the difference in elevation of the hubs.

If the difference in elevation of A and B is known, only one setup is required near either point for the adjustment.

16–6. Adjustment of the transit. The transit is designed to measure the vertical and horizontal projections of angles, and to serve as a level. To these ends, certain lines and axes must be precisely positioned. The principal lines of a transit, Fig. 16–5, are:

a. Axis of the plates. Any line through the top surface of the upper and lower plates and the center of rotation.
b. Axis of the plate bubble. The line tangent to the circular arc of the level vial at its midpoint.

c. Vertical axis. Same as for the level.
d. Axis of the standards, or the horizontal axis. The line through the center of rotation of the telescope axle and the bearings in the standards.
e. Axis of sight. Same as for the level.
f. Axis of the telescope-level bubble. Same as for the level.

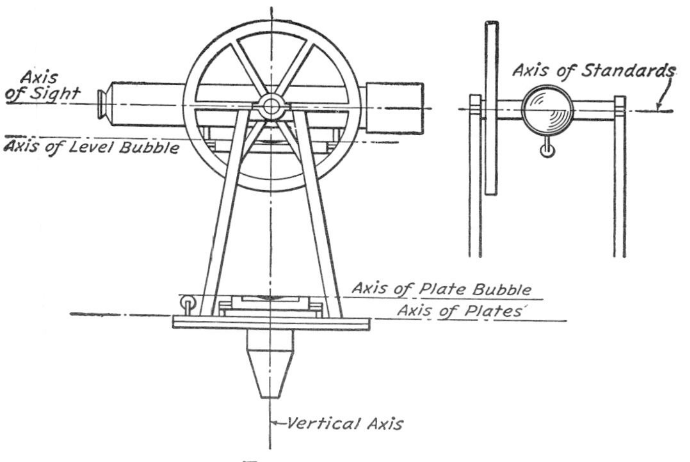

Fig. 16–5. Transit.

For correct adjustment, *a* must be parallel to *b* and perpendicular to *c*; *d* must be parallel to *a*; *e* perpendicular to *d*; and *f* parallel to *e*. To maintain these relationships, the plate-level vials, cross hairs, standards, telescope-level vial, and vertical-circle vernier must be adjustable.

Adjustment of Plate-Level Vials

Purpose. To make the axis of each plate-level bubble perpendicular to the vertical axis.

Test. Set up the instrument, bring one of the plate-level vials over two opposite leveling screws, and center it. Revolve the instrument 180° about the vertical axis to place the same level vial, turned end for end, over the same leveling screws. The distance the bubble moves from its central position is double the error.

Correction. Turn the capstan screws at one end of the level vial to move the bubble halfway back to the centered position. Level the instrument by means of the leveling screws. Repeat the test until the bubble remains centered during a complete revolution of the instrument.

Adjust the other bubble in the same manner.

Preliminary Adjustment of Vertical Cross Hair

Purpose. To place the vertical cross hair in a plane perpendicular to the horizontal axis of the instrument.

Test. Set up the transit and sight on a well-defined point with one end of the vertical cross hair. Turn the telescope on its horizontal axis so that the cross hair moves along the point. If it departs, the cross hair is not perpendicular to the horizontal axis.

Correction. Loosen all four capstan screws holding the reticle, and turn the reticle slightly until the vertical hair remains on the fixed point during the rotation of the telescope. Tighten the capstan screws and recheck the adjustment.

Adjustment of Vertical Cross Hair

Purpose. To make the axis of sight perpendicular to the horizontal axis.

Test. This test applies the double-centering procedure for prolonging a straight line. Level the transit and backsight carefully on a well-defined distant point A, Fig. 16–6, clamping the plates. Plunge the telescope and set a foresight point B at approximately the same elevation as A, and at least 600 ft away if possible. With the telescope still in the inverted position, unclamp either plate, turn the instrument on the vertical axis, backsight on the first point A again, and clamp the plate. Plunge the telescope back to its normal position and set a point C beside the first foresight point B. The distance between B and C is four times the error of adjustment because of the double reversion.

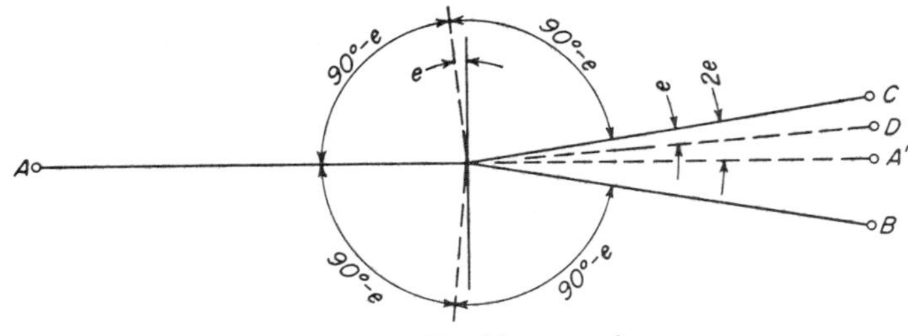

Fig. 16–6. Double reversion.

Correction. Loosen one of the side capstan screws which hold the reticle to the telescope tube and tighten the opposite screw to move the vertical hair one-fourth of the distance CB to point D. Repeat the test until the telescope sights the same point A' after reversing from the backsight A.

Adjustment of Horizontal Cross Hair

Purpose. To bring the horizontal cross hair into the optical axis of the telescope. This is necessary if the transit is to be used for leveling or for measuring vertical angles.

Test. Set up and level the transit over point A, Fig. 16–7. Line in two stakes B and C at approximately the same elevation. Stake B should be at minimum focusing distance from A, perhaps 5 or 10 ft, and stake C at least 300 ft away.

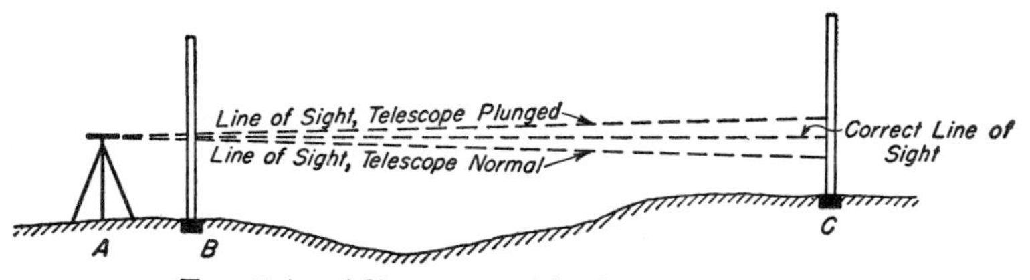

Fig. 16–7. Adjustment of horizontal cross hair.

Take readings of the horizontal cross hair upon a rod held first on B and then on C. Plunge the telescope, turn the instrument about its vertical axis, and sight on B again, setting the horizontal hair on the first rod reading. Then with the vertical and horizontal axes clamped, read the rod held on the far stake C. Any discrepancy between the two readings on the rod at C is approximately double the error.

Correction. By means of the top and bottom capstan screws holding the reticle, move the horizontal hair until it intercepts the rod halfway between the two readings on C. Repeat the test and adjustment until the horizontal-hair reading does not change for the normal and plunged sights on the far point.

Adjustment of Standards

Purpose. To make the horizontal axis of the telescope perpendicular to the vertical axis of the transit.

Test. With the transit carefully leveled parallel to the horizontal axis, sight a well-defined high point A, Fig. 16–8, at a vertical angle of at least 30°, and clamp the plates. Depress the telescope and mark a point B near the ground. Plunge the telescope, unclamp either plate, turn the instrument about the vertical axis, sight point A again, and then clamp the plate. Now depress the telescope and set another point, C, near B. Any discrepancy between B and C is the result of unequal height of the standards and represents approximately twice the error.

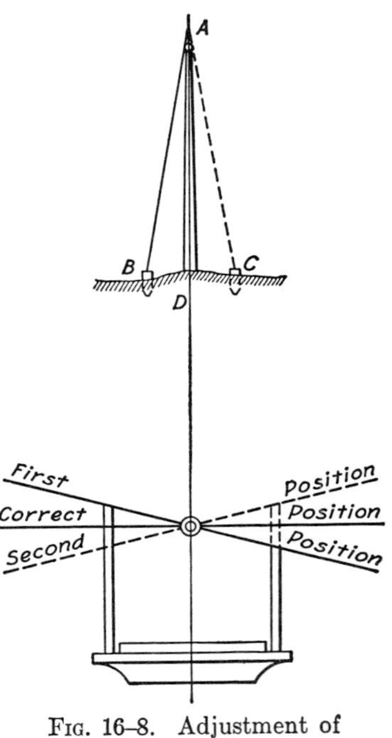

FIG. 16–8. Adjustment of standards.

Correction. Set a point D approximately halfway between B and C, and sight on it. With the plates clamped, elevate the telescope and bring the line of sight on point A by raising or lowering the movable block in one of the standards. To raise the horizontal axis, first loosen the friction screws holding the trunnion cap on the standard and then tighten the capstan screw below the block. To lower the axis, reverse the procedure. The friction screws holding the trunion cap must be set carefully to prevent the telescope from being too loose, or from binding.

Repeat the test and adjustment until the high and low points remain in the line of sight with the telescope in both the normal and inverted positions.

Adjustment of Telescope-Level Vial

Purpose. To make the axis of the level bubble perpendicular to the vertical axis and parallel to the axis of sight of the telescope.

Test. Same as for the peg adjustment of the dumpy level.

Correction. If the rod reading b_2, Fig. 16–4, indicates that an adjustment is necessary, the correct rod reading required to produce a horizontal axis of sight is set on the rod by means of the vertical circle slow-motion screw. The telescope bubble is then centered by turning the capstan screws at one end of the level vial.

Adjustment of Vertical-Circle Vernier

Purpose. To make the vernier of the vertical circle read zero when the telescope-level bubble is centered (that is, to make the index error zero).

Test. Level the instrument with both plate-level bubbles. Revolve the telescope about the horizontal axis until the telescope bubble is centered. Then read the vertical-circle vernier. The angle that is read is the index error.

Correction. Loosen the capstan screws holding the vernier plate and move it to make the zero marks of the vernier and vertical circle coincide. Then tighten the capstan screws. In making this adjustment, care should be taken to avoid leaving a gap between the vernier and vertical circle since such a space introduces errors in reading vertical angles.

16–7. Adjustment of the plane-table alidade. The adjustments of an alidade are similar to those for the transit and level, but they need not be as refined since the plane table is used for less-precise work. The importance of any slight defect of adjustment can be judged by its effect upon the map to be drawn. The most critical adjustment is that of the vertical arc, which must be used constantly to determine elevations.

Adjustment of level vials on blade to make their axes parallel to blade. Level the board, draw guide lines along the blade ends, and lift the alidade and turn it end for end, placing it between the guide lines to avoid the effect of any warping in the board. If the bubble goes off center, bring it halfway back by means of the adjusting screws. Then relevel and test again.

Adjustment of vertical cross hair to make it perpendicular to transverse axis. The procedure is the same as for the transit.

Adjustment of telescope-level vial to make its axis perpendicular to the vertical axis and thus parallel to axis of sight. For a rigidly attached vial, the procedure is the same as the peg method for the level vial on a transit telescope. For a striding level, attach the striding level to the telescope tube, center the bubble, and pick up and turn the striding level end for end. If the bubble moves off center, bring it halfway back by means of the adjusting screws.

Adjustment of reticle to make axis of sight coincide with axis of telescope sleeve. The procedure is the same as for the adjustment of the reticle in the wye level, the telescope being rotated in the sleeve.

Adjustment of vertical-arc vernier. The procedure is the same as for the transit vertical-arc vernier.

Adjustment of vernier-control level bubble. Level the telescope and set the vernier to read zero. Center the vernier-control bubble by means of the adjusting screws.

16–8. Adjustment of the Beaman stadia arc. The index mark for the V scale of a Beaman arc can be adjusted to avoid the need for an index correction.

Purpose. To make the index read 50 when the axis of sight is horizontal.

Test. Clamp the telescope with the axis of sight horizontal. The reading of the V scale should be exactly 50.

Correction. If the V scale is not perfectly set, loosen the screws holding the arc plate, move it to a position giving a reading of 50, and tighten the screws. Proper clearance must be allowed between the arc and the circle.

16–9. Adjustment of the hand level. The horizontal cross hair is the only part of a Locke hand level which can be adjusted.

Purpose. To make the axis of sight horizontal when the bubble is centered.

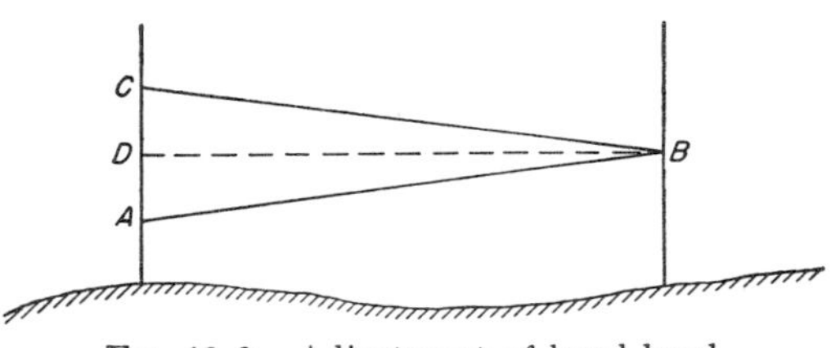

FIG. 16–9. Adjustment of hand level.

Test. With the hand level on a solid support at elevation *A*, and with the bubble centered, mark a point *B* on a post or building corner.

See Fig. 16–9. The distance AB should not be greater than 100 ft. Support the level at B, center the bubble, and note whether the line of sight strikes point A. If it does not, mark another point, C.

Correction. Bisect the distance AC and set point D. With the level at B and the bubble centered, move the cross hair to D by means of the adjusting screws.

16–10. Other instruments. Except for the bulls-eye level vials, adjustment of the self-leveling level and optical theodolites should be left to experts. Dismantling either type can result in serious damage.

16–11. Sources of error in adjustment of instruments. Some of the sources of error in adjustment of surveying instruments are as follows:

a. Shifting of the instrument—caused by settlement, pressure by the observer, or wind.
b. Adjusting screws worn or not properly set.
c. Threads of adjusting screws not fine enough for a precise setting.
d. Air boiling, instrument in direct sunlight, and sights taken through alternate sunlight and shadow.
e. Poor sights resulting from parallax.
f. Uneven wear on different parts of the instrument.

16–12. Mistakes. Typical mistakes in adjustment of surveying instruments include:

a. Arithmetical errors.
b. Adjustments not made in the proper order.
c. Failure to recheck all adjustments as the final step.

PROBLEMS

16–1. List three possible sources of error present if conditions a, b, and c given in Sec. 16–1 for adjustment do not exist.

16–2. What are the three most important rules to follow in making adjustments of surveying instruments?

16–3. State two fundamental differences between the processes involved in standardizing tapes and adjusting levels.

16–4. Tabulate three examples in which the principle of reversion is used in checking or adjusting non-surveying instruments or equipment.

16–5. Is the usual allowable error of adjustment of an engineer's level less than, equal to, or greater than errors caused by the combined effect of curvature and refraction? Explain.

16–6. What adjustments are required for a transit tripod? A tripod with a Johnson head?

16–7. Describe how the principle of reversion is used in field work, (a) with the level and (b) with the transit, to eliminate errors from inadjustments.

16–8. If the bubble is not centered at the instant of sighting in leveling, is the error personal, instrumental, natural, accidental, or systematic?

16–9. Which is the most important adjustment of the wye level? Why?

16–10. After leveling a wye level the instrumentman finds that the bubble moves off center when the telescope is rotated 180° about the vertical axis. Describe the adjustment necessary.

16–11. What defect in adjustment of the level is nullified by balancing backsights and foresights in differential-leveling operations?

16–12. Describe the proper method of adjusting for parallax when using a level.

16–13. When adjusting the wye level, why is it necessary to make the line of sight coincide with the axis of the collars before the axis of the bubble tube is made parallel to the line of sight?

16–14. If the bubble on a wye level runs when the telescope is turned between backsight and foresight, what adjustment is required? What field procedures should be used if time for adjustment is not available?

16–15. Answer the questions in problem 16–14 for a dumpy level.

16–16. Answer the questions in problem 16–14 for a transit.

16–17. Does the wye adjustment affect the accuracy of leveling? What does it accomplish?

16–18. What three precautions should be taken if work must be done with a level which is known to be out of adjustment?

16–19. Which is the more exact method of adjusting a level, the peg method or the reversal method? Why?

16–20. Compare the adjustments of a wye level with those of a dumpy level.

16–21. How does the peg adjustment of a wye level differ from that of a dumpy level? Why?

16–22. Compare the regular level vial and a reversion vial on a transit telescope with respect to advantages and adjustments.

16–23. List the adjustments of the transit, and state for each one the kind of measurement or operation in which it is of greatest importance.

16–24. List three inadjustments which will *not* affect the correct projection of a line produced by the double-centering method.

16–25. What operation of the transit requires that the horizontal axis be perpendicular to the vertical axis?

16–26. Describe, and show by a sketch with numerical values, the adjustment of a transit required to eliminate the need for double centering.

16–27. If certain observations to be made require the vertical axis of the transit to be truly plumb, which adjustments are necessary?

16–28. Which telescope axle bearing is more likely to loosen and cause the line of sight to "walk"? Why?

16–29. Describe fully the procedures for using an engineer's transit which will eliminate errors caused by lack of adjustment in measuring both horizontal and vertical angles.

16–30. Catalog the errors due to inadjustment of a transit which are eliminated by the following procedures: (a) always sighting with the intersection of the vertical and horizontal cross hairs; (b) reading vertical angles with the telescope in both the direct and reversed positions and taking the average; (c) reading horizontal angles to a high point with the telescope in both the direct and reversed positions and taking the average.

16–31. Describe the adjustment to eliminate index error.

16–32. What is the effect of the optical axis of the objective lens not coinciding with the axis of the telescope tube?

16–33. If the line of sight of a transit when revolved about the horizontal axis describes a plane which is inclined to the vertical, what adjustment is required?

16–34. Is there more than one adjustment for the hand level? Explain.

16–35. List three things which may cause a transit or level head to "freeze" on the tripod. How is the head best loosened for removal?

16–36. What is the line of collimation of a transit?

16–37. In making the peg adjustment of a dumpy level, the rod readings on points *A* and *B* were 3.25 and 4.48 ft, respectively, with the instrument set midway between the points. With the level just behind *A*, the rod reading on that point is 3.56 ft. (a) What should be the reading on point *B* if the instrument is in adjustment? (b) If the reading on *B* is 4.91 ft, explain what should be adjusted, how, and how much.

16–38. Same as 16–37 except that the given readings on *B* are 6.03 and 6.29 ft respectively.

Readings obtained in testing a wye level by the peg method are given in problems 16–39 and 16–40. What is the rod reading on *A* to give a level line of sight with the same setup at *B*? Describe the adjustment required.

16–39. Inst. at *A*: +S on *A*, 5.426; −S on *B*, 5.778.
Inst. at *B*: +S on *B*, 5.967; −S on *A*, 5.613.

16–40. Inst. at *A*: +S on *A*, 3.814; −S on *B*, 2.996.
Inst. at *B*: +S on *B*, 3.515; −S on *A*, 4.337.

16–41. With the transit at point *B*, a 600-ft backsight is taken on point *A*, the telescope plunged, and a point *C* set 400 ft away on line *AB* extended. Another backsight is taken on *A* with the telescope in its plunged position, the telescope is transited to normal, and point *C′* set. Distance *CC′* measures 0.22 ft. What is the angular error of a single plunging? Describe the adjustment required.

16–42. Similar to problem 16–41, except for an 800-ft B.S., a 600-ft F.S., and a distance *CC′* of 0.18 ft.

16–43. A horizontal angle to the right measured between a high point A and a low point B is 32°20′. To check the transit's adjustment, one sight is taken normal and one plunged on a 120-ft high tower top, the telescope depressed and points X and X' set on the ground. If the distance $XX' = 0.36$ ft and the vertical angle between A and B is approximately 45°, what error in the horizontal angle is caused by the lack of adjustment?

16–44. Similar to problem 16–43, except for a horizontal angle of 42°05′ and a distance $XX' = 0.48$ ft.

part II

chapter **17**

Determination of Meridian

17–1. General. To fix property lines and other survey courses, directions as well as lengths must be known. The direction of a line is determined by the horizontal angle between it and some reference line, usually a meridian. This meridian may be an assumed line, a magnetic north-south line, or a true meridian through the celestial poles.

The magnetic needle furnishes one means of ascertaining the north-south line. Since magnetic forces are variable in direction throughout the earth, and the compass needle is sensitive to local attraction, this method is not reliable for accurate surveying. A true north-south line determined by observations on the sun or stars is necessary if the location of points is to be permanently fixed.

North-seeking gyro attachments for use with a theodolite are now made by two instrument manufacturers. Their use by the "quick" method gives results within ± 3 minutes in about 7 minutes of observing time. A "transit" method produces results of ± 30 seconds in approximately 20 minutes of observing time.

Astronomical azimuths for first and second order triangulation must have a probable error of result not to exceed 0.3 seconds. For first, second, and third order traverse, the limits are 0.5 sec, 2.0 sec, and 5.0 sec respectively.

This chapter will present the fundamental facts of field astronomy as applied to observations on the sun and Polaris (the North Star) for azimuth. Some additional study will permit the student to make observations by using any of the stars whose positions are given in an ephemeris or the *American Nautical Almanac.* Copies of the ephemeris are made available free of charge by the various companies manufacturing surveying equipment. More-extensive tables are given in *Ephemeris of the Sun and Polaris* (published by the General Land Office), the *Nautical Almanac,* and the *American Ephemeris.*

Possession of an ephemeris will be assumed in the following discussion. (The 1968 Keuffel and Esser ephemeris has been used for reference in this chapter and the appropriate table and page number have been noted for values taken from it.)

17–2. Methods of determining azimuth. The following methods for determining an approximate north-south line will be examined briefly: (1) *Sun observations* by (a) shadow method, (b) equal-altitudes method, and (c) direct observations; (2) *circumpolar-star observations* at (a) culmination, (b) elongation, and (c) any hour angle.

Students usually find the study of field astronomy somewhat difficult because it requires many definitions, spherical trigonometry, and visualization in three dimensions. Several examples of simple methods which can be used for rough determinations of azimuth will be given before the more-technical features are discussed.

17–3. Shadow method. It is possible to establish a meridian by the shadow method without any equipment except a piece of string. In Fig. 17–1, points *A*, *B*, *C*, *D*, *E*, and *F* mark the end of the shadow of a plumbed staff or a telephone pole at intervals of perhaps 30 min of time throughout the period from 9 A.M. to 3 P.M. A smooth curve is sketched through the marks. With the staff as a center, and any appropriate radius, a circular arc is swung to obtain two intersections, *x* and *y*, with the shadow curve. A line from the staff through *m*, the midpoint of *xy*, approximates the meridian.

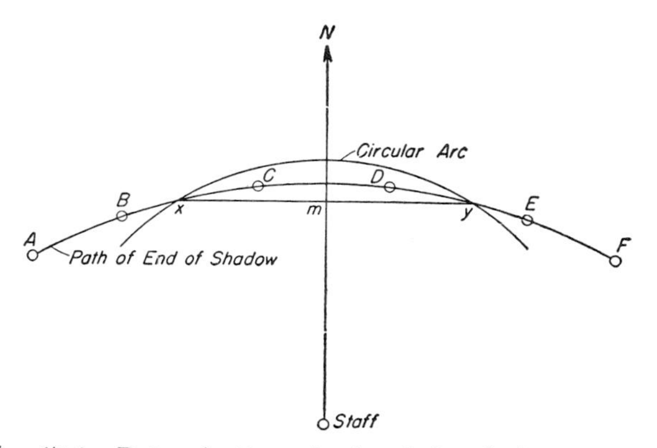

FIG. 17–1. Determination of azimuth by shadow method.

If the pole is plumb, and the ground is level, and if the shadow points are carefully marked, the angle between the line established and the true meridian should not be larger than 30 min of arc. Since the sun moves diagonally across the equator instead of traveling along a perfect east-west path, some error is inherent in the method.

17–4. Meridian by equal altitudes of the sun. Determination of the meridian by equal altitudes of the sun requires a transit, but the method is similar in principle to the shadow method. Assume that the meridian is to be passed through a point P, Fig. 17–2, over which the transit is set up. At some time between 8 A.M. and 10 A.M., say about 9 A.M., with a dark glass over the eyepiece or objective lens, the sun's disk is bisected by both the horizontal and vertical cross hairs. The vertical angle is read, the telescope is depressed, and point x is set at least 500 ft from the instrument. Shortly before 3 P.M., with the vertical angle previously read placed on the arc, the sun is followed until the vertical and horizontal cross hairs again simultaneously bisect the sun. The telescope is depressed and a point y is set at approximately the same distance from P as x. The bisector of angle xPy is the true meridian.

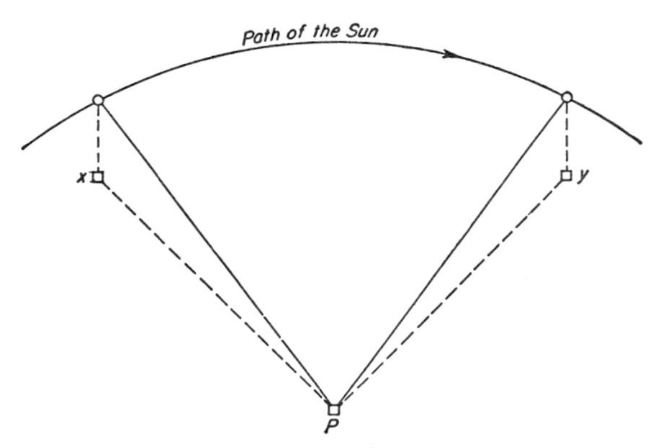

FIG. 17–2. Azimuth by equal altitudes of the sun.

As in the shadow method, perfect results cannot be attained because the changing declination of the sun gives it a diagonal path across the equator (instead of parallel with it). The declination of the sun is its distance north or south of the equator.

Other disadvantages of the method are the time required, the possibility of the sun being obscured at the time of the second sight, and the difficulty of setting the vertical angle exactly (since most arcs read only to the nearest minute). The last problem can be eliminated by setting hub x first, then elevating the telescope to observe the sun as it passes over the point, and leaving the vertical angle on the arc until the afternoon sight has been taken.

17–5. Meridian from Polaris at culmination. Figure 17–3 shows the apparent motion of Polaris as seen from the earth. The star moves in a counterclockwise direction around the north-south axis of the earth extended, but twice each day it is on the line of this axis for an observer at any location. The point at the upper limit of its travel is called *Upper Culmination* (U.C.), and the point at the lower extremity, *Lower Culmination* (L.C.). At *Eastern Elongation* (E.E.) and *Western Elongation* (W.E.) the star reaches its most distant position from the meridian.

If an observer at point O sights on Polaris at the exact instant of upper culmination or lower culmination for his longitude, he need only depress the telescope and set a hub X on the ground to obtain a true north-south line, OX. If it is more convenient to do so, the horizontal angle from a mark to the star can be measured to give the azimuth of a desired line.

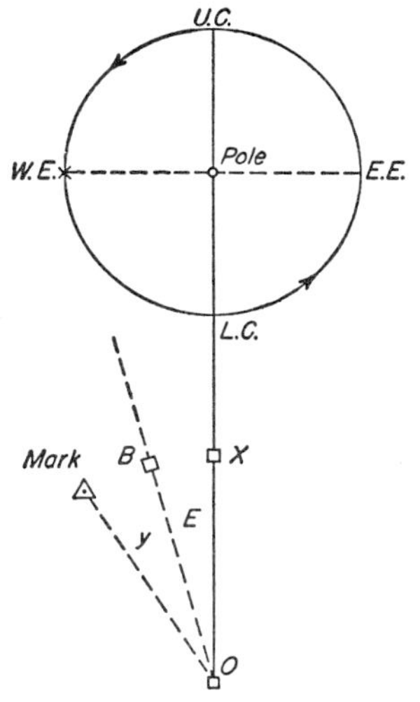

Fig. 17–3. Azimuth at culmination.

The exact time of culmination at Greenwich, England, for any date is available in an ephemeris. A few simple calculations are required to determine the time of culmination at any other location.

Some basic principles of field astronomy will now be given, along with their use in observations.

17–6. Definitions. In observations upon heavenly bodies, the sun and stars are assumed to lie on the surface of a celestial sphere of infinite radius having the same center as the earth. All stars appear to move around centers which are on the north-south axis of the celestial sphere. Figure 17–4 illustrates some of the terms used in field astronomy. Here S represents a heavenly body, as the sun or a star. Students will find it helpful to sketch the various features on a true sphere or globe.

The *zenith* is the point where the plumb line projected above the horizon meets the celestial sphere. On a diagram it is usually designated by Z. Stated differently, it is the point on the celestial sphere vertically above the observer.

The *nadir* is that point on the celestial sphere directly beneath the observer, and directly opposite to the zenith.

A *vertical circle* is any great circle of the celestial sphere passing through the zenith and nadir. As a great circle, a vertical circle is the line of intersection of a vertical plane with the celestial sphere.

An *hour circle* is any great circle on the celestial sphere whose plane is perpendicular to the plane of the celestial equator. Hour circles

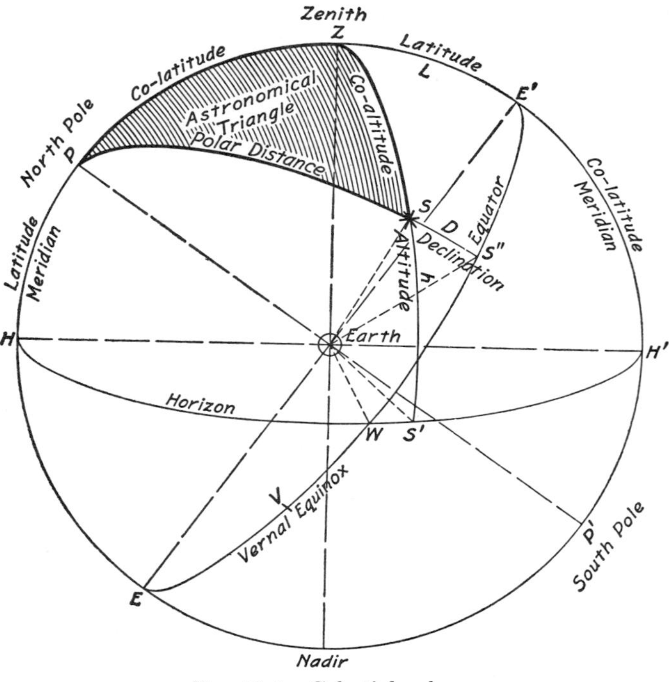

FIG. 17–4. Celestial sphere.

correspond to meridians and longitudinal lines, and are used to measure hour angles.

The *celestial equator* is the great circle on the celestial sphere whose plane is perpendicular to the axis of rotation of the earth. It is the earth's equator enlarged in diameter to that of the celestial sphere. Half of the equator is represented by $EWS''E'$.

A *celestial meridian* is the hour circle which contains the zenith. It is also defined as the vertical circle which contains the celestial pole. The intersection of the plane of the celestial meridian with the plane of the horizon is the astronomic meridian line used in plane surveying.

The *horizon* is a great circle on the celestial sphere whose plane is perpendicular to the direction of the plumb line. In surveying, the plane of the horizon is determined by the use of a spirit level. Half of the horizon is represented by $HWS'H'$.

An *hour angle* is the angle between the plane of the hour circle passing through a celestial body or point and the plane of the celestial meridian. It may be measured by the angle at the pole between an hour circle and the meridian, or by the arc of the equator which is intercepted by those circles.

The *Greenwich hour angle* of a heavenly body is the angle measured from the meridian of Greenwich westward to the meridian over which the body is passing at any moment. In the ephemeris it is designated by GHA.

The *declination* of a heavenly body is the angular distance measured along the hour circle between the body and the equator; it is plus when the body is north of the equator, and minus when south of it. Declination is usually denoted by D in formulas, and is represented by $S''S$ in Fig. 17–4. About 22 June and 22 December the sun reaches its maximum declination north and south respectively of about $23°26\frac{1}{2}'$. These are the solstices or solsticial points.

The *position* of a heavenly body with respect to the earth, at any moment, is given by its declination and Greenwich hour angle.

The *polar distance* of a body is the angular distance from the pole measured along an hour circle. It is equal to 90° minus the declination.

The *altitude* of a heavenly body is the angular distance measured along a vertical circle above the horizon. Altitude is represented by $S'S$ and in formulas is usually denoted by h. It is measured by means of the vertical arc of a transit, theodolite, or a sextant.

The *co-altitude,* or *zenith distance,* equals 90° minus the altitude.

The *azimuth* of a heavenly body is the arc of the horizon measured clockwise from either the north or south point to the vertical circle through the body. The azimuth from the south is represented by $H'S'$. It is measured as a horizontal angle by means of a transit or theodolite.

The *latitude* of an observer is the angular distance on his meridian between the equator and the zenith. It is also the angular distance between the polar axis and the horizon. Latitude is measured north or south of the equator. In formulas in this book it is denoted by L.

The *vernal equinox* is the point of intersection of the celestial equator and the ecliptic apparently traversed by the sun in passing from south to north in March. It is a fixed point on the celestial sphere (the astronomer's zero-zero point of coordinates in the sky) and moves with the celestial sphere just as the stars do. On Fig. 17–4 it is designated by V.

Refraction is the angular increase in the apparent altitude of a heavenly body due to the bending of light rays passing obliquely through the earth's atmosphere. It varies from zero for an altitude of 90° to a maximum of about 35 min at the horizon. The correction for refraction, in minutes, is roughly equal to the natural cotangent of the observed altitude. Small corrections must also be made for temperature and pressure variations.

Refraction makes observations on heavenly bodies near the horizon less reliable than those taken at high altitudes. The correction is always subtracted from observed altitudes.

Parallax results from observations being made from the surface of the earth instead of at its center. It causes a small angular decrease in the apparent altitude, hence the correction is always added. Parallax is insignificant when observing a star but must be added on solar shots. The ephemeris contains tables listing refraction and parallax corrections.

17–7. Time. Four kinds of time may be used in computing an observation.

Sidereal time. A sidereal day is the interval of time between two successive upper transits of the vernal equinox over the same meridian. Sidereal time is star time. The sidereal time at any instant is equal to the hour angle of the vernal equinox.

Apparent solar time. An apparent solar day is the interval of time between two successive lower transits of the sun. Apparent solar time is sun time, and the length of the day varies somewhat.

The average apparent solar day is 3 min 56 sec longer than a sidereal day.

Mean solar, or civil, time. This is the time kept by a fictitious sun which is assumed to move at a uniform rate. It is the basis for watch time and the 24-hr day.

The *equation of time* is the difference between mean solar and apparent solar time. Its value is continually changing as the sun gets ahead of, then falls behind, the mean sun. Values for each day of the year are given in an ephemeris.

Standard time. This is the mean time at meridians 15 deg or 1 hr apart, measured eastward and westward from Greenwich. Eastern standard time (E.S.T.) is the time of the 75th meridian and differs from Greenwich time by 5 hr (earlier, since the sun has not yet traveled from the meridian of Greenwich to the United States). Standard time was adopted in the United States in 1883 replacing 100 local times previously used. Daylight saving time (DST) in any zone is equal to standard time in the zone to the east.

In working with longitude and time zones, it is helpful to remember the following relations:

$$360° \text{ of longitude} = 24 \text{ hours}$$
$$15° = 1 \text{ hour}$$
$$1° = 4 \text{ minutes (of time)}$$

17–8. Star positions. If the pole could be seen as a definite point in the sky marked by a star, a meridian observation would require only a simple sighting. Since the pole is not so marked, observations must be made on stars—preferably stars close to the pole—whose radii of rotation and positions at given times are listed in a nautical almanac or an ephemeris. Stars appear to move counterclockwise because of the earth's clockwise rotation.

The visible star nearest the north pole is Polaris, a part of the constellation Ursa Minor, which constellation is also called the Little Dipper. The radius of rotation of Polaris, measured by a vertical angle along the meridian through the position of an observer, is its polar distance. The value of this angle changes slightly from year to year but it is approximately 0° 55′.

Polaris is located in the sky by first finding the Big Dipper, in Ursa Major. The two stars of the dipper farthest from the handle are the Pointers, as shown in Fig. 17–5. Polaris is the nearest bright

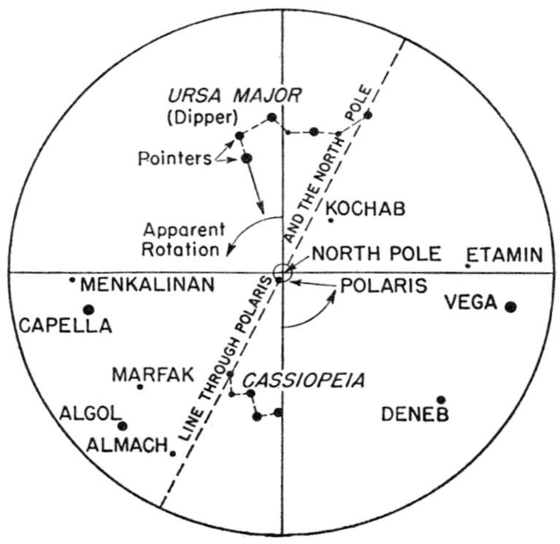

FIG. 17–5. Position at polaris.

star along the line through the pointers and is also distinguished by its position at the end of the handle of the Little Dipper. In 12,000 years, however, it will have moved away from its present position (but will return in about 25,000 years!) and a new star come close to the pole. The Southern Cross is used for observations in the Southern hemisphere since there is not a bright star near the pole.

17–9. Observation on Polaris at culmination. As previously noted, if Polaris can be observed at upper or lower culmination, the meridian is readily laid out on the ground. At culmination, however, the star appears to be moving rapidly in the east-west or west-east direction. Accurate time is therefore required to obtain a precise meridian, and only a single sight is possible for each culmination.

Since the sidereal day is 23 hr 56.1 min and the mean solar (civil) day is 24 hr, culmination occurs about 4 min earlier each day of the calendar year. Over a period of many years, however, there is very little difference in the time of culmination for the same date and location. For the example of Sec. 17–10, the time of upper culmination on 15 December 1961 was $20^h\ 21.3^m$ or only 3.3 minutes different in 7 years. Thus an old ephemeris might be satisfactory for the approximate time of elongation if the meridian determination to the nearest few minutes is good enough.

Ephemerides list the time of culmination and elongation at Greenwich for every 10th or 15th day throughout the year. It is necessary to convert these tabulations to the standard time of the place of observation. The computational methods illustrated herein are believed to provide the simplest approach to field astronomy for beginners in the subject.

An observation at culmination is simple to compute. There are three important disadvantages, however: A small error in time causes a relatively large error in azimuth; both culminations may come at awkward hours—even during daylight in the summer; and with only a single pointing on the star, any instrumental errors will cause an error in direction of the meridian when the telescope is transited.

17–10. Computations for time of culmination. Three computations or corrections must be made to convert the time of culmination listed in an emphemeris for Greenwich, England, to the standard (watch) time for culmination at any other location. These computations will be illustrated for an observation on 18 December 1968 at latitude 46° 42′ North, longitude 93° 18′ West, by taking pertinent values from an ephemeris.

1) Correction to convert the time of upper culmination tabulated in the ephemeris for Greenwich on certain dates, to the actual date of observation. The result obtained by interpolating in the table is as follows. The interpolated date or value is inset.

1968 K&E Ephem. Table 8, p. 183	15 December	$20^h\ 24.6^m$ GCT	
	18 December	$\frac{3}{10} \times 39.5 = 11.8^m$	$20^h\ 12.8^m$
	25 December	$19^h\ 45.1^m$	

2) Correction for longitude 93° 18′ west of Greenwich. A place at longitude 93° 18′ is approximately $\frac{1}{4}$ the way around the world from Greenwich, and therefore the time difference is $\frac{1}{4}$ day. The average change (decrease) in time of culmination per day (or per 360° of longitude) is 3.93 min over an entire year, but for the 10-day interval being used, the change is 3.95 min per day. The correction for 93° 18′ of west longitude is

$$3.95 \text{ min} \times \frac{93.3^\circ}{360^\circ} = -1.0 \text{ min}$$

3) Correction for longitude west of 90th meridian. A correction

must be applied for the difference in time of culmination at longitude 90° and that at 93° 18′. This means converting standard (watch) time to local time at the place of observation. Since 1° corresponds to 4 min of time, the correction is

$$3.30° \times 4 \text{ min/degree} = +13.2 \text{ min}$$

The plus indicates that culmination occurs later at longitude 93° 18′ West than at the time-zone meridian. Note that standard time at the 90th meridian is six hours earlier than Greenwich time, but it takes Polaris approximately six hours to travel from the meridian of Greenwich to the meridian of the place of observation.

The three computations show that at the moment Polaris culminates over the meridian of the observer at longitude 93° 18′, a watch set for the time zone of the 90th meridian should read

$$20^h\,12.8^m - 1.0^m + 13.2^m = 20^h\,25.0^m$$

Notice that if the observer had been on the 90th meridian a correction of only 1 min would have been necessary to get the local time of culmination after interpolating in the ephemeris.

Ephemeris tables are based on 40° latitude as an average for the United States. Since the star is on the meridian at culmination, latitude has no effect on the watch time.

17–11. Observation on Polaris at elongation. Observing Polaris at elongation instead of at culmination has several advantages: (a) The star has a negligible east-west or west-east movement and appears to travel along the vertical wire for at least 15 min before and after elongation. (b) The watch time therefore need not be precise. (c) Several sights can be taken with the telescope normal and plunged, and averaged to eliminate instrumental and personal errors. A minor disadvantage is the somewhat longer computation procedure.

The observation is made by first sighting on an illuminated fixed mark with the plates set to zero, and then loosening the upper motion and sighting on the star. The angle from mark to star, y in Fig. 17–3, is read. The telescope is plunged, a second sight taken on the star, and the angle read again. The process is repeated for the desired number of repetitions. As a final step, the original mark should be resighted to be certain the plates have not been disturbed and still read zero.

An alternative method is to sight the star at elongation with zero on the plates, depress the telescope, and set a stake *B*, Fig. 17–3, at least several hundred feet from the instrument. The procedure is repeated after plunging as a check. The direction to true north can be laid off the next day after computing the angle *E*.

Illustration 17–1

COMPUTATION OF THE BEARING OF POLARIS AT ELONGATION
Longitude 93°18′ W Latitude 46°42′ N 19 Dec. 1968

(Table 8, p. 183) Time of W. E. at Greenwich, 19 Dec. 1968		2^h 08.9^m
Correction for longitude west of Greenwich	−	1.0^m
Correction for longitude west of 90th meridian	+	13.2^m
(Table 9, p. 183) Correction for latitude other than 40°	−	0.8^m
Time of W. E. at observer's location		2^h 20.3^m

Polar distance of Polaris by interpolation from the ephemeris:

(Table 3, p. 175) 16 December	0° 52.39′	
19 December	$\frac{3}{10} \times 0.04 = 0.01$	0° 52.38′
26 December	0° 52.35′	

Bearing of Polaris at W. E. by double interpolation and extrapolation:

Polar distances	0°52.38′	0°52.50′	0°52.70′
(Table 7, p. 182) Latitude 46°	*1°15.4′*	1°15.6′	1°15.9′
Latitude 46° 42′	1°16.4′		
Latitude 47°	*1°16.8′*	1°17.0′	1°17.3′

17–12. Computations for an observation at elongation. A computation for the time of western elongation and the bearing of Polaris at elongation is given in illustration 17–1. The data (except the date) are the same as in Sec. 17–10, but an additional (fourth) correction must be made for latitude other than 40°.

The star is 1° 16.4′ to the west of north at elongation. This is the value of angle *E*, Fig. 17–3. Adding the angle *y* between the mark and the star produces the desired true bearing of the line from *O* to the mark.

Another means of computing the angle *E* between Polaris at elongation and the meridian is by the formula

$$\sin E = \frac{(\sin \text{polar distance})}{\cos \text{latitude}} \qquad (17\text{–}1)$$

or

$$\text{Bearing of Polaris (in minutes)} = \frac{\text{polar distance}}{\text{cos latitude}} \qquad (17\text{–}1a)$$

In the example,

$$\sin E = \frac{\sin 0°52.38'}{\cos 46°\ 42'} = 0.022213 \qquad E = 1°16.4'$$

or

$$E = \frac{0°52.38'}{\cos.\ 46°42'} = \frac{0°52.38'}{0.68582} = 76.4 \text{ min} = 1°16.4'$$

The mean polar distances for the years 1968 through 1975 are shown in Table 17–1.

TABLE 17–1

MEAN POLAR DISTANCES FOR THE YEARS 1968–1975

Year	Mean Polar Distance	Year	Mean Polar Distance
1968	0° 53′ 00″	1972	0° 51′ 51″
1969	0° 52′ 43″	1973	0° 51′ 34″
1970	0° 52′ 26″	1974	0° 51′ 17″
1971	0° 52′ 08″	1975	0° 51′ 00″

For observations made less than 30 min before or after elongation, the correction to the bearing at elongation can be computed by the formula

$$C = Kt^2 \qquad (17\text{–}2)$$

where C = a correction to the angle, in seconds of arc;
K = a factor depending upon the bearing at elongation = $\frac{(15)^2}{6} \times (60)^2 \times \sin 1'' \tan Z$, where Z is the azimuth at greatest elongation.
t = the time, in minutes, before or after elongation.

The value of K (which does not change with time) may be found from Table 17–2.

TABLE 17–2

CORRECTION FACTORS FOR BEARINGS AT ELONGATION

Bearing at Elongation	K
1° 00′	0.034
1° 10′	0.040
1° 20′	0.046
1° 30′	0.051
1° 40′	0.057
1° 50′	0.063

For the values previously found and an observation 30 min after elongation, the correction is

$$C = 0.046(30)^2 = 41 \text{ sec}$$

The small value of the correction demonstrates the advantage of observing Polaris at or near elongation since a large error in watch time causes only a small error in the azimuth.

17–13. Observation on Polaris at any hour angle. Frequently it is more convenient to make an observation on Polaris at any hour angle rather than at culmination or elongation. The difference between the time of observation and the time of culmination is computed, and the bearing of Polaris at that hour angle taken from the ephemeris.

A sample set of field notes and the computations for the bearing of the reference line are shown in illustration 17–2. A second horizontal angle should always be obtained with the telescope plunged, and the watch time recorded, to permit computation of an average value of the bearing.

17–14. Practical suggestions on Polaris observations. The following suggestions make observations on Polaris easier to perform:

1. It is most convenient to use eastern elongation during the late summer season in the United States.
2. Have the noteforms ready in advance of starting field work.
3. Have a watch, flashlight, reflector, and pencils. Note that the accuracy of vernier readings is lowered at night. Lighting from the side causes parallax if the plates are not at the same level. Hold the flashlight over the compass box behind the ground-glass upright piece and let the light diffuse through it.
4. The mark must be visible at night, and recognizable and definite in the daytime. It should be at least one mile away if possible to avoid instrumental errors in refocusing between mark and star, star and mark sights.
5. Set up the transit in daylight if possible. If the altitude of Polaris is known, observations can begin before dark.
6. The altitude of Polaris is equal to the latitude of the place when the star is at elongation. The star may be a degree off and out of the field of view near culmination.
7. Use a paper reflector held on the telescope by a rubber band. The light must not be too bright or the star becomes invisible.

It cannot be too dim or the cross hairs fade out. Most theodolites have a special receptacle to accommodate batteries and a bulb for lighting the cross wires, horizontal circle, micrometer scale, and vertical arc.

8. For pointings at high altitudes, the telescope objective lens may slide back and throw the instrument out of focus.

Illustration 17–2

OBSERVATION ON POLARIS AT ANY HOUR ANGLE

Data

⩚ at △ Cass. Lat. 47°22′13″ N Long. 94°32′18″ W 6 Sept. 1968

Telescope	Point Sighted	Watch Time	Watch Correction	Horizontal Angle Clockwise	Vertical Angle
	△ Snow			0° 00′	
N	Polaris	$5^h\ 44^m\ 00^s$	$0^m\ 00^s$	72° 22′ 30″	48° 08′
P	Polaris	$5^h\ 44^m\ 24^s$		72° 22′ 50″	
	△ Snow			0° 00′	
		Ave. $5^h\ 44^m\ 12^s$		72° 22′ 40″	

Hour Angle of Polaris

GCT of upper culmination 7 September 1968		$2^h\ 57.5^m$
Correction for one day at 3.91^m per day	+	3.9^m
GCT of upper culmination 6 September 1968		$3^h\ 01.4^m$
Correction for longitude 94.54° west of Greenwich	−	1.0^m
Local time of upper culmination		$3^h\ 00.4^m$
Correction for longtitude 4.54° west of 90th meridian	+	18.2^m
CST of upper culmination at longitude 94° 32.3′ W		$3^h\ 18.6^m$
Actual time of observation		$5^h\ 44.2^m$
Time difference from culmination		$2^h\ 25.6^m$
Correction for 3.91^m per day (approx. 10^s per hour)	+	0.4^m
Hour angle (west of upper culmination)		$2^h\ 26.0^m$

Bearing

(Table 10, p. 185) For latitude 47° 22.2′, tabular polar distance 0° 52.70′, and hour angle $2^h\ 26.0^m$ (36.50°), by double interpolation, bearing of Polaris (west of north)		0° 46.9′
(Table 11, p. 186) Correction (by double interpolation for polar distance 0° 52.97′ and bearing 0° 46.9′	+	0.2′
Bearing of Polaris		0° 47.1′
Clockwise angle △ Snow to Polaris		72° 22.7′
Clockwise angle △ Snow to North		73° 09.8′
Bearing △ Cass to △ Snow		N 73° 09.8′W

9. Use a sketch to eliminate errors in the relationships between the mark, north, and the star.
10. The value of 1 deg of arc in miles indicates the limit of accuracy obtainable. If the assumed latitude of the observer's position is in error by 1 mile, the resulting maximum error in azimuth at a latitude of 45° is roughly 1.5 sec.

17–15. Comparison of solar and Polaris observations. Compared with Polaris observations, sights on the sun (a) are more convenient, (b) will not give results so precise, (c) can be made while a daytime survey is in progress, and (d) have a limiting accuracy of perhaps 1 or 2 min with methods normally used.

17–16. Methods of observing the sun. Observations on the sun can be made directly by placing over the eyepiece a dark glass which is optically plane and parallel. They can be taken indirectly by focusing the sun's image on a piece of paper held behind the eyepiece. Looking at the sun directly through the telescope without a dark glass may result in permanent injury to the eye.

The best hours for observing are between 8 and 10 A.M., and between 2 and 4 P.M.

At least four observations on the sun, two normal and two plunged, with a minimum of time between them should be obtained since the sun's disc does not provide an easy target. The Roelof solar prism, when properly aligned, produces 4 intersecting suns permitting a bisection to within perhaps 5 seconds. The path of the sun can be considered a straight line over a period of about 5 min, and an average value of the readings can be used in computations. Plotting the vertical and horizontal angles against time for the four readings provides a good check. Note in illustration 17–3 that for approximately equal differences in time between observations, the changes in vertical angles are approximately equal, and the changes in horizontal angles are roughly equal.

The diameter of the sun as viewed from the earth is approximately 32 min of arc. Bisecting such a large object which is moving both horizontally and vertically is difficult, but the average person can do it with an accuracy of perhaps 1 min of arc.

A better method is to observe the border of the sun. In Fig. 17–6a the disk is brought tangent to both cross hairs first in one quadrant and then in the diagonally opposite quadrant. Averaging the two readings eliminates consideration of the semidiameter.

Illustration 17–3

SOLAR OBSERVATION FOR AZIMUTH

⊼ at △ Rover Latitude 42° 45′ N Longitude 73° 56′ W 30 July 1968

Point Sighted	Telescope	Watch Time (EDST)	Vertical Angle	Horizontal Angle (Clockwise)
△ Ridge				0° 00′
⊙ Sun	Direct	$3^h 33^m 10^s$	52° 54′	209° 33′
⊙ Sun	Direct	$3^h 34^m 20^s$	52° 43′	209° 42′
⊙ Sun	Plunged	$3^h 35^m 37^s$	52° 22′	210° 02′
⊙ Sun	Plunged	$3^h 36^m 49^s$	52° 10′	210° 11′
△ Ridge				0° 00′
Mean		$3^h 34^m 59^s$	52° 32.2′	209° 52′

Index error	=	0° 00.0′
(Tables 2, 3, 4, p. 173, 174) Refraction and parallax corr. (80°F, elev. 1200 ft) −0.73(0.97)(0.94)+0.08	= −	0.6′
True altitude	=	52° 31.6′

Correction	0^h GCT to noon	= +	12^h
Correction	To 5th time zone	= +	5^h
Correction	For daylight saving	= −	1^h
EDST of observation		=	$3^h 34^m 59^s$
Greenwich civil time of observation		=	$19^h 34^m 59^s = 19.58^h$

(Table 1, p. 166) Declination 31 July 1968 0^h GCT	=	N 18° 18.8′
Corr. for 4.42^h earlier = 0.61′ × 4.42^h	=	+ 02.7′
Declination	=	N 18° 21.5′

$$\cos Z_n = \frac{\sin D}{\cos L \cos h} - \tan L \tan h$$

$$\cos Z_n = \frac{\sin 18° 21.5'}{\cos 42° 45' \cos 52° 31.6'} - \tan 42° 45' \tan 52° 31.6'$$

$$= \frac{0.314959}{(0.734322)(0.608392)} - (0.924390)(1.30448) = -0.500857$$

$$Z_n = 120° 03.4'$$

Since the observation was made in the afternoon, this angle is counterclockwise from north. The minus sign indicates an angle greater than 90°.

Azimuth of sun	=	239° 56.6′
△ Ridge to sun	=	209° 52′
Azimuth of △ Rover to △ Ridge	=	30° 05′ (from north)

FIG. 17-6. Observation on the sun.

To avoid coordinating the movement of both tangent screws with the sun, it is simpler to follow the disk by keeping the vertical cross hair tangent to it and letting the sun come tangent to the horizontal hair, as in Fig. 17-6b. Since the altitude changes faster than the azimuth during the observing hours, it is preferable to keep the vertical hair tangent.

The transit is oriented by setting the plates to read zero and sighting along the fixed line from the observer's position to a mark. After recording the times, the vertical angles, and the horizontal angles for four or more observations, the mark is resighted to be certain the plates still read zero.

Before work is begun, the observer should have (a) the noteforms ready, (b) an ephemeris and the correct watch time, (c) the latitude and longitude of the place of observation, and (d) the index error of the instrument.

17-17. Required quantities in determining azimuth by direct solar observation. Five things must be known or determined in a solar observation for azimuth: the latitude, the declination of the sun, the time, the altitude of the sun, and the horizontal angle from some reference line to the sun.

Latitude can be taken from a map or found by a separate observation. North latitudes are considered plus in the standard formulas. For locations within the United States, an error of 1 minute in (a) the assumed latitude (i.e., approximately 1 mile), or (b) in the declination, or (c) in the measured altitude, produces an error in the computed azimuth ranging from a few seconds to a maximum of about $3\frac{1}{3}$ minutes, depending upon the time of the year and the declination on the date of observation, the hour angle, and the magnitude of the measured altitude.

Declination of the sun is given in ephemerides for each day of the year, along with the change per hour. The declination of the sun depends on its astronomical position and is independent of the observer's location. It is the same for all observers in any part of the world at the same instant. North declinations are always considered plus, regardless of the latitude of the observer.

The time of observation must be adjusted for the difference between watch time and Greenwich civil time. A correction is required only for the time zone, not for the exact longitude of the observer. Every manufacturer's ephemeris now lists the declination for Greenwich civil noon, rather than for Greenwich apparent noon, to eliminate the correction for the equation of time. Greenwich noon is 12 hours Greenwich civil time.

The apparent altitude of the sun is its angular distance above the horizon as measured by the engineer's transit. To obtain the true altitude, the observed value must be corrected for index error, for refraction and parallax, as indicated in Fig. 17–7, and for the semi-diameter of the sun if the border is sighted in only one quadrant. Errors of adjustment of the standards and cross hairs are eliminated by sighting with the telescope both normal and plunged. Careful leveling is required.

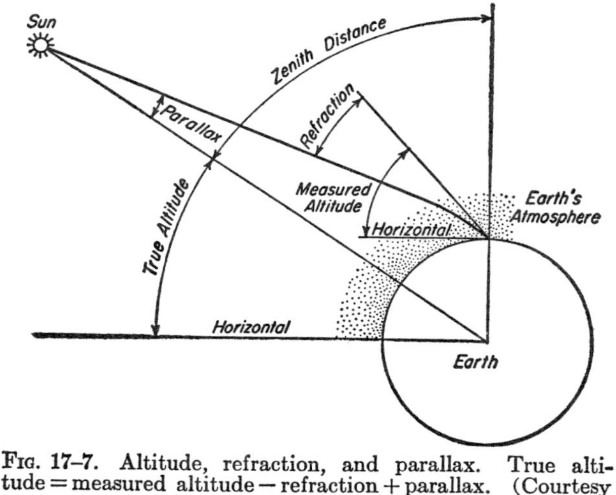

FIG. 17–7. Altitude, refraction, and parallax. True altitude = measured altitude − refraction + parallax. (Courtesy of Keuffel and Esser Company.)

The horizontal angle to the center of the sun is measured from some fixed line. This angle combined with the computed bearing of the sun at any instant gives the bearing of the reference line.

A solar attachment which costs about the same amount as a transit or lower-order theodolite gives a mechanical solution of the *PZS* triangle. By setting the latitude, declination, and hour angle on three scales and sighting the sun, a mechanical procedure places the telescope in the north-south direction.

17–18. Notes and computations for a solar observation. Determination of the azimuth of a line by observation on the sun requires the solution of a spherical triangle defined by the zenith, the north pole, and the sun. This is the *PZS* triangle, Fig. 17–4. The angle *PZS*, or Z_n, at the zenith is the bearing of the sun from the meridian through the place of observation.

Forms of the equation used in solving the *PZS* triangle are

$$\cos Z_n = \frac{\sin D}{\cos L \cos h} - \tan L \tan h \qquad (17\text{–}3)$$

and

$$\cos Z_n = \sin D \sec L \sec h - \tan L \tan h \qquad (17\text{–}4)$$

where Z_n = angle from the meridian to the sun, measured clockwise in the morning and counterclockwise in the afternoon;
D = declination of the sun at the moment of observation;

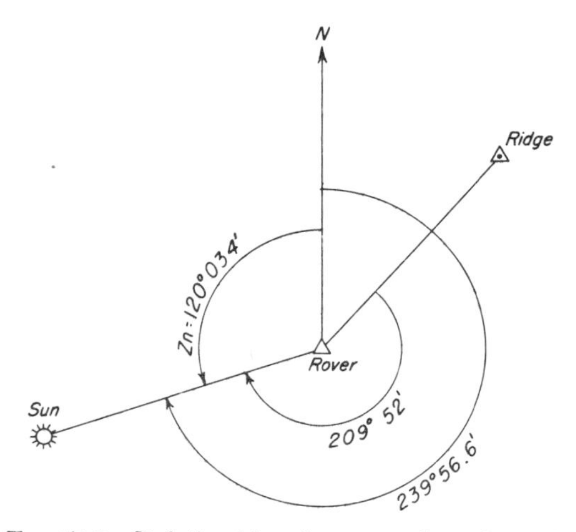

Fig. 17–8. Relationship of sun, mark, and north.

L = latitude of the place;
h = altitude of the center of the sun at the time of observation.

A different form of the equation is desirable for computations performed on a calculating machine.

An example illustrating the field data and calculations for a solar observation is given in illustration 17–3. The conditions are shown in Fig. 17–8.

17–19. Sources of error in meridian observations. Some of the sources of error in meridian observations include:

a. Transit not perfectly leveled.
b. Horizontal axis of the instrument not truly horizontal.
c. Index error not corrected.
d. Sun not bisected by both cross hairs.
e. Time not correct, or not read exactly at the moment of observation.
f. Parallax in readings taken at night.

17–20. Mistakes. Some of the more common mistakes that occur in observations for meridian are:

a. Sighting on the wrong star.
b. Using a poor signal on the reference line.
c. Computational errors.

PROBLEMS

17–1. Determine the direction of true north by the shadow method, using a flagpole or a telephone pole. Record the data obtained (times and taped or paced distances) and include a sketch. Compare the result with a meridian established by an observation on the sun or Polaris. Check the local magnetic declination by reading the bearing of the line obtained.

17–2. Draw a revised Fig. 17–1 to show a true sketch of the sun's path for a determination of the meridian in the month of May if the observer is at latitude 30° and the shadow to the south.

17–3. The coordinates of a point on the earth's surface are latitude and longitude. What are the coordinates of the sun in the same system?

17–4. A TV tower 150 ft high casts a shadow 184 ft long on level ground. What is the altitude of the sun?

17–5. Why is sun-dial time not accurate?

17–6. When is the sun's declination 0° 00′? When is it minus?

17–7. (a) What is the declination of a star as it passes through the zenith of an observer in latitude 24° 17′ S? (b) Of a star as it rises due east of an observer in latitude 38° 50′ N?

17–8. Is an observer's zenith always on his celestial meridian? Explain.

17–9. If at a given instant the GHA of Polaris is 165° 12′ what is the LHA of Polaris for an observer at longitude 70° W?

17–10. At what latitude are the altitude and declination of a celestial body equal?

17–11. Why is the vertical angle corrected for parallax in observations on the sun but not in observations on Polaris?

17–12. Why is the moon not commonly used for meridian observations?

17–13. Why does the calendar day start at midnight while the astronomer's day starts at noon?

17–14. What is the reason for having both apparent and mean solar time?

17–15. Do standard time zones follow exactly 15° intervals? Explain.

17–16. Explain the gain or loss of one day in crossing the International Date Line.

17–17. What is the central meridian for the Pacific Standard Time Zone?

17–18. How often is Apparent Solar Time the same as Civil Time?

17–19. Determine the maximum difference between apparent and mean solar time during this year.

At Greenwich noon, what is the watch reading for the locations and conditions given in problems 17–20 through 17–23?

17–20. Longitude 115° 08′ E, one hour of DST.

17–21. Longitude 84° 15′ E, two hours of DST.

17–22. Longitude 115° 08′ W, two hours of D.S.T.

17–23. Longitude 84° 15′ W, one hour of DST.

17–24. List the primary reference planes used in astronomical coordinate systems.

17–25. Explain what is meant by the Equation of Time.

17–26. Tabulate the azimuth and altitude of the (a) west point, (b) north point, and (c) celestial pole.

17–27. Give the hour angle of the (a) south point, (b) east point, and (c) zenith.

17–28. List the right ascension and declination of the vernal equinox and the autumnal equinox.

17–29. At what local civil time will the center of the sun be on the meridian 30 March of this year at longitude 75° W?

17–30. State the changes which take place in the azimuth and zenith distance of Polaris during 24 hours.

17–31. Express in time a difference of longitude of 49° 59′ 43″.

17–32. What is the watch time at Washington, D.C. (longitude $5^h08^m16^s$) when it is 9^h03^m A.M. at San Francisco (longitude $8^h09^m43^s$)?

17–33. Compute the watch time in Rome (longitude $0^h49^m30^sE$) when the time in Chicago (longitude $5^h50^m27^s$) is $11^h42^m45^s$ P.M. on 3 August.

17–34. On 6 May, for longitude 78° 00′ W at 11:00 A.M., compute the (a) standard apparent time, (b) local mean time, and (c) local apparent time if the equation of time at 0 hour GCT on 6 May is + $3^m24.3^s$ and the change per hour is +0.19 sec.

17–35. Describe and show by means of a sketch a method for determining, without the use of a table or any equipment, whether Polaris is on the meridian of the observer.

17–36. What is the most accurate method of determining the true meridian that involves the least amount of computation?

17–37. In a rough observation for true north taken on Polaris at Latitude 50° 40′ N, the movement off culmination was not considered. What maximum error in direction is possible due to this one factor?

17–38. What is the latitude of an observer, if the altitude of Polaris at U.C. is 32° 18′ and its declination is 89° 07.4′?

17–39. The vertical angle to the center of the sun measured as it crosses the observer's meridian is +42° 16′ after all corrections have been made. If the sun's declination at that time was −6° 17′, what is the observer's latitude?

17–40. Compute the hour angle of Polaris at 9:00 P.M. PST on 17 November at latitude 40° N and longitude 118° W.

17–41. The latitude of an observer is 35° 10′ 40″ and the declination of Polaris is 89° 07.4′. Compute the zenith distance of the star if it is on the observer's meridian (a) above the pole, (b) below the pole.

An observation for azimuth is made from point *X* on Polaris at Upper Culmination under the conditions listed in problems 17–42 through 17–45. Compute the four time-conversion corrections for the point.

17–42. Latitude 46° 42′ N, Longitude 93° 18′ W (Madison, Wisc.), 4 May.

17–43. Latitude 40° 50′ N, Longitude 73° 57.5′ W (New York), 6 June.

17–44. Latitude 39° 58′ N, Longitude 75° 15.2′ W (Philadelphia), 10 August.

17–45. Latitude 32° 18′ N, Longitude 106° 48′ W (Las Cruces, N.M.), 15 Nov.

17–46. Similar to problem 17–42 except for E.E. on 25 March.

17–47. Similar to problem 17–43 except for W.E. on 21 February.

17–48. Similar to problem 17–44 except for E.E. on 13 September.

17–49. In problem 17–46, if the angle to the right from a line *EF* to Polaris is 74° 46′, what is the true bearing of *EF?*

17–50. In problem 17–47, if the angle to the right from a line *KL* to Polaris is 235° 58′, what is the true bearing of *KL?*

17–51. An observer at point *X* in latitude 40° N sights on Polaris 2 hr after E.E. and reads an angle from *Y* to the star of 42° 06′ R. What is the azimuth of *XY?*

17–52. Similar to problem 17–51, but for 3 hr after W.E.

17–53. At latitude 35° N, how long will Polaris appear to climb (or descend) the vertical cross line without departing more than 20 sec in arc?

17–54. Similar to problem 17–53, except for 30 sec at latitude 50° N.

17–55. What error in azimuth is produced in an observation on Polaris at latitude 45° N if the time is off by 1 min and the star is (a) at culmination, (b) at elongation, and (c) about midway between U.C. and W.E.?

17–56. What approximate error in azimuth results from a mistake of 1 min in reading the altitude of the sun at 10 A.M. on 10 Aug. at latitude 40° N?

17–57. In making an observation on the sun for azimuth, two normal readings and two plunged readings are taken. Show by a sketch the check available within the data themselves before computations are made.

17–58. Determine the *GHA* of the sun and the declination when the standard time in N.Y. (longitude $4^h55^m50^s$) is $3^h38^m46^s$ P.M. on 10 October of this year.

17–59. What is the maximum declination of the sun at latitude 32° N, and on what date does it occur?

17–60. In observations on the sun for azimuth, what corrections must be made after recording time, vertical and horizontal angles?

17–61. List the sides of the *PZS* triangle used in computing a solar observation, and the angles opposite them.

In problems 17–62 through 17–65 compute the azimuth of line *AB* for the following data taken this year at hub *A*. The average of 4 watch times, 4 horizontal angles to the right from *AB* to the sun, and 4 vertical angles with the sun in diagonally opposite quadrants is listed.

	Date	Time	Temp.	Lat.	Long.	H Angle	V Angle
17–62.	22 May	3:00 PM	90 F	32° 18′ N	106° 48′ W	34° 00′	26° 00′
17–63.	25 Sept.	3:16 PM	75 F	36° 00′ N	117° 12′ W	53° 24′	26°54′
17–64.	3 Nov.	2:24 PM	50 F	42° 21′ N	71° 15′ W	166° 30′	22° 09′
17–65.	10 Oct.	9:06 AM	42 F	39° 41′ N	104° 58′ W	101° 36′	28° 24′

chapter **18**

Photogrammetry

PAUL R. WOLF
Associate Professor of Civil Engineering
University of Wisconsin, Madison

18–1. General. Photogrammetry is defined as the science, art, and technology of obtaining reliable measurements and qualitative information from photographs. As indicated by this definition, photogrammetry encompasses two major areas of specialization

FIG. 18–1. Phototheodolite. (Courtesy Wild Heerbrugg Instruments, Inc.)

which include (1) the determination of horizontal distances and elevations, and compilation of mosaics and topographic maps; and (2) *photographic interpretation,* the recognition and identification of objects appearing on photographs. The first area will be treated more extensively than the second although interpretation of photographs is an extremely important aspect of aerial photogrammetry.

An aerial photograph completely depicts all cultural features within the region covered. Roads, railroads, rivers, trees, factories, and cultivated lands are easily distinguished. Critical factors considered in identifying objects include their shape, size, pattern, shadow, tone, texture and site location. Photographic interpretation requires study, patience, imagination, judgment, experience, and above all an understanding of the field of endeavor for which the interpretation is being done.

Two basic types of photographs are used in photogrammetry: *terrestrial photographs* exposed with the camera on the ground, and *aerial photographs* taken wtih the camera carried by any airborne vehicle. The phototheodolite, Fig. 18–1, is a combination camera and theodolite mounted on a tripod for terrestrial work. Aerial photographs are taken with a camera of the type shown in Fig. 18–2. *Vertical photographs,* those taken with the camera axis vertical, or as nearly vertical as possible, are most commonly used. *Oblique photographs* made with the camera axis inclined intentionally at some angle between the horizontal and vertical are also utilized. A *high oblique* includes the horizon in the field of view, a *low oblique* does not.

Photography dates back to 1839 and the first attempt to use photogrammetry to fashion a topographic map occurred a year later. Photogrammetry has now emerged as the chief and almost exclusive method of topographic compilation and mapping by the U. S. Geological Survey. Cameras, films, plotting instruments, and techniques have been continually improved so that maps prepared from aerial photographs today are superior to those made by transit-stadia or plane-table methods. Other advantages of photogrammetric mapping include (a) speed of coverage of an area, (b) relatively low cost, (c) ease of obtaining topographic details especially in inaccessible areas, and (d) reduced likelihood of omitting data due to the tremendous amount of detail shown in the photographs.

Photogrammetry is presently being applied in many fields other

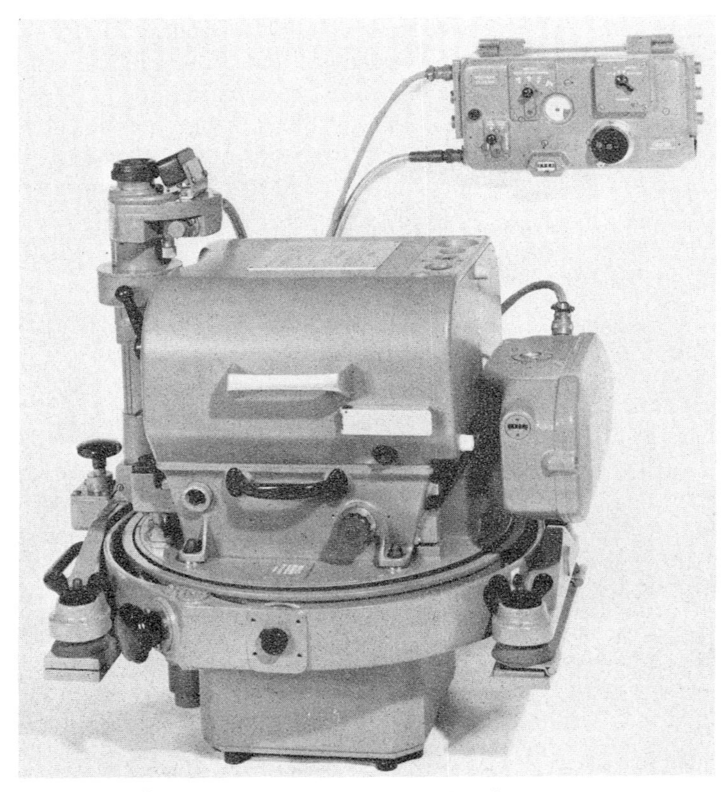

FIG. 18–2. Aerial camera. (Courtesy Wild Heerbrugg Instruments, Inc.)

than topographic mapping, e.g., geology, archeology, forestry, agriculture, conservation, planning, traffic management, military intelligence, medicine, dentistry, extra-terrestrial mapping, and engineering design. In the field of highway engineering, photogrammetry is employed extensively from reconnaissance to final design and computation of bid quantities. Its methods can provide instantaneous recordings of dynamic occurrences such as the deflections of beams under impact loads. The unique advantage of recording positions of moving objects on film for future measurements when actual physical measurements are impossible, has enabled photogrammetry to emerge as a powerful research tool. Its successful use in traffic accident investigation has the obvious advantage that photographs overlook nothing which may be needed later to reconstruct the mishap, and normal traffic flow is restored quickly since time-consuming ground measurements are eliminated.

Only basic principles of aerial photogrammetry are presented in this chapter. The geometry of aerial photographs, methods of computing elevations and horizontal distances, and ways of compiling topographic maps and constructing mosaics are discussed.

18–2. Aerial cameras. Aerial mapping cameras are precision instruments designed to take photographs from aircraft. They must be capable of exposing a large number of photographs in rapid succession while moving in an aircraft at high speed so a short cycling time, fast lens, efficient shutter, and large capacity magazine are required.

Aerial cameras are commonly classified by "type" (*single-lens frame, multi-lens frame, strip,* and *panoramic*) or by "angular field of view" (*normal angle* for fields of view up to 75°, *wide-angle* for from 75° to 100°, and *super-wide angle* for fields of view greater than 100°). Aerial cameras generally have focal lengths of 6 in., although $3\frac{1}{2}$-, $8\frac{1}{4}$-, and 12-in. focal lengths are common. The frame or format size of most mapping cameras is 9 in. by 9 in. but dimensions 7 in. by 7 in. can be obtained. A single-lens frame camera having a wide-angle field of view, 6-in. focal length, and 9-in. by 9-in. format is pictured in Fig. 18–2.

A frame aerial camera, the type most widely used, exposes the entire format simultaneously through a lens held at a fixed distance from the focal plane. Principal components of the camera are the *lens* (most important part); *shutter* to control the interval of time that light passes through the lens; *diaphragm* to regulate the size of lens opening; *filter* to reduce the effect of haze and distribute light uniformly over the format; *camera cone* to support the lens-shutter-diaphragm assembly and prevent stray light from striking the film; *focal plane* formed by the upper surface of the fiducial marks and focal-plane frame; *fiducial marks,* usually four in number to define the photographic principal point; *drive mechanism* to cock and trip the shutter, flatten the film and advance it between exposures; *camera body* to house the drive mechanism; and the *magazine* to house the film flattening device and hold the supply of exposed and unexposed film.

An aerial camera shutter can be operated manually by an operator, or by an *intervalometer* which automatically trips the shutter at a specified set time interval. A level vial attached to the camera helps keep the optical axis vertical in spite of a slight tip and tilt of the aircraft. More recently, gyroscopes have been developed to keep the camera axis approximately vertical. Acetate roll film is normally used with magazine capacities of 200 ft or more.

Images of the fiducial marks are printed on the photographs so that lines joining opposite pairs intersect at or very near the *principal point,* defined as the point where a perpendicular from the emergent nodal point of the camera lens strikes the focal plane. Fiducial marks may be located on the sides as shown in Fig. 18–3, or in the corners, Fig. 18–8.

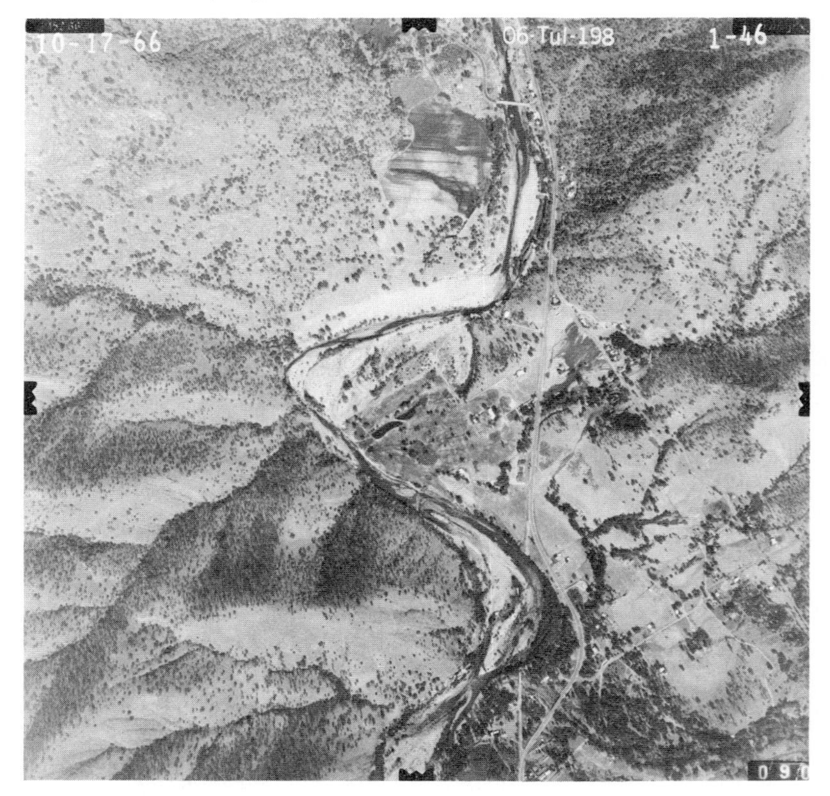

Fig. 18–3. Vertical photograph showing side fiducial marks. (Courtesy California Division of Highways.)

Aerial mapping cameras are laboratory calibrated to get precise values for the focal length, lens distortions, and format dimensions. Flatness of the focal plane, and relative position of the principal point with respect to the fiducial marks, are also specified. These calibration data are necessary for precise photogrammetric calculations.

18–3. Vertical aerial photographs. Vertical photographs, those exposed with the camera optical axis vertical or as nearly vertical as possible, are the principal mode of obtaining imagery for topographic mapping. A *truly vertical* photograph results if the axis is exactly vertical. In spite of the precautions taken, small tilts generally less than 1° and rarely greater than 3° are invariably present. *Near vertical* or *tilted* photographs contain small unintentional tilts. Photogrammetric principles and practices have been developed to handle tilted photographs so that accuracy need not be sacrificed in preparing maps from them.

A vertical photograph, although it looks like a map to laymen, is not a true orthographic projection of the earth's surface. Rather, it is a perspective projection and principles of perspective geometry must be applied in using it to prepare maps. Figure 18–4 illustrates the geometry of a vertical photograph taken at the exposure station L. The photograph, considered as a contact print positive, is a 180° exact reversal of the negative. The positive shown on Fig. 18–4 is used to develop photogrammetric equations in subsequent articles.

Distance oL, Fig. 18–4, is the camera focal length. The x and y reference axis system for measuring photographic coordinates of images is defined by straight lines joining opposite fiducial marks shown on the positive of Fig. 18–4. The x axis, arbitrarily designated as the line most nearly parallel with the direction of flight, is positive in the direction of flight. Positive y is 90° counterclockwise from positive x.

Vertical photographs for topographic mapping are taken in strips which normally run length-wise over the area to be covered. The strips or *flight lines* generally have a *sidelap* (overlap of adjacent flight lines) of 15 to 25 per cent. *Endlap* (overlap of adjacent photographs in the same flight line) is usually about 60 per cent, plus or minus 5 per cent. Figures 18–19 (a) and (b) illustrate endlap and sidelap respectively. If endlap is greater than 50 per cent, all ground points will appear in at least 2 photographs and some will

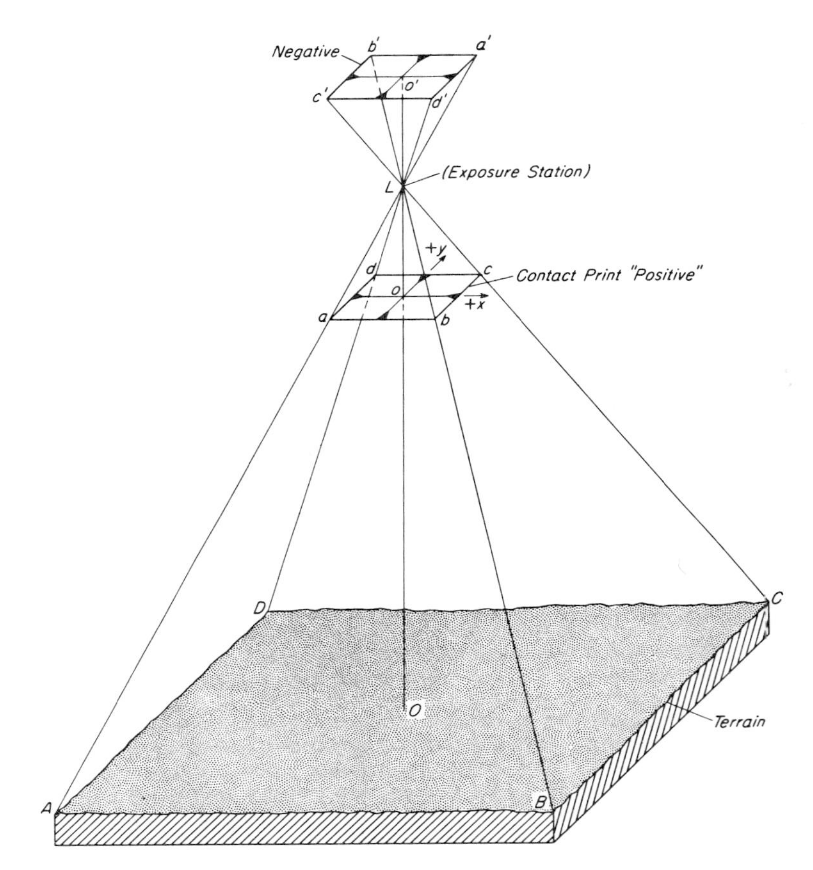

FIG. 18-4. Geometry of the vertical photograph.

show in 3. Images common to 3 photographs permit extension of control through a strip of photographs using only minimal existing control.

18-4. Scale of a vertical photograph. Scale is ordinarily interpreted as the ratio of a distance on a map to that same distance on the ground, and is uniform throughout because a map is an orthographic projection. The scale of a vertical photograph is the ratio of photo distance to ground distance. Since a photograph is a perspective view, scale varies from point to point with variations in terrain elevation.

On Fig. 18-5, L is the exposure station of a vertical photograph

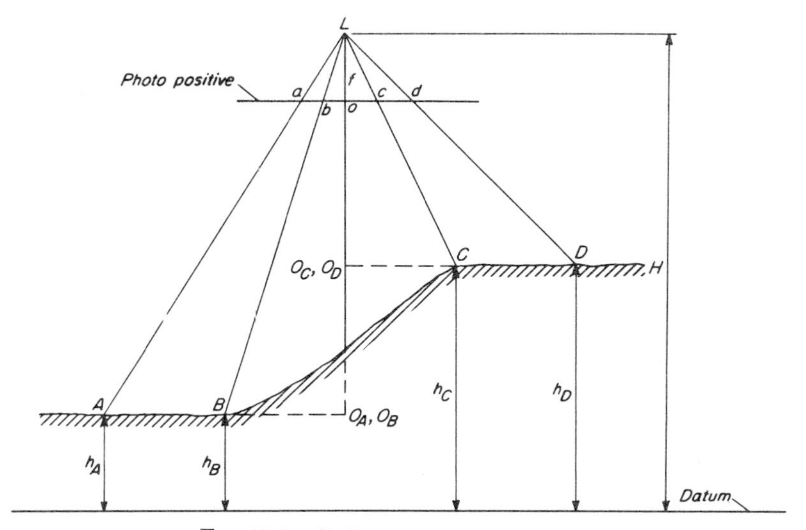

FIG. 18–5. Scale of vertical photograph.

taken at an altitude H above the datum. The camera focal length is f, and o the photographic principal point. Points A, B, C and D which lie at elevations above datum of h_A, h_B, h_C, and h_D, respectively, are imaged on the photograph at a, b, c and d. The ratio ab/AB is the photo scale at the elevation of points A and B. Scale at any point can be expressed in terms of the elevation of the point, camera focal length, and flying height above datum. From Fig. 18–5, for the similar triangles Loa and LO_AA, the following scale equation at A is written:

$$oa/O_AA = S_A = f/(H - h_A)$$

Scales at B, C, and D may be expressed similarly as

$$S_B = f/(H - h_B),\ S_C = f/(H - h_C),\ \text{and}\ S_D = f/(H - h_D)$$

It is apparent from the stated relationships that scale increases at higher elevations and decreases at lower elevations. This concept is seen graphically on Fig. 18–5. Ground lengths AB and CD are equal but photo distances ab and cd are not, cd being longer and hence at larger scale than ab due to the higher elevation of CD. In general, the scale S at any point whose elevation above datum is h may be expressed as

$$S = \frac{f}{H - h} \tag{18–1}$$

The use of an average photographic scale is frequently desirable but it must be accepted with caution as an approximation. For any vertical photograph taken of terrain whose average elevation above datum is h_{avg} the average scale S_{avg} is

$$S_{avg} = \frac{f}{H - h_{avg}} \qquad (18\text{–}2)$$

Example. The photograph of Fig. 18–5 was exposed with a 6-in. focal length camera at a flying height above mean sea level of 10,000 ft.

a) What is the photo scale at point a if the elevation of point A on the ground is 2,500 ft above mean sea level?

From Eq. 18–1

$$S_A = \frac{f}{H - h_A} = \frac{6 \text{ in.}}{10{,}000 - 2{,}500} = 1{:}15{,}000$$

b) For the same photograph, if average terrain is 4,000 ft above mean sea level, what is the average photo scale?

From Eq. 18–2

$$S_{avg} = \frac{f}{H - h_{avg}} = \frac{6 \text{ in.}}{10{,}000 - 4{,}000} = 1{:}12{,}000$$

The scale of a photograph can be determined from a map of the same area. This method does not require that the focal length and flying height be known. Rather it is necessary only to measure on the photograph a distance between two well-defined points identifiable on the map. Photo scale is then calculated from the equation

$$\text{Photo scale} = \frac{\text{photo distance}}{\text{map distance}} \times \text{map scale} \qquad (18\text{–}3)$$

In using Eq. 18–3, all distances must be in the same units and the answer is the scale at average elevation of the two points used.

Example. On a vertical photograph, the length of an airport runway measures 4.24 in. On a map plotted to a scale of 1:9,600 it extends 7.92 in. What is the photo scale at the runway elevation?

From Eq. 18–3

$$S = \frac{4.24}{7.92} \times \frac{1}{9{,}600} = \frac{1}{17{,}900} \quad \text{or } 1'' = 1{,}495'$$

Scale of a photograph can also be computed readily if lines whose lengths are common knowledge appear in the photograph. Section lines, a football or baseball field, etc., can be measured on the photograph and an approximate scale at that elevation ascertained as the ratio of measured photo distance to known ground length.

18–5. Ground coordinates from a vertical photograph. Ground coordinates of points whose images appear in a vertical photograph can be determined with respect to an arbitrary ground space axis system. The arbitrary X and Y ground axes are in the same vertical planes as the photographic x and y axes respectively, and the origin of the system is the point in the datum plane vertically beneath the exposure station. Ground coordinates of points determined in this manner are used to calculate horizontal distances, horizontal angles, and areas.

Figure 18–6 illustrates a vertical photograph taken at a flying

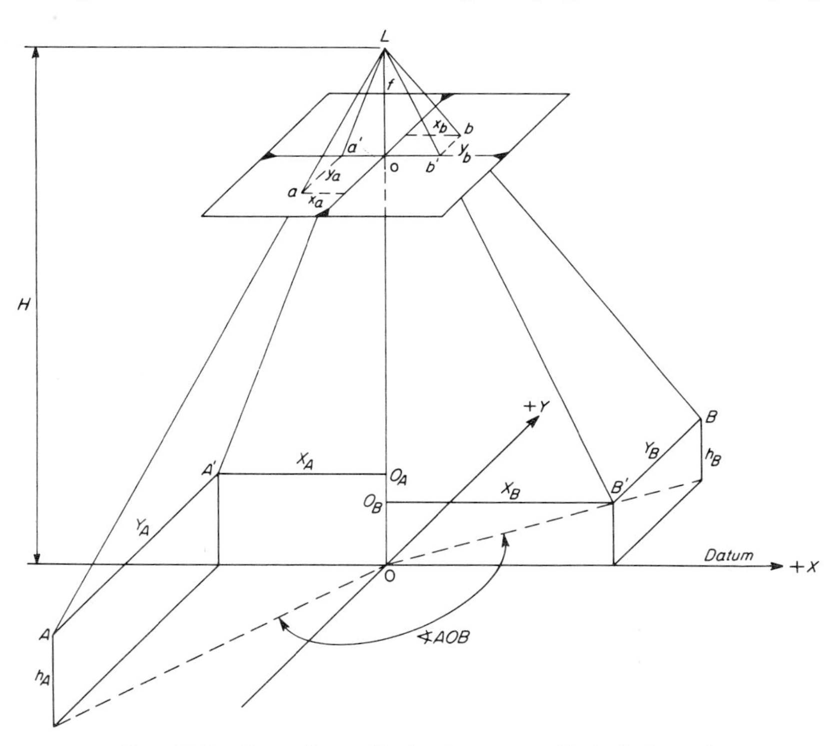

FIG. 18–6. Ground coordinates from a vertical photograph.

height H above datum. Images a and b of the ground points A and B appear on the photograph. The measured photographic coordinates are x_a, y_a, x_b, and y_b, the ground coordinates X_A, Y_A, X_B and Y_B. From similar triangles LO_AA' and Loa'

$$\frac{oa'}{O_AA'} = \frac{f}{H - h_A} = \frac{x_a}{X_A}$$

then

$$X_A = \frac{(H - h_A)\, x_a}{f} \qquad (18\text{–}4)$$

Also from similar triangles $LA'A$ and $La'a$

$$\frac{a'a}{A'A} = \frac{f}{H - h_A} = \frac{y_a}{Y_A}$$

and

$$Y_A = \frac{(H - h_A)\, y_a}{f} \qquad (18\text{–}5)$$

Similarly:

$$X_B = \frac{(H - h_B)\, x_b}{f} \qquad (18\text{–}6)$$

$$Y_B = \frac{(H - h_B)\, y_b}{f} \qquad (18\text{–}7)$$

From the X and Y coordinates of points A and B, the horizontal length of line AB can be calculated using the Pythagorean Theorem

$$AB = \sqrt{(X_B - X_A)^2 + (Y_B - Y_A)^2}$$

Also, horizontal angle AOB may be computed by

$$\text{Angle } AOB = \tan^{-1}\left(\frac{X_A}{Y_A}\right) + \tan^{-1}\left(\frac{Y_B}{X_B}\right) + 90^\circ$$

Areas are determined from the X and Y coordinates by the method discussed in Chapter 11.

18–6. Relief displacement on a vertical photograph. Relief displacement on a vertical photograph is the shift or movement of the position of an image from its theoretical datum location caused by the relief of the object, i.e., its elevation above or below the datum. Relief displacement on a vertical photograph occurs along radial lines from the principal point and increases in magnitude with increased distance of the image from the principal point.

The concept of relief displacement in a vertical photograph taken

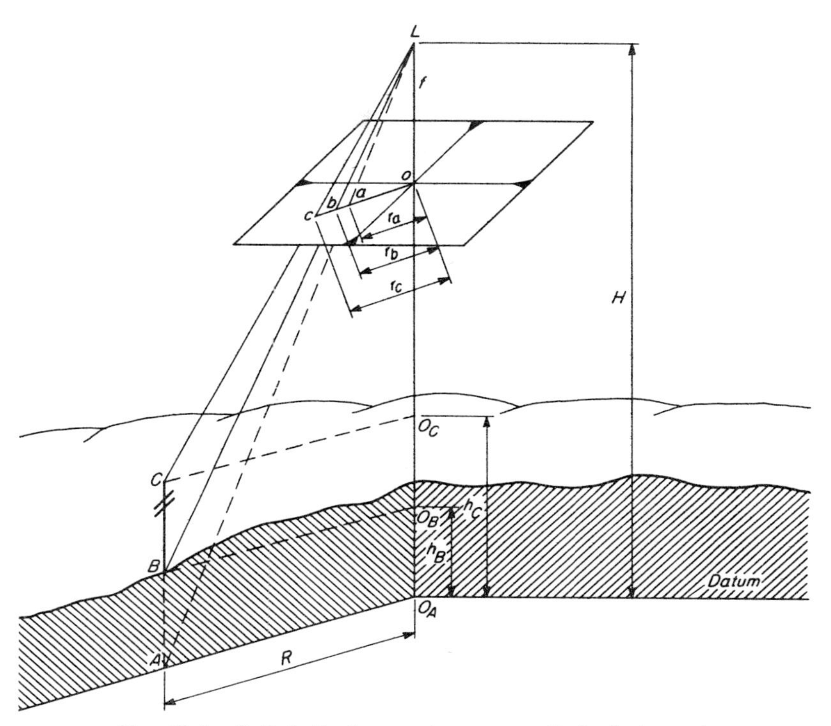

Fig. 18–7. Relief displacement on a vertical photograph.

from a flying height H above datum is illustrated in Fig. 18–7. The camera focal length is f, the principal point o. Points B and C are the base and top respectively of a power pole with images at b and c on the photograph. A is an imaginary point on the datum plane vertically beneath B with corresponding imaginary image position a on the photograph. Distance ab on the photograph is the image displacement due to h_B, the elevation of B above datum, and bc is the image displacement due to height of the power pole.

From similar triangles LO_AA and Loa an expression for relief displacement is formulated.

$$\frac{r_a}{R} = \frac{f}{H}$$

and rearranging,

$$r_aH = fR \tag{a}$$

Also from similar triangles LO_BB and Lob

$$\frac{r_b}{R} = \frac{f}{H - h_B} \qquad \text{or} \qquad r_b(H - h_B) = fR \tag{b}$$

Equating (a) and (b), $r_a H = r_b(H - h_B)$

and rearranging, $r_b - r_a = r_b h_B / H$

If $d_b = r_b - r_a =$ the relief displacement of image b, then $d_b = r_b h_b / H$ which may be written in general terms as

$$d = \frac{rh}{H'} \tag{18–8}$$

where d = relief displacement
r = photo radial distance from principal point to image of the top or high point
h = height above datum of the top or high point
H' = flying height above that same datum.

Equation 18–8 can be used to locate the photographic positions of images with respect to a common datum on a vertical photograph. True horizontal angles may then be taken directly from the photograph, and if the photo scale at datum is known, true horizontal lengths of the lines are measured directly. The datum position is located by scaling the calculated relief displacement d of a point along a radial line to the principal point (inward for a point whose elevation is above datum).

Equation 18–8 can also be applied in computing heights of vertical objects such as buildings, church steeples, radio towers, power poles, etc. To determine heights using the equation, the images of both top and bottom of the object must be visible.

Example. On Fig. 18–7, the radial distance r_b to the image of the base of a power pole is 75.23 mm and the radial distance r_c to the image of its top is 76.45 mm. Flying height H is 4,000 ft and point B is 450 ft both above mean sea level. What is the height of the pole?

The relief displacement is $r_c - r_b = 76.45 - 75.23 = 1.22$ mm. Arbitrarily selecting a datum at the pole base and applying Eq. 18 – 8

$$d = \frac{rh}{H'} \quad \text{so} \quad 1.22 = \frac{76.45\ h}{4{,}000 - 450}$$

Then $h = \dfrac{3{,}550(1.22)}{76.45} = 56.6$ ft.

The relief displacement equation is of particular value to photo interpreters who are usually interested in relative heights rather than absolute elevations.

Figure 18–8 is a vertical photograph which vividly illustrates relief displacements. Note (a) relief displacement of power poles and towers running along the photograph diagonals, (b) outward radial displacement of images from the principal point, and (c) the increased magnitude of displacements for objects farther from the

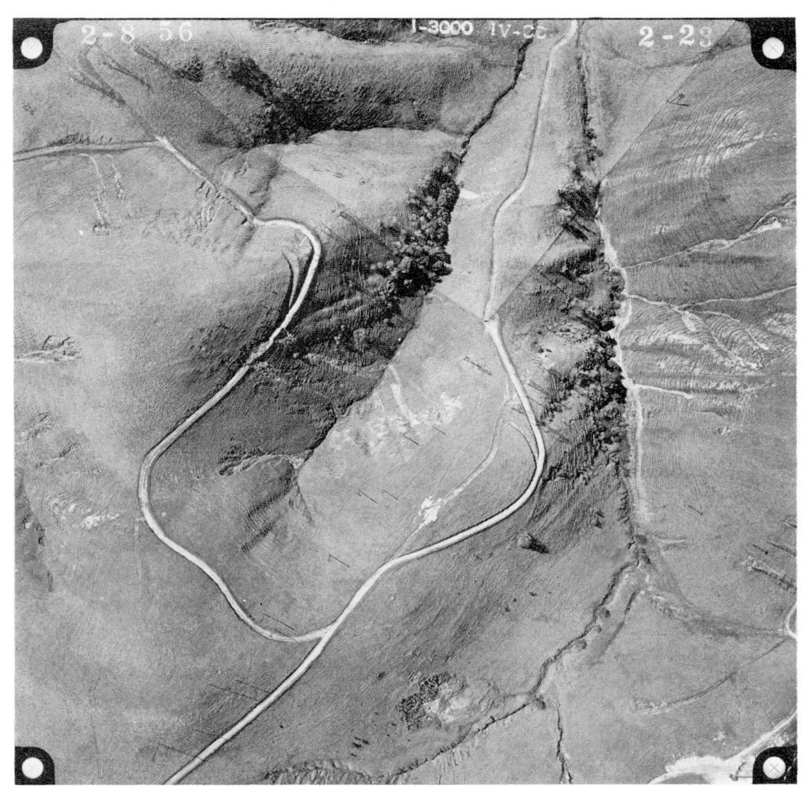

Fig. 18–8. Vertical photograph illustrating relief displacements. (Courtesy California Division of Highways.)

principal point. The shadows cast toward the lower right should not be confused with the poles and towers. Note also the apparent crookedness of an actually straight fence line, which runs from near the photograph's center toward the upper left corner, resulting from relief displacement.

18–7. Flying height of a vertical photograph. From previous articles it is apparent that flying height above datum is an important parameter in solving basic photogrammetry equations. For rough computations, flying heights can be taken from altimeter readings if available. An approximate H can be obtained also by using Eq. 18–1 if a line of known length on fairly level terrain appears on a photograph.

Example. The length of a section line on reasonably flat ground is measured on a vertical photograph as 4.15 in. Find the approximate flying height above the terrain if $f = 6$ in.

Assuming the datum lies at the section line elevation, Eq. 18–1 reduces to

$$\text{Scale} = \frac{f}{H} \qquad \text{and} \qquad \frac{4.15}{5280} = \frac{6}{H}$$

$$\text{From which} \quad H = \frac{5280 \times 6}{4.15} = 7634 \text{ ft above the terrain}$$

If the images of two ground control points A and B appear on a vertical photograph, the flying height may be determined precisely. The Pythagorean theorem can be written

$$L^2 = (X_B - X_A)^2 + (Y_B - Y_A)^2$$

Substituting Eqs. 18–4 through 18–7 in this expression

$$L^2 = \left[\frac{(H - h_B)x_b - (H - h_A)x_a}{f}\right]^2 + \left[\frac{(H - h_B)y_b - (H - h_A)y_a}{f}\right]^2 \qquad (18\text{–}9)$$

where L = horizontal length of ground line AB
H = flying height above datum
h = elevations of the control points above datum
x and y = measured photo coordinates of the control points

Equation 18–9 is a quadratic and rather formidable to solve directly so it is generally expedient to find H by a trial and error method. Compute an L^2 based upon an estimated H and compare this L^2 value with the true length. If the calculated L^2 is greater than the correct length, the first approximation of H is decreased and the computations repeated until satisfactory agreement between calculated and true L^2 is reached.

18–8. Ground control for photogrammetry. Many phases of photogrammetry depend upon ground control for suitable results. Ground control refers to any points whose planimetric positions and/or vertical elevations are known with respect to a horizontal or level datum, respectively. Images of the control points must be identifiable on photographs.

Ground control for photogrammetry can be in the form of *basic control*—triangulation, trilateration, or traverse monuments and bench marks already in existence—, or it may be *photo control*—any points whose images have been recognized on photographs and whose positions are subsequently tied to the basic control through a ground survey.

Ordinarily, photo control points are selected after photography has been finished. This insures positive identification of images of the points and selection in convenient locations. The alternative is to make a control survey prior to photography. Control points are then pre-marked with white targets so their images will show clearly on the photographs. Pre-marking with artificial targets is advisable in areas without natural objects easily recognized on photographs. Vast prairies and dense timber stands are examples of this condition.

18–9. Radial triangulation. Radial triangulation is a traditional graphical technique for extending horizontal control and compiling planimetric maps from aerial photographs. More recently, numerical methods have been used in conjunction with electronic computers to perform this operation. The photogrammetric principle in radial triangulation that relief displacement radiates from the principal point on a truly vertical photograph means that angles measured about that point in the plane of the photograph are true horizontal angles regardless of the elevations of points involved.

Radial triangulation consists basically of successive resection and intersection so it is necessary to have photographic overlap greater than 50 per cent. In addition, images of at least two (and preferably more) horizontal control points must appear somewhere within the assembly of photographs.

Radial triangulation using tracing paper overlays will be illustrated in detail as an easily visualized method. In Fig. 18–9, four photographs of a flight strip are laid out in their overlapping positions with points a and b the images of two ground control points

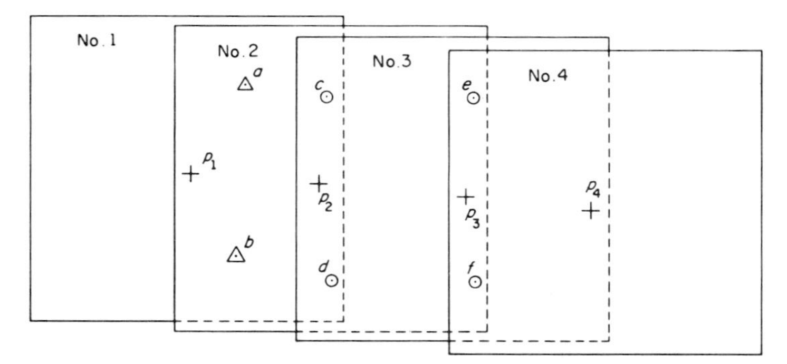

Fig. 18–9. Four overlapping photographs of a flight strip.

A and B. A tracing paper overlay is placed on photograph No. 1 and rays drawn from the principal point p_1 to a, b, and conjugate principal point p_2. Conjugate principal points are the principal points of adjacent photographs. On Fig. 18–9, for example, p_1 and p_3 are conjugate principal points of photograph No. 2. Figure 18–10

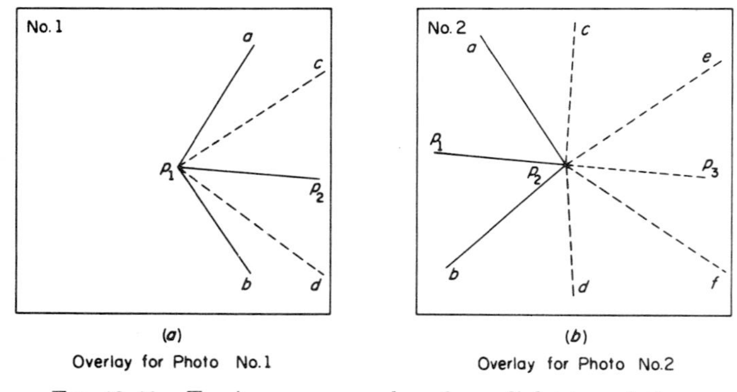

Fig. 18–10. Tracing paper overlays for radial triangulation.

(a) shows the tracing paper overlay for photograph No. 1. Angles ap_1p_2 and p_2p_1b are true horizontal angles.

A second tracing paper overlay is prepared for photograph No. 2 by drawing rays from p_2 to points a, b, and the conjugate principal point p_1. Figure 18–10 (b) is the overlay for this photograph. Ground control points A and B are plotted to the desired scale on a map sheet as shown in Fig. 18–11. Transparent overlay No. 1 is oriented on the map sheet so that rays p_1a and p_1b simultaneously pass through their respective plotted control points A and B. At the same time, photograph overlay No. 2 is oriented on the map sheet to make rays p_2a and p_2b pass through their respective plotted control points, and rays p_1p_2 and p_2p_1 coincide. The resulting locations of p_1 and p_2 define the true planimetric map positions of principal points P_1 and P_2 which are marked by pricking through the overlays with a pin. Ground coordinates of P_1 and P_2 can then be scaled from the map. This procedure for locating ground principal points is called *resection.*

With the principal points of a pair of overlapping photographs fixed on the map sheet, any number of other points whose images

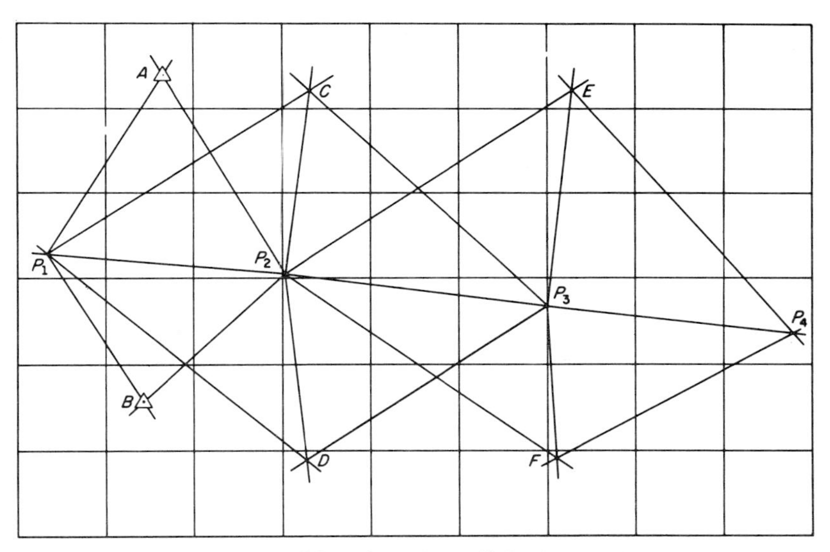

FIG. 18–11. Map sheet for radial triangulation.

appear in the overlap area of the two photographs can be established. Placing the first overlay on photograph No. 1 again, oriented as it was when rays to a, b, and p_2 were drawn, rays to new points c and d are ruled. Also, with the second overlay once more put over photograph No. 2 and properly oriented, rays are drawn from p_2 to c and d. These rays are shown dashed on Fig. 18–10. The two overlays are then spotted on the map sheet in their previously determined arrangement and the intersection of rays p_1c and p_2c settles the true planimetric position of point C. Locating new points this way is called *intersection.* Point D and any others in the overlap area can be similarly established by intersection and their ground coordinates scaled from the map.

Points C and D are called *pass points* and when their planimetric positions are located they become points of extended horizontal control. To satisfactorily serve this purpose they must be well-defined images appearing in a desirable location on three successive overlapping photographs. The ideal placements are opposite the principal and conjugate principal points, and near the center of the sidelap.

Notice on Fig. 18–9 that points c and d appear on photograph No. 3. Since the ground or map positions of these points have been determined, the map location of P_3 can be found by resection. With P_3 located, new pass points e and f are carefully chosen so that their images on photograph No. 4 will help fix P_4, etc. Radial triangulation is extended in this manner through all the photographs. Careful planning should precede a radial triangulation project and all pass points selected prior to constructing radial lines on the overlays. Rays to each control and pass point are then drawn when the tracing paper is first overlaid on the photographs.

Radial triangulation can be done by *three-point resection* but three control points are required. The procedure is basically the same as the method just described although an additional pass point is used in lieu of conjugate principal points.

Slotted templates of either metal or cardboard having precisely cut thin slots representing ray directions are frequently employed instead of tracing paper. One template is prepared for each photograph. A variation utilizes metal arms with pre-cut slots to represent the ray directions for each photograph. The arms are fastened together at the principal point by a special nut and bolt. With either the slotted or metal arm templates, the individual pieces for each photograph are assembled into a single overall unit by making the slots (rays) to common points intersect properly. A special stud at the ray intersections permits changing the scale of the entire unit by gently pulling apart or compressing the assembly.

All available horizontal control in the project area should be plotted on a map sheet and the radial triangulation fitted to it. Errors in graphical or mechanical construction will probably necessitate adjusting the entire assembly by forcing all control point rays to simultaneously intersect on their respective plotted points. This adjustment is difficult with tracing paper templates but conveniently done with slotted templates or metal arms, an important factor when extensive work is scheduled. After the overlays or templates have been fitted to all control, pass points and planimetry are finally located.

18–10. Mosaics. An aerial mosaic is an assembly of overlapping photographs. It provides a continuous representation of the terrain, contains a vast amount of detail, and is useful for many purposes, but it is not the equivalent of a map. A map shows features on a single orthographic projection whereas a mosaic is made up of a series of perspective projections.

Mosaics are generally classified as *controlled* or *uncontrolled.* A controlled mosaic is made from rectified (corrected for tilt and brought to a common scale by a rectifier) photographs and laid to accurate horizontal control. An uncontrolled mosaic is constructed with unrectified photographs and formed by simply matching the images of objects such as roads, railroads, fence lines, streams, etc. A third classification, *semicontrolled* mosaic, is sometimes used to denote those mosaics which combine some of the features of both the other two types.

Mosaics are also classified according to their purpose. An *index* mosaic, for example, is an assembly of all photographs of a particular flight laid so the photograph numbers are visible. The index mosaic is photographed and the prints filed for subsequent use in selecting the particular prints or negatives wanted from the flight. A *strip* mosaic is a compilation of the pictures on a single flight strip and suitable for qualitative study of various proposed route locations.

In preparing mosaics, single-weight paper print photographs should be used and laid on a hard nonporous surface such as masonite.

To build an uncontrolled mosaic, the center strip of photographs is first arranged by matching images and held in place by weights. A long straightedge is then placed over the entire strip so that the edge passes through as many principal points as possible. With the straightedge in this position, a fine line called an *azimuth line* is drawn on the top photograph only. Two arbitrary but widely spaced points that lie on the azimuth line of the top print are located on the second photograph and a fine line drawn on that print connecting them. This process is repeated until each photograph of the entire strip has been marked with an azimuth line The first print is then mounted on the board at a predetermined position to make the entire mosaic fit conveniently on the base. The other prints are laid in succession by matching images while maintaining the straightness of the azimuth line to prevent twisting of the flight strips which could cause serious mismatches of images near the mosaic edges.

After the first strip is completed, adjacent ones are laid in the same fashion. Trimming the photographs with a razor blade or sharp knife by cutting through the emulsion only, is necessary before mounting. The photograph is then bent along the cut and torn back on a taper of approximately one-half inch. The tapered or feathered edge is smoothed by rubbing with a fine sandpaper. Figure 18–12 shows the construction of a mosaic. Cutting lines should be chosen along the sides of roads, fence-lines, edges of woods, or other suitable delineations where natural tone changes

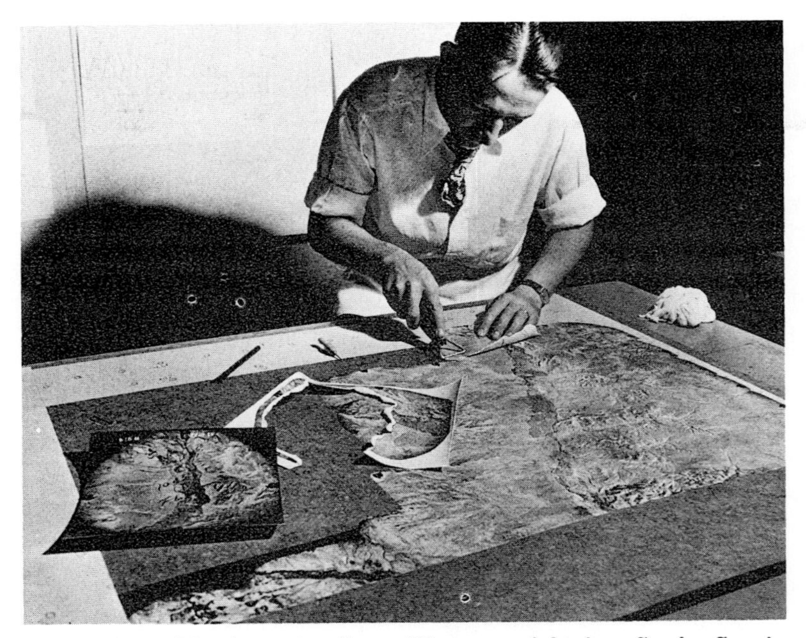

Fig. 18–12. Mosaic construction. (Photo copyright Aero Service Corp.)

occur. Cutting across fence-lines, roads, or other lineal features should be avoided as even the slightest mismatch of these features will be obvious. Only the center portion of each photograph should be used, if possible, because it contains the least relief displacement and hence minimizes difficulties in matching images.

Adhesive is applied to their backs and the photographs are pasted on the board. Excess adhesive is forced from under the prints by a movement away from the center with a squeegee, and removed with a wet sponge. Gum arabic is the most commonly used adhesive although rubber cement is also employed. The advantage of rubber cement is less shrinkage but gum arabic permits slight adjustment of the prints for a longer period of time after pasting. When completed the mosaic can be photographed and copies distributed.

18–11. Parallax. Parallax is defined as the apparent displacement of the position of an object with respect to a frame of reference due to a shift in the point of observation. For example, a person looking through the view finder of an aerial camera in an aircraft as it moves forward sees images of objects moving across his field of view. This apparent motion (parallax) is due to the changing location of the observer. Using the focal plane of the camera as a frame of reference, parallax exists for all images appearing on successive photographs due to forward motion between exposures. The greater the elevation of any point, i.e., the closer the point is to the camera, the larger will be the parallax. For 60 per cent endlap, the parallax of images on successive photographs should average approximately 40 per cent of the focal plane width.

The magnitude of parallax of a point is a function of the elevation of that point and consequently a means of calculating elevations. It is also possible to compute X and Y ground coordinates from parallax.

Movement of an image across the focal plane between successive exposures takes place in a line parallel with the direction of flight. Thus in order to measure parallax, the direction of flight must first be established. This is done by locating the positions of principal and conjugate principal points of the pair of photographs, and ruling a line on each print passing through them to get the direction of flight. This line also defines the photographic x axis where parallax is to be measured. Another line is drawn perpendicular to the flight line and passing through the principal point of each print to position the photographic y axis for starting parallax meaurements. The x coordinate of a point is scaled on each photograph with respect to the axes so constructed. Parallax of the point is then calculated from the expression

$$p = x - x_1 \tag{18–10}$$

Photographic coordinates x and x_1 are measured on the left-hand and right-hand prints respectively with due regard given for algebraic signs.

Figure 18–13 illustrates an overlapping pair of vertical photographs exposed at equal flight heights H above datum. The distance between exposure stations L and L_1 is called B the *air base*. The small inset figure shows the two exposure stations L and L_1 in superposition to make the similarity of triangles $La_1'a'$ and $LA'L_1$ more

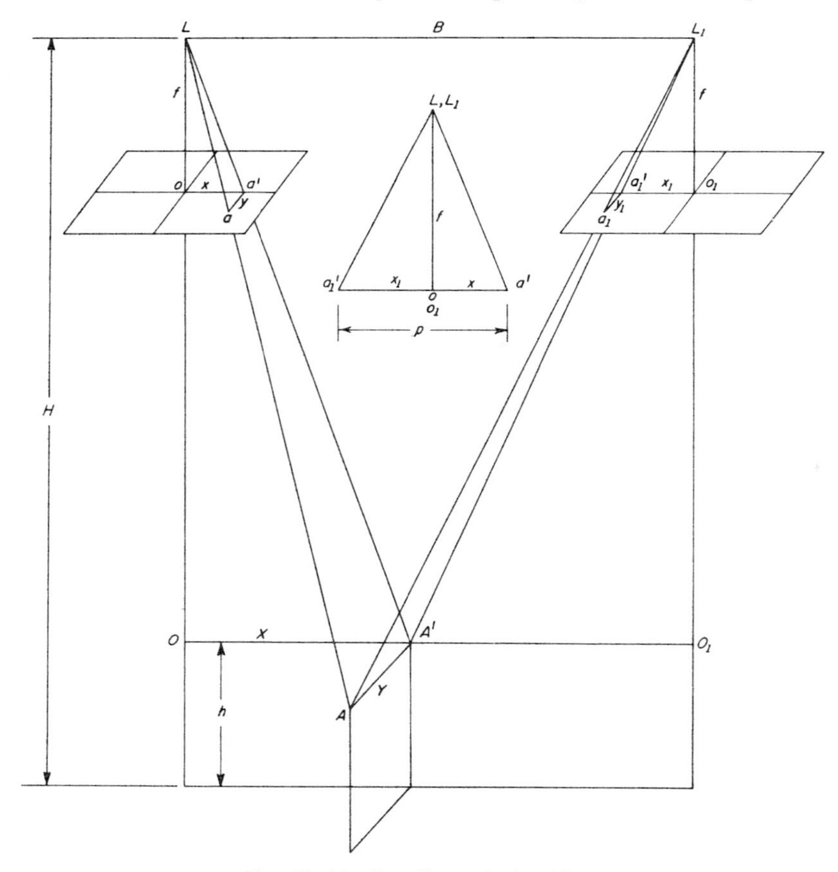

FIG. 18–13. Parallax relationships.

easily recognized. Equating these two aforementioned similar triangles there results

$$\frac{p}{f} = \frac{B}{H - h}$$

from which

$$H - h = \frac{Bf}{p} \qquad (18\text{–}11)$$

Also from similar triangles LOA' and Loa',

$$X = \frac{x}{f}(H - h)$$

Substituting Eq. 18–11 into the above,

$$X = \frac{B}{p}x \qquad (18\text{–}12)$$

Similarly from triangles LAA' and Laa',

$$Y = \frac{B}{p}y \qquad (18\text{–}13)$$

In these equations, X and Y are ground coordinates of a point with respect to an origin at a point vertically beneath the exposure station of the left photograph, positive X coinciding with the direction of flight and positive Y 90° counter-clockwise to positive X. Parallax of the point is p; x and y are photographic coordinates of the point on the left-hand print; H is the flying height above datum; h is the elevation of the point above the same datum; and f is the camera lens focal length.

Equations 18–11 through 18–13 are commonly called the parallax equations and are among the most useful ones to photogrammetrists for calculating horizontal lengths of lines and elevations of points. They also provide the fundamental basis for design and use of stereoscopic plotting instruments.

Example. The length of line AB and elevations of points A and B from two vertical photographs which contain the images of a and b are needed. Flying height above mean sea level was 4,000 ft and air base 2,000 ft. The camera had a 6-in. focal length. Measured photographic coordinates in inches on the left-hand print are $x_a = 2.10$, $x_b = 3.50$, $y_a = 2.00$, and $y_b = -1.05$; on the right-hand print $x_{1a} = -2.25$ and $x_{1b} = -1.17$.

From Eq. 18–10,

$$p_a = x_a - x_{1a} = 2.10 - (\,-2.25) = 4.35 \text{ in.}$$
$$p_b = x_b - x_{1b} = 3.50 - (-1.17) = 4.67 \text{ in.}$$

From Eq. 18–12 and 18–13

$$X_A = \frac{B}{p_a} x_a = \frac{2000 \times 2.10}{4.35} = 965.5 \text{ ft}$$

$$X_B = \frac{2{,}000 \times 3.50}{4.67} = 1{,}498.9 \text{ ft}$$

$$Y_A = \frac{B}{p_a} y_a = \frac{2{,}000 \times 2.00}{4.35} = 919.5 \text{ ft}$$

$$Y_B = \frac{2{,}000 \times (-1.05)}{4.67} = -\,449.7 \text{ ft}$$

From the Pythagorean Theorem length AB is

$$AB = \sqrt{(1498.9 - 965.5)^2 + (-449.7 - 919.5)^2} = 1{,}469.4 \text{ ft}$$

From Eq. 18–11, the elevations of A and B are

$$h_A = H - \frac{B}{p_a} f = 4{,}000 - \frac{2{,}000 \times 6}{4.35} = 1{,}241 \text{ ft}$$

$$h_B = 4{,}000 - \frac{2{,}000 \times 6}{4.67} = 1{,}430 \text{ ft}$$

18–12. Stereoscopic viewing and measuring. The term stereoscopic viewing means seeing an object in three dimensions, a process requiring a person to have normal *binocular* (two-eyed) vision. In Fig. 18–14, two eyes L and R are separated by a distance b called the *eye base.* When the eyes are focused on point A, their optical axes converge to form angle ϕ_1 and when sighting on B, ϕ_2 is produced. Angles ϕ_1 and ϕ_2 are called parallactic angles and the brain associates distances d_A and d_B with them. The depth $(d_B - d_A)$ of the objects is perceived from unconscious comparison by the brain of these parallactic angles.

If two photographs of the same subject are taken from two different perspectives or camera stations, the left print viewed with the left eye and simultaneously the right print seen with the right eye, a mental impression of a three-dimensional model results. In normal stereoscopic viewing the *eye base* gives an impression of the parallactic angles. While looking at aerial photographs stereoscopically, the exposure station spacing simulates an eye base so the viewer actually sees parallactic angles comparable with having one eye at each of the two exposure stations.

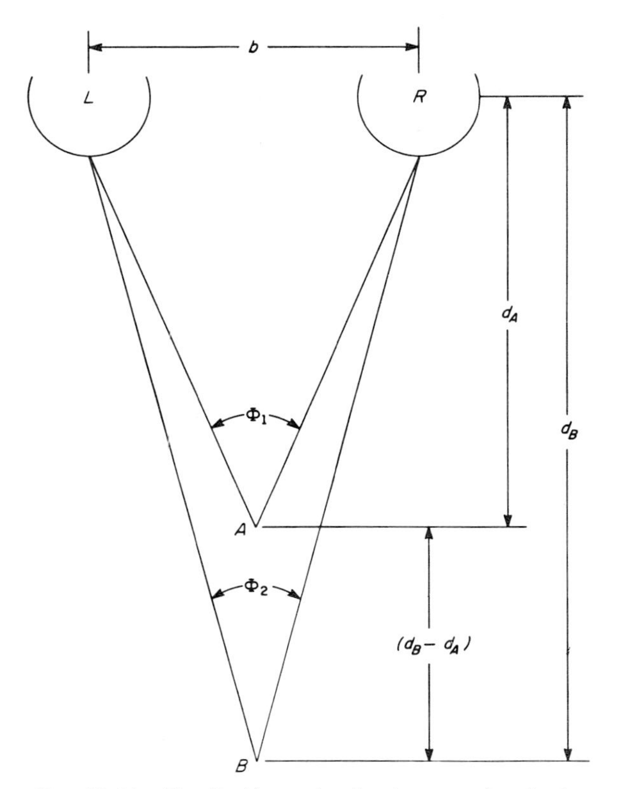

FIG. 18–14. Parallactic angles in stereoscopic viewing.

The stereoscope shown in Fig. 18–15 permits viewing photographs stereoscopically by enabling the left and right eyes to focus comfortably on the left and right prints respectively, assuming proper orientation of the overlapping pair of photographs under the stereoscope. Correct orientation requires the two photographs to be laid out in the same order they were taken, with the stereoscope so set that the line joining the lens centers is parallel with the direction of flight. Spacing of the prints is varied, carefully maintaining this parallelism, until a clear stereoscopic model is obtained.

Parallax of a point can be measured while viewing stereoscopically with the advantage of speed, and because binocular vision is used, greater accuracy. While viewing through the stereoscope, two small identical marks etched on pieces of clear glass called *half marks* may be placed over each photograph. Simultaneously

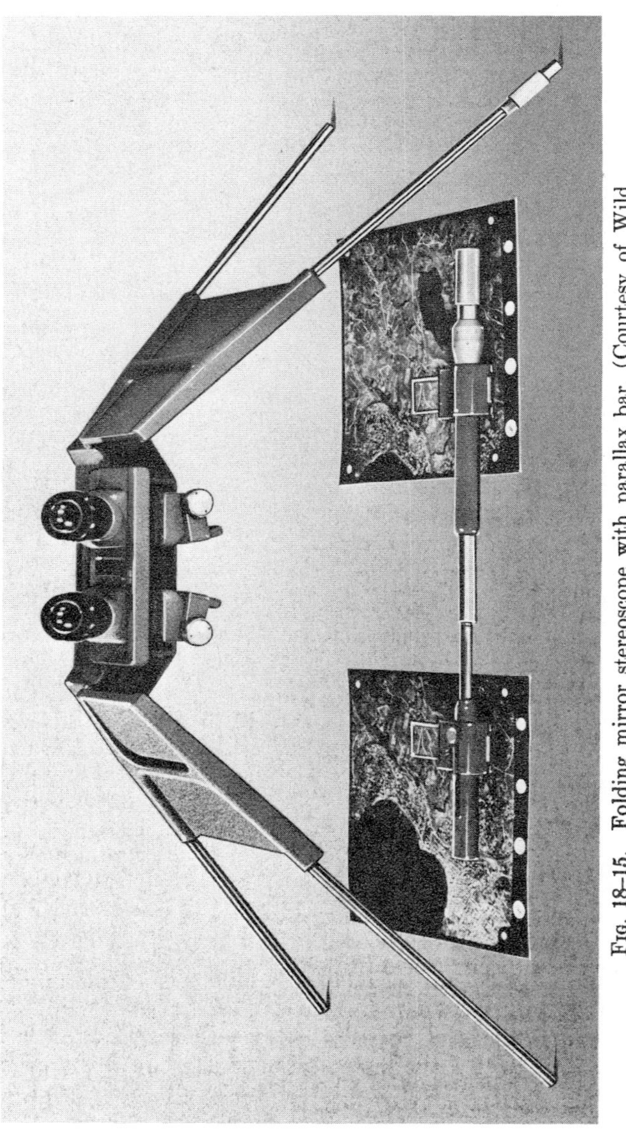

Fig. 18-15. Folding mirror stereoscope with parallax bar. (Courtesy of Wild Heerbrugg Instruments, Inc.)

seeing one mark with the left eye and the other with the right eye, their positions are shifted until they seem to fuse together as one mark which appears to lie at a certain elevation. The elevation of the mark will seem to vary or "float" as the spacing of the half marks is varied, hence it is called the *floating mark*. Figure 18–16 demonstrates this principle and also illustrates that the mark can be set exactly on particular points such as A, B, and C by placing the half marks at a and a', b and b', and c and c' respectively.

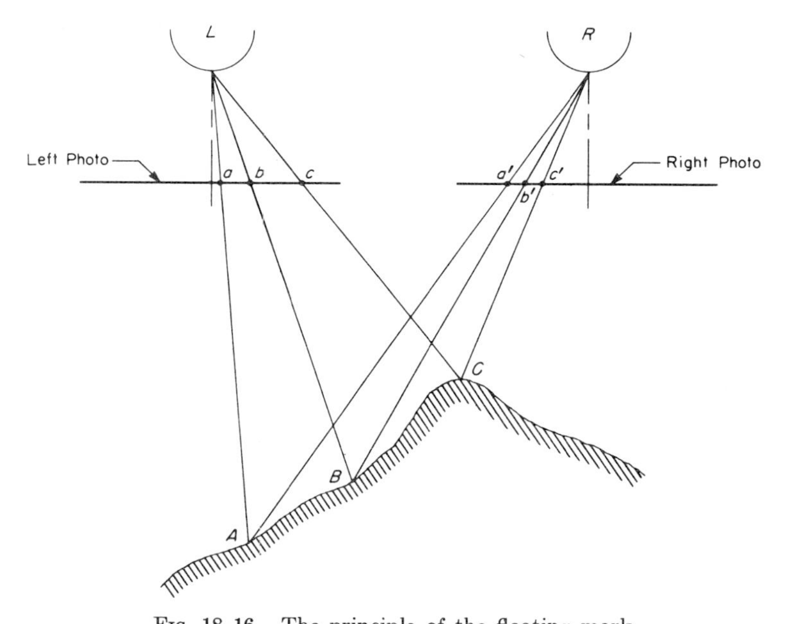

FIG. 18–16. The principle of the floating mark.

Using the floating mark principle, parallax of points is measured stereoscopically with a parallax bar as shown beneath the stereoscope in Fig. 18–15. It is simply a bar to which two half marks are fastened. The right mark can be moved with respect to the left one by turning a micrometer screw and readings taken with the floating mark set stereoscopically on various points. The micrometer readings are added to the parallax bar *set-up constant* to get parallax.

When using a parallax bar, two overlapping photographs are oriented properly for viewing under a mirror stereoscope and fastened securely with respect to each other. The parallax bar constant for the set-up is determined by measuring photo coordinates for a desired point and applying Eq. 18–10 to get its parallax. The floating mark is placed on the same point, the micrometer read, and the constant for the setup found by

$$C = p - r \tag{18–14}$$

where C = parallax bar set-up constant,
p = parallax of a point determined by Eq. 18–10,
r = micrometer reading obtained with the floating mark set on that same point.

Having determined the constant, the parallax of any other point can be computed by adding its micrometer reading to the constant and then used in the parallax equations. Each time another pair of photos is oriented for parallax measurements, a new parallax bar set-up constant must be determined.

The reading required to place the floating mark on points at any particular elevation can be calculated and set on the micrometer. With the parallax bar then linked to a pantograph and tracing pencil, the contour for that elevation is traced directly by moving the floating mark over the terrain always keeping it in contact with the ground.

EXAMPLE. Two overlapping vertical photographs are oriented under a mirror stereoscope. The x photo coordinate of point a is 65.26 mm on the left print and −5.45 mm on the right print. The parallax bar micrometer reading on a was 15.15 mm. Camera focal length = 6 in., elevation of point A is 375 ft, and flying height = 7800 ft above mean sea level. What is the required micrometer reading to trace the 500-ft contour?

a) The parallax of point a, by Eq. 18–10 is

$$p = x - x_1 = 65.26 - (-5.45) = 70.71 \text{ mm}$$

b) The air base, from Eq. 18–11 is

$$H - h = \frac{Bf}{p} \qquad 7800 - 375 = \frac{B \times 6 \times 25.4}{70.71}$$

$$B = \frac{7425\ (70.71)}{6 \times 25.4} = 3586 \text{ ft}$$

c) From Eq. 18–14, the parallax bar set-up constant is

$$C = p - r = 70.71 - 15.15 = 55.56 \text{ mm}$$

d) By Eq. 18–11, the parallax for the 500-ft contour is

$$7800 - 500 = \frac{3586\ (6)\ 25.4}{p}$$

$$p = \frac{3586\ (6)\ 25.4}{3700} = 71.92 \text{ mm}$$

e) By Eq. 18–14, the micrometer setting for the 500 ft contour is

$$r = p - C = 71.92 - 55.56 = 16.36 \text{ mm}$$

18–13. Stereoscopic plotters. The primary use of stereoscopic plotters is to compile topographic maps from overlapping aerial photographs which are developed from negatives onto $\frac{1}{8}$- or $\frac{1}{4}$-in. thick glass plates called *diapositives*. The diapositives of a pair of overlapping photographs are projected so that light rays carrying images common to both intersect to form a stereoscopic model on the table, Fig. 18–17.

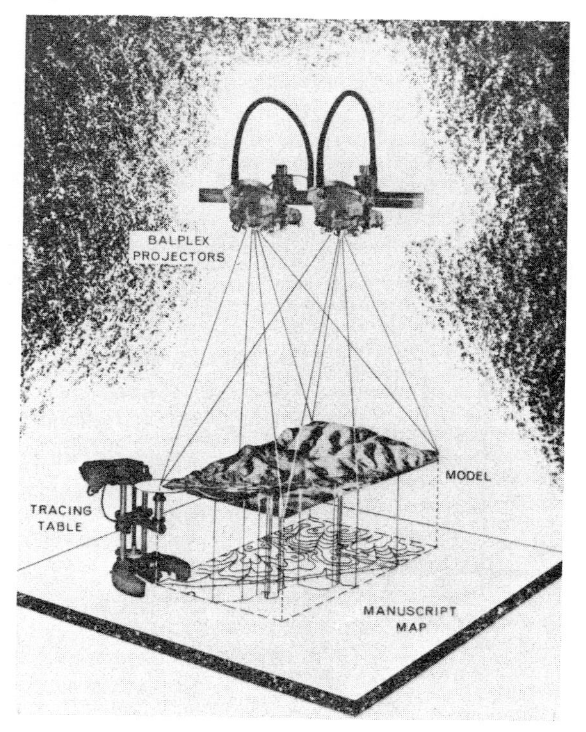

Fig. 18–17. Balplex projectors and stereoscopic model. (Courtesy Bausch and Lomb, Inc.)

There are numerous types of plotters, each having slightly different characteristics. All must include the following basic components, however: a system for (1) projection, (2) viewing, (3) measuring, and (4) tracing. These components make possible an optical-mechanical solution of the parallax equations.

Projectors used in stereoscopic plotters resemble ordinary slide projectors except that the geometry of their optics is much more precise. Also, they are capable of re-creating, at a greatly reduced scale, the exact spatial attitude of the aerial camera at the instants the overlapping photographs were exposed, hence a "true" stereoscopic model results. The Balplex plotter projection system, Fig. 18–17, consists of two projectors each having principal distances of 55 mm and an ellipsoidal reflecting surface. With the light source placed at one ellipsoid focus and the projector lens at the other, the light rays are concentrated as they pass through the lens carrying the images.

Plotter viewing systems must be designed so that the left and right eye see only the projected images of the corresponding left and right diapositive. One common method of accomplishing this is to place a blue filter in one projector and a red filter in the other. A pair of glasses with one blue and one red lens is worn by the operator so the left and right eye see only the images projected from the left and right diapositive respectively. This system of viewing stereoscopically is called the *anaglyphic* method. Another system developed recently named the *Stereo Image Alternator* (SIA) operates by means of rapidly rotating shutters located on the projectors, and a viewing eyepiece. The shutters are synchronized so that the left and right eye can see only the images from the corresponding left and right projector. The SIA offers several advantages over the anaglyphic system such as the capability of using color photographs, sharper projected images, and less light loss with no filters.

Various measuring and tracing systems have been devised for plotters but regardless of design their basic function is to measure and trace to scale the stereoscopic model. In the Balplex instrument, Fig. 18–17, light rays carrying corresponding images are intercepted on the *tracing table platen.* A small light emitted through a pin hole in the platen provides the floating mark which can be made to appear to rest exactly on a point of the model by raising or lowering the tracing table. A tracing pencil located directly below the floating mark permits positioning the point planimetrically.

A counter linked to the tracing table responds to up and down motion so the elevation for any setting of the floating mark is read directly.

When diapositives are placed in the projectors and the lights turned on, corresponding rays will not intersect properly to form a clear model because of tilt in the photographs, unequal flying heights, and of course improper projector orientation. The projectors can be moved linearly along the X, Y and Z axes and also rotated about each of them until the diapositives reproduce the exact conditions existing when the photographs were taken. This is called *relative orientation* and when accomplished, corresponding rays should intersect to form a perfect three-dimensional model.

The model is brought to required scale by making the rays of at least two ground control points intersect at their plotted positions on the manuscript map. It is leveled by adjusting the projectors so that the counter reads the correct elevations of each of a minimum of three ground control points when the floating mark is set on them. Scaling and leveling the model is called *absolute orientation.*

Figure 18–18 shows a Wild A-8 first-order plotter linked to a coordinatograph. Its projection system consists of two precisely-

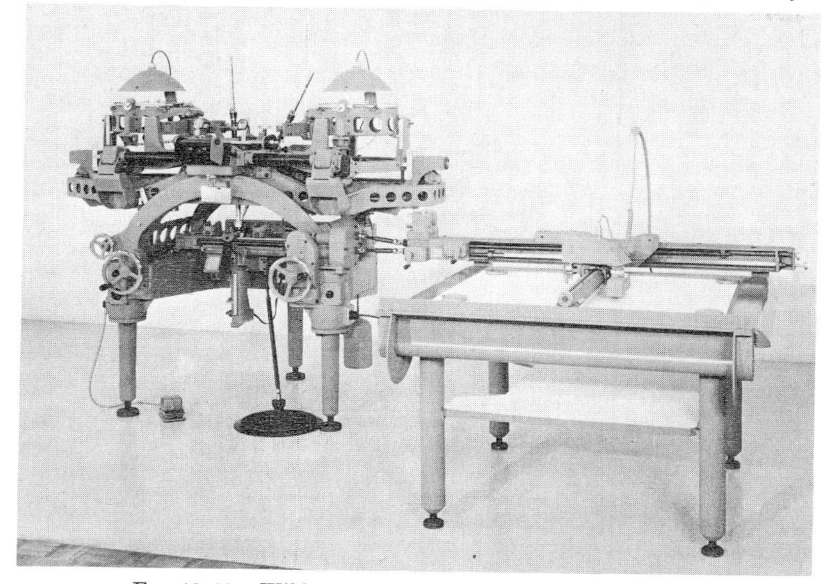

Fig. 18–18. Wild autograph A8 with coordinatograph. (Courtesy of Wild Heerbrugg Instruments, Inc.)

made metal space rods which simulate light rays. Full-size diapositives are used and viewed through binoculars via an optical train of lenses and prisms. The floating mark, composed of a pair of half marks superimposed in the optical train, is moved up or down by turning a foot disk, and impelled in the *X* and *Y* directions by means of two hand wheels. Direct linkage to the coordinatograph allows planimetry to be traced directly on the manuscript. The *X* and *Y* coordinates can also be read directly from the instrument.

18–14. Analytical photogrammetry. In recent years, with the advent of high-speed electronic computers, much of the work done with stereoscopic plotters can now be performed economically by means of analytical photogrammetry. This science embodies precise measurements of photographic coordinates of images and construction of a mathematical model which can be solved by numerical methods. The procedures of analytical-photogrammetry are particularly well adapted to establishing vertical and horizontal control through aerotriangulation.

18–15. Flight planning. Certain factors, depending generally upon the purpose of photography, must be specified to guide a flight crew in executing its primary mission of taking aerial photographs. Some of them are (a) boundaries of the area to be covered, (b) required scale of the photography, (c) camera focal length and format size, (d) endlap, (e) sidelap, and (f) aircraft ground speed in flight. Having fixed these elements, it is possible to compute the entire flight plan and prepare a flight map—a topographic map upon which the required flight lines have been delineated. The pilot flies the specified flight lines by choosing and correlating headings on existing natural features in the field shown on the flight map.

Purpose of the photography is the paramount consideration in flight planning. For example, in photography for mosaic construction it is desirable to minimize image displacement and scale variation due to diverse relief by increasing the flying height while maintaining photographic scale with a camera of long focal length. Conversely, photography taken for mapping purposes gains overall geometric strength and utility from flights at lower altitude and using a wider-angle shorter-focal-length camera.

The *C-Factor,* ratio of the flight height to the contour interval which is practical for any specific plotter, can be used to help select the flying height. It has been evaluated for various plotters and values range from approximately 500 to 1500. By this criterion, if a plotter has a C-Factor of say 1000 and a map is to be compiled with a 5-ft contour interval, then a flight height of not more than $(1000 \times 5) = 5{,}000$ ft above the terrain should be sustained.

Information ordinarily calculated in a flight plan includes (1) flying height above mean sea level; (2) interval of time between successive exposures; (3) number of photographs per flight line; (4) number of flight lines; and (5) total number of photographs. A flight plan is prepared based upon the five items.

EXAMPLE. A flight plan for an area 10 miles wide and 15 miles long is required. Average terrain in the area is 1500 ft above mean sea level. Airplane ground speed is 135 mph and the camera has a 6-in. focal length with 9-in. by 9-in. format. Endlap is to be 60 per cent, sidelap 25 per cent. Required scale of the photography is 1:12,000.

a) Flying height (from Eq. 18–1)

$$\text{Scale} = \frac{f}{H - h_{avg}} \quad \text{so} \quad \frac{1}{12{,}000} = \frac{6/12}{H - 1500}$$

$$\text{and } H = 7500 \text{ ft}$$

b) Distance between exposures.

Endlap is 60 per cent so the lineal advance per photograph is 40 per cent of the total coverage = 9 in. × 1000 ft per in. = 9000 ft. Distance between exposures = 0.40 × 9,000 = 3600 ft.

c) Intervalometer setting.

$$135 \text{ mph} \times \frac{5280 \text{ ft/mi}}{3600 \text{ sec/hr}} = 198 \text{ ft per sec}$$

$$\frac{3600 \text{ ft}}{198 \text{ ft/sec}} = 18.2 \text{ sec (use 18 seconds)}$$

Note that in rounding down to a shorter time the endlap is slightly increased thereby providing a small additional factor of safety.

$$\text{Recalculated linear advance} = 18 \text{ sec} \times 198 \text{ ft/sec} = 3560 \text{ ft}$$

d) Total number of photographs per flight line.

Length of each flight line = 15 mi × 5280 ft/mi = 79,200 ft

$$\text{Number of photos per flight line} = \frac{79{,}200 \text{ ft}}{3560 \text{ ft/photo}} = 22.3 \text{ (say 23)}$$

Add 4 photos on each end to insure complete coverage so total = 23 + 4 + 4 = 31 photos per flight line

e) Distance between flight lines.

Sidelap is 25 per cent so the side advance per flight line is 75 per cent of the total photographic coverage.

$$\text{Distance between flight lines} = 0.75 \times 9000 \text{ ft} = 6750 \text{ ft}$$

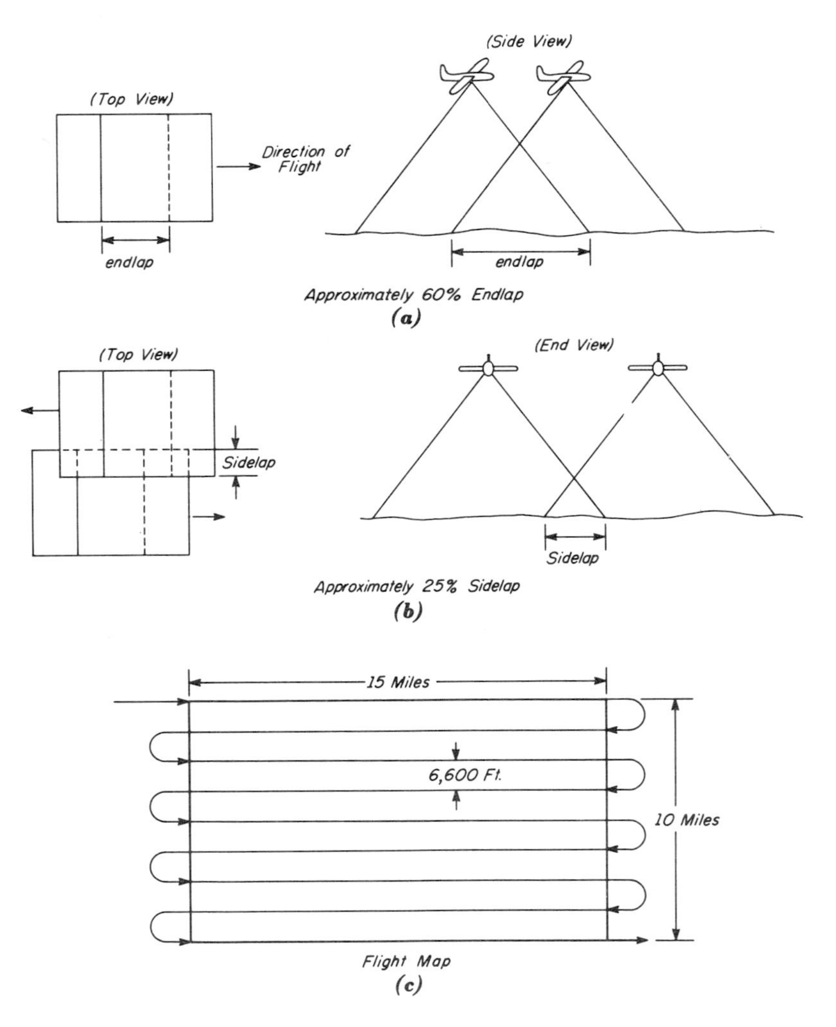

FIG. 18–19. Endlap, sidelap, and flight map.

f) Number of flight lines.

$$\text{Width of the area} = 10 \text{ mi} \times 5280 \text{ ft/mi} = 52{,}800 \text{ ft}$$

$$\text{Number of lines} = \frac{52{,}800 \text{ ft}}{6750 \text{ ft/line}} = 7.8 \text{ lines (say 8)}$$

As a factor of safety, add one flight line so that both edges of the area are overlapped by one-half the width of one flight line coverage.

$$\text{Total flight lines} = 8 + 1 = 9$$

$$\text{Actual spacing between flight lines} = \frac{52{,}800 \text{ ft}}{8 \text{ flight lines}} = 6600 \text{ ft}$$

g) Total number of photos required.

Total photos = 31 per flight line × 9 flight lines = 279 photos

Figures 18–19 (a) and (b) illustrate endlap and sidelap and Fig. 18–19 shows the flight map.

18–16. Sources of error in photogrammetry. Some of the sources of error in photogrammetric work are:

a. Measuring scale not standard length.
b. Inaccurate location of principal and conjugate principal points.
c. Failure to use camera calibration data.
d. Assuming vertical exposures when photographs are actually tilted.
e. Presuming equal flying heights when they were unequal.
f. Not considering differential shrinkage or expansion of the photographic prints.
g. Incorrect orientation of photographs under a stereoscope or in a stereoscopic plotter.
h. Faulty setting of the floating mark on a point.

18–17. Mistakes. Some mistakes which occur in photogrammetry include:

a. Reading measuring scales incorrectly.
b. Mistaking units, i.e., inches and millimeters.
c. Confusion in identifying corresponding points on different photographs.
d. Neglect of relief displacement.
e. Failure to provide proper control or using erroneous control coordinates.
f. Attaching an incorrect sign (plus or minus) to a measured photographic coordinate.
g. Blunder in computations.

PROBLEMS

The distance between two points on a vertical photograph is (X) and the corresponding ground distance is (Y). For the data given in problems 18–1 through 18–3 what is the photographic scale?

18–1. $X = 3.50$ in. $Y = 4000$ ft
18–2. $X = 2.72$ in. $Y = 2430$ ft
18–3. $X = 43.75$ mm. $Y = 1545$ ft

On a vertical photograph of flat terrain, section corners appear a distance (X) apart. If the camera focal length was (f), compute the flying height above the ground for problems 18–4 through 18–6.

18–4. $X = 2.03$ in. $f = 8\frac{1}{4}$ in.
18–5. $X = 98.55$ mm. $f = 6$ in.
18–6. $X = 51.83$ mm. $f = 88.90$ mm.

What are the scales of vertical photographs for each of the conditions given in problems 18–7 through 18–9 of flying height above sea level (H), camera focal length (f), and average ground elevation ($A.G.E.$)?

18–7. $H = 8000$ ft $f = 6$ in. $A.G.E. = 1600$ ft.
18–8. $H = 6500$ ft $f = 8\frac{1}{4}$ in. $A.G.E. = 1250$ ft.
18–9. $H = 12{,}750$ ft $f = 88.90$ mm. $A.G.E. = 4480$ ft.

On a vertical photograph of flat terrain, the scaled distance between two points is (X). In problems 18–10 through 18–12 find the scale of the photograph if the length between the same points is (Y) on a map which is plotted at a scale of (S).

18–10. $X = 3.75$ in. $Y = 4.50$ in. $S = 1{:}10{,}560$
18–11. $X = 60.25$ mm. $Y = 71.84$ mm. $S = 1{:}14{,}400$
18–12. $X = 112.48$ mm. $Y = 3.60$ in. $S = 1{:}24{,}000$

Compute the scale of a vertical photograph using the data given in problems 18–13 through 18–15.

18–13. The photographic length of a college football field is 1.75 inches from goal post to goal post.

18–14. The photographic width of an interstate highway pavement is 1.30 mm.

18–15. The photographic distance between two adjacent section corners is 3.63 inches.

For certain military reconnaissance it is necessary to obtain vertical aerial photography at a scale of (S). In problems 18–16 and 18–17 what focal length camera would be required if the lowest safe altitude for flying over enemy defenses is (H)?

18–16. $S = 1{:}10{,}000$, $H = 16{,}000$ ft.

18–17. $S = 1{:}20{,}000$, $H = 10{,}000$ ft.

Calculate the flight height above average terrain that is required in order to obtain vertical photographs for constructing a mosaic at an average scale of (X) if the camera focal length is (f), for problems 18–18 through 18–20.

18–18. $X = 1{:}14{,}400$, $f = 12$ in.

18–19. $X = 1{:}10{,}000$, $f = 8\frac{1}{4}$ in.

18–20. $X = 1{:}8000$, $f = 152.4$ mm.

In problems 18–21 through 18–23 determine the horizontal distance between two points A and B whose elevations above datum are (h_A) and (h_B) and whose images a and b have photographic coordinates x_a, y_a, x_b, and y_b on a vertical photograph. The camera focal length was 6 inches and the flying height above datum 6000 ft.

18–21. $x_a = 2.15$ in. $y_a = 1.78$ in. $x_b = -1.44$ in. $y_b = -3.10$ in.
$h_A = 350$ ft. $h_B = 210$ ft.

18–22. $x_a = 3.45$ in. $y_a = -0.76$ in. $x_b = -4.25$ in. $y_b = 2.62$ in.
$h_A = 1500$ ft. $h_B = 780$ ft.

18–23. $x_a = -74.28$ mm. $y_a = -40.77$ mm. $x_b = 102.05$ mm. $y_b = -68.13$ mm.
$h_A = 2200$ ft. $h_B = 860$ ft.

For the above problems the image c of a third point C whose elevation above datum is (h_C), has photographic coordinates of x_c and y_c. In problems 18–24 through 18–26 calculate the three horizontal angles in the triangle ABC for the data given.

18–24. The data of problem 18–22 and $x_c = 3.80$ in. $y_c = -2.95$ in.
$h_C = 100$ ft.

18–25. The data of problem 18–23 and $x_c = -2.02$ in. $y_c = 1.84$ in.
$h_C = 935$ ft.

18–26. The data of problem 18–24 and $x_c = 5.70$ mm. $y_c = 88.62$ mm.
$h_C = 1440$ ft.

Determine the height of TV towers from a vertical photograph for the conditions given in problems 18–27, 18–28, and 18–29 of flying height above the base of the tower (H), distance on the photograph from the principal point to the base of the tower (r_a) and distance from the principal point to the top of the tower (r_b).

18–27. $H = 6{,}000$ ft. $r_a = 2.775$ in. $r_b = 3.100$ in.

18–28. $H = 5{,}500$ ft. $r_a = 86.75$ mm. $r_b = 95.28$ mm.

18–29. $H = 10{,}000$ ft. $r_a = 43.92$ mm. $r_b = 48.03$ mm.

An area has an average terrain elevation of (h) feet above mean sea level and the highest point in the area is (Y) feet above average terrain. If, in problems 18–30 through 18–32, the camera focal plane opening is 9 inches by 9 inches, what flying height above mean sea level will limit relief displacement to a maximum of 0.05 inches in a vertical photograph of this area?

18–30. $h = 500$ ft $Y = 150$ ft.

18–31. $h = 1000$ ft $Y = 40$ ft.

18–32. $h = 850$ ft $Y = 65$ ft.

On a vertical photograph the images a and b of the ground points A and B have photographic coordinates x_a, y_a, x_b, and y_b respectively. The horizontal distance between A and B is (L) and the elevations above datum of A and B are h_A and h_B. In problems 18–33 through 18–35 calculate the flying height above datum for a camera having a focal length of 6 inches.

18–33. $x_a = 3.45$ in. $y_a = 2.17$ in. $x_b = -1.80$ in. $y_b = -2.32$ in.
$h_A = 425$ ft. $h_B = 570$ ft. $L = 5{,}280.5$ ft.

18–34. $x_a = -60.50$ mm. $y_a = 94.33$ mm. $x_b = -16.74$ mm. $y_b = -87.08$ mm.
$h_A = 730$ ft. $h_B = 618$ ft. $L = 4838.7$ ft.

18–35. $x_a = 0.707$ in. $y_a = -2.413$ in. $x_b = 4.316$ in. $y_b = -0.835$ in. $h_A =$ 1435.2 ft. $h_B = 1461.2$ ft. $L = 1919.4$ ft.

18–36. Discuss the advantages and disadvantages of mosaics.

18–37. How does a mosaic differ from a map?

18–38. Discuss the advantages and disadvantages of extending horizontal control by radial triangulation.

18–39. Discuss the fundamental photogrammetric principle upon which radial triangulation is based.

An air base of (B) feet exists for a pair of overlapping photographs taken at a flying height of (H) feet above MSL with a camera having a focal length (f). The photo coordinates of points A and B on the left photograph are $x_a = 40.00$ mm, $y_a = 50.00$ mm, $x_b = 3.00$ mm and $y_b = -27.00$ mm. The x photo coordinates on the right photograph are $x_a = -60.00$ mm and $x_b = -72.00$ mm. Calculate the horizontal length of the line AB for problems 18–40 through 18–42.

18–40. $B = 3000$ ft $H = 9000$ ft $f = 8\frac{1}{4}$ in.

18–41. $B = 7200$ ft $H = 15{,}000$ ft $f = 6$ in.

18–42. $B = 4500$ ft $H = 5500$ ft $f = 88$ mm.

In problems 18–43 through 18–45 calculate the elevations above datum of points A and B from the overlapping photographs.

18–43. For problem 18–40.

18–44. For problem 18–41.

18–45. For problem 18–42.

A pair of overlapping vertical photographs are oriented and secured under a mirror stereoscope and a parallax bar reading is obtained on point a of $(r_a) = 14.23$ mm. The x photo coordinate of point a is 2.45 inches on the left photograph and -1.24 inches on the right photograph. The camera focal length is 6 inches, flying height is 10,000 feet above datum, and the air base is 4,000 feet. In problems 18–46 through 18–48 what is the difference in elevation between points B and C if their parallax bar readings are (r_b) and (r_c) respectively?

18–46. $r_b = 15.75$ mm. $r_c = 15.02$ mm.

18–47. $r_b = 14.63$ mm. $r_c = 16.28$ mm.

18–48. $r_b = 13.94$ mm. $r_c = 17.86$ mm.

For the overlapping pair of photographs in problems 18–46 through 18–48, what parallax bar reading is required to trace a contour of the elevations given in problems 18–49 through 18–51?

18–49. 3500 ft.

18–50. 4000 ft.

18–51. 2500 ft.

Aerial photography is to be taken of an area that is (X) miles square. Flying height will be (H) feet above average terrain and the camera will have a focal length (f). If the focal plane opening is 9 inches by 9 inches and minimum sidelap is 30 per cent, how many flight lines will be needed to cover the area for the data given in problems 18–52 through 18–54?

18–52. $X = 10$ $H = 6000$ $f = 6$ in.

18–53. $X = 15$ $H = 12{,}000$ $f = 209.8$ mm.

18–54. $X = 140$ $H = 8000$ $f = 3\frac{1}{2}$ in.

Aerial photography was taken at a flying height (H) ft above average terrain. If the camera focal plane dimensions were 9 in. by 9 in., the focal length (f), and the spacing between adjacent flight lines (X) ft, what is the per cent sidelap for the data given in problems 18–55 through 18–57?

18–55. $H = 4000$ $f = 6$ in. $X = 4500$.

18–56. $H = 6500$ $f = 88$ mm. $X = 12{,}000$.

18–57. $H = 10{,}750$ $f = 8\frac{1}{4}$ in. $X = 7000$.

Photographs at a scale of (S) are required to cover an area (X) miles square. The camera has a focal length (f) and focal plane dimensions of 9 inches by 9 inches. If endlap is 60 per cent and sidelap 30 per cent, how many pictures will be required to cover the area for the data given in problems 18–58 through 18–60?

18–58. $S = 1{:}12{,}000$ $X = 10$ $f = 6$ in.

18–59. $S = 1{:}14{,}400$ $X = 20$ $f = 3\frac{1}{2}$ in.

18–60. $S = 1{:}7200$ $X = 35$ $f = 152.4$ mm.

chapter 19

Boundary Surveys

19–1. General. The earliest surveys were made to locate or relocate boundary lines of property. Trees and other natural objects, or stakes placed in the ground, were used to identify the corners. As property increased in value and owners disputed rights to land, the importance of permanent monuments and written records became evident. Property titles now are transferred by written documents called *deeds,* which contain a description of the boundaries of the property.

Property is described in several ways: (a) by metes and bounds, (b) by block-and-lot system, (c) by coordinate values for each corner, or (d) by township, section, and smaller subdivision. The first three methods will be discussed briefly in this chapter. Subdivision of the public lands into townships and sections will be covered in Chapter 20.

Practically all property surveys today are wholly or partly resurveys rather than originals. In retracing old lines a surveyor must exercise acute judgment based on practical experience and a knowledge of land laws, and be precise in his measurements. The necessary mathematics and the proper use of the transit, tape, and level can be learned in a relatively short time. This background must be bolstered by tenacity in searching the records of all adjacent property as well as studying the description of the land in question. In the field, a surveyor must be untiring in his efforts to find points called for by the deed. Often it is necessary to obtain testimony from old settlers or others having knowledge of accepted land lines and the location of corners, reference points, fences, and other evidence of the lines.

A land surveyor may be confronted with defective surveys; incompatible descriptions and plats of common lines for adjacent tracts; lost or obliterated corners and reference marks; discordant stories by local residents; questions of riparian rights; and a multitude of legal decisions on cases involving property boundaries. His duty to a client is to sift all available evidence and try to obtain a meeting of minds among persons involved in any property-line dispute. In this task he is doing professional work although he has no legal authority to force a compromise or settlement of any kind. Fixing of title boundaries must be done by agreement of adjacent owners or by court proceedings. To serve legally as an expert witness in proceedings to establish boundaries, a surveyor must be registered.

Many municipalities have rigid laws covering subdivisions. Regulations may specify the minimum size of lot; the allowable closures for surveys; the types of corner marks to be used; the minimum width of streets, and the procedure for dedicating them; the rules for registry of plats; and other matters. The mismatched street and highway layouts of today could have been eliminated by suitable subdivision regulations in past years.

An estimate made of the surveying involved in the 41,000-mile Interstate Highway System shows that approximately 750,000 separate parcels of land will have to be surveyed and acquired. Some parcels will be completely taken, and only one survey will be needed. Many others will be only partial takings, and therefore two or three separate surveys, closures, and area computations will be required. Electronic digital computers will be used for this volume of work.

Boundary surveying is learned only by practice in the field and study of the laws pertaining to land. Discussion in an elementary text must of necessity be brief. For more-extensive coverage the reader is referred to several excellent books on the subject listed in the References, page 603.

19–2. Basis of land titles. In the eastern part of the United States, individuals acquired the first land titles by gift or purchase from the English Crown. Surveys and maps were completely lacking or inadequate, and descriptions could be given only in general terms. The remaining land in the Thirteen Colonies was transferred to the states at the close of the Revolutionary War. Later this land was parceled out to individuals, generally in irregular tracts. The boundary lines were described by metes and bounds (directions by magnetic bearings, and lengths in Gunter's chains, poles, or rods).

Many original transfers, and subsequent ownerships and subdivisions, were not recorded. Those that were legally registered usually had scanty or defective descriptions since land was cheap and abundant. The trees, rocks, and natural landmarks defining the corners were soon disturbed. The intersection of two property lines might be described only as "the place where John killed a bear," or "the bend in a footpath from Jones' cabin to the river."

Numerous problems in land surveying stem from the confusion engendered by early property titles, descriptions, and compass surveys. The locations of thousands of corners have been established by compromise after resurveys, or by court interpretation of all available evidence pertinent to their original or intended positions. Other corners have been fixed by *squatters' rights, adverse possession,* and *riparian changes.* Many boundaries still are in doubt, particularly in areas having marginal land where the cost of a survey ("retracement of history") exceeds the value of the property.

The fact that the four corners of a field can be found, and that the distances between them agree with the calls in a description, does not necessarily mean they are in the proper place. Title or ownership is complete only when the land covered by a deed is positively identified and located on the ground.

Years of experience in a given area are needed by a land surveyor in order for him to become familiar with local conditions, basic reference points, and legal interpretations of boundary problems. The method used in one state for prorating differences between recorded and measured distances may not be acceptable in another state. Different interpretations are given locally to the superiority or definiteness of one distance over another associated with it; to the position of boundaries shown by occupancy; to the value of corners in place in a tract and its subdivisions; and to many other factors. Registration of land surveyors is therefore required in most states to protect the public interest.

19–3. Property description by metes and bounds. Descriptions by metes and bounds have a *point of beginning,* such as a stake, fence post, road intersection, or some natural feature. In recent years artificial permanent monuments long enough to reach below the frost line have been used. They consist of metal pipes, steel pins, and concrete posts. Lengths and bearings of successive lines from the point of beginning are given. Any lengths in chains, poles, and rods are being replaced by measurements in feet and decimals secured by means of the surveyor's tape. Bearings may be magnetic or true, the latter being preferable of course.

In relocating an old survey, precedence (weight of importance) is commonly assigned as follows: (1) Marks or monuments in place;

(2) calls for boundaries of adjoining tracts; (3) courses and distances shown in the original notes or plat.

The importance of permanent monuments is evident—in fact, long iron pins and/or concrete markers are required at all property corners before surveys will be accepted for recording in some areas. Actually, almost anything can be called for as a monument. A map attached to the description clarifies it and scaling provides a rough check on the angles and distances.

Property descriptions are written by surveyors and lawyers. A single error in transcribing a numerical value, or one incorrect or misplaced word or punctuation mark, may result in litigation for a generation or more, since the intentions of *grantor* (person selling property) and *grantee* (person buying property) are not fulfilled.

In order to increase the precision of property surveys, large cities and some states have established control monuments to supplement the triangulation stations of the United States Coast and Geodetic Survey. Property corners can then be tied to these control points, and boundary lines can be relocated with assurance.

A description of a piece of property in a deed should always contain the following information in addition to the recital:

a) *Point of beginning.* This should be identifiable, permanent, well-referenced, and near the property. If the coordinates are known, they should be given.

b) *Definite corners.* Clearly defined points, with coordinates if possible.

c) *Lengths and directions of the sides of the property.* All lengths in feet and decimals, and directions by angles or true bearings, must be given to permit computation of the error of closure. Omitting the length or bearing of the closing line to the point of beginning and substituting a phrase "and thence to the point of beginning" is no longer permitted. The date of the survey must be included. This is particularly important if the bearings are referred to magnetic north.

d) *Names of adjoining property owners.* These are given to avoid claims for land in case an error in the description leaves a gap.

e) *Area.* The included area is normally given as an aid in the valuation and identification of the property. Areas of rural land are given in acres; those of city lots, in square feet.

An example of a metes-and-bounds description of a suburban lot in the eastern section of the United States follows:

Beginning at the center of a stone monument located on the South property line of Maple Street in the Town of Snowville, New York, about 452 feet from the east line of Adams Road, at the northeast corner of land now owned by James Brown; thence along east line of Brown S 5° 10′ E, two hundred fifty-two and eight tenths (252.8) feet to an iron pipe on the north line of property owned by Henry Cook; thence along a fence on north line of Cook's land S 82° 41′ E one hundred ninety-seven and three tenths (197.3) feet to center of stone wall on the west line of Charles Williams; thence N 46° 10′ E one hundred thirty-four and five tenths (134.5) feet along center line of the stone wall to an iron pipe set at the end of the wall; thence N 5° 10′ W two hundred ten and no tenths (210.0) feet to a stone monument on the south line of Maple Street; thence S 84° 50′ W three hundred (300.0) feet along south line of Maple Street to place of beginning, containing 1.64 acres more or less. The bearings given are true bearings.

A typical old description of a city lot follows (note the lack of comparable precision in angles and distances):

Beginning at a point on the west side of Beech Street marked by a brass plug set in a concrete monument located one hundred twelve and five tenths (112.5) feet southerly from a city monument No. 27 at the intersection of Beech Street and West Avenue; thence along the west line of Beech Street S 15° 14′ 30″ E fifty (50) feet to a brass plug in a concrete monument; thence at right angle to Beech Street S 74° 45′ 30″ W one hundred fifty (150) feet to an iron pin; thence at right angles N 15° 14′ 30″ W parallel to Beech Street fifty (50) feet to an iron pin; thence at right angles N 74° 45′ 30″ E one hundred fifty (150) feet to place of beginning; bounded on the north by Norton, on the east by Beech Street, on the south by Stearns, and on the west by Weston.

19–4. Property description by block-and-lot system. In subdivisions and in large cities it is more convenient to identify individual lots by *block and lot number*, by *tract and lot number*, or by *subdivision name and lot number*. Examples are:

Lot 34 of Tract 12314 as per map recorded in book 232 pages 23 and 24 of maps, in the office of the county recorder of Los Angeles County.

Lot 9 except the North 12 feet thereof, and the East 26 feet of Lot 10, Broderick's Addition to Minneapolis. [Parts of two lots are included in the parcel described.]

That portion of Lot 306 of Tract 4178 in the City of Los Angeles, as per Map recorded in Book 75 pages 30 to 32 inclusive of maps in the office of the County Recorder of said County, lying Southeasterly of a line extending Southwesterly at right angles from the Northeasterly line of said Lot, from a point in said Northeasterly line distant Southeasterly 23.75 feet from the most Northerly corner of said Lot.

Map books in the city or county recorder's office give the location and dimensions of all the blocks and lots. It is now standard practice to require subdividers to file a map with the proper office showing the type and location of monuments, size of lots, and other pertinent information such as the dedication of streets. It is evident that if the boundary lines of a tract are in doubt, the individual lot lines must be questioned also.

The block-and-lot system is a short and unique means of describing property for tax purposes as well as for transfer. Identification by street and house number is satisfactory only for tax-assessment records.

19–5. Property description by coordinates. The advantages of state plane coordinate systems in improving the accuracy of local surveys, and in facilitating the relocation of lost and obliterated corners, have led to their legal acceptance in property descriptions. The coordinate description of corners may be used alone, or as an alternative method. Wider use of the coordinate system will be made as more reference points become available to the local surveyor.

An example of a description in which coordinates are used will be given by quoting a complete deed of easement.

DEED OF EASEMENT

This indenture, made this 7th day of February, 1953, between JOHN S. DOE, party of the first part, and CITY OF ALAMEDA, a municipal corporation, party of the second part,

Witnesseth:

That the said party of the first part, for and in consideration of the sum of One ($1.00) Dollar to him in hand paid by the party of the second part, in lawful money of the United States, at or before the execution and delivery of these presents, the receipt whereof is hereby acknowledged, does hereby grant, bargain, sell, transfer and convey unto the said party of the second part a permanent easement and right of way, at any time, or from time to time, to construct, maintain, operate, replace, remove and renew an underground sanitary sewer and storm drain in, through, under and along that certain piece or strip of land in the City of Alameda, State of California, particularly described as follows, to-wit:

BEING a strip of land six feet (6.00′) wide,

COMMENCING at United States Coast and Geodetic Survey Monument "Otis Mo" having the coordinates y = 461,113.18 feet and x = 1,496,962.81 feet as based on the California Coordinate

System Zone 3, as are all coordinates, bearings and distances in this description;

THENCE South 75° 38′ 32″ West 40.21 feet to the intersection of the center lines of Otis Drive (formerly Front Street) and Mound Street as shown on the Map of Alameda, surveyed and drawn by J. T. Stratton and filed in the Alameda County Recorder's Office July 24, 1879, on Page 1 of Book 6 of Maps, whose coordinates are y = 461,103.21 feet and x = 1,496,923.86 feet and which intersection is shown on Drawing 3667, Case 54, City Engineer's files, Alameda, California, which drawing is hereby made a part of this instrument as Exhibit "A" attached;

THENCE along the center line of Otis Drive South 47° 41′ 56″ East 295.10 feet to the intersection of the center lines of Court Street and Otis Drive as shown on the aforesaid map by J. T. Stratton, and whose coordinates are y = 460,904.59 feet and x = 1,497,142.12 feet;

THENCE along the center line of Court Street South 42° 20′ 33″ West 135.00 feet;

THENCE North 47° 41′ 56″ West 30.00 feet to the east corner of the parcel as deeded to John S. Doe and recorded in Book 6598 of Official Records at Page 353, in the Alameda County Recorder's Office, and which point is the TRUE POINT OF BEGINNING, having the coordinates y = 460,825.00 feet and x = 1,497,029.00 feet;

THENCE along the northeast line of the said parcel North 47° 41′ 56″ West 58.77 feet to the north corner.

THENCE along the Northwest property line South 42° 20′ 33″ West 6.00 feet;

THENCE South 47° 41′ 56″ East 58.77 feet to the Southeast line of said parcel;

THENCE along the said line North 42° 20′ 33″ East 6.00 feet to the TRUE POINT OF BEGINNING.

TO HAVE AND TO HOLD the above mentioned and described easement, together with the appurtenances, unto the said party of the second part, its successors and assigns forever for the sole object and purpose of at any time, or from time to time, to construct, maintain, operate, replace, remove and renew an underground sanitary sewer and storm drain, as above set out and for no other purpose, and should the said real property herein described be at any time used for any other purpose by said party of the second part than for the easement and right of way for the purposes hereinabove set out, or should the use of said real property for said purposes be permanently terminated or permanently discontinued, then this easement and right of way hereby granted and conveyed shall immediately lapse and become null and void and said easement and right of way shall immediately revert to the grantor.

IN WITNESS WHEREOF, the said party of the first part has hereunto set his hand and seal, the day and year first hereinabove written.

(Signature) ______________________________

Approved as to form:

____

City Attorney of the City of Alameda

Description approved:

____

City Engineer of the City of Alameda

Earthquakes in Alaska, California, etc., have caused ground shifts which move corner monuments and thereby change their coordinates. The monuments, rather than coordinates, will then have greater weight in ownership rights.

19–6. Field work. The first task in the field is to locate the property corners. Here a most valuable instrument to the land surveyor—the shovel—frequently comes into use. In many cases one or more lines may have to be run from control points some distance away to check or establish the location of a corner. If two points are available with known coordinates on a state-wide or local system, a connecting line or traverse is used to transfer true bearings or coordinates to the property boundaries, as indicated in Fig. 19–1.

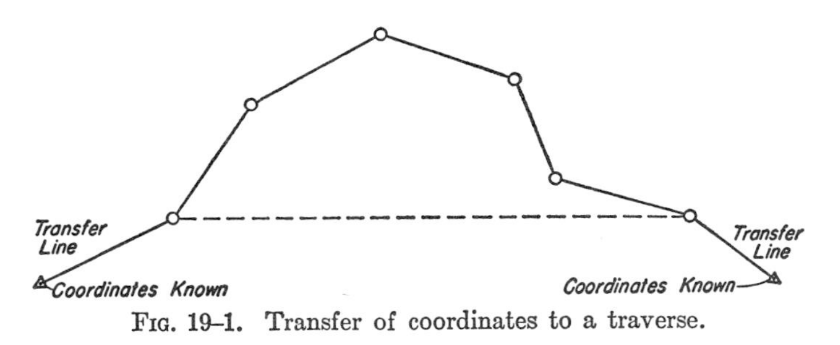

FIG. 19–1. Transfer of coordinates to a traverse.

Generally a closed traverse is run around the property, all corners being occupied if possible. Fences, trees, shrubbery, hedges, party walls, and other obstacles may necessitate a traverse that is inside or outside the property. From measurements to the corners, it is possible to calculate their coordinates and the lengths and bearings of the sides (see Sec. 10–11). All measurements should be made with a precision suited to the value of the land. If coordinates on a grid system are to be used in the description, the accuracy of the system itself should be maintained if possible.

Measurements made with a standardized tape and corrected for temperature and tension may not agree with distances on record or those between the marks. This provides a real test for the surveyor. Perhaps his tape is different in length from the one used in the original survey; the marks may have been disturbed, or be the wrong ones; the monuments may not check with others in the vicinity believed or known to be correct; and several previous surveys may not agree with each other.

Bearings, and angles between adjacent sides, may not fit those called for. The discrepancy could be due to a faulty original compass survey, incorrect corner marks, or other causes.

A surveyor's job is to relocate or establish boundary markers in the exact position where they were originally set. If discrepancies exist between distances of record and measured values for found monuments, the data should be used to get a calibration value for the tape which relates it to the original surveyed distances. In relocating monuments, this calibration is employed to lay off the actual distances used by the original surveyor.

19–7. Registration of title. To remedy the difficulties arising from inaccurate descriptions and disputed boundary claims, some states provide for the registration of title to property under rigid rules. The usual requirements include marking each corner with standardized monuments referenced to established points, and recording a plat drawn to scale and containing specified items. Titles are then guaranteed by the court under certain conditions.

A number of states have followed Massachusetts' example and maintain separate *land courts* dealing exclusively with land titles. As the practice spreads, precision of property surveys will be increased and the transfer of property simplified.

A comparable service is offered by *title insurance companies* which search, assemble, and interpret official records, laws, and court decisions affecting ownership of land. The title company insures a purchaser against loss by guaranteeing that its findings regarding defects, liens, encumbrances, restrictions, assessments, and easements are correct. Defense against lawsuits is provided by the company against these threats to a clear title if the claims are shown in the public records and are not exempted in the policy. The location of corners and lines is not guaranteed. Hence it is necessary to establish on the ground the exact boundaries called for by the deed and the title policy.

Many technical and legal problems are considered before title insurance is granted. In Florida, for example, title companies refuse to issue a policy covering a lot if the fences in place are not on the property line. Occupation and use of land belonging to a neighbor but outside his apparent boundary line as defined by a fence may lead to a claim of adverse possession.

Adverse rights are obtained against all except the public by occupying a parcel of land for a period of years specified by law and performing certain acts. Possession must be (a) actual, (b) exclusive, (c) open and notorious, (d) hostile, (e) continuous, and (f) under color of title. In some states all taxes must be paid. The time required to establish a claim of adverse possession varies from a minimum of 7 years in Florida to a maximum of 60 years for urban property in New York. The customary period is 20 years.

The continuous use of a street, driveway, or footpath by an individual or the general public for a specified number of years results in establishment of a right-of-way privilege which cannot be withheld by the original owner.

19–8. Sources of error. Some of the sources of error in boundary surveys follow:

a. Corner not defined by a unique point.
b. Unequal precision of angles and distances.
c. Length measurements not made with a standardized tape and corrected for tension, temperature, and slope.
d. Magnetic bearings not properly corrected to the date of the new survey.

19–9. Mistakes. Typical mistakes in connection with boundary surveys are the following:

a. Use of wrong corner marks.
b. Failure to check deeds of adjacent property as well as the description of the parcel in question.
c. Ambiguous deed descriptions.
d. Omission of the length or bearing of the closing line.
e. Failure to close on known control.

PROBLEMS

19–1. Determine from one of the publications listed in the References the meaning of eminent domain; riparian rights; color of title; deed; and title.

19–2. Determine, from the County courthouse or other local records, the number of types of property descriptions being used in the area. Copy one example of each type available.

19–3. What must the surveyor do in interpreting obscure descriptions?

19–4. List the essential points covering the problem which arises when corners in place do not agree with deed descriptions.

19–5. What is the first job of a surveyor employed to survey a farm located in an unfamiliar area?

19–6. List in their order of importance the following items used in legally interpreting deed descriptions: written and measured lengths; written and measured bearings; monuments; witness corners and ties; areas; and testimony of living and dead witnesses. Justify your answer.

19–7. Submit a copy of a typical recital obtained from the deed description of a local piece of property.

19–8. Compute and plot the metes-and-bounds description of the suburban lot given in Sec. 19–3. Is the accuracy satisfactory today for an average town of 5,000 people?

19–9. Similar to problem 19–8, except for the city lot described in Sec. 19–3.

19–10. In a description of land by metes and bounds, what purpose or purposes may be served by the statement "more or less" added to the acreage?

19–11. Telephone poles are set on two corners of a rectangular lot. Explain how you would make a survey of the lot to locate its boundaries and to determine the area.

19–12. Starting with a house number and street name, obtain a copy of the deed description of the property by searching the records in the County courthouse. Determine whether the corners described can be located readily on the ground.

19–13. Write a metes-and-bounds deed description for the house and lot where you live, assuming courses and distances where necessary. Draw a sketch map of the property and buildings, showing all information which should appear on a plan to be filed with the deed description at the registry of deeds.

19–14. Determine from the County Recorder's office, or a local surveyor, how the assignment of block, tract, subdivision, and lot numbers is made and by whom.

19–15. Sketch the property described as part of lots 9 and 10 in Sec. 19–4.

19–16. Sketch the portion of lot 306 described in Sec. 19–4.

19–17. Determine the size of strip and the form of easement used by the local electric and gas utilities, and submit a copy of one example.

19–18. Similar to problem 19–17 for the telephone company or other facility.

19–19. Compute and check the coordinates and closure of the easement described in Sec. 19–5.

19–20. A farmer owning 200 acres of rolling terrain near a city hires a surveyor to subdivide the property for sale as lots. Outline the steps to be taken by the surveyor.

19–21. A survey of a city block starting from proven corner monuments shows that the block is 2.62 ft longer than the plat distance. Lot boundaries were never staked. Show on a sketch how the excess is distributed, and explain.

19–22. Use the conditions given in problem 19–6, but assume that the lots have been staked for one half of the block.

19–23. Two neighbors have a boundary dispute and hire a surveyor to check the line. Outline the authority of the surveyor, (a) assuming that the line he establishes is satisfactory to his clients and (b) assuming that it is unsatisfactory to one or both of them.

19–24. A surveyor, running the metes-and-bounds boundary lines of an area prior to subdividing it, is unable to find one of the original corners, so he sets his own. The subdivision is completed and the lots are sold. Another surveyor, two years later, proves that the new corner is 8 ft east of the original corner. Which is the legal corner? Why? What should be done about the other corner?

19–25. In establishing or re-establishing property lines or corners, what judicial authority does a licensed surveyor possess if his locations do not agree with those of another licensed surveyor or individual?

19–26. An error has been made on a plat of a subdivision and duly recorded. How can this defect be remedied and by whom?

19–27. List all types of pertinent information or data which should appear on a completed plan of a property survey.

19–28. Outline and sketch the procedure for surveying and monumenting the boundary of a piece of land which is to be split into three irregular parts. Using assumed data, write a description for the deed of one part.

19–29. What period of time is required to establish a claim for adverse rights in your city and state?

19–30. Secure a sample or make a copy of the pertinent features included in the policy of a title insurance company for a city lot.

19–31. A deed calls for a property line to run "S 62° 45′ E, 255.47 feet to a concrete monument set flush with the ground." A resurvey shows that the monument is hidden by 2 ft of fill and the bearing and distance are S 62° 43′ E and 254.47 ft; but it is definitely established as the monument called for. (a) What governs? (b) If a later monument set exactly on the point at N 62° 45′ E and 255.47 ft during a resurvey is found, what governs? (c) If a fence has been built on the later line and the property on both sides of it has been occupied and used for

22 years on the assumption that the fence was on the correct line, what is the result?

19–32. List the reasons why every purchaser of a farm, city lot, or home should require a survey before final payment is made.

19–33. Of the 5 required items necessary in a deed description, Sec. 19–3, which do you consider the most important? Why? What else might well be included?

19–34. Which of the four methods of property description do you consider best? Why?

19–35. What is the purpose of the "hostile" requirement in the process of acquiring adverse rights to real property?

NOTE: Study of some of the references listed on page 603 is helpful in solving the problems of this chapter.

chapter 20

Surveys of the Public Lands

20–1. General. The term *public lands* has been applied broadly to the areas which have been subject to administration, survey, and transfer of title under the public-land laws of the United States. These lands include those turned over to the Federal Government by the Colonial States, and those larger areas acquired by purchase from (or treaty with) the native Indians, or foreign powers that had previously exercised sovereignty.

Thirty states, including Alaska, comprise the public domain which has been, or will be, subdivided into rectangular tracts (see Fig. 20–1). The area represents approximately 72 per cent of the United States.

> The title to the vacant lands, therefore the direction over the surveys, within their own boundaries, was retained by the Colonial States, the other New England and Atlantic Coast States (excepting Florida), and later by the states of West Virginia, Kentucky, Tennessee, and Texas, in which areas the United States public land laws have not been applicable.
>
> The beds of navigable bodies of water are not public domain and are not subject to survey and disposal by the United States. The sovereignty is in the individual states.
>
> In 17 of the 29 states . . . the swamp and overflowed lands, though public domain, pass to the States upon identification by public land survey, and approved selection, the title being subject to the disposal by the States.[1]

Survey and disposition of the public lands was governed originally by two factors:

a. A recognition of the value of a grid system of subdivision, based on experience in the Colonies and Europe.

[1] *Manual of Instructions for the Survey of the Public Lands*, 1947 edition, United States Government Printing Office.

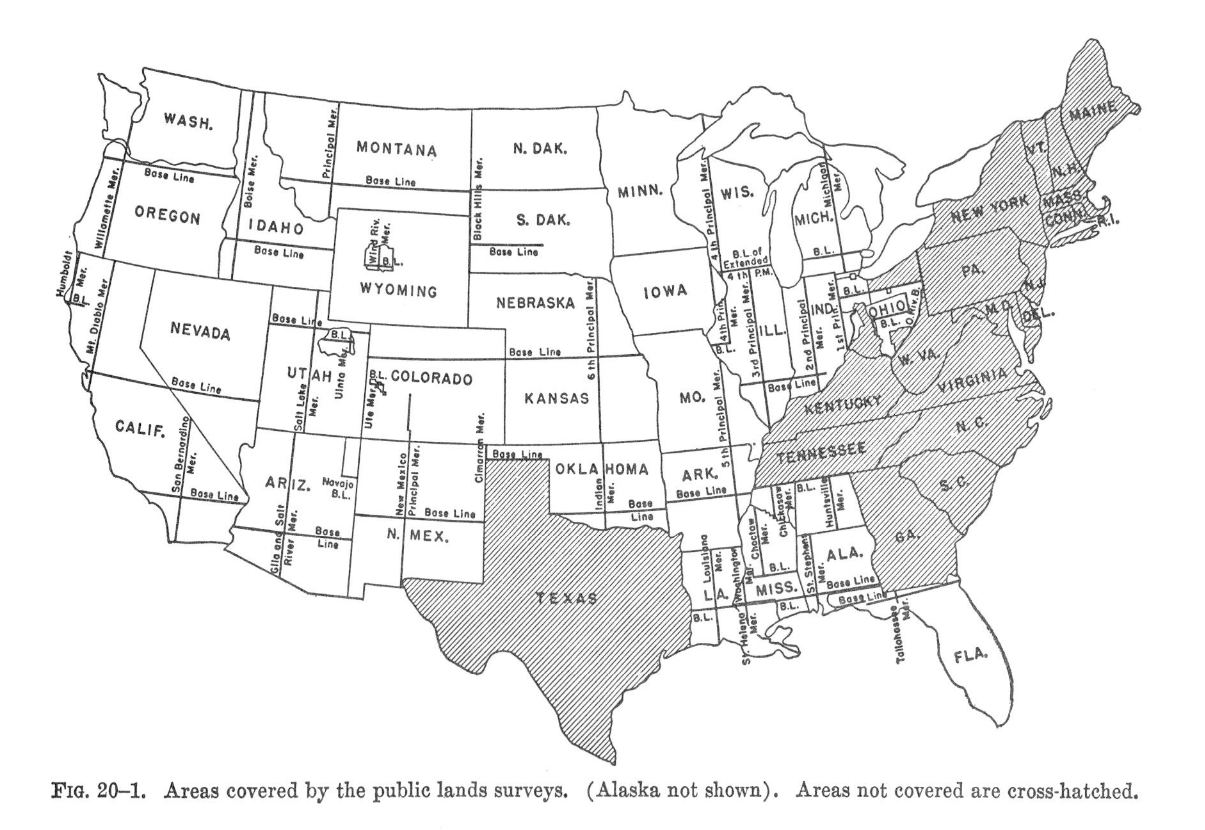

FIG. 20–1. Areas covered by the public lands surveys. (Alaska not shown). Areas not covered are cross-hatched.

b. The need of the Colonies for revenue from the sale of the public lands. Monetary returns from sale of the public lands were disappointing, but the farsighted vision of the planners of the grid system of subdivision deserves commendation.

20–2. Instructions for survey of the public lands. The United States system of public-lands surveys was inaugurated in 1784, with the territory northwest of the Ohio River as a test area. Sets of instructions for the surveys were issued in 1785 and 1796. Manuals of instructions were issued in 1855, 1881, 1890, 1894, 1902, 1930, and 1947.

In 1796 a Surveyor General was appointed and the numbering of sections was changed to the system now in use. Supplementary rules were promulgated by each local Surveyor General "according to the dictates of his own judgment" until 1836, when the General Land Office was reorganized. Copies of changes and instructions for local use were not always preserved and sent to Washington. As a result, no office in the United States has a complete set of instructions under which the original surveys were supposed to have been made. Although the same general method of subdivision was followed, detailed procedures were altered in surveys made at different times in various areas of the country.

Most of the later public-land surveys have been run by the procedures to be described, or variations of them. The job of the present-day surveyor consists of retracing the original lines and perhaps further subdividing sections. To do so, he must be thoroughly familiar with the rules, laws, equipment, and field conditions governing the work of his predecessors in a given area.

Basically, the rules of survey stated in the 1947 *Manual of Instructions* are as follows:

> The public lands shall be divided by north and south lines run according to the true meridian, and by others crossing them at right angles, so as to form townships six miles square. . . .
>
> The corners of the townships must be marked with progressive numbers from the beginning; each distance of a mile between such corners must be also distinctly marked with marks different from those of the corners.
>
> The township shall be subdivided into sections, containing as nearly as may be, six hundred and forty acres each, by running through the same, each way, parallel lines at the end of every two miles; and by marking a corner on each of such lines at the end of every mile. The sections shall be numbered, respectively, beginning with the number one in the

northeast section, and proceeding west and east alternately through the township with progressive numbers till the thirty-six be completed.

Authority to establish section lines at intervals of one mile was given in 1800 and is included in all manuals of instructions since that date.

Additional rules of survey covering field books, subdivision of sections, adjustment for excess and deficiency, and other matters, are given in the manuals. Surveys were made by private surveyors, who were paid $2 per mile of line run until 1796 and $3 per mile run thereafter, on a contract basis. Sometimes the amount was adjusted in accordance with the importance of the line, the terrain, the location, and other factors. From his meager fee the surveyor had to pay and feed his party of at least four men while on the job, and in transit to and from distant points. He had to brush out and blaze the line, set corners and other marks, and provide satisfactory notes and one or more copies of completed plats. The contract system was discarded in 1910. Public-lands surveyors are now appointed.

Since meridians converge, it is evident that the requirement that lines shall conform to the true meridians, and townships shall be six miles square, is mathematically impossible. An elaborate system of subdivision was therefore worked out as a practical solution.

It should be noted that two principles furnish the legal background for stabilizing land lines:

a. Boundaries of public lands established and returned by duly appointed surveyors are unchangeable.
b. Original township and section corners established by surveyors must stand as the true corners which they were intended to represent, whether in the place shown by the field notes or not.

In general, the procedure for the survey of the public lands provides for the following subdivisions:

a. Division into quadrangles, or tracts, approximately 24 miles square.
b. Division of tracts into townships (16), approximately 6 miles on a side.
c. Division of townships into sections (36), approximately 1 mile square.
d. Subdivision of sections (usually by the local surveyor).

It will be helpful to keep in mind that the purpose of the grid system was to obtain sections 1 mile on a side. To this end, all discrepancies were thrown into the sections bordering the north and west township boundaries to get as many *regular sections* as possible.

20–3. Initial point. Subdivision of the public lands became necessary in any area as settlers moved in and mining or other land claims were filed. The original hope that surveys would precede settlement was not fulfilled.

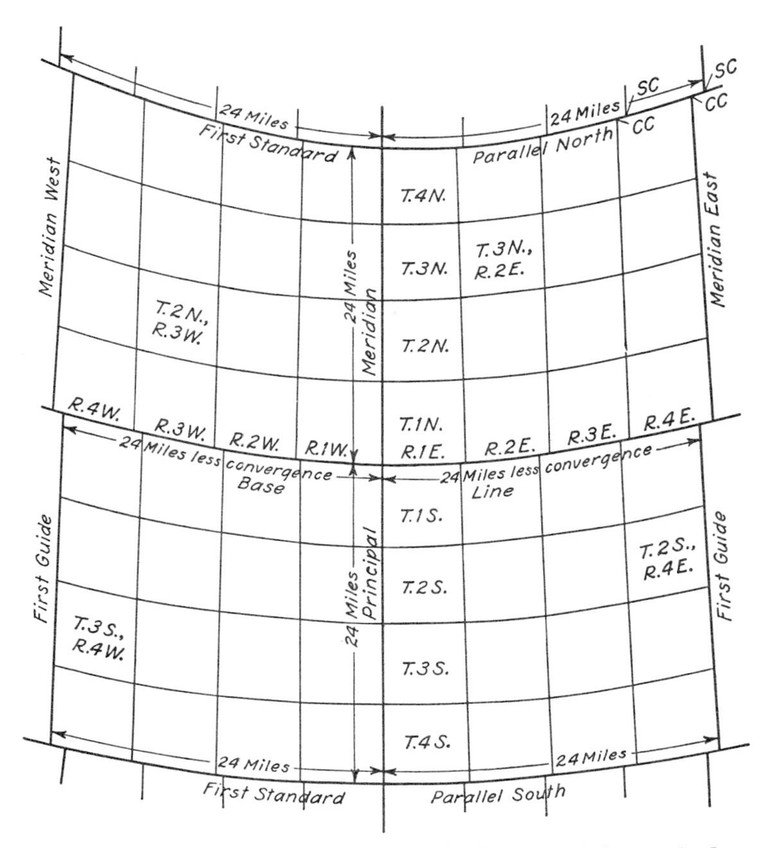

FIG. 20–2. Survey of quadrangles. (Only a few of the standard corners and closing corners are identified.)

In each area an initial point was established and located by astronomical observations. The manual of 1902 was the first to specify an indestructible monument, preferably a copper bolt, firmly set in a rock ledge if possible, and witnessed by rock bearings.

Thirty-five initial points are available, three of them being in Alaska where additional points probably will be required.

A principal meridian and a base line were passed through each initial point, such as the point in the center of Fig. 20–2.

20–4. Principal meridian. From each initial point, a true north-south line called a principal meridian (Prin. Mer. or P.M.) was run north and/or south to the limits of the area to be covered. Generally a solar attachment—a device for solving mechanically the mathematics of the astronomical triangle—was used. Monuments were set for section and quarter-section corners every 40 chains, and at the intersections with all meanderable bodies of water (streams 3 ch. or more in width, and lakes covering 25 acres or more).

The line was supposed to be within 3 min of the cardinal direction. Two independent sets of linear measurements were required to check within 20 lk. (13.2 ft) per 80 ch., which corresponds to a ratio of error of only 1/400. The allowable difference between sets of measurements is now limited to 14 lk.

20–5. Base line. From the initial point, the base line was extended east and/or west as a true parallel of latitude to the limits of the area to be covered. As required for the principal meridian, monuments were set for section and quarter-section corners every 40 ch., and at the intersections with all meanderable bodies of water. Permissible closures were the same as those for the principal meridian.

Base lines were run as circular curves with chords of 40 ch. by (a) the solar method, (b) the tangent method, or (c) the secant method.

Solar method. An observation is made with the solar attachment to determine the direction of true north. A right angle is then turned off and a line extended 40 ch., where the process is repeated. The series of lines so established, with a slight change in direction every half-mile, closely approaches a true parallel. Obviously if the sun is obscured, the method cannot be used.

Tangent method. This method of laying out a true parallel is illustrated in Fig. 20–3. A 90° angle is turned to the east or the west, as may be required, from a true meridian, and corners are set every 40 ch. At the same time, proper offsets are taken from tables and measured north from the tangent to the parallel. In the example shown, the offsets in links are 1, 2, 4, $6\frac{1}{2}$, . . . 37. The error resulting from taking right-angle offsets instead of offsets along the converging lines is negligible.

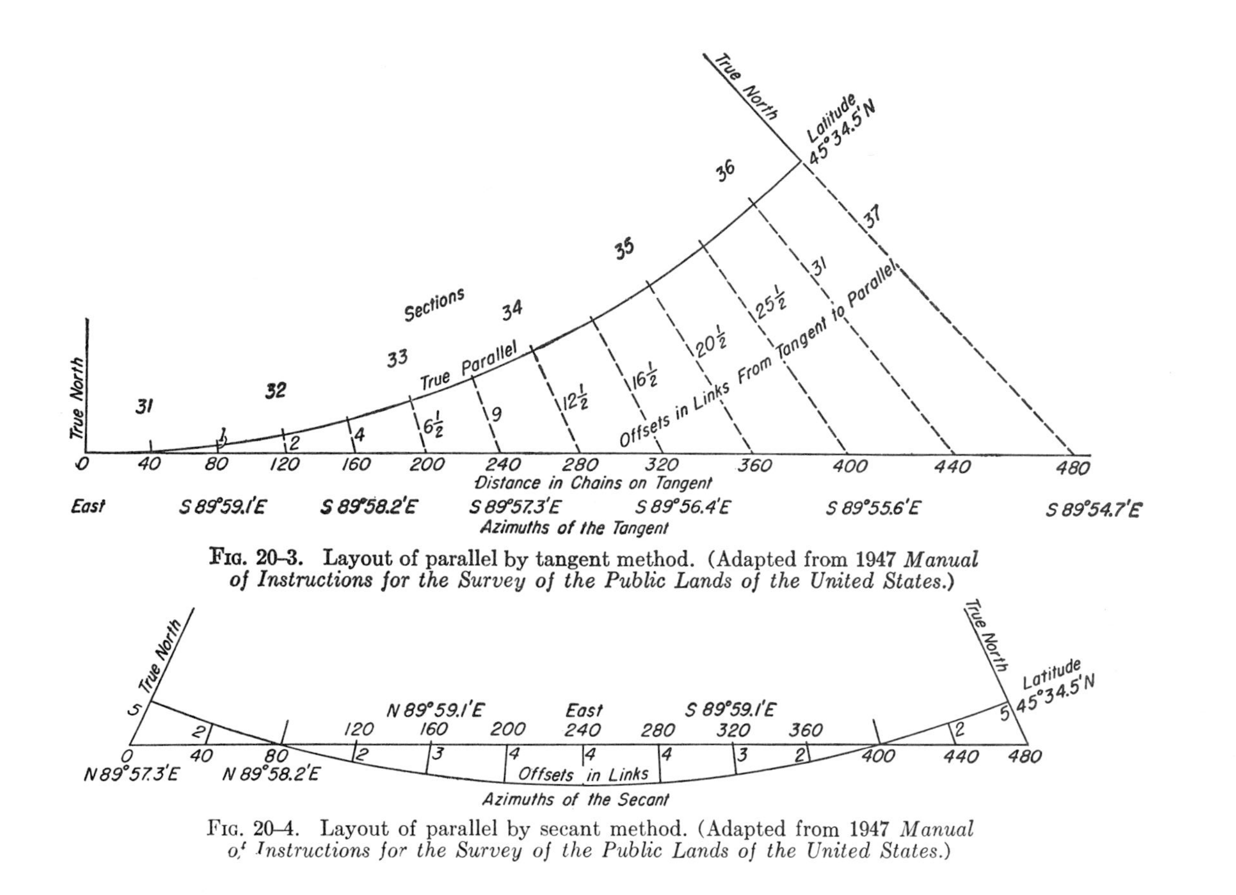

FIG. 20–3. Layout of parallel by tangent method. (Adapted from 1947 *Manual of Instructions for the Survey of the Public Lands of the United States.*)

FIG. 20–4. Layout of parallel by secant method. (Adapted from 1947 *Manual of Instructions for the Survey of the Public Lands of the United States.*)

The main objection to the tangent method is that the parallel departs considerably from the tangent and therefore two lines must be brushed out.

Secant method. This method of laying out a true parallel is shown in Fig. 20–4. It actually is a modification of the tangent method in which a line parallel to the tangent at the 3-mile (center) point is passed through the 1-mile and 5-mile points to obtain minimum offsets, as shown in Tables V and VI in Appendix B.

Field work consists in establishing a point on the true meridian, south of the beginning corner, at a distance taken from a table for the latitude of the desired parallel. The proper bearing angle from the same table is turned to the east or west from the true meridian to define the secant, which is then projected 6 miles. Offsets are measured north or south from the secant to the parallel.

The advantages of the secant method are its simplicity, the fact that the offsets are small and can be measured perpendicular to the secant without error, and the reduced amount of clearing required.

20–6. Standard parallels (correction lines). After the principal meridian and base line have been run, standard parallels (Stan. Par. or S.P.), also called correction lines, are run as true parallels of latitude 24 miles apart in the same manner as was the base line. All 40-ch. corners are marked. In some of the early surveys, standard parallels were placed at intervals of 30 or 36 miles.

Standard parallels are numbered consecutively north and south of the base line; examples are First Standard Parallel North and Third Standard Parallel South.

20–7. Guide meridians. Guide meridians (G.M.) are run due north from the base line and standard parallels at intervals of 24 miles east and west of the principal meridian, in the same manner as was the principal meridian, and with the same limits of error. Before the work is started, the chain or tape must be checked by measuring 1 mile on the base line or standard parallel. All 40-ch. corners are marked.

Because of convergence of the meridians, a *closing corner* (CC) is set at the intersection of each guide meridian and a standard parallel or base line (see Fig. 20–2). The distance from the closing corner to the *standard corner* (SC), which was set when the parallel was run, is measured and recorded in the notes as a check. Any error in the 24-mile length of the guide meridian is put in the northernmost half-mile.

The guide meridians are numbered consecutively east and west of the principal meridian; examples are First Guide Meridian West and Fourth Guide Meridian East.

20–8. Township exteriors. Meridional (range) lines and latitudinal (township) lines. Division of a quadrangle, or tract, into townships is accomplished by running range (R.) lines and township (T. or Tp.) lines.

Range lines are true meridians through the standard township corners previously established at intervals of 6 miles on the base line and standard parallels. They are extended north to intersect the next standard parallel or base line, and closing corners are set (see Figs. 20–2 and 20–5).

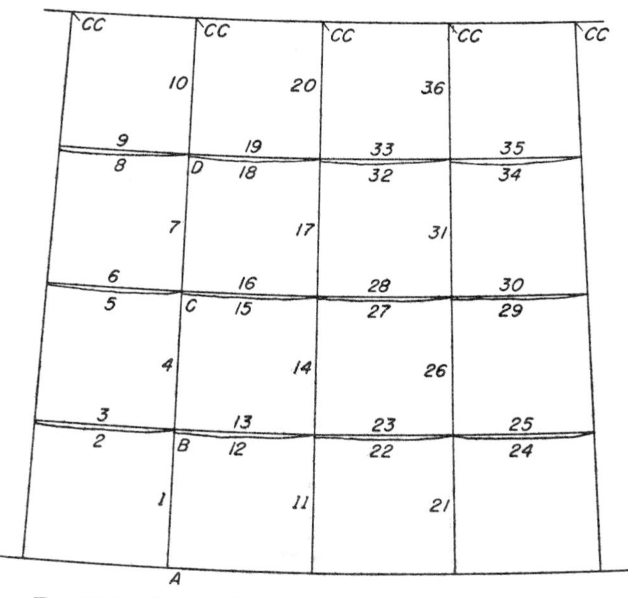

FIG. 20–5. Order of running lines for the subdivision of a quadrangle into townships.

The formulas for convergence of meridians (derived in various texts on geodesy, with results given in Table IV) are as follows:

$$\theta = 52.13d \tan \phi \qquad (20\text{–}1)$$

$$c = \tfrac{4}{3} Ld \tan \phi \qquad \text{(slight approximation)} \qquad (20\text{–}2)$$

where θ = angle of convergence, sec;
d = distance between meridians, miles, on a parallel;
ϕ = mean latitude;
c = linear convergence, ft;
L = length of meridians, miles.

Township lines connect township corners previously established at intervals of 6 miles on the principal meridian, guide meridians, and range lines.

20–9. Designation of townships. A township is identified by a unique description based upon the principal meridian governing it.

A north-and-south row of townships is called a *range*. Ranges are numbered in consecutive order east and west of the principal meridian, as indicated in Fig. 20–2.

An east-and-west row of townships is called a *tier*. Tiers are numbered in order north and south of the base line. By common practice, the term "tier" is usually replaced by "township" in designating the rows.

An individual township is identified by its serial number north or south of the base line, followed by the number east or west of the principal meridian. An example is Township 7 South, Range 19 East, of the Sixth Principal Meridian. Abbreviated, this becomes T 7 S, R 19 E, 6th P.M.

20–10. Subdivision of a quadrangle into townships. The method to be used in subdividing a quadrangle into townships is fixed by regulations in the *Manual of Instructions*. Under the old regulations, township boundaries were required to be within 21 min of the cardinal direction. Later this was reduced to 14 min in order to keep the interior lines within 21 min of the cardinal direction.

The detailed procedure for subdividing a quadrangle into townships can best be described as a series of steps designed to produce the maximum number of regular sections with a minimum amount of unproductive travel by the field party. The order of running the lines is shown by consecutive numbers in Fig. 20–5. Some of the details are described in the following steps:

a) Begin at the southeast corner of the southwest township, point A, after checking the chain or tape against a 1-mile measurement on the standard parallel.

b) Run north on the true meridian for 6 miles, setting alternate section and quarter-section corners every 40 ch. Set township corner *B*.

c) From *B*, run a random line *2* due west to intersect the principal meridian. Set temporary corners every 40 ch.

d) If the random line has an excess or deficiency of 3 ch. or less (allowing for convergence), and a falling north or south of 3 ch. or less, the line is accepted. It is then corrected back, line *3*, and all corners are set in their proper positions. Any excess or deficiency is thrown into the most-westerly half-mile. The method of correcting a random line having an excess of 1 ch. and a north falling of 2 ch. is shown in Fig. 20–6.

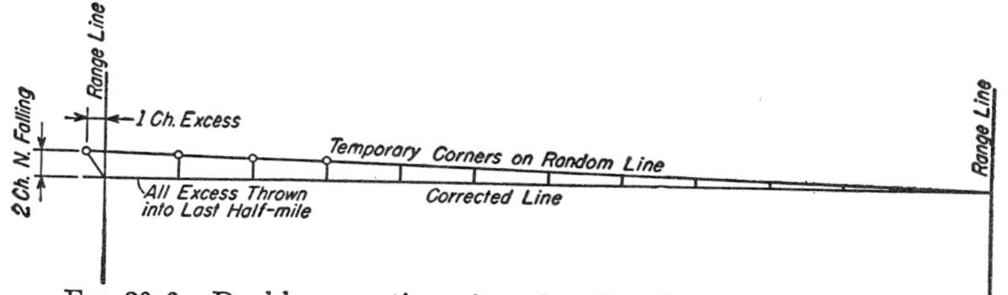

Fig. 20–6. Double correction of random line for excess and falling.

e) If the random line misses the corner by more than the permissible 3 ch., all four sides of the township must be retraced.

f) The same procedure is followed until the southeast corner, *D*, of the most-northerly township is reached. From *D* the range line *10* is continued as a true meridian to intersect the standard parallel or base line, where a closing corner is set. All of the excess or deficiency in the 24 miles is thrown into the most-northerly half-mile.

g) The second and third ranges of townships are run in the same way, beginning at the south line of the quadrangle.

h) While the third range is being run, random lines are also projected to the east and corrected back, and any excess or deficiency is thrown into the most-westerly half-mile. (All points may have to be moved diagonally to the corrected line, instead of just the last point as in Fig. 20–6.)

20–11. Subdivision of a township into sections. The detailed procedure for subdividing a township can be described most readily as a series of steps designed to produce the maximum number of regular sections 1 mile on a side. Sections are numbered from 1 to 36, beginning in the northeast corner of the township and ending in the southeast corner, as shown in Fig. 20–7.

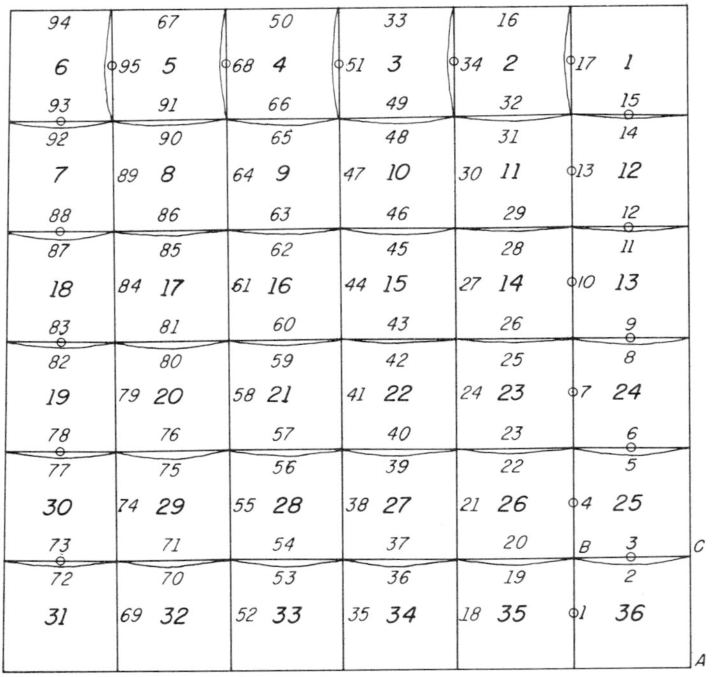

FIG. 20–7. Order of running lines for the subdivision of a township into sections.

a. Set up at the southeast corner of the township, point *A*, and observe the meridian. Retrace the range line northward, and the township line westward, for 1 mile to compare the meridian, the needle readings, and the tape with those of the previous surveyor.
b. From the southwest corner of Sec. 36, run north *parallel* with the east boundary of the township. Set quarter-section and section corners on line *1*, Fig. 20–7.
c. From the section corner just set, run a random line *parallel* with the south boundary of the township eastward to the range line. Set a temporary quarter-corner at 40 ch.

d. If the 80-ch. distance on the random line is within 50 lk., falling or distance, the line is accepted. The correct line is calculated, and the quarter-corner is located at the *midpoint* of the line *BC* connecting the previously established corner *C* and the new section corner *B*.

e. If the random line misses the corner by more than the permissible 50 lk., the township lines must be rechecked and the cause of the error determined.

f. The east range of sections is run in similar manner until the southwest corner of Sec. 1 is reached. From this point a random line is run northward to connect with the section corner on the north township line. The quarter-corner is set 40 ch. from the south section corner (on line 16 corrected back by later manuals). All discrepancies in the 6 miles are thrown into the last half-mile.

g. Successive ranges of sections across the township are run until four have been completed. All meridional lines are parallel with the east side of the township, and all east-west lines are parallel with the south boundary.

h. When the fifth range is being run, random lines are projected to the west as well as to the east. The quarter-corners in the west range are set 40 ch. from the east side of the section, all excess or deficiency resulting from errors and convergence being thrown into the most-westerly half-mile.

i. If the north side of the township is a standard parallel, instead of running a random line to the north, lines parallel with the east township boundary are projected to the correction line and closing corners are set. The distance to the nearest standard corner is measured and recorded.

j. True bearings of the interior range lines for any latitude can be obtained by applying corrections from tables for the convergence at a given distance from the east boundary.

By throwing the effect of convergence of meridians into the westernmost half-mile of the township and all errors to the north and west, twenty-five regular sections 1 mile square are obtained. Also, the south half of Secs. 1, 2, 3, 4, and 5, and the east half of Secs. 7, 18, 19, 30, and 31, are normal size.

20–12. Subdivision of sections. To divide a section into quarter-sections, straight lines are run between opposite quarter-section corners previously established or re-established. This rule holds whether or not the quarter-section corners are equidistant from the adjacent section corners.

To divide a quarter-section into quarter-quarter-sections, straight lines are run between opposite quarter-quarter-section corners established at the midpoints of the four sides. The same procedure is followed to obtain smaller subdivisions.

If the quarter-sections are on the north or west side of the township, the quarter-quarter-section corners are placed 20 ch. from the south or east quarter-section corners—or a proportional distance if the total length on the ground is not equal to that on record.

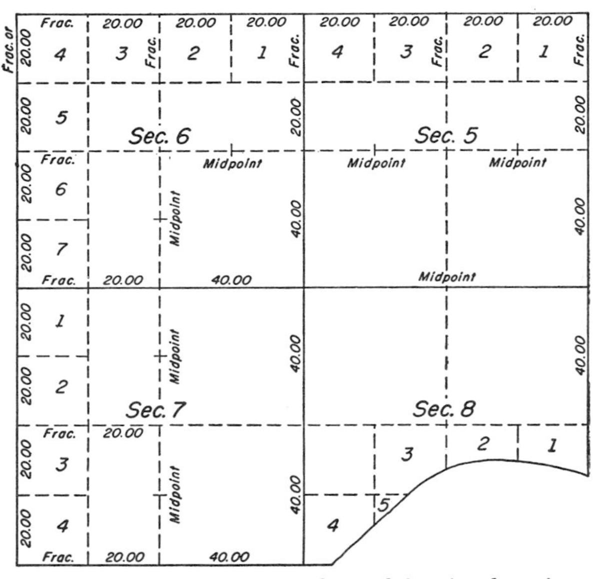

Fig. 20–8. Subdivision of regular and fractional sections.

20–13. Fractional sections. The quarter-sections along the north and west boundaries of a township made irregular by discrepancies of measurements and convergence of the range lines are usually numbered and sold as lots, as indicated in Fig. 20–8. One such quarter-section in a Wisconsin township contains 640 acres!

In sections made fractional by rivers, lakes, or other bodies of water, lots are formed bordering on the body of water and numbered consecutively through the section (see Sec. 8 in Fig. 20–8). The boundaries of the lots usually follow the quarter-section and quarter-quarter-section lines, but extreme lengths and narrow widths are avoided, as are areas of less than 5 acres or more than 45 acres.

Lot lines are not actually run in the field. Like the quarter-section lines, they are merely indicated on the plats by protraction. The areas by which the lots are sold are computed from the plats.

20–14. Notes. Specimen field notes for each of the several kinds of lines to be run are shown in various instruction manuals. Actual field notes had to follow closely the model sets.

The original notes, or copies of them, are maintained in a land office in each state, for the benefit of all interested persons.

TABLE 20–1

SUBDIVISION STEPS

Item	Subdivision of a Tract	Subdivision of a Township
Starting point	SE corner of SW township	SW corner of SE section (36)
Meridional lines		
Name	Range line	Section line
Direction	True north	North, parallel with east range line
Length	Six miles = 480 chains	One mile = 80 chains
Corners set	Quarter-section and section corners at 40 and 80 chains alternately	Quarter-section corner at 40 chains. Section corner at 80 chains
Latitudinal lines		
Name	Township line	Section line
Direction of random	True east–west parallel	East, parallel with south side of section
Length	Six miles less convergence	One mile
Permissible error	Three chains, length or falling	Fifty links, length or falling
Distribution of error		
Falling	Corners moved proportionately from random to true line	Corners moved proportionately from random to true line
Distance	All error thrown into west quarter-section	Error divided equally between quarter-sections

[*Work repeated until north side of area is reached. Subdivision of last area on the north of the range of townships and sections follows.*]

TABLE 20-1—*Continued*

Case I. When Line on the North Is a Standard Parallel

Item	Subdivision of a Tract	Subdivision of a Township
Direction of line......	True north	North, parallel with east range line
Distribution of error in length........	Placed in north quarter-section	Placed in north quarter-section
Corner placed at end .	Closing corner	Closing corner
Permissible errors....	Specified in Manual of Instructions	Specified in Manual of Instructions

Case II. When Line on the North Is Not a Standard Parallel

Item	Subdivision of a Tract	Subdivision of a Township
Direction of line......	No case	Random north and correct back to section corner already established
Distribution of error in length........	....	Same as Case I

[*Other ranges of townships and sections continued until all but two are laid out.*]

Item	Subdivision of a Tract	Subdivision of a Township
Location of last two ranges..........	On east side of tract	On west side of township
Next-to-last range subdivided..........	As before	As before
Last range		
Direction of random	True east	Westerly, parallel with south side of section
Nominal length....	Six miles less convergence	One mile less convergence
Correction for temporary corners...	Corners moved proportionately from random to true line	Corners moved proportionately from random to true line
Distribution of error of closure........	Corners moved westerly (or easterly) to place error in west quarter-section	Corner is placed on the true line so that error falls in west quarter-section

20-15. Outline of subdivision steps. Pertinent points in the subdivision of quadrangles into townships, and townships into sections, are summarized in Table 20-1.

20-16. Marking corners. Various materials were approved and used for monuments in the original surveys. These included pits and mounds; stones; posts; charcoal; and broken bottles. A wrought-iron pipe 2 in. in inside diameter and 30 in. long is now standard except in rock outcrop, where a brass tablet $3\frac{1}{4}$ in. in diameter is specified.

Stones and posts were marked with one to six notches on one or two faces. The arrangements identify a monument as a particular section or township corner. Each notch represents 1 mile of distance to a township line or corner. Quarter-sections were marked with the fraction "$\frac{1}{4}$" on a single face.

In the prairie country, where large stones and trees were scarce, a system of pits and mounds was used to mark corners. Different groupings of pits and mounds, 12 in. deep and 18 in. square, designated corners of the several classes. Unless perpetuated by some other type of mark, these corners were lost in the first ploughing.

20-17. Witness corners. Whenever possible, monuments were witnessed by several adjacent objects such as trees and rock outcrops. Bearing trees were blazed on the side facing the corner and marked with scribing tools.

When a regular corner fell in a creek, pond, swamp, or other place where it was impracticable to place a mark, *witness corners* (WC) were set on all lines leading to the corner. The letters WC were added to all other marks normally placed on the corner, and the corner was in turn witnessed in the usual manner.

20-18. Meander corners. A meander corner (MC) was established on survey lines intersecting the bank of a stream having a width greater than 3 ch., or a lake, bayou, or other body of water of considerable extent. The distance to the nearest section corner or quarter-section corner was measured and recorded in the notes. A monument was set and marked MC on the side facing the water, and the usual witnesses were noted. If practicable, the line was carried across the stream or other body of water by triangulation to another corner set in line on the farther bank.

A traverse joining successive meander corners along the banks of streams or lakes was begun at a meander corner and followed as closely as practicable the sinuosities of the bank until the next meander corner was reached. The traverse was checked by calculating the position of the new meander corner and comparing it with its known position on a surveyed line.

Meander lines follow the mean high-water mark and are used for plotting and protraction of area only. They are *not* boundaries defining the limits of property adjacent to the water.

20–19. Lost and obliterated corners. A common problem in resurveys of the public lands is the replacement of lost or obliterated corners. This difficult task requires a combination of experience, hard work, and ample time to reproduce the location of a wooden stake or post monument incorrectly set 75 years ago on an undependable section line, and with all witness trees long-since cut or burned by apathetic owners.

An *obliterated corner* is one at whose point there are no remaining traces of the monument, or its accessories, but whose location has been perpetuated or may be recovered beyond reasonable doubt. The corner may be restored from the acts or testimony of interested landowners, surveyors, qualified local authorities, or witnesses, or from written evidence. Satisfactory evidence has value in the following order:

a. Evidence of the corner itself.
b. Bearing trees or other witness marks.
c. Fences, walls, or other evidence showing occupation of the property to the lines or corners.
d. Testimony of living persons.

A *lost corner* is one whose position cannot be determined, beyond reasonable doubt, either from traces of the original marks or from acceptable evidence or testimony that bears on the original position. It can be restored only by rerunning lines from one or more independent corners (existing corners that were established at the same time, and with the same care as the lost corner). Usually single- or double-proportionate measurements are necessary, the latter being used for township corners common to 4 sections and section corners common to 4 sections on interior of township. Thus the reliability of corner locations at the centers of sections and quarter sections also depends upon these measurements.

Proportionate measurements allot the discrepancies between the original and later recordings among the several parts of a line in the proportions in which the original recorded measurements were distributed. The method should be used only as a last resort.

20–20. Accuracy of public-lands surveys. The accuracy required in the early surveys was of a very low order. Frequently it fell below that which the notes showed. A small percentage of the surveys were made by men drawing upon their imagination in the comparative comfort of a tent. Obviously no monuments were set and the notes serve only to confuse the situation for present-day surveyors and landowners. Some surveyors threw in an extra chain-length at intervals to assure a full measure!

The poor results obtained in many areas were due primarily to the following reasons:

a. Lack of training of personnel. Some contracts were given to men without any technical background.
b. Poor equipment.
c. Surveys made in unsettled and apparently valueless areas.
d. Marauding Indians, swarms of insects, and dangerous animals.
e. Lack of appreciation for the need of accurate work.
f. Surveys made in piecemeal fashion as the Indian titles and other claims were extinguished.
g. Work done by contract at low prices.
h. Absence of control points.
i. Field inspection not provided until 1850, and not actually carried out until 1880.
j. Magnitude of the problem.

In general, considering the handicaps listed, the work was reasonably well done in most cases.

20–21. Descriptions by township, section, and smaller subdivision. Description by the sectional system offers a means of defining boundaries uniquely, clearly, and concisely. Several examples of acceptable descriptions are listed.

Sec. 6, T 7 S, R 19 E, 6th P.M.
Frac. Sec. 34, T 2 N, R 5 W, Ute Prin. Mer.
The SE¼, NE¼, Sec. 14, Tp. 3 S, Range 22 W, S.B.M. [San Bernardino Meridian].
E½, NE¼, Sec. 20, T 15 N, R 10 E, Indian Prin. Mer.
E 80 acres of the NE¼ of Sec. 20, T 15 N, R 10 E, Indian Prin. Mer.

Note that the last two descriptions do not necessarily describe the same land. A California case in point occurred when the owner of a SW¼ section, nominally 160 acres but actually 162.3 acres, deeded the westerly portion as "the West 80 acres" and the easterly portion as "the East ½."

Sectional land which is privately owned may be partitioned in any manner at the option of the owner. The metes-and-bounds form is preferable for irregular parcels. In fact, metes and bounds are required to establish the boundaries of mineral claims, and various grants and reservations.

Differences between the physical and the legal (or record) ground locations and areas may result because of departures from accepted procedures in description writing; loose and ambiguous statements; or dependence upon the accuracy of early surveys.

20–22. Sources of error. Some of the many sources of error in retracing the public-lands surveys follow:

a. Discrepancy between the length of the chain of the early surveyor and that of the modern tape.
b. Changes in the magnetic declination and/or the local attraction.
c. Lack of agreement between field notes and actual measurements.
d. Changes in watercourses.
e. Nonpermanent objects which were used for corner marks.
f. Loss of witness corners.

20–23. Mistakes. Typical mistakes in the retracement of boundaries in public-land surveys are as follows:

a. Failure to follow the general rules of procedure governing the original survey.
b. Neglecting to check the tape used against distances on record for marks in place.
c. Resetting corners without exhausting every means of relocating the original corners.

PROBLEMS

20–1. Why are the boundaries of public lands established by duly appointed surveyors unchangeable, even though set incorrectly in the first surveys?

20–2. What differences exist between the original and present instructions and practice in spacing corners and running section lines?

20–3. Describe the method of running a principal meridian. State the original and present accuracy required.

20–4. List the advantages and disadvantages of each of the three methods of running base lines and standard parallels.

Determine the offsets from, and the azimuths of, a secant to lay out a base line at the latitudes listed in problems 20–5 through 20–7. Draw a sketch showing the values.

20–5. At latitude 32°30′N.

20–6. At latitude 41°00′N.

20–7. At latitude 46°30′N.

What is the convergence, in ft, of two meridians for the conditions given in problems 20–8 through 20–10?

20–8. Meridians originally 24 miles apart, after 24 miles, latitude 32°30′N.

20–9. Meridians originally 18 miles apart, after 12 miles, latitude 41°00′N.

20–10. Meridians originally 6 miles apart, after 6 miles, latitude 46°30′N.

20–11. At latitude 45°N, what is the maximum offset from the secant to the parallel, and the maximum difference of the azimuth of the secant from the cardinal direction?

What is the nominal distance in miles between the points given in problems 20–12 through 20–14?

20–12. Second Guide Meridian West and east range line of Range 32 West.

20–13. Third Guide Meridian West and west range line of Range 15 East.

20–14. First Guide Meridian East and west range line of Range 19 East.

20–15. In latitude 45°N, at what distance from the Principal Meridian will the quarter-quarter corner of Sec. 36, T 1 N, theoretically fall exactly opposite a section corner of Sec. 6, T 1 S, due to convergence?

20–16. Show on a sketch where closing corners are located on township lines and the areas they govern.

20–17. Outline the steps in subdividing a quadrangle into townships. Show on a sketch the order of the work by numbering the lines consecutively.

20–18. Similar to problem 20–12, for subdividing a township into sections.

20–19. What steps in public-lands subdivision are left to the local surveyor?

Show on a sketch; label the principal meridian, township lines, etc.; give distances where they are specified by law; and compute the nominal area of the pieces of property described in problems 20–20 through 20–23.

20–20. SW$\frac{1}{4}$, NE$\frac{1}{4}$, Sec. 15, T 2 S, R 1 E, N.M. P.M.

20–21. S$\frac{1}{2}$, NW$\frac{1}{4}$, Sec. 26, T 3 N, R 5 W, Boise P.M.

20–22. NW$\frac{1}{4}$, SE$\frac{1}{4}$, SW$\frac{1}{4}$, Sec. 9, T 4 N, R 2 E, 6th P.M.

20–23. NE$\frac{1}{4}$, SE$\frac{1}{4}$, NE$\frac{1}{4}$, Sec. 8, T 1 N, R 2 W, 3rd P.M.

In problems 20–24 and 20–25 how many rods of fence are required to enclose the areas?

20–24. Sections 16, 17, 20, 21 and 28.

20–25. Sections 9, 16, 21, 22 and 27.

20–26. What allowable error in distance is permitted in the location of the northeast corner of T 1 N, R 1 E?

20–27. What are the two causes for sections in a normal township being smaller than 640 acres? Which sections usually are not regular?

20–28. Which lines in the public-lands surveys are laid out parallel to other lines?

20–29. Which section line is run as a random line?

20–30. The owner of the NW¼ of the SE¼ of Sec. 8 wishes to have his corners monumented. Assume that all section corners and quarter-section corners originally set by the public-lands surveyors are in place. Outline the procedure, using a sketch to show all lines run and the new corners set.

20–31. The east line of the NE¼ of Sec. 4 has a record distance of 40.24 ch., and a length measured in the field of 40.10 ch. At what distance from the northeast corner of Sec. 4 would you set the southeast corner of lot 1? Give the reason for your answer.

20–32. The exterior dimensions of Sec. 6 on the west, north, east, and south sides are 79 ch., 78 ch., 82 ch., and 81 ch., respectively. Explain with the aid of a sketch how to divide the section into quarters, giving distances used to establish the required points and lines.

20–33. If you found a section corner set by the Bureau of Land Management and marked as shown in the figure, in what direction would you go to find the nearest established section corner, and how would it be marked?

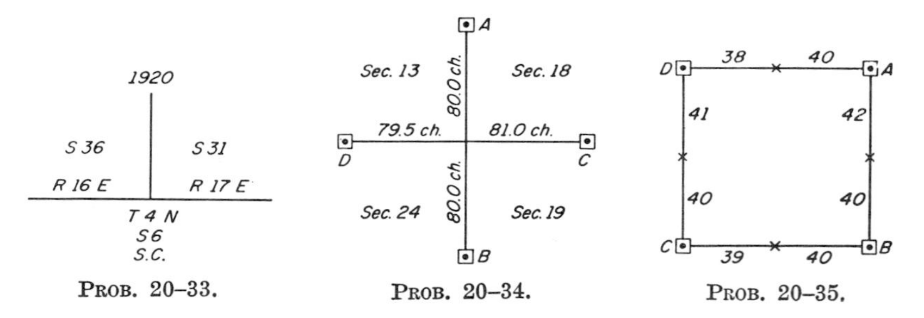

PROB. 20–33. PROB. 20–34. PROB. 20–35.

20–34. The distances given on the accompanying sketch are the original government survey distances. The southeast corner of Sec. 13 is lost. Corners *A, B, C,* and *D* are in. If the measured distance *AB* is 160.6 ch. and *DC* is 170.0 ch., explain how to establish the location of the southeast corner of Sec. 13.

20–35. Section 6 is to be subdivided into quarters. The original survey distances in chains are shown on the sketch. Corners *A, B, C,* and *D* are found, but all other corners are lost. The measured resurvey distances are: *AB* = 82.4 ch., *BC* = 79.6 ch., *CD* = 81.8 ch., and DA = 78.2 ch. Explain how to subdivide the section into quarters, giving the actual distances to be used.

20–36. Using the data and figure of problem 20–35, explain how the NW¼, NW¼ of Sec. 6 would be staked out. Give actual distances to be used.

20–37. A man owns part of lot 4 in Sec. 5. Make up necessary data to give a complete description of any part of this lot. Show its location by a sketch.

20–38. Show by a dimensioned sketch how a county surveyor would set the corners for the SE¼ of the NE¼ of Section 10.

20–39. The remeasured distance from the section corner for Secs. 9, 10, 15 and 16 to the quarter-section corner between Secs. 9 and 10 is found to be 41.32 chains. At what distance from that section corner should the quarter-quarter-corner be set on this line in subdividing the section?

20–40. Which classes of corners are relocated by (a) single proportionate measurement, and (b) double proportionate measurement?

20–41. The lengths of the boundaries of Sec. 3 in a normal township, as returned in the original field notes are: south, 80.32 ch.; west, 82.26 ch.; east, 82.48 ch.; and north, 80.00 ch. Sketch the subdivision of this section by protraction as it would be on the official plat.

20–42. If the southern boundary of a township lies on a standard parallel in latitude 39°N, what is the theoretical length of the north boundary of the township?

20–43. Determine the difference between the bearing of the east boundary of a township and that of the line between Secs. 19 and 20 in latitude 45°N. If the bearing of the township boundary is N0°06′E, what should be the bearing of the section line?

20–44. Why are meander lines not the boundaries defining the area of ownership of tracts adjacent to the water?

20–45. A meanderable river follows a winding course from southwest to northeast across a section. The position of the regular southwest corner of the section falls in the water. Draw a sketch illustrating this condition, and indicate the position of the meander and witness corners and the meander lines.

20–46. Write a complete legal description for a parcel of land consisting of a quarter-section in New Mexico whose south boundary line is approximately 52 miles north of the base line and whose east line is about 31 miles west of the principal meridian.

20–47. What advantage would a stadia-interval factor of 132 (instead of the usual 100) have in public-lands surveys?

20–48. Explain how metes and bounds can be used in areas covered by the public-lands surveys.

chapter **21**

Construction Surveys

21–1. General. In recent years the construction industry has become the largest single industry in the United States. Construction surveying, as the basis for all construction and part of it, has therefore become increasingly important. It is estimated that 60 per cent of all surveying man-hours are spent on location-type surveys giving line and grade. Nevertheless, sufficient attention has not always been given to surveys and maps of underground utilities and some types of construction projects. Surveys for a subdivision usually cost less than 1 per cent of the expenditure for surfacing streets and laying water and sewer lines.

A topographic survey of the area is the first requirement in locating or positioning a structure. Reference points to control the construction stakes and to aid in checking the progress of work are needed.

A few of the most common types of construction surveys, with which every engineer should be familiar, will be described briefly. Construction surveying is best learned on the job by adapting basic principles to the project at hand. Because each project introduces individual problems, textbook coverage of construction surveying is somewhat limited.

Laser beams are now being used to give line and grade on various kinds of construction projects such as tunnels, underwater dredging, etc.

21–2. Staking out a pipeline. The flow in water lines is usually under pressure, but most sewers have gravity flow. Alignment and grade must therefore be carefully watched on sewer pipes. Larger water lines also have definite grades because blowoffs are needed at low points and air releases are required at high points.

Construction stakes, sometimes set on the center line at 50-ft stations when the ground is reasonably uniform, disappear on the first ditcher or bulldozer pass so parallel offset lines are desirable.

Marks should be closer on horizontal and vertical curves and for pipes of large diameter. On hard surfaces where stakes cannot be driven, points are marked by paint, spikes, drill holes, or other means.

Figure 21–1 shows the arrangement of *batter boards* for a sewer line. Batter boards are usually 1″ × 6″ boards nailed to 2″ × 4″ posts which have been pointed and driven into the ground. The top of the batter board is placed a full number of feet above the invert (lower inside surface), or above the flow line, of the pipe. Nails are driven into the tops of the boards so that a string stretched between them will define the center of the pipeline. A graduated stick is used to measure the required distance from the string to the pipe invert, or to the flow line. Thus the string gives both line and grade. It can be kept taut by hanging a weight on each end after wrapping it around the nails.

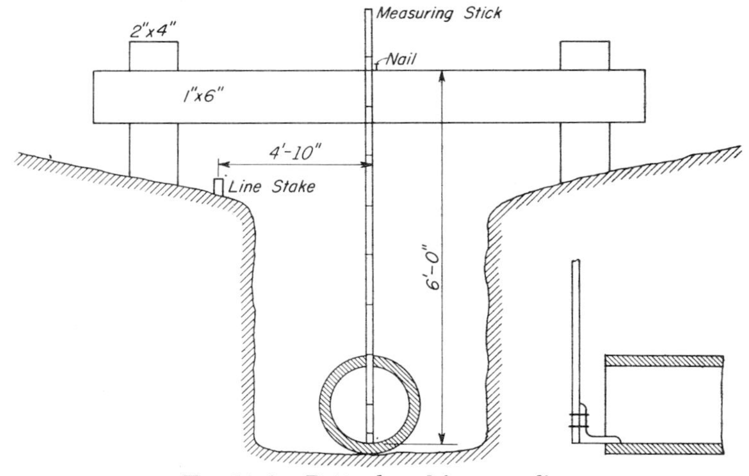

FIG. 21–1. Batter board for sewer line.

In Fig. 21–1, instead of a fixed batter board, a 2″ × 4″ carrying a level vial can be placed on top of the offset-line stake whose elevation is known. Measurement is made from the underside of the leveled 2″ ×4″ with a tape or graduated stick to establish the flow line.

On some jobs the engineer's level is set up in the ditch to give line and grade.

Suitable grades for trenches to avoid excessive cut and fill, and to permit connections with other facilities, are determined from a profile such as that in Fig. 5–25.

21–3. Staking out grades. Staking out grades is the reverse of taking profiles, although in both operations the center line should first be marked and stationed in horizontal location.

Grade elevations are determined by a leveling process. A level is set up and its H.I. found by reading a plus sight on a bench mark. The difference between the H.I. and the grade elevation at any station is the *grade rod.* The *ground rod* at each station is obtained by holding the rod on successive stations. Grade rod minus the ground rod at a station is the cut or fill at that station. The cut or fill is marked on an offset-line stake on the side facing the center line. The station is given on the other side of the stake. If stakes are driven to grade, the tops are marked with blue keel, and commonly called "blue tops."

Note that the interval between the middle crosshair and the upper (or lower) stadia wire gives a $\frac{1}{2}$ per cent grade; between the two stadia wires there is a 1 per cent grade. After setting one cross hair on the grade rod reading at a station, $\frac{1}{2}$ or 1 per cent grades are readily set even though the telescope is not perfectly level.

As an example, assume that the H.I. of a level is 292.5 ft, the grade elevation of station 0 + 00 is 280.0 ft, the rod readings on the ground at station 0 + 00 and 1 + 00 are 10.4 and 8.3 ft, respectively, and a +1.00 per cent grade is to be laid out.

At station 0 + 00 the grade rod is 292.5 − 280.0 = 12.5 ft. The cut is 12.5 − 10.4 = 2.1 ft, and the stake is marked C 2.1.

At station 1 + 00 the grade elevation is 280.0 + 1.0 = 281.0 ft and the grade rod is 292.5 − 281.0 = 11.5 ft. The cut is therefore 11.5 − 8.3 = 3.2 ft, and this stake is marked C 3.2.

If a target rod is used, the target can be set at the proper elevation to put the base of the rod at grade, or an even number of feet above or below it. Stakes are then driven until the horizontal line of sight of the level coincides with the center line of the target when the rod is held on top of each stake.

21–4. Staking out a building. The first task in staking out a building is to locate it properly on the correct lot by making measurements from the property lines. Most cities have an ordinance establishing setback lines to improve appearance and fire-protection.

Stakes may be set initially at the exact building corners as a visual check on the positioning of the structure, but obviously such stakes are lost when work is begun on the footings. A set of batter boards and reference stakes, located as shown on Plate A–12 in Appendix A, is therefore placed near each corner but out of the way of construction. The boards are nailed a full number of feet above the bottom of the footing, or above the elevation of the first floor. Corner stakes and batter-board points for rectangular buildings are checked by measuring the diagonals for comparison with each other and the computed values. A bench mark (two or more on large projects) beyond the construction area but within easy sight distance is necessary to control elevations.

Nails are driven into the tops of the batter boards so that strings stretched between them will define the outside wall line of the building. Again, the boards give line and grade.

Permanent foresights may be helpful to establish the principal lines of the structure. Targets or marks on nearby existing buildings are commonly used if movement due to thermal effects and settlement is considered negligible. On concrete structures, such as retaining walls, offset lines are used because the outside wall line is obstructed by forms.

Plate A–12 shows the location of batter boards and the steps to be followed in setting them for a small building. The positions of such things as interior footings, anchor bolts for columns, and special piping and equipment may first be marked by steel pins and later by the actual bolts required, survey discs, or scratches on the bolts or concrete surfaces. Exact location and alignment of machines and equipment in the building can be achieved by the precise surveying instruments and methods discussed in Chapter 25, Industrial Applications.

21–5. Staking out a highway. Location stakes for a highway are set on the center line and on an offset line at full 100-ft stations, at the beginning and end of horizontal and vertical curves, and at other critical points. A profile is run on the center line and the elevation is determined at each of the stakes.

To guide the contractor in making excavations and embankments, *slope stakes* are set at the intersection of the ground and each side slope, as shown in Fig. 21–2. Note that actually there is no cut at the slope stake—the cut marked is the distance down to grade. Slope stakes are located by a trial-and-error method based upon mental calculations involving the H.I., the grade rod, the ground rod,

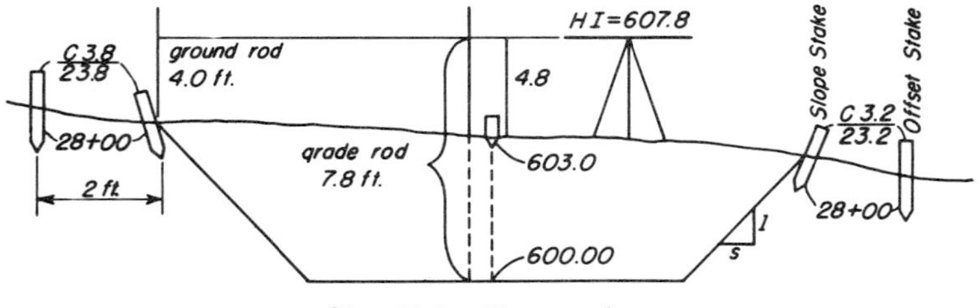

FIG. 21–2. Slope stakes.

half the width of the roadway, and the side slope. One or two trials are generally sufficient to fix the position of the stake within an allowable error of 0.3 ft to 0.5 ft.

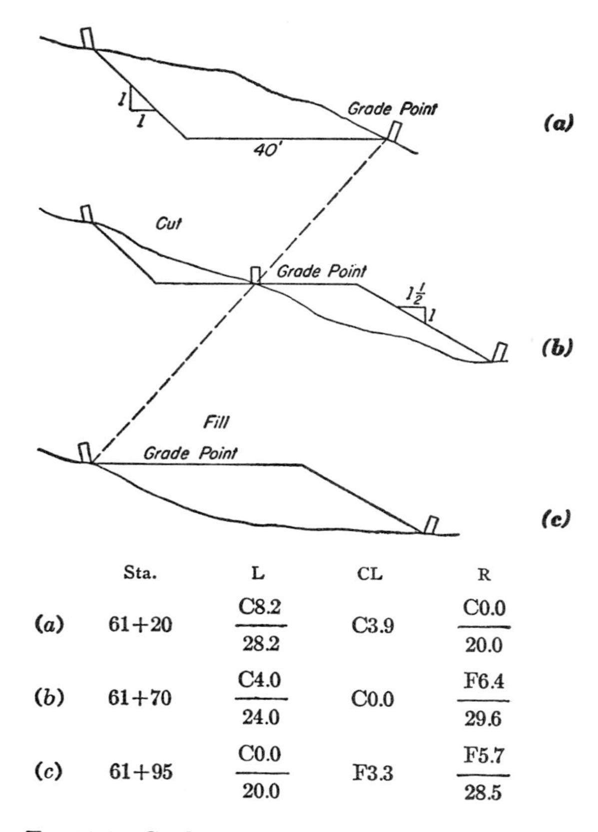

	Sta.	L	CL	R
(*a*)	61+20	$\frac{C8.2}{28.2}$	C3.9	$\frac{C0.0}{20.0}$
(*b*)	61+70	$\frac{C4.0}{24.0}$	C0.0	$\frac{F6.4}{29.6}$
(*c*)	61+95	$\frac{C0.0}{20.0}$	F3.3	$\frac{F5.7}{28.5}$

FIG. 21–3. Grade points at transition sections.

Grade stakes are set at the intersection of grade and ground. Three transition sections occur in passing from cut to fill, and a grade stake is set in each one, as indicated in Fig. 21–3. A line connecting the grade stakes defines the change from cut to fill.

Steps to be taken in slope staking are listed in sequence, assuming a roadway of 40 ft and side slopes in cut of 1 to 1.

1. Compute the cut at center line stake from profile and grade elevations (603.0 − 600.0 = C 3.0 in Fig. 21–2). Check in field by grade rod minus ground rod = 7.8 − 4.8 = C 3.0 ft.

2. Estimate the difference in elevation between the left-side slope stake point (20+ ft out) and the center stake. Apply the difference, say +0.5 ft, to the center cut and get an estimated cut of 3.5 ft.

3. Mentally calculate the distance out to the slope stake, 20 + 1(3.5) = 23.5 ft.

4. Hold zero end of a cloth tape at the center stake while the rodman goes out at right angles (turned by prism or extended arm method) with the other end and holds the rod at 23.5 ft.

5. Forget all previous calculations to avoid confusion of too many numbers and remember only the grade rod value.

6. Read the rod with an engineer's level set anywhere, or by hand level while standing at center stake. Get cut from grade rod—ground rod, perhaps 7.8 − 4.0 = C 3.8 ft.

7. Compute required distance out for this cut, 20 + 1(3.8) = 23.8 ft.

8. Check tape to see what is actually being held and find it is 23.5 ft.

9. Distance is within a few tenths of a ft and close enough. Move out to 23.8 ft if ground is level and drive stake. Move farther out if the ground slopes up since a greater cut would result and thus the slope stake must be beyond the compute distance, or not so far if the ground has begun to slope down which gives a smaller cut.

10. If the distance has been missed badly, make a better estimate of cut, compute a new distance out, take a reading to repeat the procedure.

11. In going out on the other side, the rodman lines up the center stake and the left-hand slope stake to get his right-angle direction.

12. To locate grade stakes at road edge, one man carries the zero end of the tape along the center line while the rodman walks parallel

with him holding the 20-ft mark until the required ground rod reading is found by trial. Note that the grade rod changes during the movement but can be computed readily at 5- or 10-ft intervals. The notekeeper should have the grade rod listed in his book for quick reference at full stations and other points where slope stakes are to be set.

13. Grade points on the center line are located using a starting guess determined by comparing the cut and fill at back and forward stations.

Location-staking of railways and canals follows the methods used for highways.

21–6. Sources of error. Some of the sources of error in construction surveys follow:

a. Movement of stakes and marks.
b. Failing to check the diagonals of a building.
c. Lack of foresight as to where construction will void points.
d. Failing to use tacks for proper line when justified.
e. Notation for cut or station on stake not checked.
f. Wrong datum for cuts, as cut to finish grade instead of cut to subgrade.

21–7. Mistakes. Typical mistakes that are often made in construction surveys include the following:

a. Arithmetic mistakes, generally due to lack of a check.
b. Use of incorrect elevations, grades, and stations.
c. Carrying out computed values beyond the accuracy possible in the field (a good hundredth is worth all the bad thousandths).
d. Reading the rod on top of the center-line stakes instead of on the ground beside them.
e. Failure to compute midpoint grades by successive increments so that any error will be evident if the last grade elevation does not check,

PROBLEMS

21–1. How far apart should stakes be set on a sewer job on a flat grade? On a steep grade? On a sharp curve?

21–2. To grade a rolling sloping surface it is required to set grade points on stakes at stations 0+52, 0+96, and 1+74 on a straight grade between the tops of stakes at station 0+00, elevation 64.25 and station 2+42, elevation 68.82. Arrange notes in suitable form and compute rod settings for the grade points, if the B.S. on 0+00 is 7.56 ft.

21–3. Are batter boards ever placed inside a building? Explain.

21–4. By means of a sketch, show how and where batter boards should be set for an I-shaped building 60 ft by 45 ft, all wings being 15 ft wide.

21–5. Same as problem 21–4 except for an H-shaped building 54 ft by 54 ft with wings 18 ft wide.

21–6. List two or more ways of stilling the swing of a plumb-bob line extending several stories outside a building wall.

21–7. Can the corner of a building be plumbed by running the vertical cross line of a transit up and down the wall line? Explain.

21–8. What tolerance is reasonable for the plumbing of elevator shafts?

21–9. Compare the accuracy required on a survey for a tall office building with that for an ordinary property survey. What tolerances are reasonable for wall alignment and floor elevations?

21–10. What tolerances are reasonable in positioning the anchor bolts for a bridge? For an elevated water tank?

21–11. What method can be used to check the deviations from the vertical of steel tubes driven to form cast-in-place concrete piles?

21–12. Arrange in proper order the following sequence of giving construction lines: (a) for foundation, (b) for excavation, (c) for main structure, (d) for subdivisions of the structure.

21–13. Discuss the suitability of a 0.00% grade for a street.

21–14. On a construction contract, is the owner or the contractor responsible for the layout accuracy?

21–15. What are the advantages of the coordinate system for an overpass layout?

21–16. Outline a suitable method for giving grade for (a) a parking lot, (b) a reinforced-concrete culvert, and (c) a footing for a bridge pier.

21–17. If a pipeline is parallel to and near the curb, would it be preferable to put the offset line on the curb or on the road surface?

21–18. What is a reasonable chord length for a circular curve having $I = 60°$ and $D = 20°$ to lay out a 10-in. water line?

21–19. An engineer's level is set up so that sighting on the h.i. with the middle cross hair gives the proper elevation of the invert at the start of a sewer line. The grade elevations are then obtained by sighting with the lower cross hair from the same set up. What grade is being used?

21–20. Approximately how close should the levels check on a survey for a concrete-surfaced highway to be one mile long?

21–21. Are distances measured by stadia satisfactory for preliminary highway surveys?

21–22. What are the most important things to be considered in laying a grade line on a ground profile?

21–23. A highway survey is run by deflection angles. Which is the most important transit adjustment to check?

21–24. Show on a sketch the definition of grade rod for a cut of 2.2 ft, a fill of 3.5 ft, and a grade point.

21–25. State briefly the surveys you would make on which to base the layout of an irrigation system.

21–26. When and where are offset lines used?

chapter 22

Circular Curves

22–1. General. Straight (tangent) sections of most types of transportation routes, such as highways, railroads, and pipelines, are connected by curves in both the horizontal and vertical planes. An exception is a transmission line, in which a series of straight lines is used with direct angular changes.

Two types of horizontal curves are employed—circular arcs and spirals. A *simple curve*, Fig. 22–1a, is a circular arc connecting two tangents. A *compound curve*, Fig. 22–1b, is composed of two circular arcs of different radii tangent to each other, with their centers on the same side of the common tangent. The combination of a short length of tangent connecting two circular arcs having centers on the same side, as in Fig. 22–1c, is called a *broken-back curve*. A *reverse curve*, Fig. 22–1d, consists of two circular arcs tangent to each other, the centers being on opposite sides of the common tangent. Reverse, compound, and broken-back curves are unsuitable for modern high-speed highway and railroad traffic.

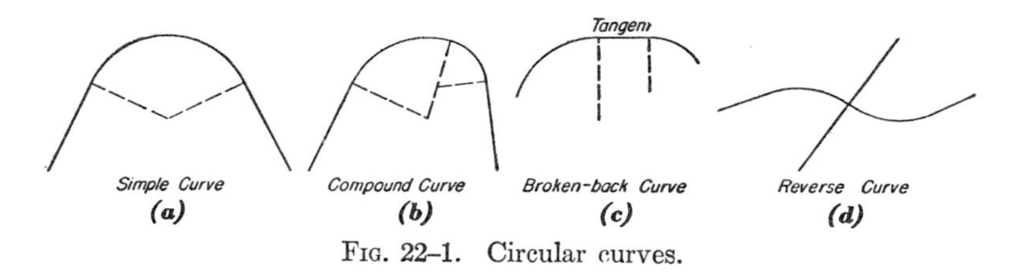

FIG. 22–1. Circular curves.

Easement curves are desirable to lessen the sudden change in curvature at the junction of a tangent and a circular curve. A *spiral* makes an excellent easement curve because its radius decreases uniformly from infinity at the tangent to that of the curve it meets. Spirals are used to connect a tangent with a circular curve, a tangent with a tangent (double spiral), and a circular curve with a circular curve. Figure 22–2 illustrates these arrangements.

The effect of centrifugal force on a vehicle passing around a curve can be balanced by *superelevation* of the outer rail of a track and the outer edge of a highway pavement. The correct superelevation on a spiral increases uniformly with the distance from the beginning of the spiral, and is in inverse proportion to the radius at any point. Properly superelevated spirals insure smooth and safe riding qualities with less wear on equipment. For detailed coverage of the spiral and superelevation, the reader is referred to one of the books on route surveying listed in the References, page 603.

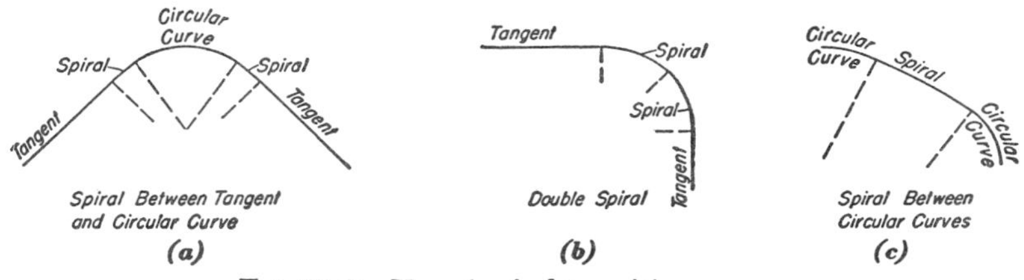

FIG. 22–2. Use of spiral transition curves.

Circular arcs and spirals are used for curves in the horizontal plane because they are readily laid out in the field by transit and tape.

Grade lines are joined in the vertical plane by parabolic curves, discussed in Chapter 23. Elevations on parabolic curves are easily computed and can be established on the ground by leveling.

22–2. Degree of curve. In European practice and some American highway work, circular curves are designated by their radius; for example, "1000-ft curve" and "3200-ft curve." American railroads and various highway departments prefer to identify curves by their *degree*, employing either the chord definition or the arc definition.

In railroad practice, the degree of curve is the angle at the center of a circular arc subtended by a chord of 100 ft. This is the *chord definition* and is indicated in Fig. 22–3a. In some highway work, the degree of curve is the angle at the center of a circular arc subtended by an arc of 100 ft. This is the *arc definition,* as illustrated in Fig. 22–3b. Formulas relating the radius R and the degree D are shown beside the illustrations.

Radii of chord- and arc-definition curves for values of D from 1° to 6° are given in Table VII. Although the differences appear to be small in this range, they have some significance in computations.

The chord-definition curve is consistent in using chords for computation and layout. Its disadvantages are: (a) R is not directly proportional to the reciprocal of D; (b) the formula for length is slightly approximate; (c) corrections are necessary in using the short-cut formulas for tangent distance and external distance; (d) greater difficulty is encountered in checking sharp curves.

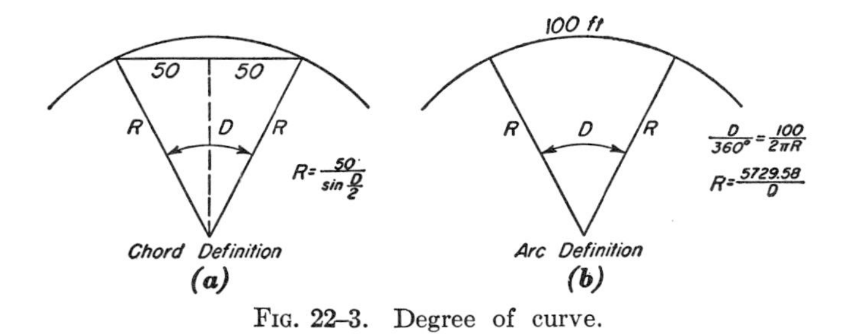

FIG. 22–3. Degree of curve.

The arc-definition curve has the disadvantage that most measurements between stations are less than a full tape length. Computations are facilitated since exact answers for the radius, tangent distance, and external distance are obtained from tabular values for a 1° curve by dividing by the degree D. Also, the formula for length is exact, this being an advantage in preparing right-of-way descriptions.

The arc and chord definitions give practically the same result when applied to the flat curves common on modern highways and railroads.

Circular curves are usually laid out in the field by deflection angles and taped chords.

22–3. Derivation of formulas. Circular-curve elements are shown in Fig. 22–4. The *point of intersection* of the tangents (P.I.) is also called the *vertex* (V). The *beginning of the curve* (B.C.) and the *end of the curve* (E.C.) are also termed the *point of curvature* (P.C.) and the *point of tangency* (P.T.). Other expressions for these points are *tangent to curve* (T.C.) and *curve to tangent* (C.T.).

The distance from the B.C. to the P.I., and from the P.I. to the E.C., is the *tangent distance* (T). The line connecting the B.C. and the E.C. is the *long chord* (L.C.). The *length of curve* (L) is the distance from the B.C. to the E.C. measured along the curve for the arc definition, or along 100-ft chords for the chord definition.

The *external distance* (E) is the distance from the vertex to the curve on a radial line. The *middle ordinate* (M) is the (radial) distance from the midpoint of the long chord of a circular curve to the midpoint of the curve.

The change in direction of the two tangents is the deflection angle I, which is equal to the central angle.

By definition, and by inspection of Fig. 22-4, relations for the *arc definition* follow:

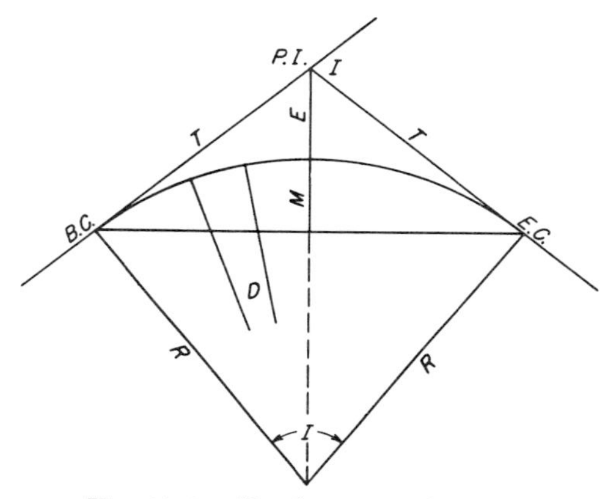

FIG. 22-4. Circular curve elements.

$$\frac{D^\circ}{360^\circ} = \frac{100}{2\pi R} \quad \text{and} \quad R = \frac{100 \times 360}{2\pi D} = \frac{5729.58}{D} \tag{22-1}$$

$$T = R \tan \frac{I}{2} \tag{22-2}$$

$$T_a = \frac{T_{1^\circ}}{D_a} \tag{22-3}$$

$$L = 100 \frac{I}{D} \tag{22-4}$$

$$\text{L.C.} = 2R \sin \frac{I}{2} \tag{22-5}$$

$$\frac{R}{R + E} = \cos \frac{I}{2};\ E = R\left(\sec \frac{I}{2} - 1\right) = R \text{ exsec } \frac{I}{2} \tag{22-6}$$

$$E_a = \frac{E_{1^\circ}}{D_a} \tag{22-7}$$

$$\frac{R - M}{R} = \cos\frac{I}{2};\ M = R\left(1 - \cos\frac{I}{2}\right) = R \text{ vers } \frac{I}{2} \qquad (22\text{–}8)$$

The formulas for T, L, E, and M apply also to a chord-definition curve. Small corrections must be added to the values of T_a and E_a found by formulas 22–3 and 22–7 respectively.

The formula relating R and D for a chord-definition curve is as follows:

$$R = \frac{50}{\sin D/2} \qquad (22\text{–}9)$$

22–4. Sample computation. Assume that field measurements show $I = 8° 24'$ and the station of the P.I. is 64 + 27.46, and that terrain conditions require use of the maximum degree of curve permitted by the specifications, which is, say, 2° 00′. Then, for an *arc-definition* curve,

$$R = \frac{5729.58}{2} = 2864.79 \text{ ft}$$

$$T = 2864.79 \times 0.07344 = 210.39 \qquad \text{or} \qquad T = \frac{420.8}{2} = 210.4^*$$

$$L = 100 \times \frac{8.40}{2} = 420.00$$

$$E = 2864.79 \times 0.002693 = 7.71 \qquad \text{or} \qquad E = \frac{15.4}{2} = 7.7^*$$

$$M = 2864.79 \times 0.002686 = 7.69 \text{ ft}$$

Station P.I. =	64 + 27.46
T =	2 + 10.39
Station B.C. =	62 + 17.07
L =	4 + 20.00
Station E.C. =	66 + 37.07

The computations for T and E marked with an asterisk are short-cut methods based upon the use of tables which list the tangent and external distances of 1° curves for various values of I. Division of these values by D gives R, T, and E directly for the arc-definition curve. Small corrections must be added for a chord-definition curve.

Computations for the stations of the B.C. and E.C. should be arranged as shown. Note that the station of the E.C. *cannot* be obtained by adding the tangent distance to the station of the P.I., although the position of the E.C. on the ground is determined by measuring the tangent distance from the P.I. The points representing the B.C. and E.C. must be carefully marked, and placed exactly on the tangent lines at the correct distance from the P.I., so that other computed values will fit their fixed positions on the ground.

After the curves have been inserted, the stationing must follow the actual line to be constructed. An *equation of chainage* is therefore applied to correct the original stationing. This equation represents the difference between the sum of the two tangents and the length of the curve. Its numerical value increases after each curve in the route.

22–5. Curve layout. Except for unusual cases, such as street railways, the radii of curves in route surveys are generally too large to permit swinging an arc from the curve center. Circular curves are therefore laid out by (a) deflection angles and chords, (b) tangent offsets, (c) chord offsets, (d) middle ordinates, and (e) other methods. Layout by deflection angles from the tangents is the standard method and one of two procedures which will be discussed in this text.

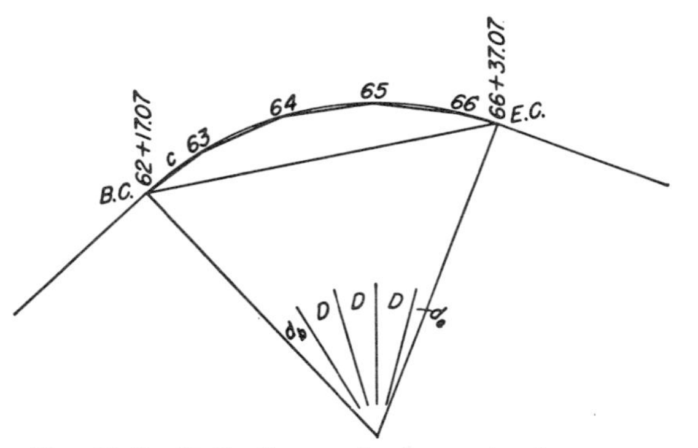

Fig. 22–5. Deflection angles for a circular curve.

In Fig. 22–5, assume that the transit is set up over the B.C. (station 62 + 17.07 in the preceding example). The first point to be marked on the curve is station 63 + 00, since cross sections are normally taken, construction stakes are set, and computations of earthwork are made at full stations and critical points. The distance from the B.C. to the next full station is called the *subchord, c.*

The *deflection angle* from a tangent for a full 100-ft arc is $D/2$, since an angle formed by a tangent and a chord is measured by one-half the intercepted arc. The *subdeflection angle* $d/2$ from the B.C. to station 63 + 00 is computed by the proportion

$$\frac{d}{2} : \frac{D}{2} = c : 100$$

from which

$$\frac{d}{2} = \frac{cD}{200} \tag{22–10}$$

A small subdeflection angle is expressed in minutes, rather than in degrees, and formula 22–10 then becomes

$$\frac{d}{2} = 0.3cD \tag{22–11}$$

where D is in degrees and the subdeflection angle is in minutes.

In this example,

$$\frac{d}{2} = 0.3 \times 82.93 \times 2 = 49.76 \text{ min}$$

Deflection angles are normally carried out to several decimal places for checking purposes, and to avoid accumulating small errors when D is an odd value, such as $3° \, 17.24'$.

An alternate method of computing deflection angles is to multiply the deflection angle per foot of arc or chord by the length of arc or chord being laid out. In the present example, the deflection angle per foot of arc is $(D/2)/100 = 1° \, 00'/100 = 0.60$ min/ft. Then the deflection angle $d/2$ from the B.C. to station 63+00 is 0.60 min × 82.93 = 49.76 min.

With the transit set up over the B.C., it is oriented by backsighting on the P.I., or on a point along the back tangent, with $0° \, 00'$ on the plates. The subdeflection angle of $49' \, 45''$ is then turned. Meanwhile the 17-ft mark of the tape is held on the B.C. The 100-ft end of the (subtracting) tape is swung until the line of sight hits the point 0.07 ft back from the 100-ft mark. This is station 63 + 00.

The rear tapeman next holds the zero mark on station 63, and the forward tapeman sets station 64 by direction from the instrumentman, who has placed an angle of $1° \, 49' \, 45''$ on the plates.

Deflection angles to stations following 63 are found by adding $D/2 = 1° 00'$ for each full chord. The subdeflection angle $d/2$ for the final subchord equals 22.24 min. As a check, the deflection angle from the B.C. to the E.C. must equal $I/2$. In the field it is essential to read and check the deflection angle to the E.C. (when visible from the B.C.) before starting to run the curve. An error in I, T, or L, or in the stationing, will then be discovered without wasting time running an impossible curve.

Field notes for the curve of this example are recorded in Plate A–13 as they would appear in a field book. The notes run up the page to permit sketching in a forward direction.

For a 2° 00′ curve, the lengths of arcs and chords are the same to two decimal places. On sharper curves the chords would be shorter than the corresponding arc lengths. For example, the chord measurement to lay out a 100-ft arc for a 6° 00′ curve is 99.95 ft.

In many cases it is desirable to *back in* a curve by setting up over the E.C. instead of the B.C. One setup is thereby eliminated and the long sights are taken on the first measurements. In precise work it is better to run in the curve from both ends to the center, where small errors can be adjusted more readily.

The curve used in a particular situation is selected to fit ground conditions and specification limitations on maximum D or minimum R. Normally the value of I and the station of the P.I. are available from field measurements on the preliminary line. Then a value of D up to the maximum for the railroad line, or an R suitable for the highway type, is chosen. Sometimes the value of E or M required to miss a stream or steep slope outside or inside the P.I. is measured, and D or R is computed. The tangent distance governs infrequently; the length of curve, practically never.

22–6. Curve layout by offsets. For short curves, when a transit is not available, for checking purposes, and if coordinates are desired, one of four offset-type methods may be used for circular curves: tangent offsets, chord offsets, middle ordinates, and ordinates from the long chord. Figure 22–6 shows the relationship of chord offsets, tangent offsets, and middle ordinates. Visually and by formula comparison, the chord offset (chord definition)

$$\text{C.O.} = 200 \sin D/2 = (100)^2/R = 2\ \text{T.O.} = (\text{approx.})\ 8\ \text{M.O.}$$

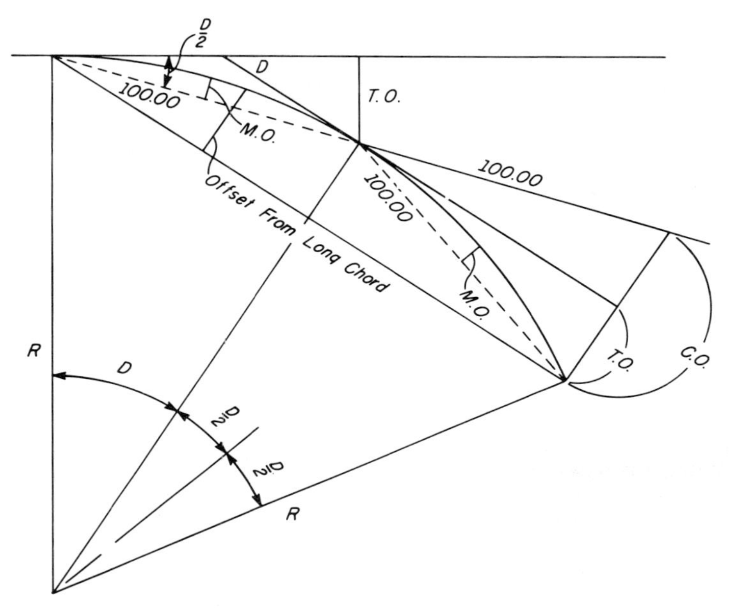

FIG. 22–6. Curve offsets.

A useful equation in laying out or checking curves in place is

D (in degrees) $= m$ (in inches) for a 62-ft chord (approx).

A sample computation of tangent offsets for a circular curve laid out from the B.C. and E.C. to a common station near the midpoint—thereby avoiding long measurements and providing a check point where small adjustments can be made if necessary—is given in illustration 22–1 for Fig. 22–7. Ordinates and abscissas are computed

Illustration 22–1

TANGENT OFFSETS FOR CIRCULAR CURVE

Point	Station	α	$c \sin\alpha$	Total Ordinate	$c \cos\alpha$	Total Abscissa
E.C.	80 + 00	0°00′	0.000	0.000	0.000	0.00
	79 + 50	1°15′	1.090	1.09	49.998	50.00
	79 + 00	3°45′	3.270	4.36	49.893	99.89
	79 + 00	4°45′	4.140	6.28	49.828	119.79
	78 + 50	2°15′	1.963	2.14	49.961	69.96
	78 + 00	0°30′	0.1746	0.17	19.999	20.00
B.C.	77 + 80	0°00′	0.000	0.00	0.000	0.00

from successive values of angle α where α = *central angle to the beginning of a chord + the deflection angle for that chord.*

Given data: $D = 5°00'$ (chord definition), B.C. = 77 + 80.00 and E.C. = 80 + 00.00. Then $L = 220.00$ ft, $I = LD/100 = 11°00'$, and $d_1/2 = \alpha_1 = 0.3 \times 20 \times 5 = 30$ min. Angle α_2 to 78 + 50 = 1°00′ (deflection angle for a 40-ft chord) + 1°15′ (deflection angle for a 50-ft chord) = 2°15′.

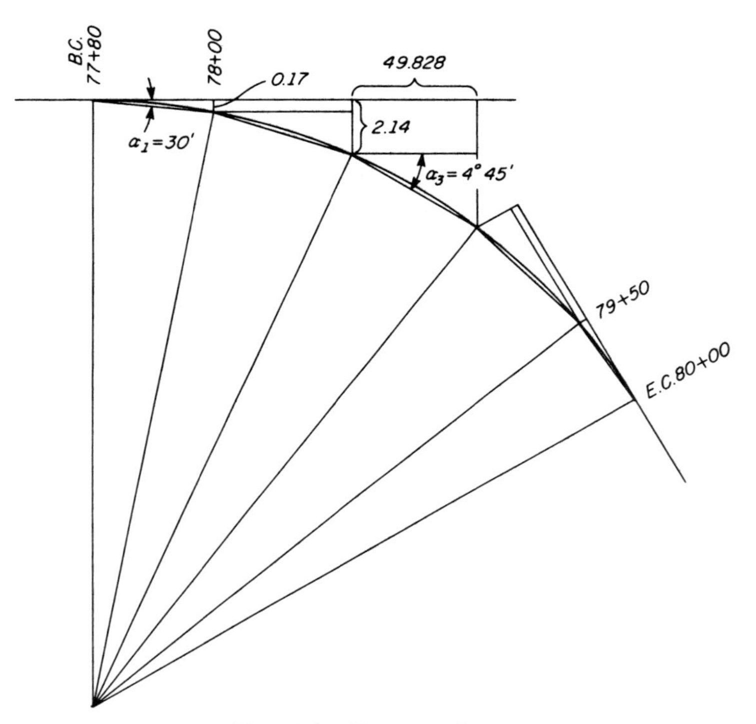

FIG. 22–7. Tangent offsets.

22–7. Setups on the curve. Obstacles and extremely long sight distances sometimes make it necessary to set up on the curve. The simplest procedure to follow is one which permits use of the same notes computed for running the curve from the B.C.

In this method the transit is backsighted on any station on the curve, with the telescope inverted and the plates set to *the deflection angle for that station from the B.C.* The telescope is plunged to the normal position, and the deflection angles previously computed for the various stations from the B.C. are used.

In the example of Secs. 22–4 and 22–5, if a setup is required at station 65, place 0° 00′ on the plates and sight to the B.C. with the telescope inverted. Plunge, set the plates to read the deflection angle 3° 49′ 45″, and set station 66. Or, if the B.C. is not visible, set 0° 49′ 45″ on the plates, sight on station 63, plunge, set 3° 49′ 45″ on the instrument, and locate station 66.

A simple sketch will make clear the geometry basic to this procedure.

22–8. Sight distance on horizontal curves. Highway safety requires certain minimum sight distances in zones where passing is permitted, and in nonpassing areas to assure a reasonable stopping distance if there is an object on the roadway. Specifications and tables list suitable values based upon vehicular speeds, the perception and reaction times of the average individual, the braking distance for a given coefficient of friction during deceleration, and the type and condition of the pavement.

A minimum sight distance of 600 ft is desirable for speeds as low as 30 miles per hr.

An approximate formula for sight distance can be derived by referring to Fig. 22–8, in which the clear sight distance past an

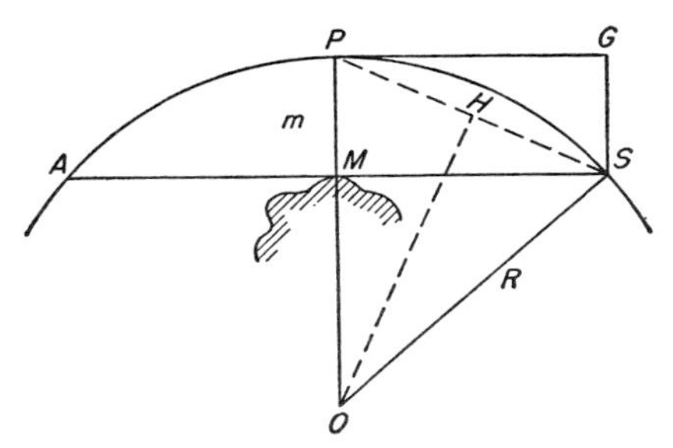

FIG. 22–8. Sight distance.

obstruction is the length of the long chord *AS*, denoted by C; and the required clearance is the middle ordinate *PM*, denoted by m. Then in triangles *SPG* and *SOH*,

$$m : SP = \frac{SP}{2} : R \qquad \text{and} \qquad m = \frac{(SP)^2}{2R}$$

Usually m is small compared with R, and SP may be assumed equal to $C/2$. Then

$$m = \frac{C^2}{8R} \tag{22-12}$$

If the distance m from the center line of the highway to the obstruction is known or can be measured, the available sight distance C is calculated from the formula. Actually the car travels along the curve on either the inside or outside lane; so the sight distance AS is not exactly the true stopping distance. The computed distance is on the safe side and is satisfactory for practical use.

22–9. Compound and reverse curves; special problems. Compound and reverse curves are combinations of two or more circular curves. They should be used only for low-speed traffic ways, and in terrain where simple curves cannot be fitted to the ground without excessive construction costs.

Special formulas have been derived to facilitate computations for such curves and are demonstrated in texts on route surveying. A compound curve is run by setups at the B.C. and E.C., or perhaps by one setup at the *point of compound curvature* (C.C.). Reverse curves are handled in similar fashion.

Many special problems in connection with curves arise in office design and in field layout. It may be necessary to pass a curve through a given point such as an underpass or bridge crossing; to relocate a new tangent a certain distance inside or outside of the old tangent; or to replace a broken-back curve with a simple curve. These problems are solved by applying a basic understanding of curve fundamentals and referring to a route-surveying text.

22–10. Sources of error. Some of the sources of error in curve computations and layout are:

a. Inability to set on the plates the required subdivisions of a minute for the deflection angles.
b. Poor intersections between tape line and sight line on flat curves.
c. Use of less than full tape lengths on arc-definition curves.
d. The large numbers which must be used to obtain answers with six significant figures.

22–11. Mistakes. Some typical mistakes that occur in laying out a curve in the field are the following:

a. Failure to double or quadruple the deflection angle at the P.I. before computing or laying out the curve.
b. Adding the tangent distance to the station of the P.I. to get the station of the E.C.
c. Using 100.00-ft chords to lay out arc-definition curves having D greater than 2°.
d. Taping subchords of nominal length for chord-definition curves having D greater than 5° (a nominal 50-ft subchord for a 6° curve requires a measurement of 50.02 ft).

PROBLEMS

22–1. A circular curve has a 1600-ft radius. What is its degree (a) by arc definition and (b) by chord definition?

22–2. Similar to problem 22–1, but for a 2800-ft radius.

22–3. List the advantages and disadvantages of the chord and arc definitions for the degree of a circular curve.

Compute T, L, and stations B.C. and E.C. for a circular curve in problems 22–4 through 22–7.

22–4. Highway curve with R = 572.96 ft, I = 35°20′, sta. V = 74 + 28.7.

22–5. Highway curve with D = 2°00′, I = 11°00′, P.I. = sta. 62 + 48.1.

22–6. Railroad curve with D = 3°00′, I = 19°24′, sta. V = 78 + 64.22.

22–7. Railroad curve with R = 1011.51 ft, I = 26°20′, V = 48 + 03.06.

22–8. Compute E and M for problem 22–4.

22–9. Compute E and M for problem 22–5.

22–10. Compute E and M for problem 22–6.

Compute and tabulate the curve data R, D, T, L, E, M, B.C., E.C., and deflection angles to lay out the curves given in problems 22–11 through 22–18.

22–11. Railroad curve with D = 2°30′, I = 8°12′R, sta. V = 82 + 09.56.

22–12. Highway curve with R = 2000 ft L, I = 16°40′, sta. V = 12 + 80.5.

22–13. Railroad curve with D = 4°20′, I = 10°20′, sta. V = 37 + 28.20.

22–14. Highway curve with R = 1200 ft, I = 10°20′, sta. V = 37 + 28.2.

22–15. Railroad curve with L = 300 ft, D = 5°00′, sta. V = 37 + 65.18.

22–16. Highway curve with L = 3200 ft, R = 6000 ft, sta. V = 80 + 70.4.

22–17. Railroad curve with T = 600 ft, D = 3°15′, B.C. = sta. 63 + 95.06.

22–18. Highway curve with T = 900 ft, R = 3600 ft, B.C. = Sta. 41 + 83.2.

Calculate and tabulate all data required to lay out the simple circular curves of problems 22–19 through 22–22.

22–19. The D for a highway curve will be rounded off to the nearest multiple of 30 min. Field conditions show E should be as near 48 ft as possible, and I = 25°30′, P.I. = 51 + 47.3 ft, $E_{1°}$ = 144.9 ft, $T_{1°}$ = 1296.5 ft.

22–20. The R for a highway curve will be rounded off to the nearest multiple of 100 ft. Field conditions indicate that M should be approximately 31.5 ft. $I = 14°24'$, and P.I. = 58 + 04.7 ft.

22–21. Similar to problem 22–19 except that E should be 54 ft and D rounded off to the nearest 10 minutes.

22–22. Similar to problem 22–20 except that R should be rounded off to the nearest 50 ft, and M is approximately 36 ft.

22–23. The P.I. of a highway survey line falls in a stream so a cutoff line AB = 400.00 ft is run between the tangents. In the triangle formed by points A and B and the vertex V, angle $VAB = 18°\ 00'$ and angle $VBA = 20°\ 00'$. The station of A is 36+47.10. Compute and tabulate the curve notes to run a 4° 00′ curve to connect the tangents AV and BV.

22–24. Similar to problem 22–23 except that $AB = 320.00$ ft and the bearing of line AV from A through the vertex is N 50°00′E, the bearing of VB is S 85°00′E, and station of point B is 52 + 16.32 ft.

22–25. What are the disadvantages of running circular curves by using single deflection angles?

22–26. Why is the vertex angle usually measured by repetition?

In problems 22–27 through 22–29, after a B.S. on the B.C. with 0°00′ on the plates, what is the deflection angle to the curve point noted?

22–27. Setup at midpoint, deflection to the E.C.

22–28. Setup at the ⅜ point of the curve, deflection to the ¾ point.

22–29. Setup at the ⅝ point of the curve, deflection to the ¼ point.

What sight distance is available if there is an obstruction on the inside of a curve as noted in problems 22–30 through 22–32?

22–30. For problem 20–16, obstacle 15 ft from curve on radial line to P.I.

22–31. For problem 20–17, obstacle 20 ft inside curve.

22–32. For problem 20–18, obstacle 22 ft inside curve.

For problems 22–33 through 22–35 compute and tabulate the coordinates to stake out a circular railroad curve by tangent offsets. Regular and half stations are to be set and approximately half of the curve laid out from each tangent.

22–33. Station of B.C. = 46 + 72.40, E.C. = 50 + 18.90, and $D = 5°40'$.

22–34. Length = 400 ft, $D = 6°00'$, and station of B.C. = 76 + 50.00.

22–35. Station of E.C. = 85 + 00.00, $D = 4°40'$, and $L = 530$ ft.

22–36. List in their order of importance the influence of R, D, L, T, E, M, and LC in selecting curves to fit ground conditions. Justify your listing.

22–37. Prepare a set of field notes for running a circular curve of 100-ft radius for a street-car track making a 90° right turn on to a cross street. Which definition of curve should be used?

22–38. Compute the length of chord (in feet) for which the middle ordinate (in inches) equals twice the degree of curve.

22–39. A race track must have exactly one mile along its centerline with semi-circles and two tangents, and ⅓ of its total length has to be in the two curves. Find the length, radius, and degree of the curves.

22–40. Similar to problem 22–39, except that ¼ of the total length is in the two curves.

22–41. A compound curve consists of 5 equal-length circular curves of $D = 2°$, 4°, 6°, 4° and 2° in that sequence and connects two tangents having an intersection angle of 72°00′. If the station of the B.C.C. is 51 + 48.6, compute the stations of the C.C. and E.C.C.

chapter **23**

Parabolic Curves

23–1. General. Parabolic curves are used in the vertical plane to provide a smooth transition between grade lines on highways and railroads. They are also employed in the horizontal plane on landscaping work and for other functions which require a curve of pleasing appearance that can be laid out with a tape. The parabolic curves used in route-surveying practice differ only slightly from large radius circular curves.

Parabolic curves can be computed by the *tangent-offset method* and by the *chord-gradient method.* Both systems are based upon the following property of a parabola:

> *Offsets from a tangent to a parabola are proportional to the squares of the distances from the point of tangency.*

In this brief treatment only a single arithmetic method will be demonstrated. It is applicable to curves having either equal tangents or unequal tangents. Another property of the parabola supplements the tangent-offset rule in this method:

> *The center of a parabola is midway between the vertex and the long chord; that is, the external distance always equals the middle ordinate.*

A vertical curve must (a) fit the grade lines it connects, (b) have a length sufficient to meet the specification covering the maximum rate of change of grade per station (0.05 to 0.10 per cent on railroads), and (c) provide the required sight distance. Generally the vertex is placed at a full station in railroad work but it may be at a plus station on a highway layout.

23–2. Computations for an equal-tangent curve. In Fig. 23–1 a -2.00 per cent grade meets a $+1.60$ per cent grade at station $87 + 00$ and at elevation 743.24 ft. Assume that an 800-ft curve fits the ground conditions fairly well and satisfies the specifications.

Grade elevations on the tangents are determined by computing

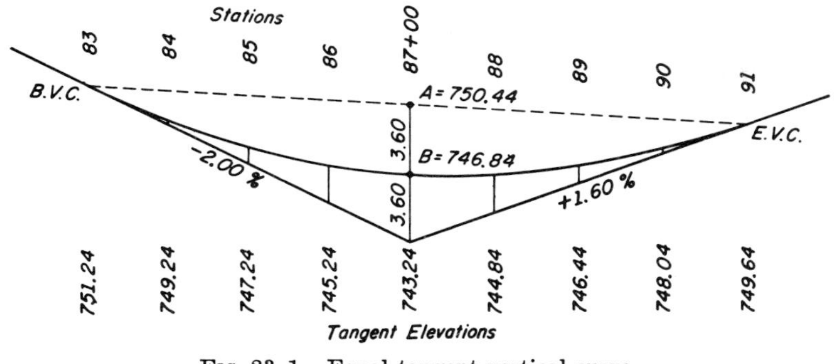

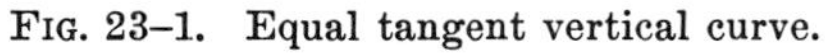

FIG. 23–1. Equal tangent vertical curve.

forward and backward from the vertex. The elevation of point A (the center of the long chord), found by averaging the elevations of the B.V.C. and the E.V.C., is 750.44 ft. Point B, the midpoint of the curve, is halfway between point A and the vertex, and is at elevation 746.84.

The tangent offset at the center of the curve is therefore 3.60 ft. Offsets from the tangents at the other full stations are

$$\left(\tfrac{1}{4}\right)^2 \times 3.60 = 0.225 \text{ ft}$$
$$\left(\tfrac{1}{2}\right)^2 \times 3.60 = 0.90 \text{ ft}$$
$$\left(\tfrac{3}{4}\right)^2 \times 3.60 = 2.025 \text{ ft}$$

Illustration 23–1

EQUAL-TANGENT VERTICAL CURVE

Station	Point	Tangent Elevation	Tangent Offset	Elevation on Curve	First Diff. of Offsets	Second Diff. of Offsets
92		751.24				
91	E.V.C.	749.64	0.00	749.64		
					0.225	
90		748.04	0.225	748.26		0.45
					0.675	
89		746.44	0.90	747.34		0,45
					1.125	
88		744.84	2.025	746.86		0.45
					1.575	
87	P.V.I.	743.24	3.60	746.84		
86		745.24	2.025	747.26		
85		747.24	0.90	748.14		
84		749.24	0.225	749.46		
83	B.V.C.	751.24	0.00	751.24		
82		753.24				

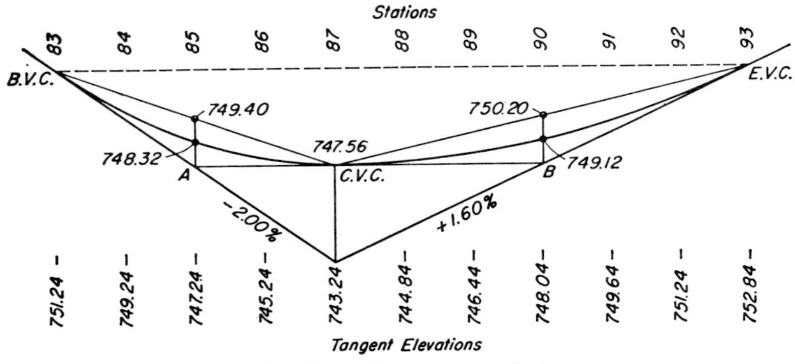

FIG. 23–2. Unequal tangent vertical curve.

Final elevations on the curve are shown in the usual form in illustration 23–1. Offsets can be computed from one tangent for the full length of the curve but it is simpler to use symmetrical offsets on both sides of the vertex with the known tangent elevations.

A check on the tangent offsets is obtained by computing the first and second differences between the offsets, as shown in the right-hand columns of illustration 23–1. Unless disturbed by rounding off, all the second differences ("rate of change") should be equal. The grades shown in Figs. 23–1 and 23–2 are greatly exaggerated. Actually there is no significant difference between vertical lines and offsets perpendicular to the tangents for the light grades used in practice. Verticals are more convenient and give a smooth curve. Figures 23–1 and 23–2 show *sag curves.* If the grades are reversed, *summit curves* are formed.

23–3. Computations for an unequal-tangent curve. In Fig. 23–2, assume that for the two grades of the previous example a 400-ft vertical curve is to be extended back from the vertex, and a 600-ft curve run in forward to better fit the ground conditions.

Connect the midpoints of the two curves, stations 85 and 90, to obtain line *AB*. Draw lines from the B.V.C. and E.V.C. to point C.V.C., station 87. Compute the elevation of the C.V.C., 747.56, by proportion from the known elevations of *A* and *B*.

Now compute two vertical curves, one from the B.V.C. to the C.V.C. and another from the C.V.C. to the E.V.C., by the method of Sec. 23–2. Since both curves are tangent to the same line *AB* at point C.V.C., they will be tangent to each other and form a smooth curve. Critical elevations are shown in Fig. 23–2.

Vertical curve computations by themselves are quite simple, hardly a challenge to the electronic computer. But when vertical curves are combined with horizontal curves, spirals, and superelevation on complex highway interchange coordinate calculations, programming can save time.

23–4. High or low point on a vertical curve. To investigate drainage conditions, clearance beneath overhead structures, cover over pipes, and sight distance, it may be necessary to determine the elevation and location of the low (or high) point on a vertical curve. At the low or high point, a tangent to the curve will be horizontal and its slope will be equal to zero. Based on this fact, the following formula is readily derived:

$$x = \frac{g_1 L}{g_1 - g_2} \tag{23–1}$$

x = distance, B.V.C. to the high or low point of the curve;
g_1 = tangent grade through the B.V.C.;
g_2 = tangent grade through the E.V.C.;
L = length of the curve, in stations.

If g_2 is substituted for g_1 in the numerator, the distance x is measured from the E.V.C.

In the problem of Fig. 23–1, $g_1 = -2.00$ per cent, $g_2 = +1.60$ per cent, and $L = 8$ stations. Then

$$x = \frac{-2.00 \times 8}{-2.00 - (+1.60)} = 4.444 \text{ stations}$$

The elevation at this point (station 87 + 44.4) is

$$(743.24 + 0.444 \times 1.60) + \left(\frac{3.56}{4}\right)^2 \times 3.60 = 746.80 \text{ ft}$$

23–5. Sight distance. The formula for sight distance S with the vehicle on a vertical curve, and S less than the length L, is

$$S^2 = \frac{8Lh}{g_1 - g_2} \tag{23–2}$$

where S = sight distance, in stations;
L = length of curve, in stations;
h = height of driver's eye, and the object sighted, above the roadway (by AASHO recommendation, 3.75 ft).

Then for a summit curve having a length of 800 ft, and grades of +2.00 per cent and −1.60 per cent, if h = 3.75 ft,

$$S = \sqrt{\frac{8 \times 8 \times 3.75}{2.00 - (-1.60)}} = 8.16 \text{ stations} = 816 \text{ ft}$$

Since this distance is greater than the length of curve and thus not in agreement with the assumption used in deriving the formula, a different expression must be employed.

If the vehicle is off the curve and on the tangent to it, the formula for sight distance which is applicable and derived in a similar manner is

$$S = \frac{L}{2} + \frac{4h}{g_1 - g_2} \tag{23-3}$$

In the preceding example, with $h = 3\frac{3}{4}$ ft,

$$S = \frac{8}{2} + \frac{4 \times 3.75}{2.00 - (-1.60)} = 8.17 \text{ stations}$$

For a combined horizontal and vertical curve, the sight distance is the smaller of the two values computed independently for each curve.

23–6. Sources of error. Some of the sources of error in vertical-curve work are:

a. Carrying out grade percentages beyond 0.01 per cent. Multiples of 0.05 per cent or 0.10 per cent are desirable for highways.
b. Carrying out computed elevations to less than 0.01 ft.
c. Selecting the vertex at other than a full station.

23–7. Mistakes. Typical mistakes that are made in computations for vertical curves include the following:

a. Arithmetical errors.
b. Using an incorrect offset at the vertex.
c. Subtracting the offsets from the tangents for a sag curve, or adding them for a summit curve.
d. Failing to check the computed values for curve elevations, to be

certain they fit those of the tangents, the P.V.I., the B.V.C., and the E.V.C.

PROBLEMS

23–1. A manhole is 12 ft from the center line of a street that is 48 ft wide and has a 6-in. parabolic crown. If the center of the street is at elevation 241.30 ft, what is the elevation of the manhole cover?

23–2. Width of a street pavement is 22 ft and average parabolic crown from the center to each edge is $\frac{1}{8}$ in. per ft. What is the drop in pavement surface from the street center to a point 6 ft from the edge?

Tabulate elevations of the stations on a vertical curve for the conditions stated in problems 23–3 through 23–6.

23–3. A −4.00% grade meets a +2.00% grade at station 65 + 00 and elevation 361.44 ft; 800-ft curve.

23–4. A +3.00% grade meets a +1.00% grade at station 78 + 25, elevation 400.00 ft; 600-ft curve.

23–5. A 250-ft curve, 50-ft stations, grades of +0.60% and −1.00%, P.V.I. at station 42 + 70 and elevation 820.00 ft.

23–6. A 1000-ft curve, grades of +6.00% and −4.00%, P.V.I. at station 33 + 00, elevation 710.00 ft.

Field conditions require a highway curve to pass through a fixed point. Compute a suitable curve and elevations for problems 23–7 through 23–9.

23–7. Grades of +0.00% and −4.00%, P.V.I. elevation 400.00 ft and station 70 + 00. Fixed point elevation 397.00 at station 70 + 00.

23–8. Grades of 0.00% and +2.00%, P.V.I. elevation 350.00 ft at station 30 + 00. Fixed point elevation 352.00 ft at station 30 + 00.

23–9. Grades of −2.00% and +1.00%, P.V.I. station 66 + 00, elevation 800.00 ft. Fixed point elevation 801.50 ft at station 66 + 00.

23–10. A proposed highway has an ascending grade of 0.40% followed by a descending grade of 2.60% meeting at station 15 + 00 and elevation 625.50 ft. An existing cross road intersects the planned one at right angles at station 15 + 80 and elevation 622.55 ft. What length of vertical curve should be used?

23–11. A +2.00% grade meets a 0.00% grade at station 72 + 00 and elevation 900.00. The 0.00% grade in turn joins a −3.00% grade at station 77 + 00. Compute and tabulate the notes for a 1000-ft vertical curve to fit the given conditions.

Compute and tabulate station elevations for an unequal-tangent vertical curve to fit the conditions noted in problems 23–12 through 23–14.

23–12. A +2.00% grade meets a −4.00% grade at station 24 + 00 and elevation 230.24 ft. Length of first curve 200 ft, second curve 600 ft.

23–13. Grade $g_1 = -3.00\%$, $g_2 = +4.00\%$, P.V.I. at station 48 + 00 and elevation 324.60 ft, $L_1 = 800$ ft, and $L_2 = 400$ ft.

23–14. The P.V.I. of -1.20% and $+2.40\%$ grades is at station 26 + 00 and elevation 150.00 ft. Lengths of curves are 600 ft and 800 ft.

23–15. In determining sight distances on vertical curves, how does the computer determine whether the cars or objects are on the curve or tangents?

23–16. What sight distance is available in problem 23–6?

23–17. What sight distance is available in problem 23–7?

23–18. Determine the low point of the curve in problem 23–3.

23–19. Calculate the high point of the curve in problem 23–5.

What is the minimum length of vertical curve to provide the required sight distance under the conditions given in problems 23–20 through 23–22?

23–20. A summit curve with grades of $+4.00\%$ and -1.60%. Sight distance 600 ft.

23–21. Grades of $+3.60\%$ and -2.40%. Sight distance of 800 ft.

23–22. Sight distance of 1000 ft, grades of $+1.80\%$ and -3.20%.

23–23. Why are the second differences of the tangent offsets equal for a parabolic curve?

23–24. What is meant by the rate of change on vertical curves?

23–25. Why is a parabola rather than a circular arc used for vertical curves?

23–26. Write a computer program for the vertical curve of illustration 23–1.

23–27. When is it advantageous to use an unequal tangent vertical curve instead of an equal tangent layout?

23–28. Check the elevation given on Fig. 23–2 by the method of second differences and explain the results.

23–29. A highway to be built will cross a pipe culvert at right angles at station 42+75. Elevation of top of pipe is 368.2 ft. The P.I. of a 500-ft parabolic curve will be located at station 42+00 and elevation 371.40. Grades of forward and backward tangents are -1.40% and $+1.10\%$ respectively. What is the depth of cover over the pipe?

23–30. A $+1.00\%$ grade meets a 0.00% grade at station 18+50 and elevation 200.00 ft. The 0.00% grade then meets a -2.00% grade at station 20+00 and elevation 200.00. Compute an unequal tangent vertical curve to pass through elevation 199.75 at station 18+50 and also through elevation 199.75 at station 20+00.

23–31. In laying out preliminary grades on a highway line, what consideration must be given to the later introduction of vertical curves?

chapter **24**

Volumes

24–1. General. Surveyors are called upon to measure volumes of earthwork and concrete for various types of construction projects. Volume computations are required to determine the capacity of bins, tanks, reservoirs, and buildings, and to check the quantities in stockpiles.

Much of the field work formerly involved in running preliminary center lines, getting cross-section data, and making slope-stake and other measurements on long route surveys is now being done more efficiently by photogrammetry. Computation of earthwork for route surveys of more than a few stations is, or will be, carried out on electronic computers. It is not intended to discuss photogrammetric or electronic-computer methods herein. Rather, basic field and office procedures for determining and calculating volumes on small-scale work—and merely carried out faster and more economically by the newer methods—will be presented briefly.

The unit of volume is a cube having edges of unit length. Both the cubic foot and the cubic yard are used in surveying calculations but the latter is the standard unit for earthwork. One cubic yard equals 27 cubic feet.

24–2. Methods of measurement. Direct measurement of volumes is rarely made in surveying, since it is difficult to actually apply a unit of the material involved. Indirect measurements are obtained by measuring lines and areas which have a relation to the volume desired.

Three principal methods are used: (a) the cross-section method, (b) the unit-area, or borrow-pit, method, and (c) the contour-area method.

24–3. The cross-section method. When the cross-section method is used for computing volumes, vertical cross sections are taken (usually at right angles to the center line) at intervals of 50 or 100 ft. The areas of these sections can be found by calculation from the field

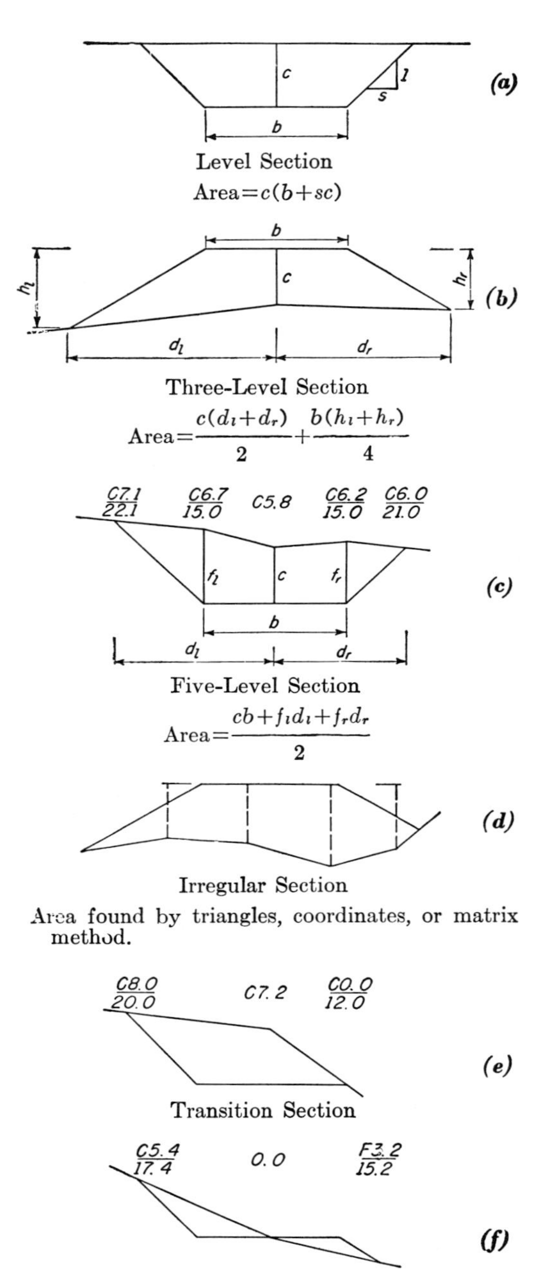

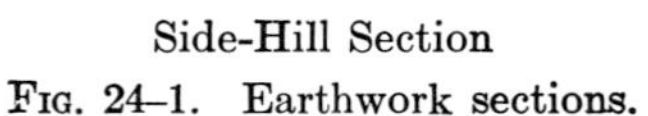
FIG. 24-1. Earthwork sections.

data, or by planimetering the plotted cross sections. The volume is computed by the *average-end-area formula* or by the *prismoidal formula.*

24–4. Types of cross sections. The types of cross sections commonly used on route surveys are shown in Fig. 24–1. In flat terrain the *level section* in (*a*) is suitable. The *three-level section* in (*b*) is generally employed where ordinary ground conditions prevail. Rough topography may require a *five-level section,* as in *(c),* or more practically an *irregular section,* as in *(d).* A *transition section,* as in *(e)* and a *side-hill section,* as in *(f),* occur in passing from *cut* (excavation) to *fill* (embankment), and on side-hill locations.

The width b of the base or finished roadway is fixed by the project requirements. The base is usually wider in cuts than on fills, to provide for drainage ditches. The side slope s depends upon the type of soil encountered. Side slopes in fill must be flatter than those in cuts where the soil remains in its natural state.

Cut slopes of 1 to 1 and fill slopes of $1\frac{1}{2}$ (horizontal) to 1 (vertical) are satisfactory for ordinary loam soils.

Formulas for the areas of the different sections are readily derived and listed beside some of the sketches in Fig. 24–1.

24–5. End-area formulas. The volume between two vertical cross sections A_1 and A_2 is equal to the average of the end areas multiplied by the horizontal distance L between them. Thus

$$V_e = \frac{L(A_1 + A_2)}{2 \times 27} \quad cu\ yd \tag{24–1}$$

This formula is approximate and gives answers which generally are slightly larger than the true prismoidal volumes. It is used in practice because of its simplicity. Increased accuracy is obtained by decreasing the distance L between the sections. When the ground is irregular, cross sections must be taken close together.

24–6. Prismoidal formula. The prismoidal formula applies to the volumes of all geometric solids which can be considered prismoids. Most earthwork volumes fit this classification but relatively few of them warrant the precision of the prismoidal formula, which is

$$V_p = \frac{L(A_1 + 4A_m + A_2)}{6} \tag{24-2}$$

where V_p = volume;
A_1 and A_2 = areas of successive cross sections taken in the field;
A_m = area of a section midway between A_1 and A_2;
L = horizontal distance between A_1 and A_2.

Various handbooks give tables and diagrams to facilitate computations of areas and volumes.

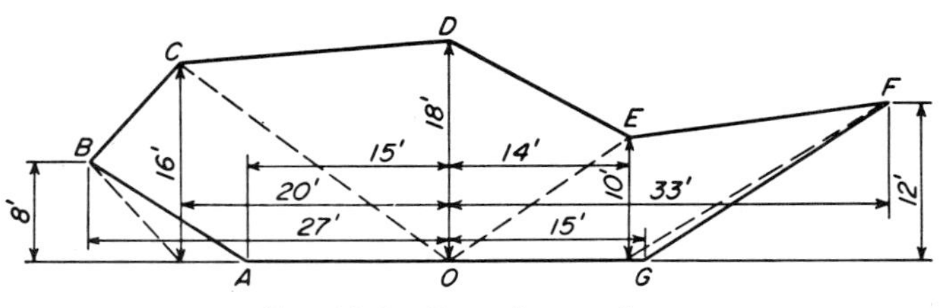

FIG. 24–2. Irregular section.

To illustrate the computation of area for the section shown in Fig. 24–2, assume that the following set of notes was obtained in the field (the letters A, B, C, D, E, F, and G, and the values at A and G were not recorded in the field; they are added for this discussion), or compiled from the contours on a suitable-scale topographic map.

Station	*A*	*B*	*C*	*D*	*E*	*F*	*G*
24 + 00	$\frac{0}{15}$	$\frac{C\ 8}{27}$	$\frac{C\ 16}{20}$	$\frac{C\ 18}{0}$	$\frac{C\ 10}{14}$	$\frac{C\ 12}{33}$	$\frac{0}{15}$

The numerators are cuts, in feet, and the denominators are distances out from the center line. Fills are noted by the letter F. Using C instead of plus for cut, and F instead of minus for fill, eliminates confusion. On many projects the measurements are carried out to tenths of a foot. In computing areas by coordinates, cut values are considered plus, distances to the right of the center line are plus, and distances to the left of the center line are minus.

The area of the cross section in Fig. 24–2 will be found in three ways: (a) by coordinates, (b) by dividing the area into simple figures (triangles), and (c) by the matrix system.

Area by Coordinates

$$
\begin{aligned}
0(-27 - 15) &= 0\\
8(-20 + 15) &= -\ 40\\
16(\ \ 0 + 27) &= +432\\
18(\ 14 + 20) &= +612\\
10(\ 33 + \ 0) &= +330\\
12(\ 15 - 14) &= +\ 12\\
0(-15 - 33) &= 0\\
2)\ &+1346\\
\text{Area}\ &673
\end{aligned}
$$

Area by Triangles

$$
\begin{aligned}
18(34) &= 612\\
10(33) &= 330\\
12(\ 1) &= 12\\
16(27) &= 432\\
-8(\ 5) &= -\ 40\\
2)&1346\\
\text{Area}\ &673
\end{aligned}
$$

Area by matrix system. The matrix system for computing earthwork can be used for any type of section and has many engineering applications.

Using the notes obtained in the field for points B, C, D, E, and F, place half the roadway width beneath zero on both ends of the data. This produces the same expression used in the coordinate method, since the coordinates of A and G are 0/15.

Then beginning at the center, multiply diagonally downward and add the results as shown. Again, beginning at the center, multiply diagonally upward and add the results. Subtract the upward sum from the downward sum to obtain the double area. Divide by 2. The computations may be arranged as follows:

Downward:

$$18(20) + 16(27) + 8(15) + 18(14) + 10(33) + 12(15) = 1674$$

Upward:

$$0(16) + 20(8) + 27(0) + 0(10) + 14(12) + 33(0) = 328$$

$$2)1346$$

$$673$$

It is necessary to make separate computations for cut and fill if they occur in the same section, since they must always be tabulated independently for pay purposes.

To find the volume between this section at station 24 and another at the next station, 25, which is 100 ft distant, assume that the area of the section at station 25 is 515 sq ft. Then the volume in cubic yards, by the end-area formula, is

$$V = \frac{100}{2 \times 27}(A_1 + A_2) = \frac{100}{54}(673 + 515) = 2200 \text{ cu yd}$$

In order to use the prismoidal formula it is necessary to know the area of the section halfway between the stations. This area is found by the usual computation after averaging the heights and widths of the end sections. Obviously the middle area is *not* the average of the end areas, since there would then be no difference between the results of the end-area formula and the prismoidal formula.

The prismoidal formula generally gives a volume less than that found by the end-area formula. For example, the volume of a pyramid by the prismoidal formula is $Ah/3$, whereas by the average-end-area method it is $Ah/2$. An exception occurs when the center height is great but the width narrow at one station, and the center height is small but the width large at the adjacent station. Figure 24–3 illustrates this condition. The difference between the volume

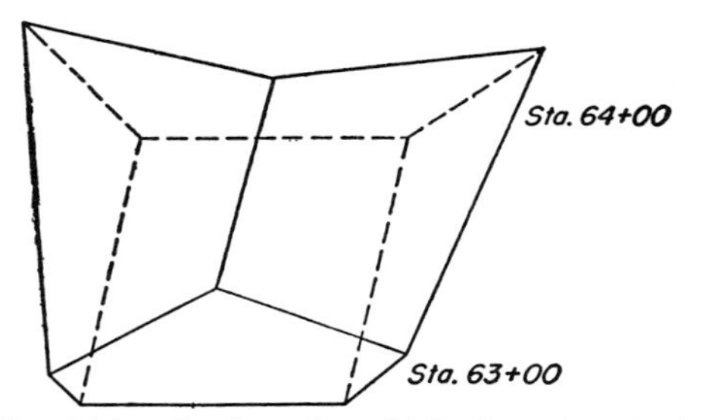

Fig. 24–3. Sections for which the prismoidal correction is added to the end area volume.

obtained by the average-end-area formula and that by the prismoidal formula is called the *prismoidal correction, Cp.*

Various books on route surveying give formulas and tables for computing prismoidal corrections which can be applied to the average-end-area volumes to get prismoidal volumes. Except in rock-excavation and in concrete work, the use of the prismoidal formula is not normally justified by the low precision of the field data. Frequently it is easier to compute the end-area volume and the prismoidal correction than to calculate the prismoidal volume directly. For projects with more than a few cross sections, computer programs are available and commonly used but a surveyor and engineer still must understand the method of computation. Earthwork calculations from photogrammetric data should check good field survey results within 1 per cent in open country free of big trees or high brush, and have received legal and contractor acceptance.

Mass diagrams are used to analyze the movement of earthwork quantities. Total mass (cut minus fill up to the station) is plotted at each station starting at the project's beginning. Horizontal (balance) lines then determine the limit of economic haul and direction of movement of material.

24–7. Unit-area, or borrow-pit, method. Calculations of volumes by the unit-area, or borrow-pit, method were discussed in Sec. 5–29.

Greater accuracy can be obtained in rough terrain by using triangular areas instead of rectangular blocks. The volume is then equal to the area of each triangle times the average of its three corner heights. Thus

$$V = \frac{A(a + b + c)}{3} \tag{24–3}$$

24–8. Contour-area method. Volumes based on contours can be obtained from contour maps by planimetering the area enclosed by each contour and multiplying the average of areas for adjacent contours by the contour interval, i.e., using formula 24–1. For example, if in the plan view of Fig. 13–5, page 279, the area within the 10-ft contour is 19,650 sq ft and that within the 20-ft contour is 12,720 sq ft, the average end-area volume is 10(19,650 + 12,720)/2(27) = 5994 cu yd. Use of the prismoidal formula is seldom, if ever, justified on this type of computation.

24–9. Sources of error. Some common errors in computing areas of sections and volumes of earthwork are:

a. Carrying out areas of cross sections beyond the nearest square foot, or beyond the limit justified by the field data.
b. Carrying out volumes beyond the nearest cubic yard.
c. Failing to correct for the effect of curvature when a section on a horizontal curve has cut on one side of the center line and fill on the other side.

24–10. Mistakes. Typical mistakes made in earthwork calculations include:

a. Errors in arithmetic.
b. Using the prismoidal formula when end-area volumes are sufficiently accurate.
c. Using end-area volumes for pyramidal or wedge-shaped solids.
d. Mixing cut quantities and fill quantities.

e. Not considering transition sections when passing from cut to fill, or from fill to cut.

PROBLEMS

24–1. Why is the roadway in cut normally wider than roadway in fill?

24–2. Discuss comparative side slopes in cut and fill.

24–3. Why must cut and fill volumes be totaled separately?

Draw the cross sections and compute V_e for the data given in problems 24–4 through 24–8.

24–4. Two level sections at 100-ft stations having center heights of 2.0 and 5.0 ft in cut. Base width = 30 ft, side slopes are 1 to 1.

24–5. Two level sections at 50-ft stations with center heights 3.2 ft and 4.4 ft. Base width = 24 ft, side slopes 1½ to 1 (fill).

24–6. Notes for station 47 + 00.
Base = 20 ft, s.s. = 1 : 1.
Area at sta. 47 + 80 = 140 sq ft.

$$\frac{C\,4.0}{14.0} \qquad C\,5.0 \qquad \frac{C\,6.0}{16.0}$$

24–7. A drainage canal with b = 12 ft and side slopes of 1 to 1

Sta. 18	$\frac{C\,2.6}{8.6}$	C 3.0	$\frac{C\,2.2}{8.2}$
Sta. 17	$\frac{C\,1.8}{7.8}$	C 2.0	$\frac{C\,1.4}{7.4}$

24–8. Same cuts as problem 24–7 but base = 10 ft, side slopes are 2 : 1.

24–9. Compute the section areas in problem 24–4 by the matrix method.

24–10. Find the section area in problem 24–6 by the matrix method.

24–11. Calculate the section area in problem 24–7 by coordinates.

24–12. Compute the C_p and V_p for problem 24–4.

24–13. Determine C_p and V_p for problem 24–7.

24–14. Find the base width and side slopes for Fig. 24–1e and compute the section area by three methods.

24–15. Rewrite the field notes and compute section area by matrix for Fig. 24–1c using the same cuts but a base of 36 ft, s.s. = 1¼:1.

24–16. Calculate V_e and V_p between the sections shown in Figs. 24–1e and 24–1f for a length of 75 ft.

24–17. Complete the notes and calculate V_e and V_p for sections similar to Fig. 24–3 using a base = 40 ft and side slopes of 1 to 1.

Station 53 + 00	$\frac{C\,5.6}{\quad}$	C 4.2	$\frac{C\,4.8}{\quad}$
Station 52 + 00	$\frac{C\,3.0}{\quad}$	C 6.0	$\frac{C\,4.4}{\quad}$

24–18. Similar to problem 24–17 except the base = 32 ft, s.s. = 1½:1.

24–19. Compute the volume in problem 5–41 using triangular prisms.

24–20. Compute the volume in problem 5–42 using triangular prisms.

Find V_e, C_p and V_p for slope stake notes of problems 24–21 and 24–22.

24–21. Base in fill = 16 ft
Base in cut = 24 ft
Side slopes 1½ to 1

Station				
6 + 40	$\frac{F\ 2.0}{11.0}$	$\frac{0.0}{3.0}$	C 1.4	$\frac{C\ 2.0}{15.0}$
6 + 10	$\frac{F\ 4.0}{14.0}$		0.0	$\frac{C\ 1.6}{14.4}$

24–22. Determine width of base in cut and fill, and side slopes as well as V_p.

Station				
9 + 90	$\frac{0.0}{8.0}$		F 4.2	$\frac{F\ 6.2}{17.3}$
9 + 70	$\frac{C\ 0.4}{12.6}$	$\frac{0.0}{6.0}$	F 3.0	$\frac{F\ 3.0}{F\ 12.5}$
9 + 60	$\frac{C\ 1.6}{14.4}$		0.0	$\frac{F\ 4.0}{14.0}$

For problems 24–23 and 24–24 compute the capacity of a reservoir between highest and lowest contours using planimetered areas taken from a topographic map.

24–23.

Elevation	200	205	210	215	220	225
Area in sq ft	620	854	1016	1195	1423	1682

24–24.

Elevation	860	880	900	920	940
Area in sq ft	154	376	588	732	1015

24–25. Calibration of a polar planimeter using a 4-in. diameter circle gives an average reading of 1.382 revolutions of the roller. Ten-ft contours planimetered on a reservoir site map to a scale of 1 in. = 400 ft give the results shown. Compute the reservoir volume in acre-ft.

Contour	640	650	660	670	680	690
Planimeter reading	0.000	0.816	1.128	2.860	3.721	5.696

24–26. Why is it often easier to compute prismoidal volume by means of V_e—C_p instead of calculating V_p directly?

24–27. List several things that can be learned from a mass diagram.

chapter 25

Industrial Applications

25–1. General. The increase in the size of manufactured items, and the decrease in tolerance limits in both individual parts and final assembly in mass production, necessitate precise measurements. The toolmaker's old stand-bys—taut wires and hanging plumb bobs, trammels and steel scales, insulated micrometer rods, gages, surface plates, and jo blocks—are no longer good enough. A new and rapidly developing field, *optical alignment,* is providing the equipment capable of meeting the close tolerances specified for large mock-ups, jigs, and end products (actually optics were used around 1925 to align machine tools).

Shop production operations require answers to four basic questions: (1) Is it straight? (2) Is it flat? (3) Is it plumb? (4) Is it square? Exact alignment, precise linear and angular measurements, and perfect level and plumbness for vertical control can be obtained with a properly adjusted surveyor's level and transit. When measurements to thousandths of a foot were sufficient for shop practice, standard surveying equipment could meet the requirements. Today, however, tolerances of only a few thousandths of an inch and seconds of arc have resulted in refinements of even the better surveying instruments.

In fact, efforts are being continued to split an inch into 10,000,000 parts, because industry says that this refinement is necessary if technology is to move ahead. Standard gage blocks are not sufficiently accurate to meet tolerance demands in precision bearings and elsewhere. The gage block must be ten times as precise as the tolerance being measured. The inch has been split into one million parts for over twenty years by scientists using an interferometer, but industry did not utilize this accomplishment until about the mid-1950's. By contrast, a difference of less than 0.0005 in. in a pair of blocks is generally not discernible by human touch, and the thickness of the average human hair is approximately 0.003 in.

The fundamental theory of measurements discussed in previous chapters applies directly to optical alignment in shop work. Optics provide a line of sight that is absolutely straight, has no weight, and serves as a perfect base from which to make accurate measurements.

Layout in a shop requires the establishment of (a) a reference line, (b) lines parallel with or perpendicular to the reference line, and (c) consistent elevations over large surfaces. These are the equivalent of line and grade in construction surveying.

25–2. Basic Equipment. Four basic instruments are used in industrial applications of surveying methods:

a. Alignment telescope, or jig alignment telescope.

b. Jig transit.

c. Precise level.

d. Alignment laser.

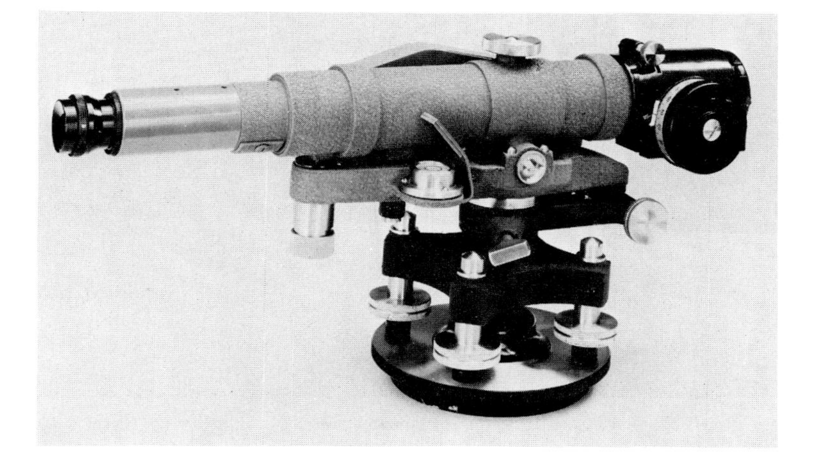

FIG. 25–1. Alignment telescope. (Courtesy of Keuffel and Esser Company.)

The alignment telescope, Fig. 25–1,[1] provides permanent reference lines on a jig or other structure. The jig transit, Fig. 25–7, establishes

[1] The illustrations in this chapter, and some of the text material, are from *Optical Alignment Equipment,* published by Keuffel and Esser Company, and are reprinted here by permission.

an absolutely vertical plane exactly where desired—on line with two marks, or precisely at right angles to any other line of sight. The precise (tilting) level, Fig. 5–10 and Fig 5–11, fixes a true horizontal plane at any desired height. All three instruments are self-checking, can be tested quickly, and adjusted exactly.

Two *optical micrometers*, built into the alignment telescope as indicated in Figs. 25–1 and 25–2, permit the line of sight to be moved

Fig. 25-2. Optical alignment of compressors. (Courtesy of Keuffel and Esser Company.)

parallel to itself sideways or vertically. The jig transit and precise level also can be equipped with an optical micrometer. The motions are controlled by micrometer knobs that show the extent of the movement in thousandths of an inch.

Measurements from the line of sight are made with a precision *optical alignment scale*, one type of which is represented in Fig. 25–12. This scale provides a definite target at every 0.1 in. The remaining decimal part is measured with the optical micrometer to 0.001 in.

25-3. Sighting telescope. The fundamental element of the alignment telescope, jig transit, and tilting level is the sighting telescope. It must be skillfully designed and perfectly manufactured to combine proper resolving power, definition, magnification, eye distance, size of pupil, and field of view. Some of these features were discussed in Sec. 5-8.

The telescope on a jig transit and on a tilting level is mounted in the same manner as that on standard surveying equipment. The alignment telescope usually is supported by the jig itself, on an instrument stand as in Fig. 25-2, or on a tooling bar.

25-4. Alignment telescope. The alignment telescope, Fig. 25-1, can be focused for any distance from practically zero (the point sighted being actually in contact with the front end of the telescope), to infinity. It can be mounted directly within a jig frame, since it requires a minimum of space. The telescope shown has a field of view of approximately 37 min at infinity focus; a resolving power of 3.4 sec; a magnification varying automatically from about 4 power at zero focus to 46 power at infinity focus; an open cross-line pattern for the reticle; and a 90° prismatic-eyepiece attachment that can be rotated through 360°. It is $17\frac{3}{4}$ in. long.

The support provided is either a permanently attached sphere or a spherical adapter which slides over the telescope tube and can be removed to permit use of the telescope in special fixtures. Typical supports are shown in Fig. 25-3. The adapter gives greater flexibility in positioning the telescope to clear jig parts. Also used is a tooling bar consisting of a track and a carriage for the optical instrument, with provision for accurately positioning it.

25-5. Use of alignment telescope. The alignment telescope provides a long, absolutely straight, permanent optical reference line for a jig or other structure on which measurements are to be made. A bracket to support a socket for mounting the telescope or a target is built at each end of the jig, as indicated in Fig. 25-3. The telescope, mounted in a sphere, is held in one socket. A target, centered in another sphere, is held by the second socket. A special device with adjusting screws, called an aligning bracket, is supplied for aiming the

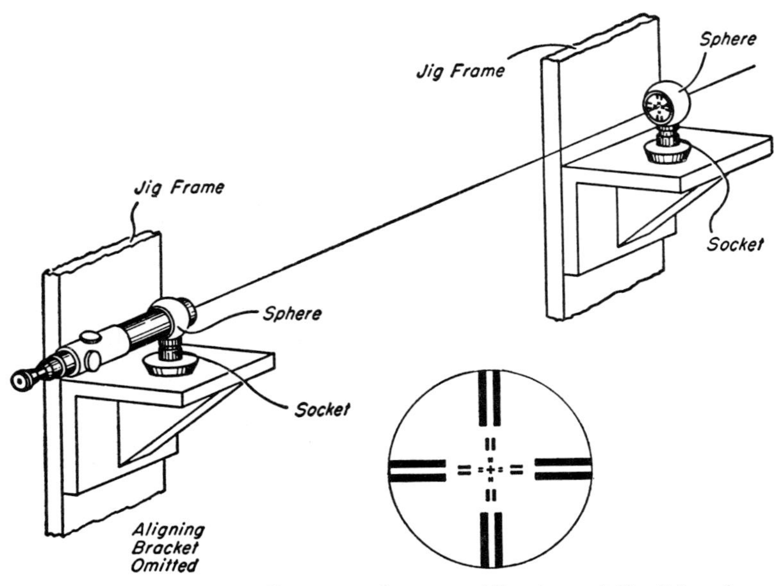

FIG. 25-3. Use of alignment telescope. (Courtesy of Keuffel and Esser Company.)

telescope. The optical reference line is then established by setting the cross lines of the telescope on the target. This line can be recovered readily as long as the sockets remain on the jig.

The sockets have adjustable mounts by which the line of sight is positioned originally. Accurate setting is necessary only when more than one line of sight is to be used, or when the line of sight must be level.

Usually it is convenient to have the line of sight exactly level. To accomplish this, a sphere with a target is placed in each socket and one of the sockets is adjusted to place the two targets at exactly the same elevation determined by means of a tilting level. One target is then replaced by the alignment telescope, which is tilted until the other target appears to coincide with the cross lines when the micrometers are set at zero. The line of sight generally is located to establish either the center line or a line parallel with it.

The line of sight is picked up by targets wherever needed. A target consists of a glass circle mounted in a steel ring ground to a specified diameter. Etched on the glass is a design which marks the center of the ring and which may include scales giving distances from the center. Usually the target is illuminated from behind. Station positions—distances parallel to the line of sight—are measured, with an inside micrometer and measuring rods, from a reference button on the socket support to a reference button on the piece to be positioned. A precise tape is used for longer distances.

25-6. Positioning a part. Parts may be positioned by (a) two targets on line, (b) autoreflection, and (c) measuring from the line of sight.

Two targets on line. When a small jig part is to be positioned on or near the line of sight, as indicated in Fig. 25-4, it is built with a bracket to hold a square-setting aligning tube containing a target in each end; a bracket to hold a shop level perpendicular to the line of sight; and a button to give the correct station. The part is positioned by adjusting it until the two targets are on line, the bubble of the shop level centers when placed on the bracket, and the button is at the correct station.

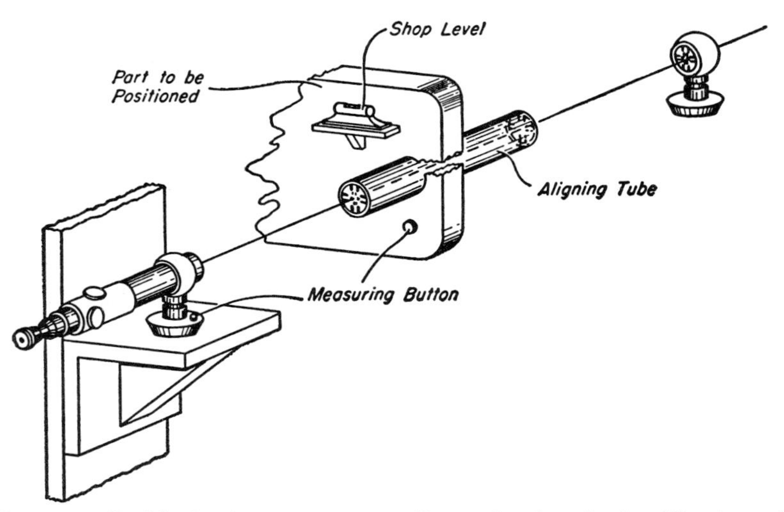

Fig. 25-4. Positioning by two targets on line and a shop level. (Courtesy of Keuffel and Esser Company.)

Autoreflection. When greater accuracy is necessary, an optically flat mirror target is mounted on the part to be positioned so that its reflecting surface is parallel to the proper reference plane on the part and it is also in the line of sight of the alignment telescope. The part is positioned, as indicated in Fig. 25–5, by placing the mirror target on line; setting the button at the proper station; and turning and tilting the part until the cross lines in the alignment telescope appear to coincide with the reflection of a target mounted on the end of the telescope, this reflection being produced by the mirror target.

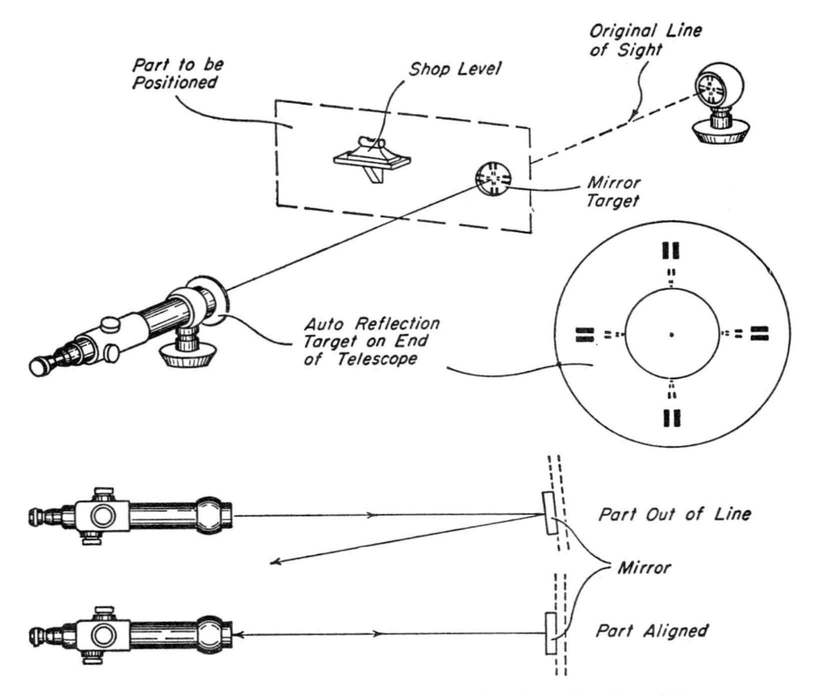

FIG. 25–5. Positioning by autoreflection and a shop level. (Courtesy of Keuffel and Esser Company.)

Measuring from the line of sight. When the optical-micrometer knobs of the alignment telescope are turned, the line of sight is moved parallel to itself by the number of thousandths of an inch shown on the graduated drums. Optical micrometers are used to determine any error in positioning the target, and thus to check the tolerances. Tolerance stops are provided for repeated checking of the same tolerance. Under most circumstances, however, measurements of distances greater than the range covered by the micrometers should be made from the line of sight with a precision scale, a special vernier caliper, or a height gage that has a target attached to the movable section.

Possible arrangements are illustrated in Fig. 25–6. Sometimes a target is placed on the line of sight and measurements are made with an inside micrometer and measuring rods.

A special device, Fig. 25–6, may be built with a target at one end; a measuring button on the other end at a predetermined distance from the target center; and a bracket to hold a shop level. The bracket holding the level can be placed to accommodate measurements vertically, horizontally, or at any desired angle in a plane perpendicular to the line of sight.

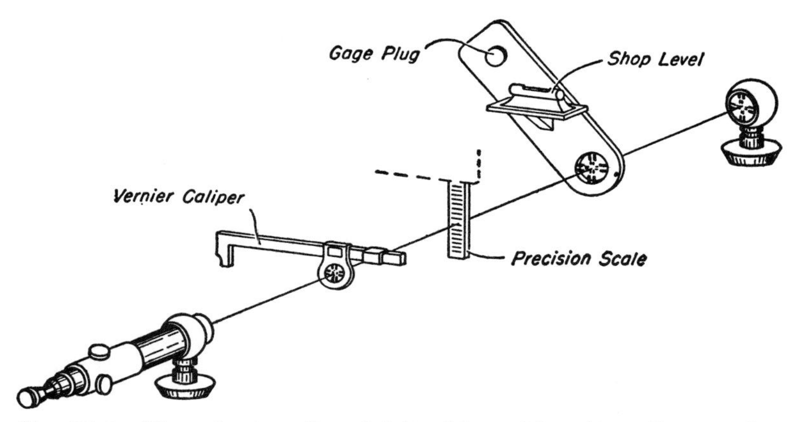

Fig. 25–6. Measuring from line of sight with special vernier caliper, precise scale, and special fixture. (Courtesy of Keuffel and Esser Company.)

25–7. The jig transit. The jig transit, Fig. 25–7, is designed especially for optical alignment. Its basic operation and functions are like those of an engineer's transit, but the instrument does not have a circle or arc for measuring angles. Particular features essential to optical tooling are listed.

Telescope level. The telescope level is as sensitive (20 sec per 2-mm division) as that on a standard transit.

Special telescope. The telescope has a magnification varying automatically from 20x at 8 in. to 30x at infinity, a resolving power of 4 sec, and an erecting eyepiece.

Adjustable line of sight. The vertical plane of the line of sight can be adjusted laterally to make it pass through the vertical axis. It is then possible to reverse the instrument without shifting the vertical plane of the line of sight horizontally.

Right-angle eyepiece. This facilitates sighting to high points. Also, it may be rotated 360° to permit use of the instrument near an obstruction.

Additional attachments that are usually furnished on a jig transit include the following:

1) An optical micrometer, with counterweight, to reduce the time of setup when *bucking-in,* and to measure the actual position of any

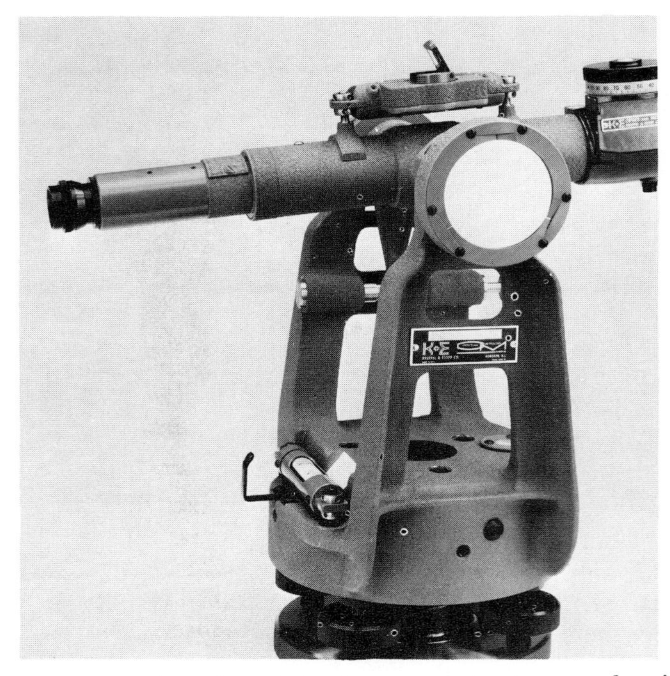

FIG. 25–7. Jig transit with optical micrometer and telescope-axle mirror. (Courtesy of Keuffel and Esser Company.)

part of the jig, both vertically and horizontally, in any plane perpendicular to the line of sight.

2) Striding level and collars.

3) Finder sights on the striding level to assist in positioning the jig transit at right angles to the line of sight of the alignment telescope.

4) A telescope-axle mirror, which has an optically flat front surface and can be attached on either end of the telescope axle so as to be parallel to the line of sight. With this device the plane of the line of

sight of the jig transit can be set perpendicular to any optical line by means of autoreflection, as indicated in Fig. 25–8.

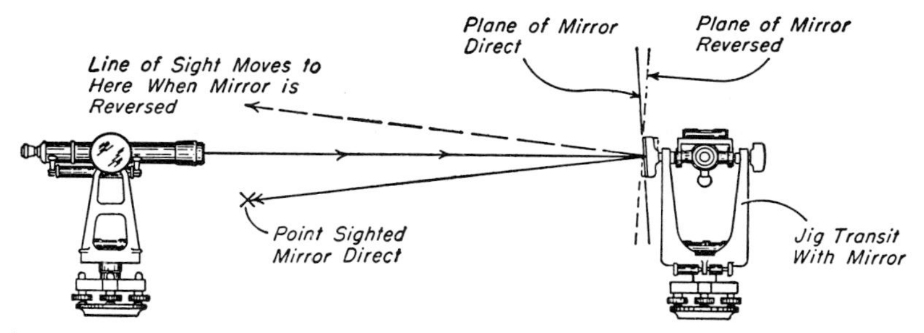

FIG. 25–8. Setting a jig transit at right angles to a sight line. (Courtesy of Keuffel and Esser Company.)

25–8. Use of the jig transit. The jig transit is designed to establish a precise vertical plane wherever desired, and to serve as a level when ordinary accuracy is sufficient. It can be used for a variety of purposes, both in original layout and in precise control.

The jig transit should be mounted on a mechanical lateral adjuster supported on a rigid stand, or on an adjustable mount. An optical micrometer greatly facilitates operation of the instrument. When the micrometer is not used, the sunshade should be placed on the telescope to reduce the unfocused light that tends to dim the image.

It is essential that the jig transit be set up with the telescope axle exactly horizontal and with the line of sight in the desired plane. When in this position, the line of sight will sweep the vertical plane required.

Usually the desired vertical plane is marked by two scribed points. The transit is placed in line with them by a process called bucking-in, the procedure being illustrated in Fig. 25–9. After the instrument is set up in line as judged by eye, the procedure is as follows:

a. Loosen a pair of adjacent leveling screws and rotate the leveling head until a pair of opposite screws is in line with the vertical plane to be established.
b. Center the two plate bubbles.
c. Aim at the far point.
d. Aim at the near point without turning the instrument about its vertical axis, and note where the line of sight falls.

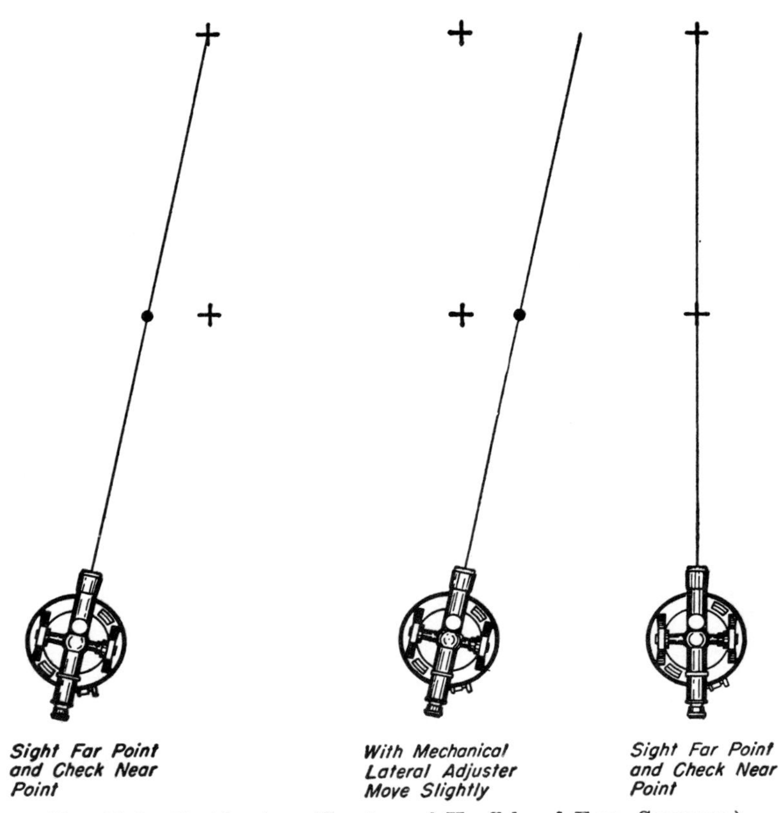

FIG. 25-9. Bucking-in. (Courtesy of Keuffel and Esser Company.)

e. Move the instrument with the mechanical lateral adjuster until the line of sight has moved in the direction of, and slightly beyond, the near point.
f. Center the plate bubbles.
g. Aim at the far point.
h. Check the near point.
i. Continue until the line of sight checks on both points. Level the instrument each time it is moved.

The striding level and optical micrometer also can be used for bucking-in, but the methods will not be described here.

25-9. The tilting level. The tilting level was described in Chapter 5. Essentially the same type is used in shop work for various purposes, including the determination of elevations at grid corners established on large plates and objects to check their flatness.

25–10. Alignment laser. The laser alignment system, Fig. 25–10, provides a *continuous visible line of sight* which permits a higher degree of accuracy than other optical alignment systems—about

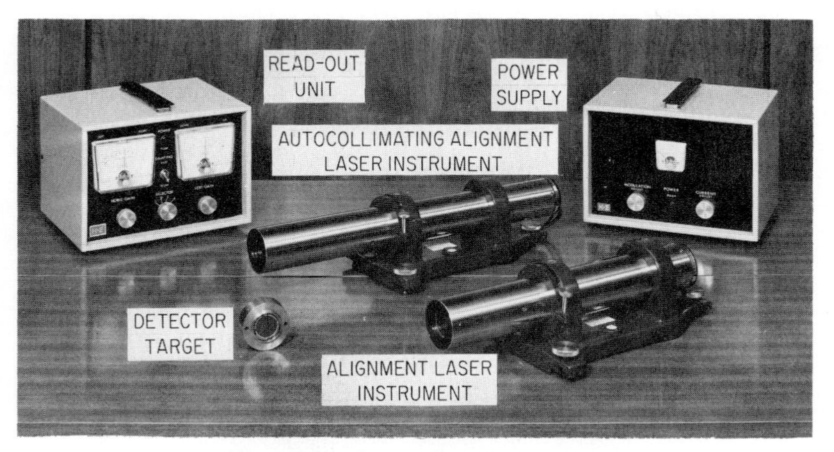

Fig. 25–10. Laser alignment system.
(Courtesy of Keuffel and Esser Company.)

0.001 inch up to 150 ft, and 0.002 inches from 150 to 300 ft, in lateral displacement. Human error is also reduced by the visible direct-reading meters.

Displacement is measured with compact photo-electric detector targets connected by wires to the read-out unit. The position of the laser beam falling on the target is registered on two large meters on the front panel of the read-out unit. The left-hand meter shows horizontal displacement of the beam from the target's center directly in thousandths of an inch. The right-hand meter gives vertical displacement to the same accuracy and range of ± 0.025 inch. One standard instrument with accessories can be used to determine straightness, flatness, levelness, and squareness.

The laser tubes are low power, $\frac{1}{4}$ to $\frac{1}{2}$ milliwatt, helium-neon (red color) continuous lasers and can be replaced in the user's plant.

25–11. Optical micrometers. Optical micrometers are manufactured as attachments for levels and transits to be used for precise leveling and alignment in shop work. Measurements to 0.001 in. can be made with them.

An optical micrometer consists of a disk of optical glass with flat parallel faces; such a disk is called a *planoparallel plate.* It is designed to permit precise tilting by moving a graduated drum, located as shown in Fig. 25–11. The device is mounted on the instrument in place of the sunshade, with the plate in front of the objective lens. The lower right-hand screw clamps the micrometer in position. A knob and a drum control the planoparallel plate. Figure 25–11 shows the line of sight moved 0.081 in. to the right as read on the drum.

FIG. 25–11. Optical micrometer. (Courtesy of Keuffel and Esser Company.)

25–12. Use of the optical micrometer. Figure 25–12 illustrates schematically the use of an optical micrometer for measuring a vertical distance. The sights are taken on an accurately graduated steel scale divided into tenths of an inch. Figure 25–12a shows the line of sight falling between 2.6 and 2.7 in. for the zero position of the

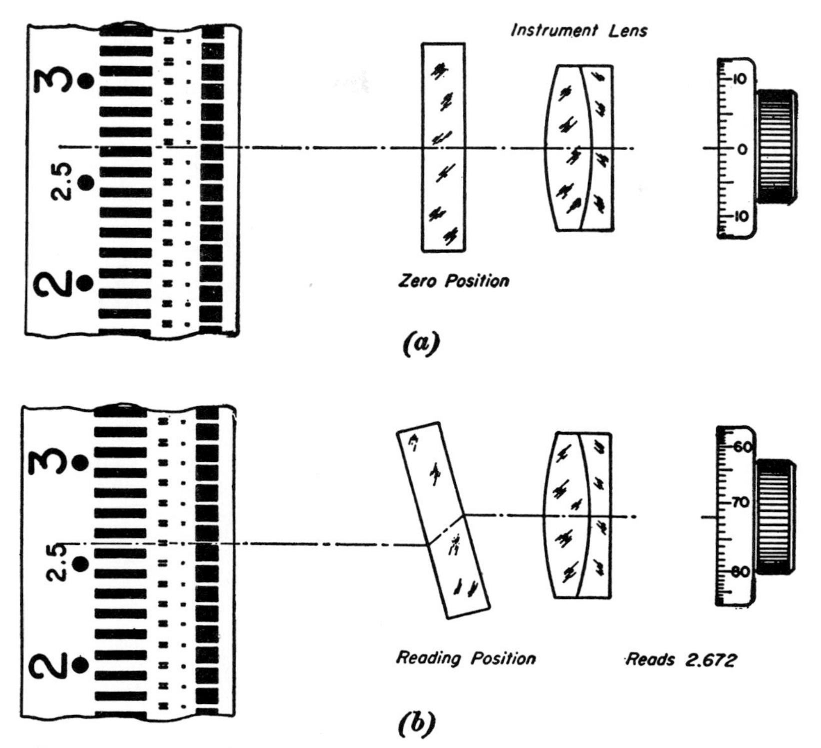

FIG. 25–12. The principle of the optical micrometer. (Courtesy of Keuffel and Esser Company.)

micrometer. The drum is graduated in both directions from 0 to 100. To avoid turning in the wrong direction, the drum is first set to zero and then rotated to make the line of sight move toward the graduation with the lesser value on the steel scale (2.6 in. in Fig. 25–12b). The drum always records the movement of the line of sight from its zero position. Figure 25–12b shows a movement of 72.0 thousandths, thus making the reading $2.6 + 0.072 = 2.672$ in.

25–13. Other equipment. Additional items of equipment used in optical alignment include the following:

Optical alignment scales. The design of one type of scale, based upon extensive tests, is shown in Fig. 25–12. Each 0.1-in. graduation and separation is positioned correctly to within ±0.001 in. at 68 F. The design is based upon the principle that a cross line can be set most accurately by centering it between two black lines on a white surface, provided that the white areas between the lines and the cross line are the optimum width. Accordingly the spacing is selected to serve over a certain range of distances for the width of cross lines in a particular telescope. Other types of scales are shown in Fig. 25–13.

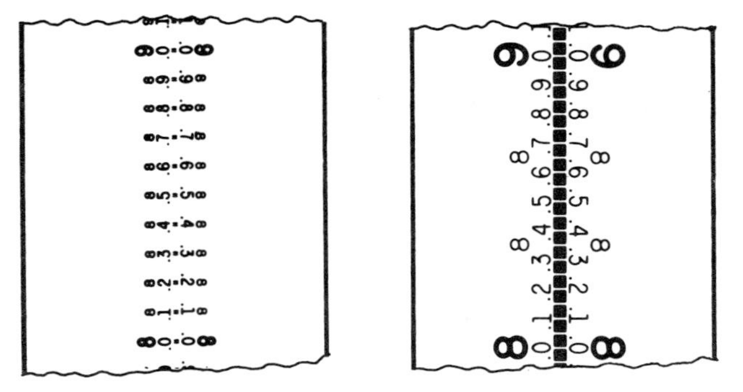

FIG. 25–13. Optical alignment scales. (Courtesy of Keuffel and Esser Company.)

Optical alignment targets. A typical arrangement of graduations on optical alignment targets is shown in Fig. 25–3.

Spherical adapters. A fixed or movable sphere is useful in positioning alignment telescopes and targets. Figure 25–3 illustrates one type of arrangement of spherical adapters.

Trivet. A trivet is a special device for supporting a transit or level near the floor when the line of sight must be low. Steel-pointed shoes on the three short legs can be removed to allow the trivet to be bolted directly to a jig or other fixture.

Autoreflection angle mirror. An autoreflection angle mirror reflects the line of sight of a telescope equipped with an autoreflection target at any appropriate angle up to about 160°. The device consists of a mirror with an optically flat (to within one-fourth the wavelength of light) front surface mounted vertically on the base of an engineer's transit. The mirror is adjusted to make the plane of its surface pass exactly through the vertical axis of the transit. It can therefore be turned through any horizontal angle like a transit.

Rapid developments in the field of optical tooling have produced many instruments which cannot be described in this text. Manufacturers' catalogues and other technical publications are sources of information on the design and use of such equipment.

25–14. Practical applications. A few present applications of optical alignment, which will suggest other possibilities, include:

1) Ship building—aligning shaft bearings, keel, and propulsion shaft, establishing witness lines for boring operations; and controlling machining of steam catapults.

2) Paper industry—leveling sole plates for large machines; laying crosslines for column erection on the sole plates; and aligning and leveling rolls.

3) Large lathes—aligning and leveling the lathe, such as a gun-boring lathe with a 10-ft face plate and a bed 200 ft long.

4) Compressors—aligning bearings, shafts, and other parts (Fig. 25–2).

5) Aircraft industry—layout of jigs for airframes and other parts.

6) Missiles—laying out launching sites for intercontinental ballistic missiles, such as at the Titan underground site where guide rails over 125 ft long had to be accurate in position to within 0.005 in. at one end and true over the entire length to $\frac{1}{64}$ in.

25–15. Sources of error in optical alignment. Some sources of error in optical tooling are:

a. Using poorly graduated scales.
b. Bubble not exactly centered at the time of sighting.
c. Targets not perfectly placed.

25–16. Mistakes. Typical mistakes in using optical alignment equipment include the following:

a. Movement of targets before sights have been completed.
b. Line of sight incorrectly relocated.
c. Micrometer read in wrong direction.

PROBLEMS

25–1. What are the advantages and disadvantages of using a jig transit or an alignment telescope instead of an engineer's transit to align equipment shafts in a shop?

25–2. By means of a sketch, show how the jig transit and tilting level can be set up outside a work area to align and position parts readily without interfering with the erection personnel.

25–3. Check the flatness, straightness, and parallel alignment of the planers in the school shop, and prepare a suitable set of notes.

25–4. Sketch and outline a procedure for checking the straightness of a machine part 10 ft long by using three optical alignment scales and a jig transit. (NOTE: First align two of the scales with the transit.)

25–5. By sketch and tabulation, show a method of using the precise level to check the flatness of a plate 10 ft by 50 ft.

25–6. Check the four corners of a bedplate and the bedways on a machine for level.

25–7. Sketch the position of optical alignment equipment which can be employed to check the main shafts of electric generators for levelness.

25–8. Sketch and outline a procedure for checking the 90° angle between a vertical column face and the run of the bedways on a large machine by using a jig transit and a precise level.

25–9. Check a grinder in the shop for level of bed, horizontal travel of tool guide bar, and vertical straightness of the machine faces.

25–10. Sketch and outline a method of placing guided-missile launching tracks within 0.001 inch in alignment and grade over a 2000-ft length.

25–11. Explain the purpose and principles embodied in the design of the optical alignment scale shown in Fig. 25–12.

25–12. What is meant by collimation?

25–13. List three advantages of laser alignment over any other type of alignment system.

25–14. List the main differences between the jig transit and engineer's transit.

25–15. Give three or more types of operating equipment not mentioned in this chapter which require precise alignment during installation.

25–16. Describe three situations where offset sight lines from true center lines may be needed, and a means of establishing them.

25–17. How would you determine whether an aluminum sheet 10 ft by 10 ft is within thickness tolerances of ± 0.003 inches?

appendix A

SUGGESTED ORDER OF FIELD ASSIGNMENTS

(Based on Three-Hour Field Periods and Three-Man Parties)

Period	Problem
1.	Measuring distances with a steel tape. Referencing hubs.
2.	Measuring distances with a steel tape. Pacing.
3.	Differential leveling between two bench marks.
4.	Differential leveling. Reciprocal leveling.
5.	Profile levels.
6.	Closing the horizon. Measurement and layout of angles with a tape.
7.	Compass and pacing survey of a five-sided area.
8.	Double direct angles, and bearings, of a closed traverse.
9.	Prolonging a line by double centering. Prolonging a line past an obstacle.
10.	Double deflection angles of a closed traverse (same traverse as in Period 8).
11.	Determination of stadia interval factor.
12.	Azimuth stadia traverse.
13, 14.	Topographic details by transit-stadia (planimetric).
15.	Trigonometric leveling.
16.	Staking out a building.
17.	Layout and leveling of shop equipment.
18.	Borrow-pit leveling.
19.	Layout of simple circular curve.
20.	Contours by transit-stadia.
21.	Topographic details by plane table.
22.	Topographic details and three-point location by plane table.
23.	Property survey.
24.	Observation on the sun for azimuth. Observation on Polaris for azimuth.
25.	Field test on use of equipment.
26, 27.	Mapping.

NOTES

a) This list of field problems covers sufficient material for a two-quarter or one-semester program. Assignments 1 through 18 are adequate for the typical first course in surveying given to civil and non-civil engineering students. A few deletions may be desirable to fit the particular group involved.

b) The suggested order permits a quick start on drafting-room computations and mapping if inclement weather is experienced early in the school term.

c) Many of the assignments can be carried out on the traverse used during the first field period and thus provide a sustained project rather than a series of unrelated problems.

d) Pacing, and closing the horizon, are done individually while the other two men in the party are taping.

e) The compass and pacing survey, and the layout of shop equipment, are appropriate for certain students.

STATE PLANE COORDINATES

PAUL R. WOLF
Associate Professor of Civil Engineering
University of Wisconsin, Madison

A–1. Introduction.[1] Most surveys of small areas are based on the assumption that the earth's surface is a plane. For large-area surveys, however, it is necessary to consider the earth's curvature. In the past, horizontal positions of widely spaced geodetic stations were listed in terms of the spherical coordinates, latitude, and longitude. Unfortunately practicing surveyors often were not familiar with this method of referencing points. Clearly a system of listing geodetic stations using plane rectangular coordinates in feet and decimals was required. The United States Coast and Geodetic Survey fulfilled this need by developing a State Plane Coordinate System for each state.

A State Plane Coordinate System provides a common datum for referencing horizontal control of all surveys in a large area in the same way that mean sea level furnishes a single datum for vertical control. It eliminates having individual surveys in an area based on different assumed coordinates which are not related to those employed in other work. At present, State Coordinates are widely used in photogrammetric mapping, highway construction projects, and property surveying. In some localities new subdivision layouts must include State Coordinates. More extensive use is likely in the future.

The earth's curved or mean sea level surface is closely approximated by a *spheroid* which is a mathematical surface of revolution. To convert spherical coordinates of a portion of the earth's surface to plane rectangular coordinates, it is necessary to mathematically project points from the spheroid to some imaginary *developable* surface—a solid surface which can be developed or "unrolled and laid out flat" onto a plane surface without distortion of shape or size. A rectangular grid can be superimposed on the developed plane surface and the positions of points in the plane specified with respect to X and Y grid axes.

[1] The suggestion for inclusion of this chapter, and the example problem, came from Professor Porter W. McDonnell, Jr., Pennsylvania State University, Mont Alto Campus.

Two basic projections are employed in State Coordinate systems, the *Lambert Conformal Conic Projection* and the *Transverse Mercator Projection*. The former utilizes an imaginary cone and the latter an imaginary cylinder as their developable surfaces, Figures A–1 (a) and (b) respectively. The cone and cylinder are secant

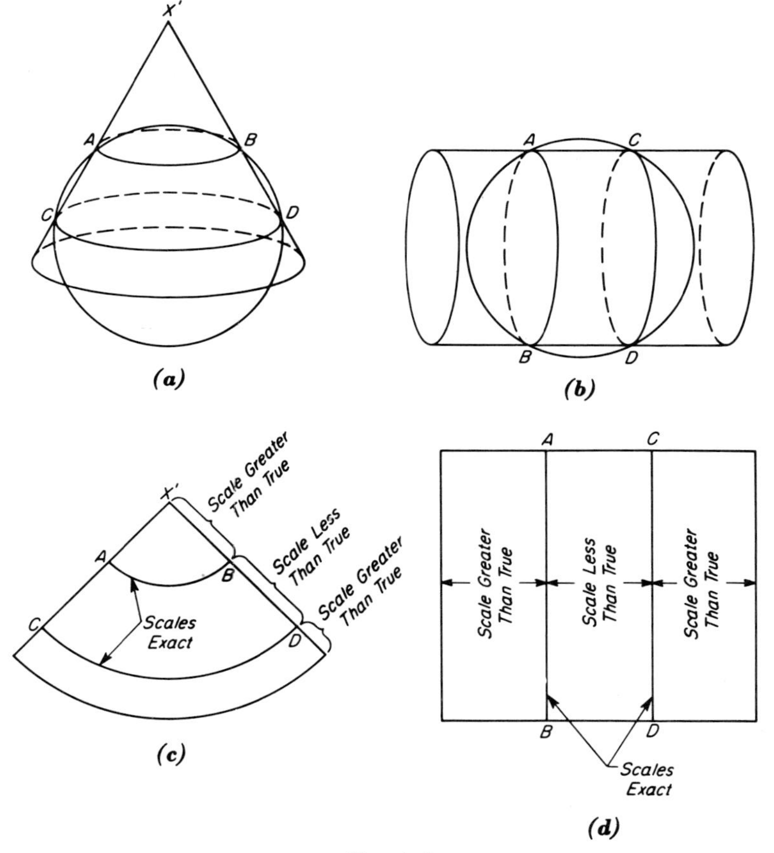

Fig. A–1

to the spheroid in the State Coordinate systems, i.e., they intersect the spheroid in two places along lines *AB* and *CD* as shown. Figures A–1 (c) and (d) illustrate plane surfaces developed from the cone and cylinder respectively.

In computing projections, points are projected mathematically from the spheroid along radial lines from some point near the earth's center to the surface of the imaginary cone or cylinder. Figure A–2 illustrates this projection of points diagrammatically and also

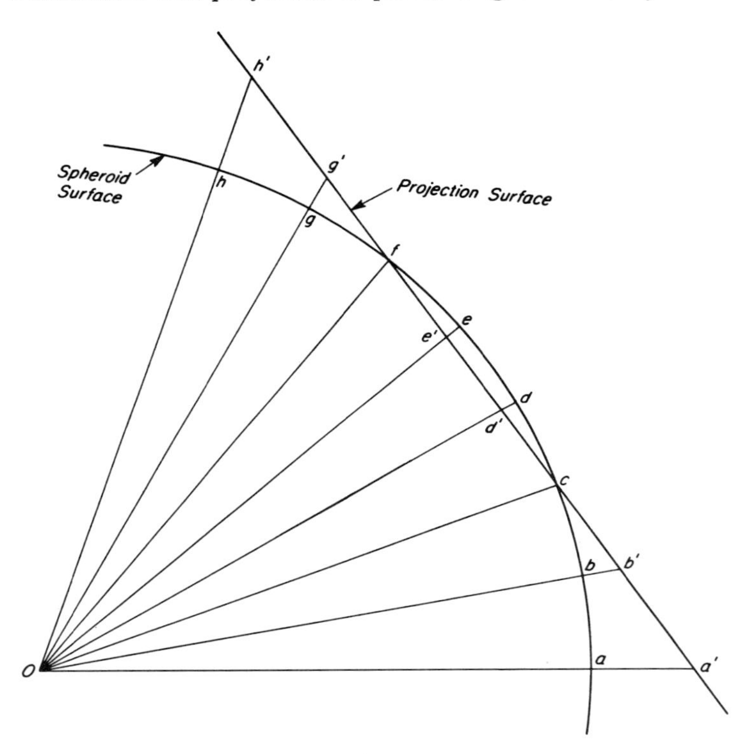

FIG. A–2. Method of projection.

displays the relationship between the length of a line on the spheroid and the length of that line when projected onto the surface of either the cone or cylinder. Note that the distance $a'b'$ on the projection is greater than ab on the spheroid, and similarly $g'h'$ is longer than gh. From this observation it is clear that map projection scale is larger than the true spheroid scale where the cone or cylinder is outside the spheroid. Conversely, distance $d'e'$ on the projection is less than de on the spheroid and thus map scale is less than true spheroid scale when the projection surface is inside the spheroid. Points c and f occur at the intersection of projection and spheroid surfaces and therefore map scale equals true spheroid scale along the lines of intersection. These relationships of map scale to true spheroid scale for various positions on the two projections are indicated in Figs. A–1 (c) and (d).

It is impossible to project points from the spheroid to any developable surface without introducing distortions in the lengths of lines or in the shapes of areas. These distortions are held to a minimum, however, by selected placement of the cone or cylinder secant, and also by limiting the zone size or extent of coverage of the earth's surface for any one projection. If the width of zones is held to a maximum of 158 miles, distortions are kept to one part in 10,000 or less, the accuracy intended by the USC&GS in its development of the State Coordinate systems.

A-2. The Lambert Conformal Conic Projection. The Lambert Conformal Conic Projection, as its name implies, is a projection onto the surface of an imaginary cone. The term conformal means that true angular relationships are retained around all points. This projection is used in 31 of the 50 states. The scale on a Lambert projection varies from north to south as shown in Fig. A–1 (c) so the projection is ideal for mapping areas extending great distances in an east-west direction, for example Kentucky, Pennsylvania, and Tennessee.

In the Lambert projection the cone intersects the spheroid along two parallels of latitude, called *standard parallels,* at one-sixth of the zone width from the north and south zone limits. On the projection, Fig. A–3, all meridians are straight lines converging at X', the apex of the cone, and all parallels of latitude are arcs of concentric circles whose centers are at the apex of the cone. The projection is centered in the zone in an east-west direction by selecting a *central meridian* whose longitude is near mid zone. The direction of the central meridian on the projection establishes *grid north.* All lines parallel with the central meridian on the projection point in the direction of grid north. Except at the central meridian, directions of true north and grid north do not coincide.

Given the latitude and longitude of any point P, its X and Y State Coordinates in the Lambert projection are readily calculated. Consider the developed plane of the Lambert projection illustrated in Fig. A–3. Point X' is at the apex of the cone, point O is the origin of rectangular coordinates. Line $X'A$ is the central meridian of the projection. A constant C, usually 2,000,000, is adopted to offset

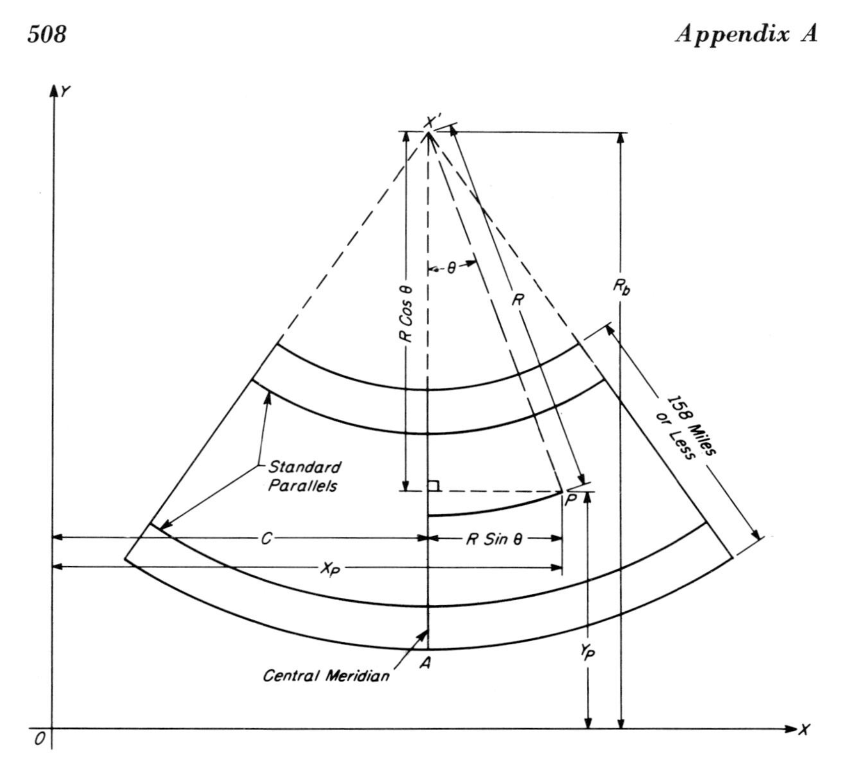

FIG. A–3. The Lambert Conformal Conic projection.

the central meridian from the Y grid axis so that the X coordinates of all points are positive. Line $X'P$ represents a portion of the meridian through point P with its length designated as R. The angle θ between the central meridian and the meridian $X'P$ is termed the *mapping angle.*

The USC&GS has computed and published projection tables for every state.[2] For any point P, the value of R is listed in the tables versus the latitude of P, and θ is recorded versus longitude P. A constant R_b is also given for any particular zone. From Fig. A–3, and appropriate projection tables, the following equations may be written for the X and Y coordinates of P:

$$X_p = R \sin \theta + C$$
$$Y_p = R_b - R \cos \theta$$

[2] Projection Tables for every state are available at a nominal fee from the Superintendent of Documents, Government Printing Office, Washington, D.C. 20402.

Note that if θ is to the left of the central meridian its sign is negative, if to the right its sign is positive. Except where a line of reference azimuth exceeds 5 miles in length, grid azimuth may be calculated with sufficient accuracy from geodetic azimuth using the following equation:

$$\text{Grid Azimuth} = \text{Geodetic Azimuth} - \theta$$

A–3. The Transverse Mercator Projection. The Transverse Mercator Projection is also a conformal projection and based upon an imaginary secant cylinder as its developable surface. Because its scale varies in an east-west direction, it is used to map areas of 22 states long in a north-south direction such as Illinois and Indiana.[3]

The axis of the imaginary cylinder of the Transverse Mercator projection lies in the plane of the earth's equator. The cylinder cuts the spheroid along two small circles equidistant from the central meridian. On the developed plane surface, Fig. A–4, all parallels of latitude and all meridians except the central meridian are curves as shown in light broken lines. A central meridian establishes the direction of grid north and the X and Y coordinates of points are measured perpendicular and parallel to the central meridian respectively.

Referring to Fig. A–4 and appropriate Transverse Mercator projection tables, the following equations may be written for the X and Y coordinates of any point P:

$$X'_p = H \times \Delta\lambda'' \pm ab$$
$$X_p = X'_p + K$$
$$Y_p = Y_o + V\left[\frac{\Delta\lambda''}{100}\right]^2 \pm c$$

In the equations, X'_p is the distance that point P is either east or west of the central meridian. The difference in seconds between the longitude of the central meridian and longitude of point P is $\Delta\lambda''$, its algebraic sign being negative if P is west, and positive if P is east of the central meridian. The constant K is 500,000 for most states. Values of H, a, Y_o and V are tabulated versus the latitude of point P in the projection tables, and b and c are listed versus $\Delta\lambda''$. If the sign of the product ab is negative it decreases

[3] Both the Lambert Conformal Conic and the Transverse Mercator projections are used in Alaska, Florida and New York.

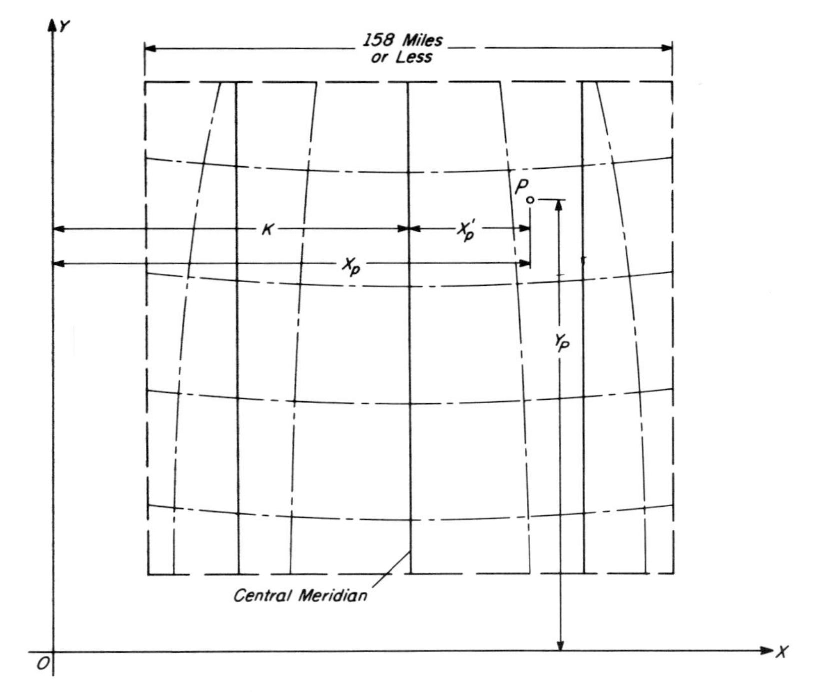

FIG. A–4. The Transverse Mercator projection.

numerically the product of $H \times \Delta\lambda''$, and if positive it increases $H \times \Delta\lambda''$.

Again, except where a line of reference azimuth exceeds 5 miles in length, grid azimuth may be calculated to sufficient accuracy from the geodetic azimuth using the following equation:

$$Grid\ Azimuth = Geodetic\ Azimuth - \Delta\alpha''$$

In the equation, $\Delta\alpha'' = \Delta\lambda''\ sin\ (Lat_p) + g$, where g is listed versus $\Delta\lambda''$ in the projection tables and Lat_p is the latitude of point P.

Ordinarily, State Coordinates and the grid azimuth to a nearby reference mark are published by the USC&GS for all geodetic stations of the U.S. network and therefore need not be calculated. If computations are necessary, it is a simple task when projection tables are used. Besides including the tabulated data described, the tables also provide detailed example problems. If work involving State Coordinates is anticipated, projection tables for your state and adjacent ones should be obtained since a survey may cover several zones (double sets of coordinates are tabulated near zone edges).

A–4. Computing State Coordinates of Traverse Stations. Placing a survey on the State Coordinate grid normally requires traversing (or triangulating or trilaterating) to start and end on existing stations which have known State Coordinates and from which known grid azimuth lines have been established. Generally these coordinates and azimuths are available for immediate use but if not they can be calculated as indicated above if latitude and longitude are known.

It is important to note that if a survey begins with a given grid azimuth and ties into another, all intermediate azimuths will automatically be grid azimuths. Thus azimuth corrections for convergence of meridians are not necessary when the State Coordinate system is used throughout the survey.

A simple traverse computation in the State Coordinate system is presented to illustrate its simplicity. The traverse consists of two sides, starts at Irwin Triangulation Station in Ohio, and ends at B.M. 1705. Station A in between is the only new point in the survey and its coordinates will be determined.

Measured lengths of the traverse lines must be reduced to distances at mean sea level and then to lengths at their positions on the State Coordinate grid. The following equation is used for reduction of measured lengths to mean sea level distances:

$$L_s = L_m \frac{R_e}{R_e = h}$$

where L_s is the sea level length of the line,
L_m is the measured length of the line,
R_e is the mean radius of the earth (approx. 20,906,000 ft),
h is the average elevation of the measured line above MSL

The ratio of $[R_e/(R_e + h)]$ is commonly called the *sea level factor*. For this example problem, average elevation = 687 ft, the factor is 0.999966.

The length of line at sea level is next modified by multiplying it by the *scale factor* obtained from the projection tables and corresponding to a particular area of the zone in which the line falls. The amount of this reduction or increase in sea level length varies from zero along the two lines of exact scale, to maximum and minimum values determined by the zone size. In Connecticut, for example, the correction is never more than one part in 40,000. In other states it may reach one part in 10,000, or slightly more in a few cases. For the illustrative problem, the average latitude of the traverse location obtained from published data on the two control stations is used as the argument to enter the table and find a scale factor of 0.999941.

The product of sea level factor and scale factor is commonly called the *grid factor*—in this problem 0.999966 times 0.999941 = 0.999907.

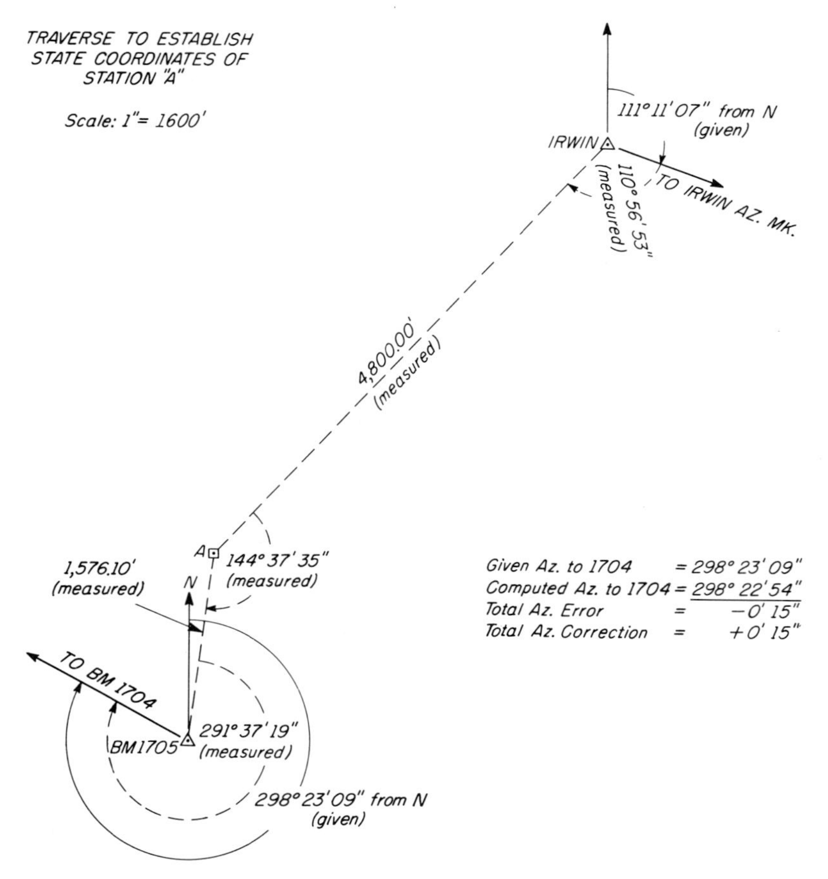

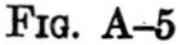
FIG. A-5

When reduction of numerous traverse lengths is imperative and elevations are uniform throughout the survey area so a single sea level factor can be used for the entire traverse, a common grid factor reduces the calculations. In many cases the grid factor is so close to 1.000000 it can be ignored, but should be checked if a measured ground distance disagrees slightly with the State Plane Coordinates.

Figure A–5 shows the given or fixed stations and azimuth lines for the example problem. Measured angles and courses were added with dashed lines. Published data for the two control stations is as follows:

Irwin Triangulation Station: E or x = 1,367,887.24
N or y = 442,126.54

Azimuth to Irwin Azimuth Mark = 111°11′07″ from grid North
Approximate elevation = 883 ft

B.M. 1705: E or x = 1,364,481.50
N or y = 437,001.53

Azimuth to B.M. 1704 = 298°23′09″ from grid north
Approximate elevation = 492 ft

TABLE A–1

Angular Closure and Adjustment

Station	From	To	Prelim. Azimuth / Angle to Right / Prelim. Azimuth	Corr. / Cum. Corr.	Final Azimuth / Angle to Right / Final Azimuth
	Irwin	Irwin Az.	111° 11′ 07″		111° 11′ 07″ Given
IRWIN	Irwin Az.	A	110° 56′ 53″	+5	110° 56′ 58″
	Irwin	A	222° 08′ 00″	+5	222° 08′05″
	A	Irwin	42° 08′ 00″		42° 08′ 05″
A	Irwin	1705	144° 37′ 35″	+5	144° 37′ 40″
	A	1705	186° 45′ 35″	+10	186° 45′ 45″
	1705	A	6° 45′ 35″		6° 45′ 45″
BM 1705	A	1704	291° 37′ 19″	+5	291° 37′ 24″
	1705	1704	298° 22′ 54″	+15	298° 23′ 09″ Given

TABLE A-2

Traverse Closure and Adjustment (Compass Rule)

Station	Grid Bearing	Measured Distance	Grid Factor	Grid Distance	Sine / Cosine	Departure	Latitude	Grid Coordinates: Prelim. X / Corr. / Final X	Grid Coordinates: Prelim. Y / Corr. / Final Y
Irwin							(Given)⟶	1,367,887.24	442,126.54
	S 42° 08′ 05″ W	4800.00	0.999907	4799.55	0.6708761	−3219.90		1,364,667.34	438,567.34
					0.7415694		−3559.20		
								−0.20	−0.62
A								1,364,667.14	438,566.72
	S 6° 45′ 45″ W	1576.10	0.999907	1575.95	0.1177540	− 185.57		1,364,481.77	437,002.36
					0.9930428		−1564.98		
								−0.27	−0.83
BM 1705							(Given)⟶	1,364,481.50	437,001.53
							Error of closure =	+0.27	+0.83

Linear error of closure $= \sqrt{(0.27)^2 + (0.83)^2} = 0.87$ ft.

Relative error of closure $= \dfrac{0.87}{6376} = \dfrac{1}{7330}$

Compass Rule:

$$X - \text{correction} = \frac{-0.27}{6376} = -0.042 \text{ ft. per 1,000 ft. of cumulative distance}$$

$$Y - \text{correction} = \frac{+0.83}{6376} = -0.130 \text{ ft. per 1,000 ft. of cumulative distance}$$

The traverse computation has been performed in 5 steps.
1) Distribution of the angular error of closure to get corrected grid azimuths for all traverse sides. Angles-to-the-right were measured. Closing error (difference between given azimuth at BM 1705 and the azimuth computed by using the 3 measured angles and given azimuth at Irwin) is – 0°00′15″. This error was distributed equally among the 3 angles as shown in Table A–1. Final grid azimuths shown in the last column are converted to grid bearings in Table A–2.
2) Reduction of measured distances to grid distances is done in Table A–2 by multiplying values in the third column by those in the fourth column, or by subtracting a correction of 0.093 ft per 100 ft (which is less than 1 part in 10,000).
3) Computation of latitudes and departures.
4) Computation of preliminary grid coordinates listed in the last column of Table A–2. Errors of closure in the x and y directions are found by subtracting the given, or fixed, coordinates of BM 1705 from those obtained by traversing. The relative error of closure is 1:7,330.
5) Traverse adjustment by the Compass Rule. Corrections for preliminary coordinates are computed in proportion to the accumulated traverse distances up to any given station by slide rule and applied in the last columns to get final State Plane Coordinates for station A. Full corrections (–0.27 and –0.83) applied at BM 1705, of course, give the fixed values written in previously.

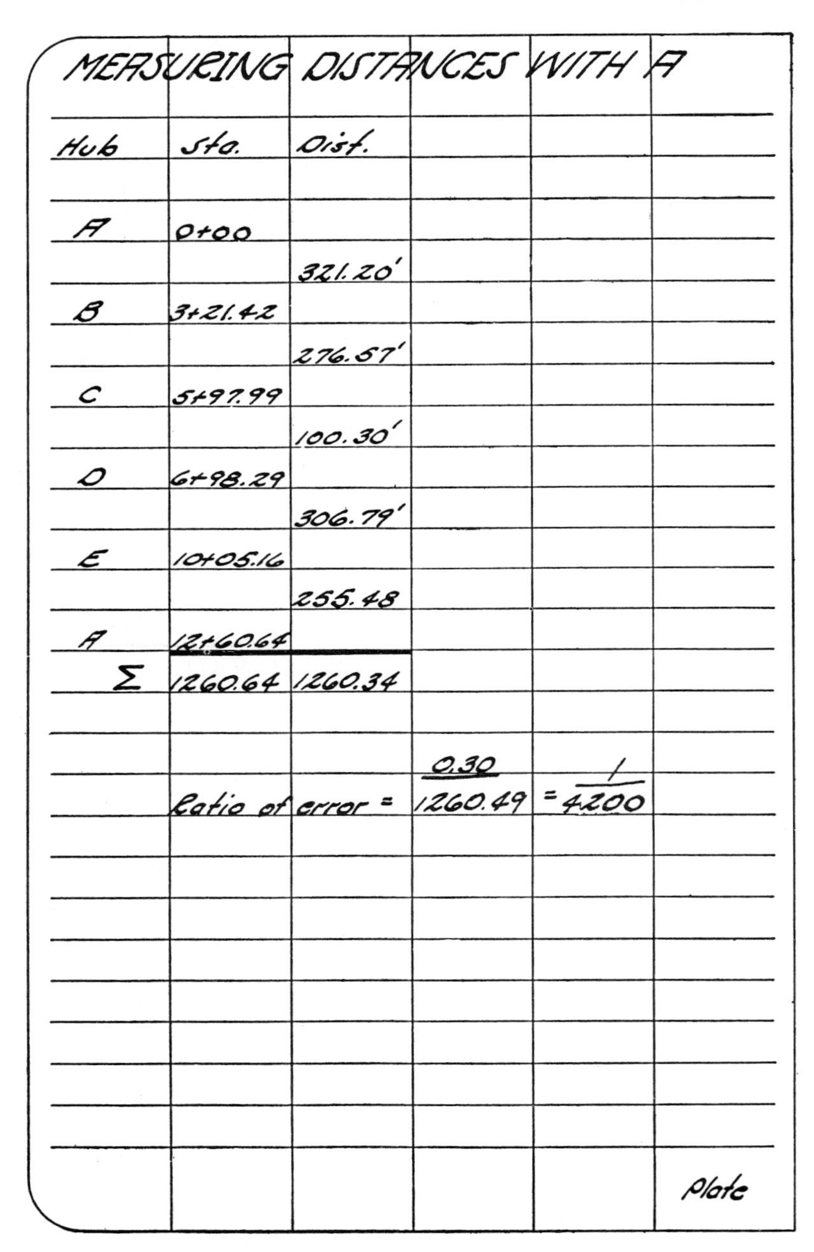

MEASURING DISTANCES WITH A

Hub	Sta.	Dist.			
A	0+00				
		321.20′			
B	3+21.42				
		276.57′			
C	5+97.99				
		100.30′			
D	6+98.29				
		306.79′			
E	10+05.16				
		255.48			
A	12+60.64				
Σ	1260.64	1260.34			
			0.30	1	
	Ratio of	error =	1260.49	= 4200	
					Plate

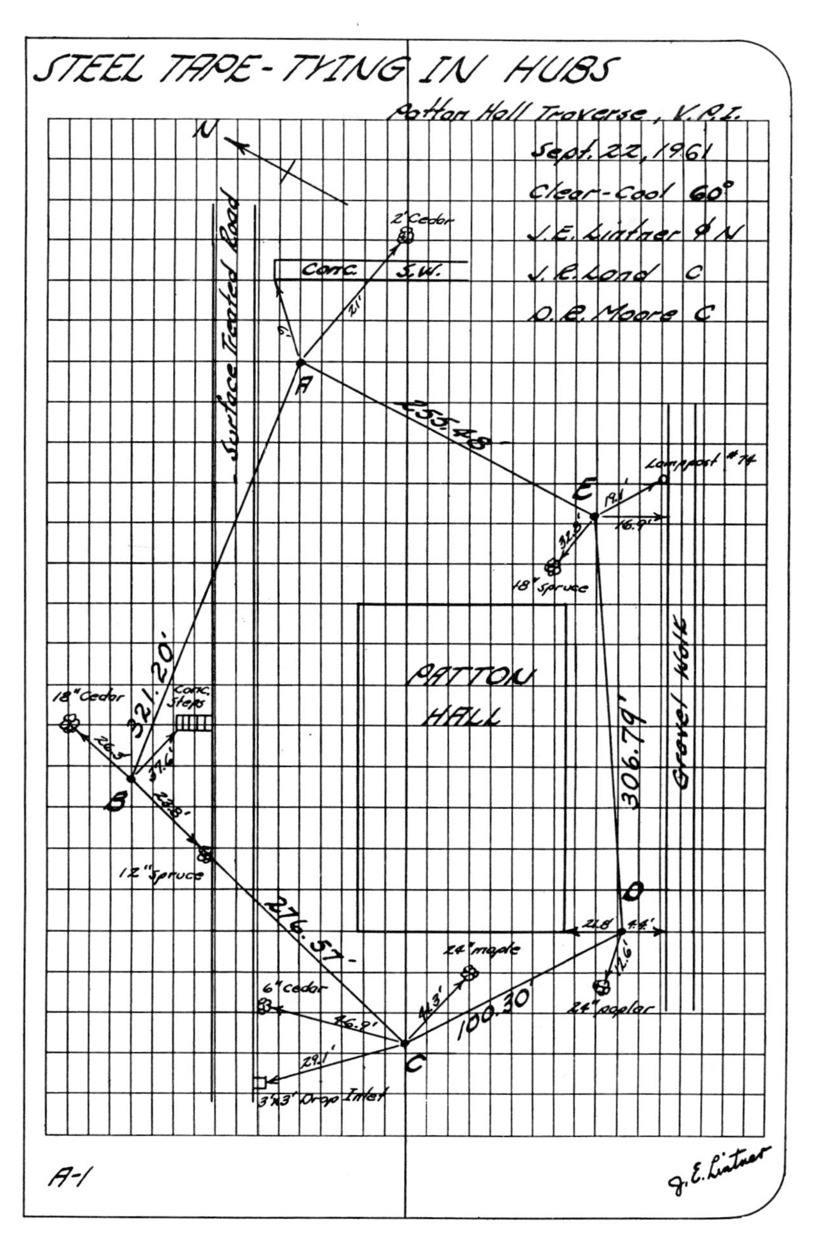
STEEL TAPE-TYING IN HUBS
Patton Hall Traverse, V.P.I.
Sept. 22, 1961
Clear-Cool 60°
J.E. Lintner
J.R. Lond C
D.R. Moore C
N
Surface Treated Road
2" Cedar
Conc. S.W.
21'
6'
A
255.48'
E
Lamppost #74
19.1'
16.9'
32.5'
18" Spruce
PATTON HALL
Gravel Walk
306.79'
321.20'
18" Cedar
Conc. Steps
26.3'
37.6'
B
23.8'
12" Spruce
276.57'
D
22.8'
4.4'
12.6'
24" maple
24" poplar
6" cedar
46.9'
14.3'
100.30'
29.1'
C
3'x3' Drop Inlet
A-1
J. E. Lintner

DISTANCES

No. of Paces	Direction	Taped Dist.
154	S	400′
155	N	400′
155	S	400′
156	N	400′
155	Average	

Length of pace = $\frac{400}{155}$ = 2.58′

No. of paces per 100′ = 39−

Plate

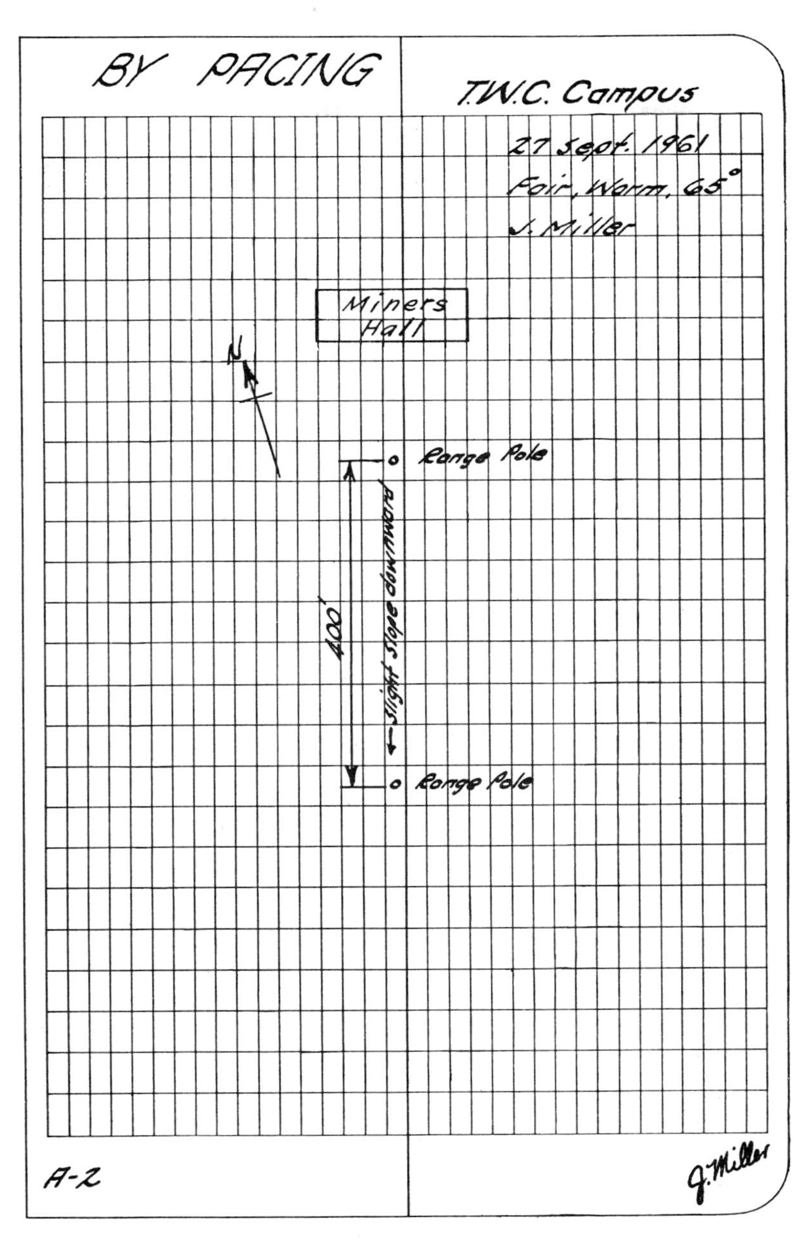
BY PACING
T.W.C. Campus
27 Sept. 1961
Fair, Warm, 65°
J. Miller
Miners Hall
N
Range Pole
400'
Slight Slope downward
Range Pole
A-2
J. Miller

DIFFERENTIAL LEVELS

Sta.	+ sight	H.I.	− sight	Elev.	Dist.
B.M. Mil.				100.00	
	1.33	101.33			150
T.P. 1			8.37	92.96	150
	0.22	93.18			135
T.P. 2			7.91 ~~8.91~~	85.27	135
	0.96	86.23			160
T.P. 3			11.72	74.51	160
	0.46	74.97			160
B.M. Rutgers			8.71	66.26	160
	2.97		36.71		1210
B.M. Rutgers				66.26	
	11.95	78.21			180
T.P. 1			2.61	75.60	180
	12.55	88.15			180
T.P. 2			0.68	87.47	180
	12.77	100.24			155
B.M. Mil.			0.21	100.03	155
	37.27		3.50		1030
B.M. Rutgers	True elev. above MSL			2053.18	
	Elev. diff.			33.75	
B.M. Mil.	MSL elev.			2086.93	
					Plate

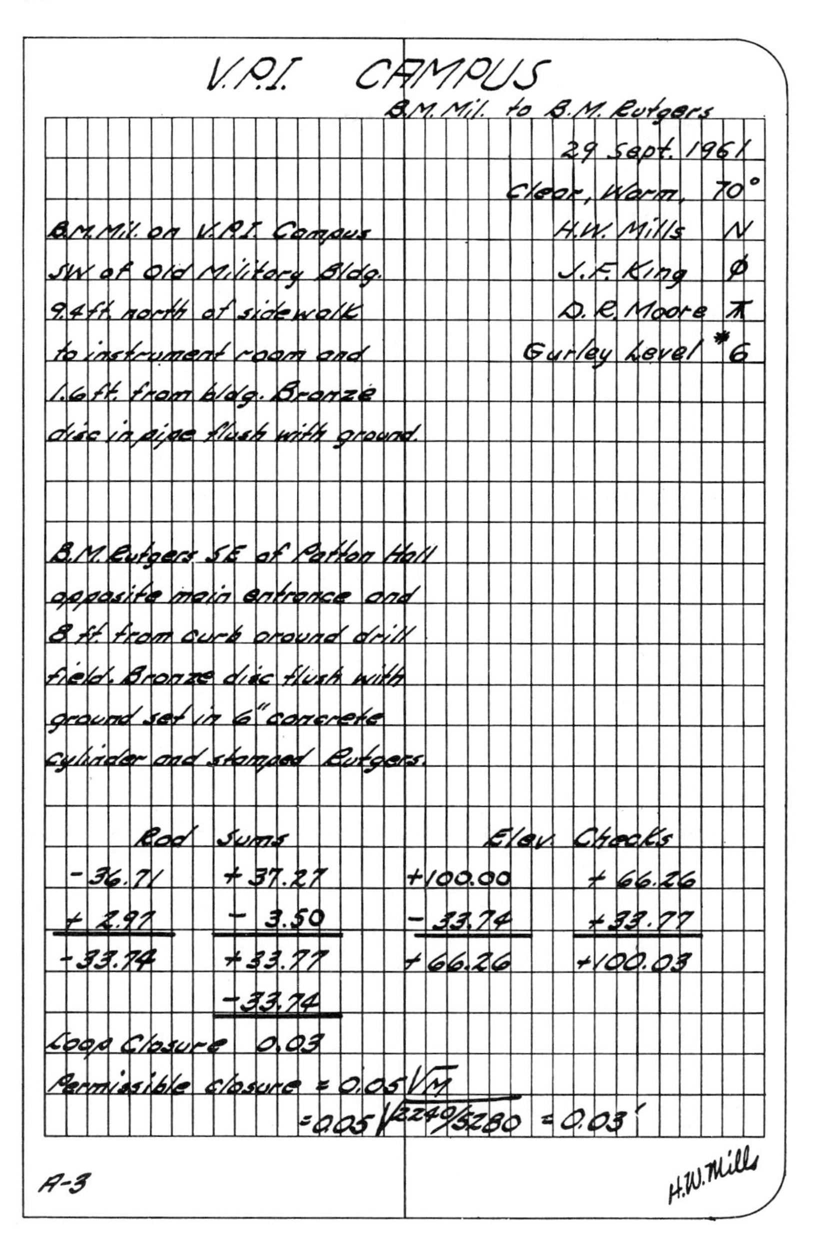
V.P.I. CAMPUS

B.M. Mil. to B.M. Rutgers

29 Sept. 1961

Clear, Warm, 70°

H.W. Mills N

J.F. King ϕ

D.R. Moore ⊼

Gurley Level #6

B.M. Mil. on V.P.I. Campus SW of Old Military Bldg. 9.4 ft. north of sidewalk to instrument room and 1.6 ft. from bldg. Bronze disc in pipe flush with ground.

B.M. Rutgers SE of Patton Hall opposite main entrance and 8 ft. from curb around drill field. Bronze disc flush with ground set in 6" concrete cylinder and stamped Rutgers.

Rod	Sums	Elev.	Checks
−36.71	+37.27	+100.00	+66.26
+2.97	−3.50	−33.74	+33.77
−33.74	+33.77	+66.26	+100.03
	−33.74		

Loop Closure 0.03

Permissible closure $= 0.05\sqrt{M}$

$= 0.05\sqrt{2240/5280} = 0.03'$

A-3

H.W. Mills

RECIPROCAL LEVELING

Station	+ Sight		− Sight	Elev. Diff.	Elev.
B.M. Rutgers	2.605				2053.182
B.M. Eggle.			12.304		
			12.302		
			12.293		
			12.297		
B.M. Eggle.		Aver.	12.299	9.694	2043.488
B.M. Eggle.	11.203				2043.488
B.M. Rutgers			1.528		
			1.517		
			1.519		
			1.522		
B.M. Rutgers		Aver.	1.522	9.681	2053.169
Σ	13.808		13.821		
			13.808		
		Closure	0.013		
		Mean		9.688	
B.M. Eggle.		Mean	Elev.		2043.494
					Plate

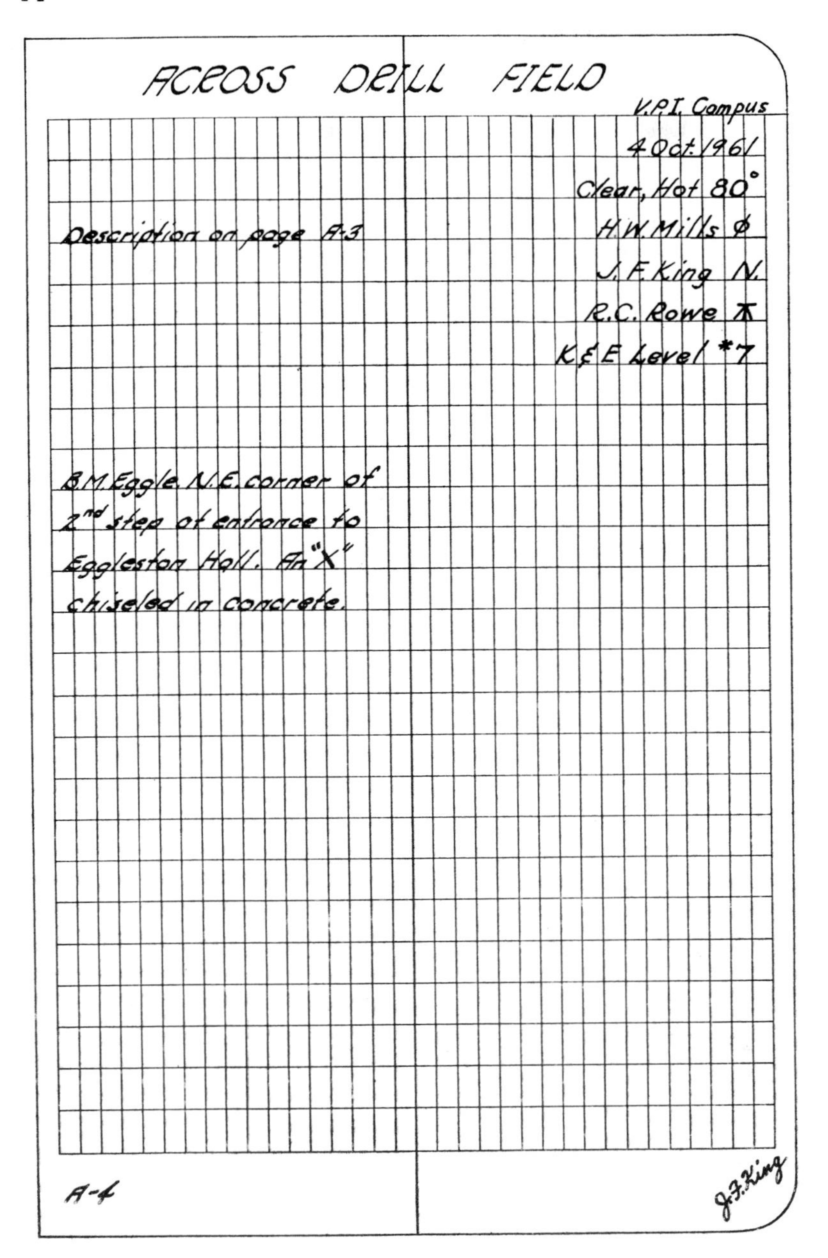
ACROSS DRILL FIELD
V.P.I. Campus
4 Oct. 1961
Clear, Hot 80°
Description on page A-3
H.W. Mills ϕ
J.F. King N.
R.C. Rowe ⊼
K&E Level *7
B.M. Eggle. N.E. corner of
2nd step of entrance to
Eggleston Hall. An "X"
chiseled in concrete.
A-4
J.F. King

PROFILE LEVELS

Station	+ sight	H.I.	— sight	Int. sight	Elev.
B.M. Road	10.14	370.62			360.48
0+00				9.36	361.26
0+20				9.8	360.8
1+00				6.5	364.1
2+00				4.3	366.3
2+60				3.7	366.9
3+00				7.1	363.5
3+90				11.7	358.9
4+00				11.2	359.4
4+35				9.5	361.1
T.P. 1	7.33	366.48	11.47		359.15
5+00				8.4	358.1
5+54				11.08	355.40
5+74				10.66	355.82
5+94				11.06	355.42
6+00				10.5	356.0
7+00				4.4	362.1
T.P. 2	2.55	363.77	5.26		361.22
8+00				1.2	362.6
9+00				3.9	359.9
9+25.2				3.4	360.4
9+25.3				4.6	359.2
9+43.2				2.2	361.6
B.M. Store			0.76		363.01
Σ	20.02		17.49		Plate

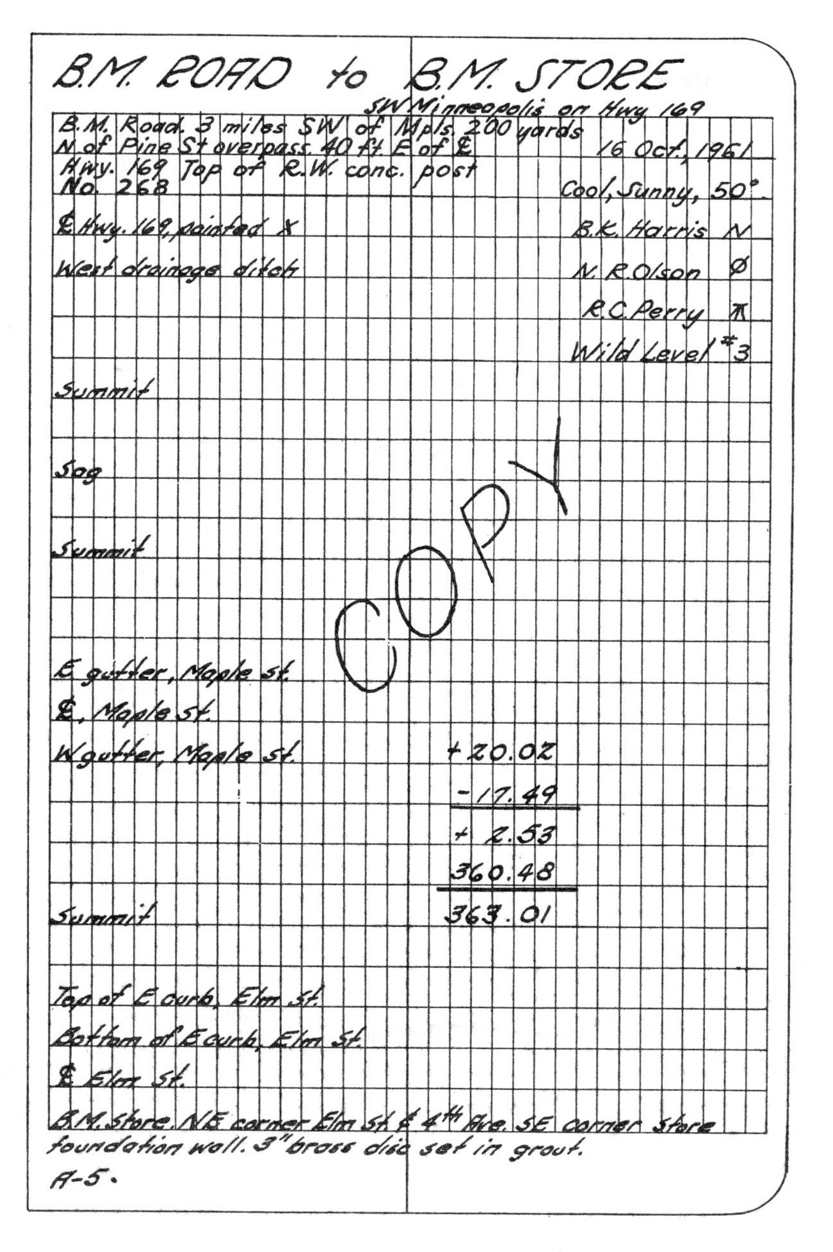
B.M. ROAD to B.M. STORE
SW Minneapolis on Hwy 169
B.M. Road. 3 miles SW of Mpls. 200 yards N. of Pine St overpass. 40 ft. E of ℄ Hwy. 169 Top of R.W. conc. post No. 268
16 Oct. 1961
Cool, Sunny, 50°
℄ Hwy. 169, painted X
B.K. Harris N
West drainage ditch
N. R. Olson Ø
R.C. Perry ⊼
Wild Level #3
Summit
Sag
Summit
COPY
E gutter, Maple st.
℄, Maple st.
W gutter, Maple st.
+20.02
-17.49
+2.53
360.48
Summit
363.01
Top of E curb, Elm St.
Bottom of E curb, Elm St.
℄ Elm St.
B.M. Store. NE corner Elm St. & 4th Ave. SE corner Store foundation wall. 3" brass disc set in grout.
A-5.

BORROW PIT LEVELING

Point	+ sight	H.I.	− sight	Elev.	Cut
B.M. Road	4.22	364.70		360.48	
A,0			5.2	359.5	1.5
B,0			5.4	359.3	1.3
C,0			5.7	359.0	1.0
D,0			5.9	358.8	0.8
E,0			6.2	358.5	0.5
A,1			4.7	360.0	2.0
B,1			4.8	359.9	1.9
C,1			5.2	359.5	1.5
D,1			5.5	359.2	1.2
E,1			5.8	358.9	0.9
A,2			4.2	360.5	2.5
B,2			4.7	360.0	2.0
C,2			4.8	359.9	1.9
D,2			5.0	359.7	1.7
A,3			3.8	360.9	2.9
B,3			4.0	360.7	2.7
C,3			4.6	360.1	2.1
D,3			4.6	360.1	2.1
A,4			3.4	361.3	3.3
B,4			3.7	361.0	3.0
C,4			4.2	360.5	2.5
B.M. Road	4.23				
					Plate

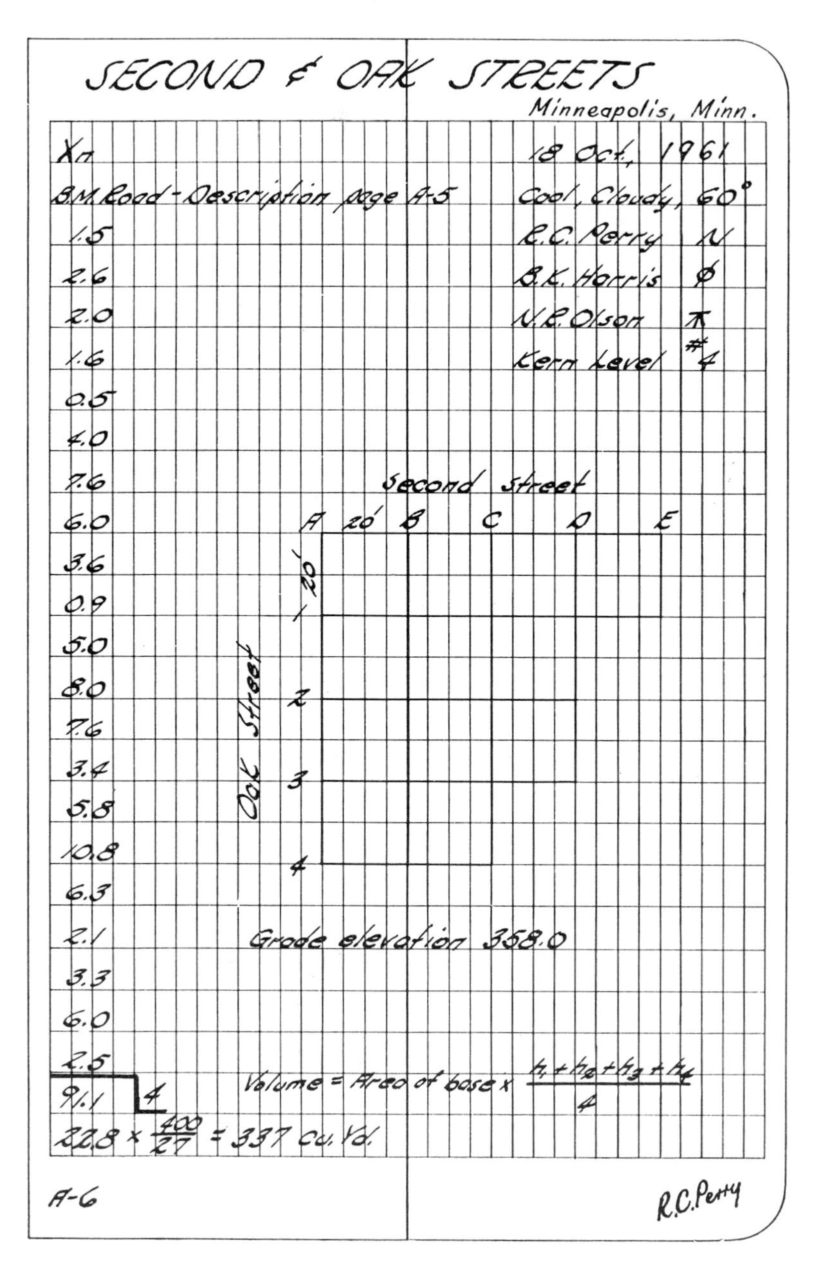
SECOND & OAK STREETS
Minneapolis, Minn.
Xn
18 Oct., 1961
B.M. Road - Description page A-5
Cool, Cloudy, 60°
1.5
R.C. Perry N
2.6
B.K. Harris Ø
2.0
N.R. Olson ⊼
1.6
Kern Level #4
0.5
4.0
7.6
Second Street
6.0
A 20′ B C D E
3.6
20′
0.9
5.0
8.0
Oak Street
2
7.6
3.4
3
5.8
10.8
4
6.3
2.1
Grade elevation 358.0
3.3
6.0
2.5
91.1 4
Volume = Area of base × (h1 + h2 + h3 + h4)/4
22.8 × 400/27 = 337 Cu. Yd.
A-6
R.C. Perry

CLOSING THE HORIZON					
	⊼ at △ A				
	All angles read clockwise				
Object	Vern. A.	Vern. B	Mean	Unadj Angle	Sta. Adj. Angle
	Reading △B to △C				
△B	0° 00′ 00″	180°00′00″	0°00′00″		
3 Rep N	126°36′20″	306°36′20″	(Read for prelim. check)		
3 Rep P	253°13′00″	73°13′00″	253°13′00″	42°12′10″	42°12′09″
	Reading △C to △D				
3 Rep N	(Reading not required)				
3 Rep P	(612°) 252°53′40″	72°54′00″	252°53′50″	59°56′48	59°56′47″
	Reading △D to △B				
3 Rep N	(Reading not required)				
3 Rep P	(2160°) 0°00″20″	180°00′20″	0°00′20″	257°51′05″	257°51′04″
				360°00′03″	360°00′00″
	Vernier	Closure	0°00′20″		
	Horizon	Closure	0°00′03″		
	Sta. Adjustment		0°00′01″		
					Plate

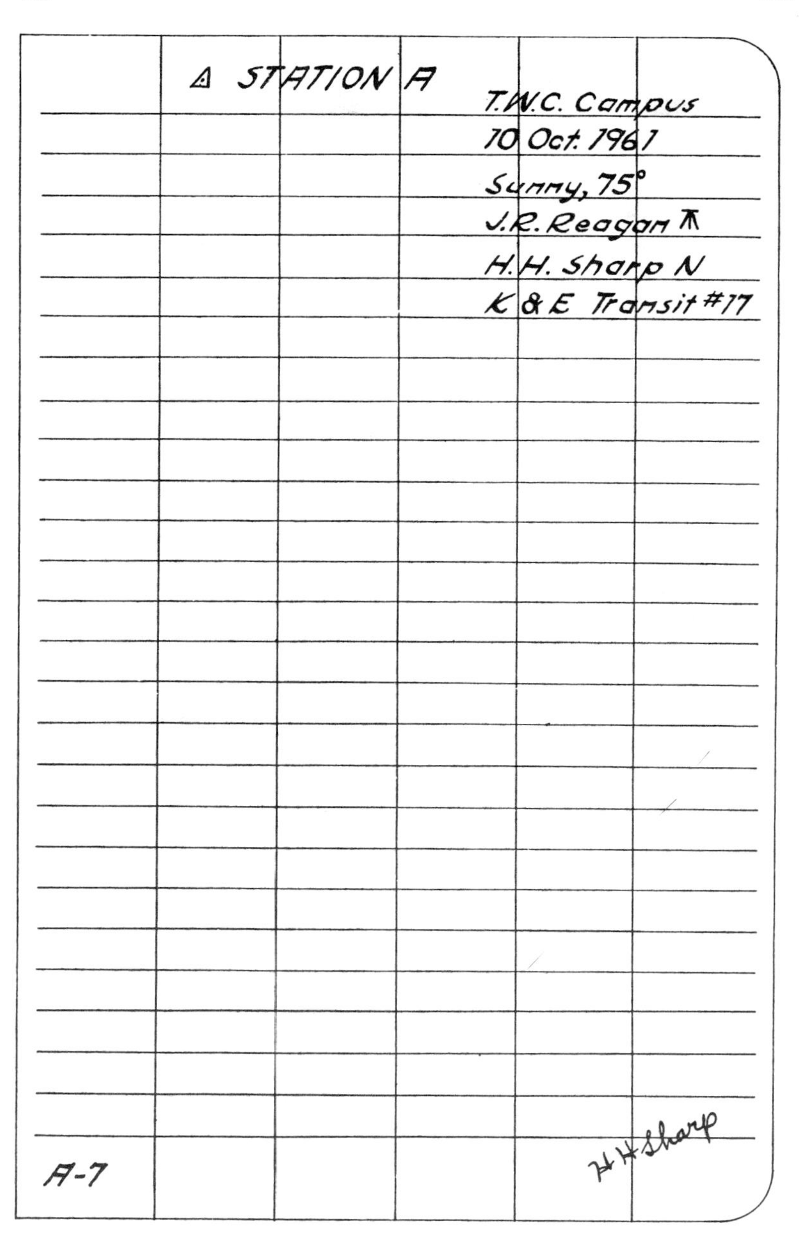
△ STATION A
T.W.C. Campus
10 Oct. 1961
Sunny, 75°
J.R. Reagan ⊼
H.H. Sharp N
K & E Transit #17
A-7
H H Sharp

DOUBLE DIRECT ANGLES

Hub	Dist.	Single ∠	Double ∠	Aver. ∠	Mag. Bearing
A		38°-58'-00"	77°-56'-40"	38°-58'-20"	
	321.31'				S71°-30'W
B		148°-53'-40"	297°-47'-00"	148°-53'-30"	
	276.57'				N41°-00'E
C		84°-28'-00"	168°-56'-00"	84°-28'-00"	
	100.30'				S56°-00'E
D		114°-40'-20"	229°-21'-00"	114°-40'-30"	
	306.83'				N58°-30'E
E		152°-60'-00"	306°-58'-00"	152°-59'-00"	
	255.48'				N32°-00'E
A					
Σ	1260.49'			539°-59'-20"	
			Closure	0°-00'-40"	

Σ interior ∠s $= (N-2)180°$

$= (5-2)180°$

$= 540°\text{-}00'$

Permissible closure $= \sqrt{N} \times$ least count of vernier

$= \sqrt{5} \times 1/3$ min.

$= 0.7$ min.

Plate

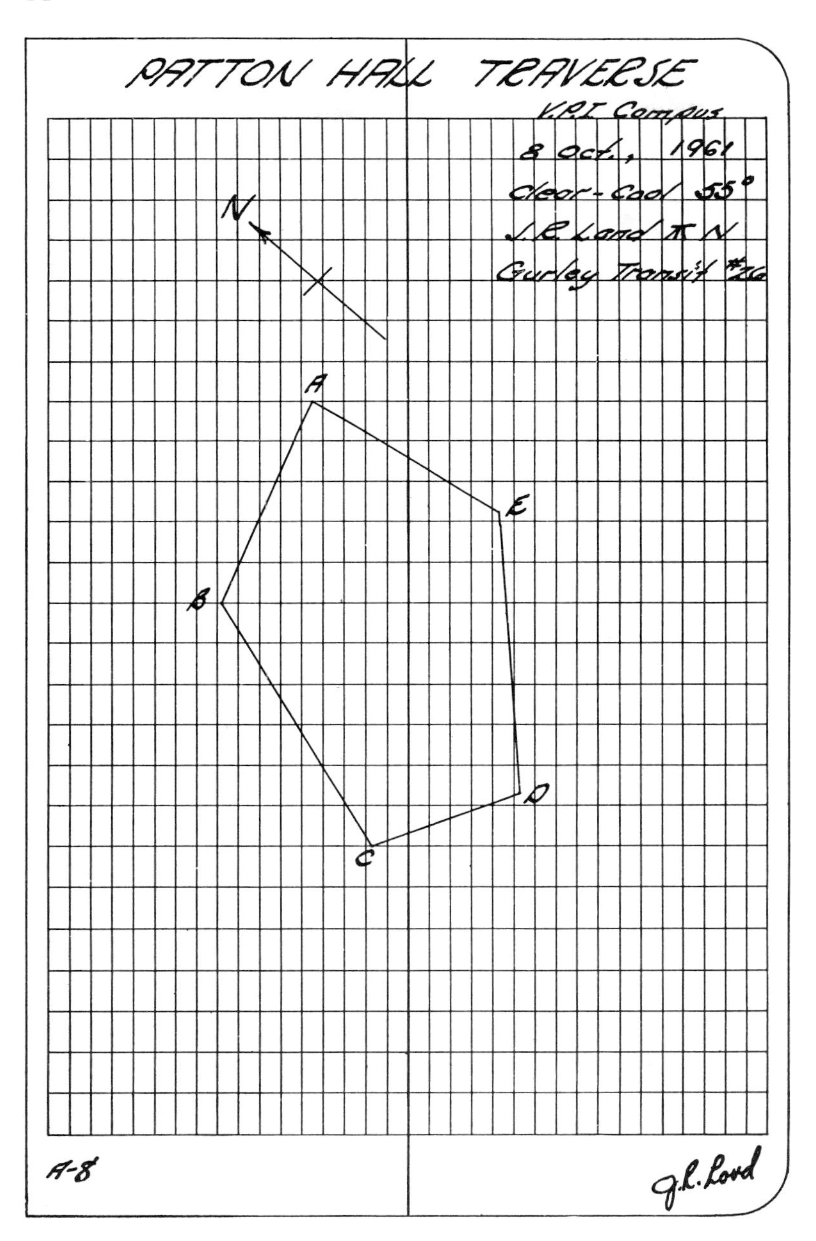
PATTON HALL TRAVERSE
V.P.I. Campus
8 Oct., 1961
Clear - Cool 55°
J.R. Land ⊼ N
Gurley Transit #26
N
A
B
C
D
E
A-8
J.R. Lord

DETERMINATION OF

i = Dist. between upper & lower cross-hairs

f = Focal length

f_1 = Dist. from objective lens to plane of cross-hairs

f_2 = Dist. from objective lens to rod

$\frac{1}{f_1} + \frac{1}{f_2} = \frac{1}{f}$ Practically, $\frac{1}{f_2}$ small & ∴ $f_1 = f$

C = Dist. from plumb bob to lens

S = Dist. from focal point to rod

D = Dist. from plumb bob to rod

R = Rod interval

$S = D - (c+f)$, $\frac{f}{i} = \frac{S}{R}$ or $S = R\frac{f}{i}$

No.	D	c	f	c+f = C	Rod upper
1	100.00′	0.48	0.69	1.17	0.49
2	200.00′	0.49	0.70	1.19	0.99
3	300.00′	0.49	0.70	1.19	1.49
4	400.00′	0.49	0.70	1.19	2.02
5	500.00′	0.49	0.70	1.19	2.52
Means or Sums				1.19	

Sample computation of No. 3

D = 300.00′

c+f = 1.19

S = 298.81

R = 3.00

$\frac{f}{i} = \frac{S}{R} = \frac{298.81}{3.00} = 99.60$

Plate

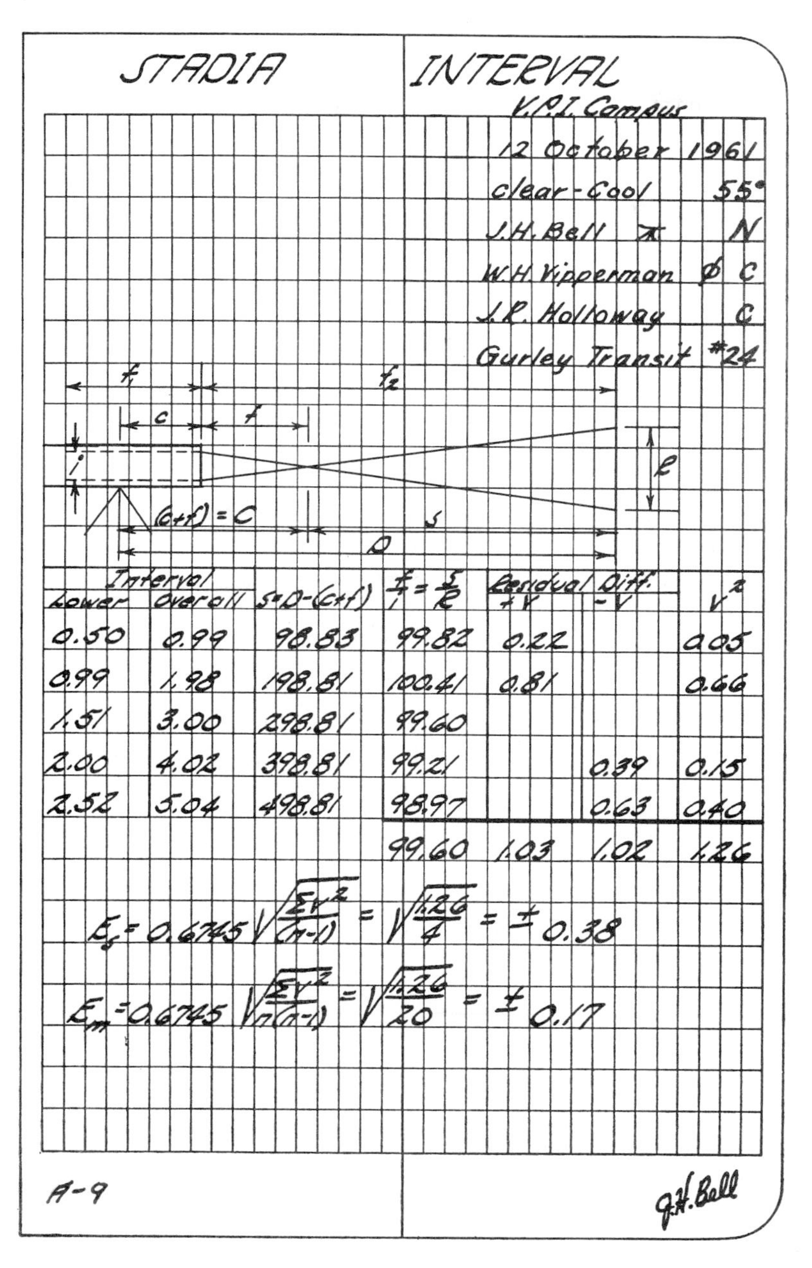

STADIA INTERVAL

V.P.I. Campus

12 October 1961

clear - Cool 55°

J.H. Bell ⊼ N

W.H. Kipperman ϕ C

J.R. Holloway C

Gurley Transit #24

Interval Lower	Overall	$s = D-(c+f)$	$\frac{f}{i} = \frac{s}{R}$	Residual Diff. +V	−V	V^2
0.50	0.99	98.83	99.82	0.22		0.05
0.99	1.98	198.81	100.41	0.81		0.66
1.51	3.00	298.81	99.60			
2.00	4.02	398.81	99.21		0.39	0.15
2.52	5.04	498.81	98.97		0.63	0.40
			99.60	1.03	1.02	1.26

$$E_s = 0.6745\sqrt{\frac{\Sigma v^2}{(n-1)}} = \sqrt{\frac{1.26}{4}} = \pm 0.38$$

$$E_m = 0.6745\sqrt{\frac{\Sigma v^2}{n(n-1)}} = \sqrt{\frac{1.26}{20}} = \pm 0.17$$

A-9

J.H. Bell

	STADIA		SURVEY		
Station	Stadia Inter.	Azimuth	Vert Angle or Rod R.	Horizontal Distance	Elev.
	⊼@⊡B,	Elev.	177.42	h.i.=5.0	
⊡A	6.74	148°04′	−0°34′	675	170.7
⊡C	4.21	60°12′	−1°35′	422	165.8
1	0.91	90°45′	8.3	92	174.1
2	1.66	120°20′	−2°12′	167	171.0
3	3.15	126°30′	−2°06′	316	165.9
4	4.60	143°45′	−1°23′	461	166.0
5	7.85	141°30′	−0°38′	786	168.7
6	2.47	167°20′	8.6	248	173.8
7	1.97	172°20′	9.6	198	172.8
8	4.99	181°15′	3.2	500	179.2
9	5.79	221°45′	+1°02′	580	187.8
10	3.47	256°00′	+1°50′	348	188.5
11	1.17	342°05′	2.4	118	180.0
12	1.71	350°15′	2.4	172	180.0
⊡D	4.49	6°10′	5.2	450	177.2
⊡A	6.74	148°04′	−0°34′	675	170.7
	⊼@⊡C,	Elev.	165.77	h.i.=5.2	
⊡B	4.20	240°12′	+1°34′	421	177.3
13	3.21	286°00′	+2°01	322	177.1
14	2.36	32°05′	8.5	237	162.5
15	2.60	41°50′	10.0	261	161.0
16	4.59	68°30′	−1°22′	460	154.8
					Plate

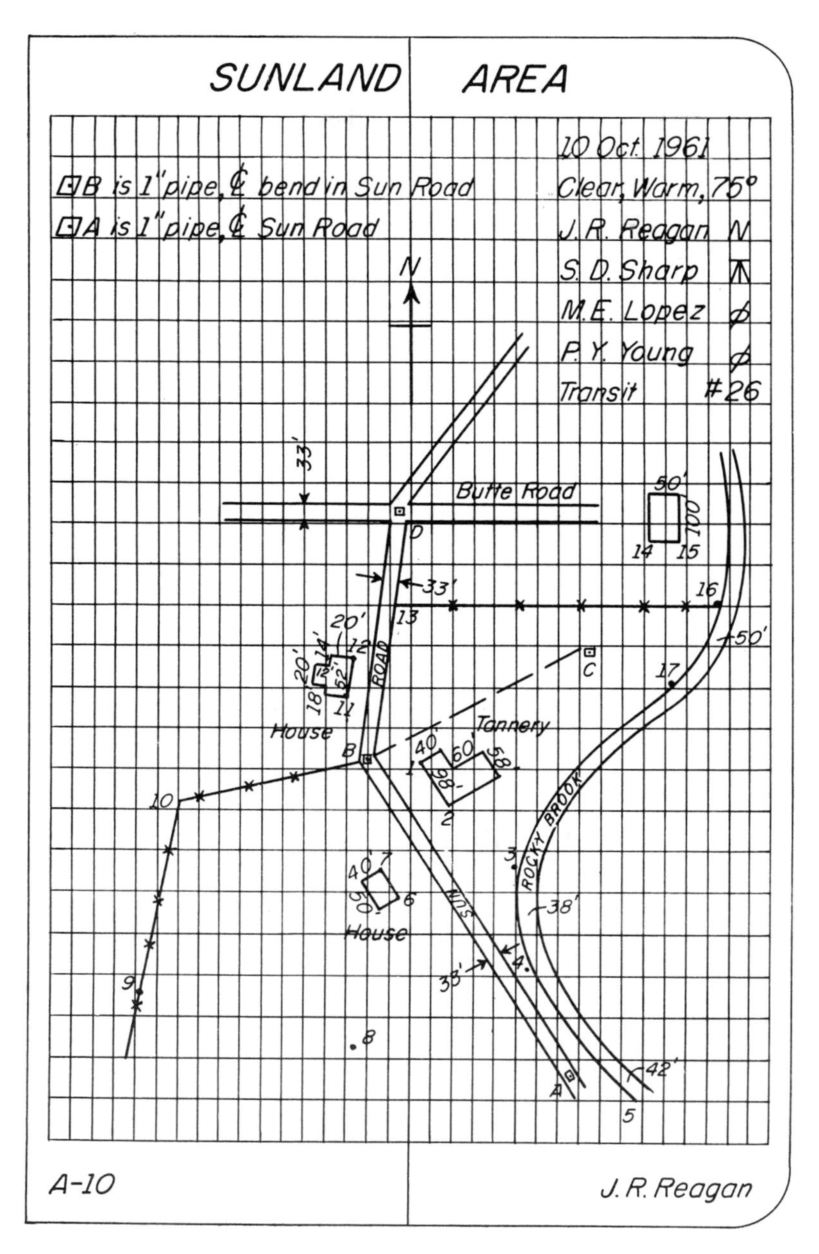
SUNLAND AREA
10 Oct. 1961
B is 1" pipe, ℄ bend in Sun Road
A is 1" pipe, ℄ Sun Road
Clear, Warm, 75°
J. R. Reagan
S. D. Sharp
M. E. Lopez
P. Y. Young
Transit #26
N
33'
Butte Road
D
50'
100'
14
15
16
33'
13
20'
14'
12
20'
12'
52'
18
11
ROAD
C
17
50'
House
B
Tannery
40
60
58
1
98
2
10
ROCKY BROOK
3
40
7
50
6
House
SUN
38'
4
33'
9
8
A
42'
5
A-10
J. R. Reagan

CROSS SECTION LEVELING

Sta.	+ sight	H.I.	− sight	Elev.	
5+00			9.5		
4+00			12.6		
T.P.-1	10.25	106.61	1.87	96.36	
3+00			2.1		
2+50			5.8		
2+00			7.4		
1+35			9.7		
1+00			5.6		
0+50			7.6		
0+00			8.5		
B.M. Pod	8.51	98.23		89.72	
					Plate

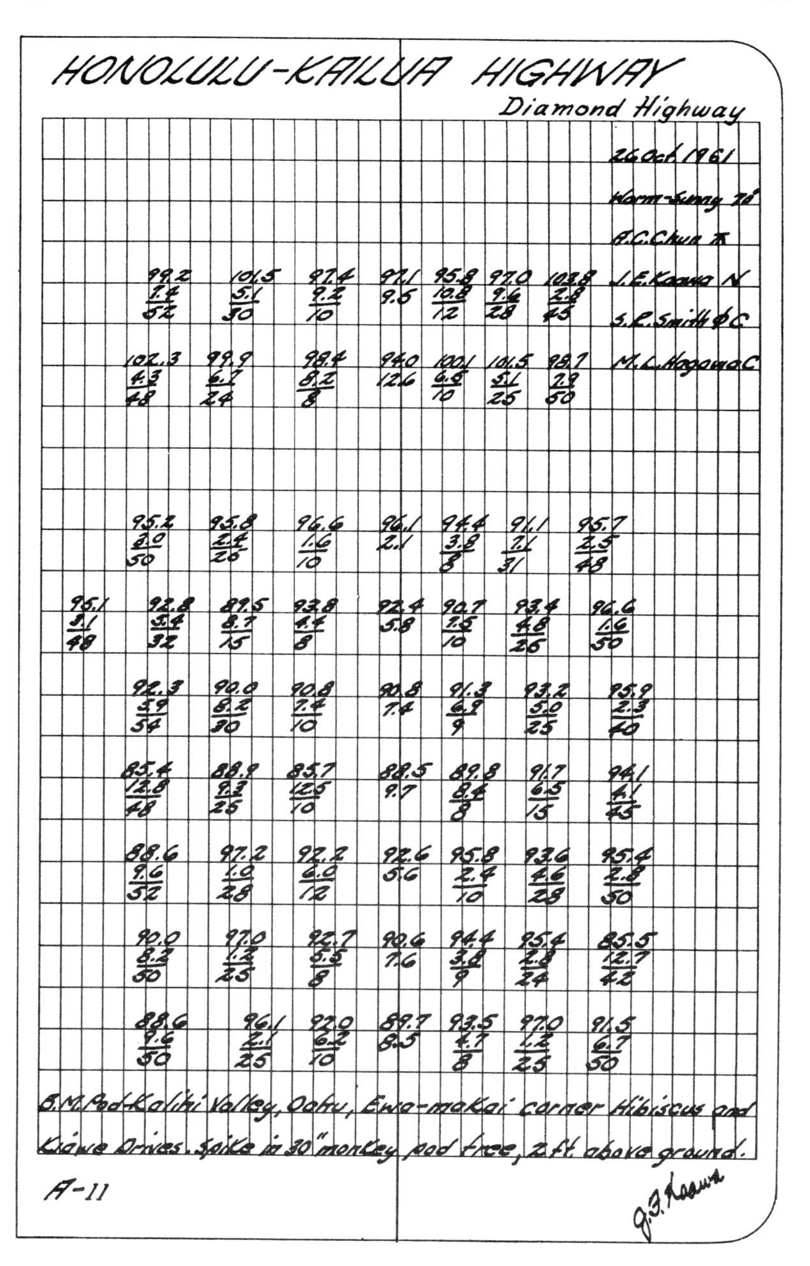

HONOLULU-KAILUA HIGHWAY

Diamond Highway

26 Oct. 1961

Warm-Sunny 70°

A.C. Chun R

J.E. Kaawa N

S.L. Smith ϕC

M.L. Hagawa C

99.2 / 7.4 / 52 101.5 / 5.1 / 30 97.4 / 9.2 / 10 97.1 / 9.5 95.8 / 10.8 / 12 97.0 / 9.6 / 28 103.8 / 2.8 / 45

102.3 / 4.3 / 48 99.9 / 6.7 / 24 98.4 / 8.2 / 8 94.0 / 12.6 100.1 / 6.6 / 10 101.5 / 5.1 / 25 98.7 / 7.9 / 50

95.2 / 3.0 / 50 95.8 / 2.4 / 25 96.6 / 1.6 / 10 96.1 / 2.1 94.4 / 3.8 / 8 91.1 / 7.1 / 31 95.7 / 2.5 / 48

95.1 / 3.1 / 48 92.8 / 5.4 / 32 89.5 / 8.7 / 15 93.8 / 4.4 / 8 92.4 / 5.8 90.7 / 7.5 / 10 93.4 / 4.8 / 26 96.6 / 1.6 / 50

92.3 / 5.9 / 54 90.0 / 8.2 / 30 90.8 / 7.4 / 10 90.8 / 7.4 91.3 / 6.9 / 9 93.2 / 5.0 / 25 95.9 / 2.3 / 40

85.4 / 12.8 / 48 88.9 / 9.3 / 25 85.7 / 12.5 / 10 88.5 / 9.7 89.8 / 8.4 / 8 91.7 / 6.5 / 15 94.1 / 4.1 / 45

88.6 / 9.6 / 52 97.2 / 1.0 / 28 92.2 / 6.0 / 12 92.6 / 5.6 95.8 / 2.4 / 10 93.6 / 4.6 / 28 95.4 / 2.8 / 50

90.0 / 8.2 / 50 97.0 / 1.2 / 25 92.7 / 5.5 / 8 90.6 / 7.6 94.4 / 3.8 / 9 95.4 / 2.8 / 24 85.5 / 12.7 / 42

88.6 / 9.6 / 50 96.1 / 2.1 / 25 92.0 / 6.2 / 10 89.7 / 8.5 93.5 / 4.7 / 8 97.0 / 1.2 / 25 91.5 / 6.7 / 50

B.M. Pod-Kalihi Valley, Oahu, Ewa-makai corner Hibiscus and Kiawe Drives, spike in 30" monkey pod tree, 2 ft. above ground.

A-11

J.J. Kaawa

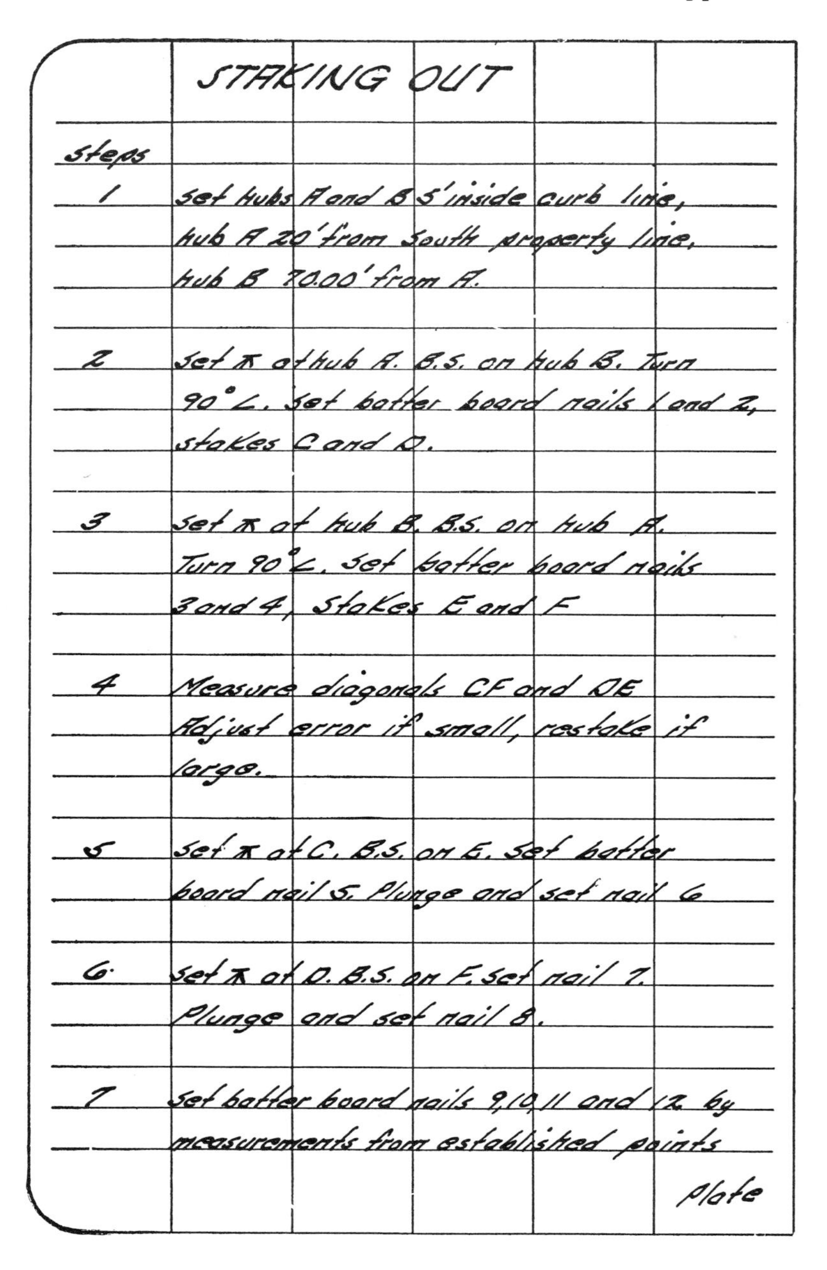
STAKING OUT

steps

1 set hubs A and B 5' inside curb line, hub A 20' from South property line, hub B 70.00' from A.

2 Set ⊼ at hub A. B.S. on hub B. Turn 90° L. Set batter board nails 1 and 2, stakes C and D.

3 Set ⊼ at hub B. B.S. on hub A. Turn 90° L. Set batter board nails 3 and 4, Stakes E and F

4 Measure diagonals CF and DE Adjust error if small, restake if large.

5 Set ⊼ at C. B.S. on E. Set batter board nail 5. Plunge and set nail 6

6 Set ⊼ at D. B.S. on F. Set nail 7. Plunge and set nail 8.

7 Set batter board nails 9, 10, 11 and 12 by measurements from established points

Plate

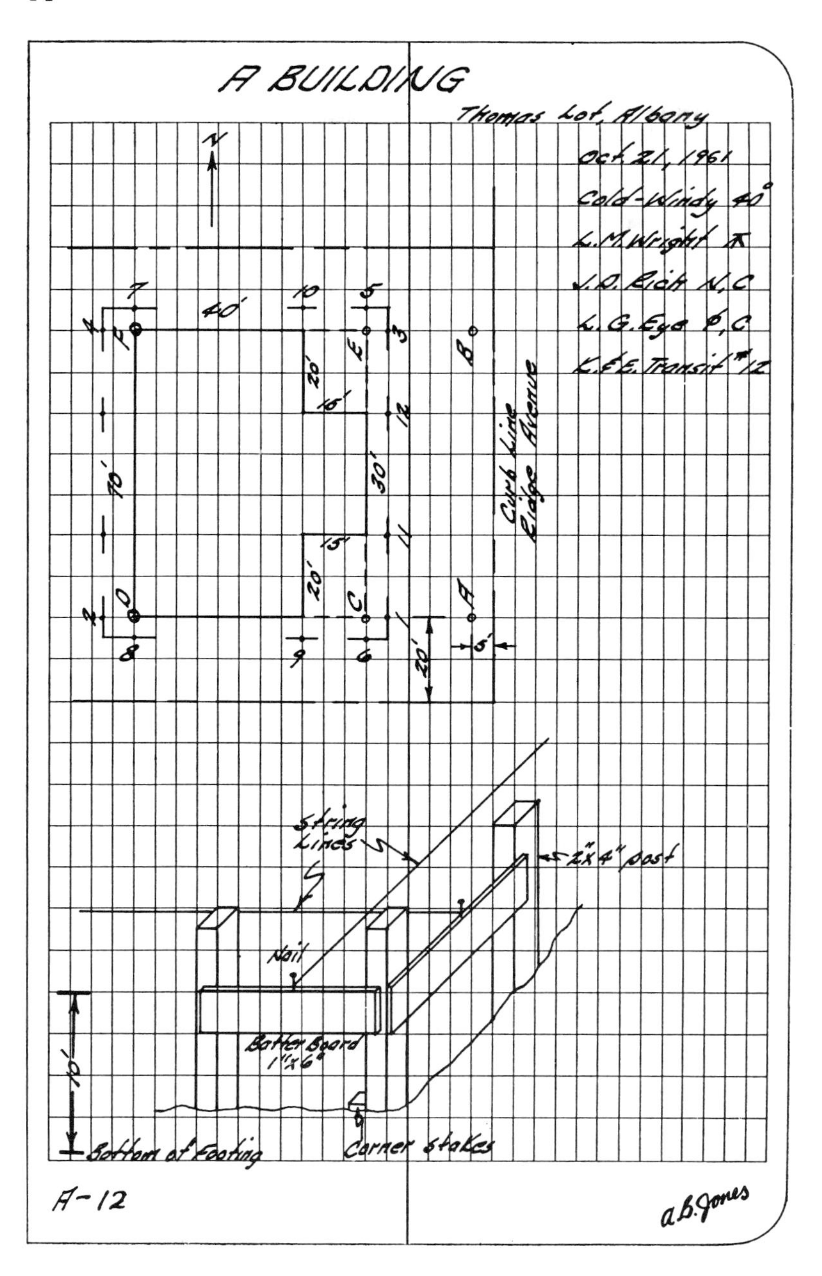
A BUILDING
Thomas Lot, Albany
Oct. 21, 1961
Cold-Windy 40°
L.M. Wright
J.D. Rich N.C
L.G. Eye F.C
K.&E. Transit #12
N
40'
70'
20'
15'
30'
15'
20'
20'
5'
Curb Line
Ridge Avenue
String Lines
2"x4" post
Nail
Batter Board 1"x6"
10'
Bottom of Footing
Corner stakes
A-12
a.B. Jones

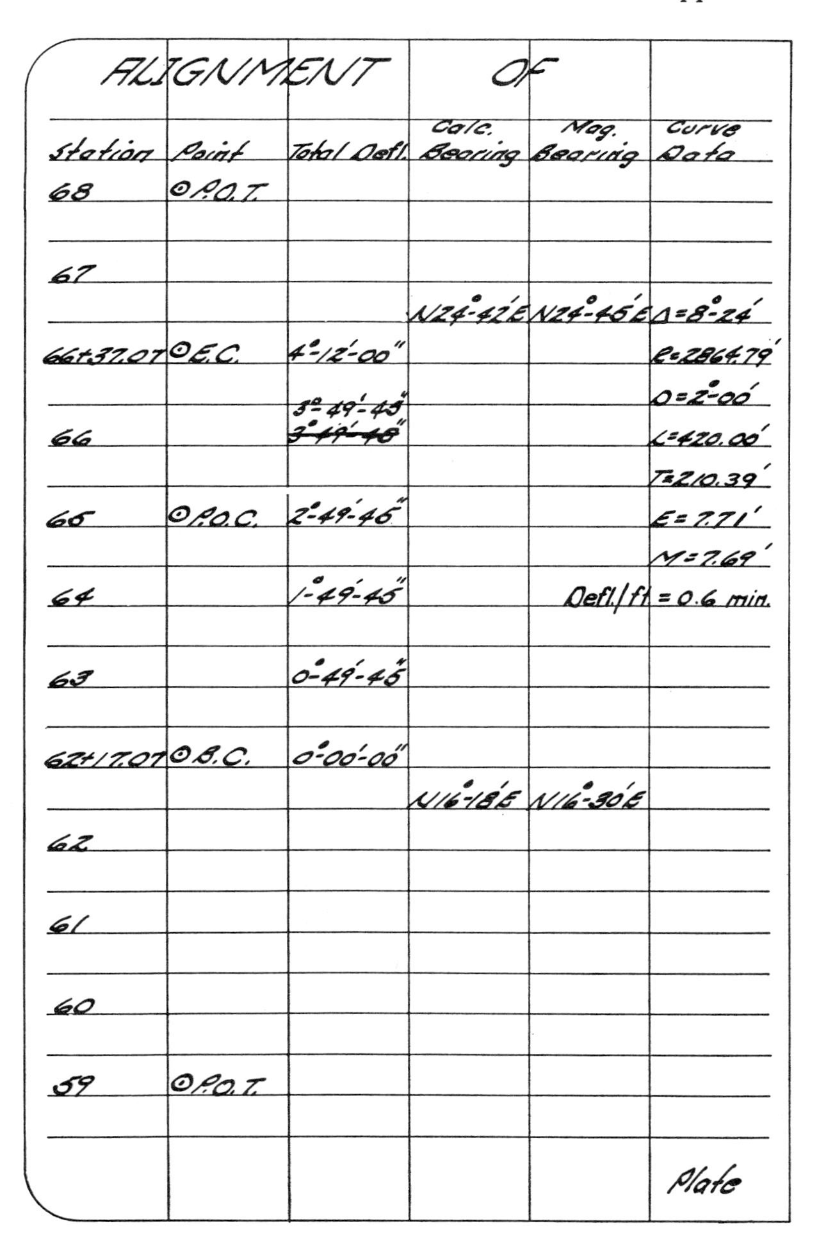

ALIGNMENT OF

Station	Point	Total Defl.	Calc. Bearing	Mag. Bearing	Curve Data
68	⊙P.O.T.				
67					
			N24°42'E	N24°45'E	Δ=8°24'
66+37.07	⊙E.C.	4°12'00"			R=2864.79'
					D=2°00'
66		3°49'45" ~~3°49'48"~~			L=420.00'
					T=210.39'
65	⊙P.O.C.	2°49'46"			E=7.71'
					M=7.69'
64		1°49'45"		Defl./ft	= 0.6 min.
63		0°49'45"			
62+17.07	⊙B.C.	0°00'00"			
			N16°18'E	N16°30'E	
62					
61					
60					
59	⊙P.O.T.				
					Plate

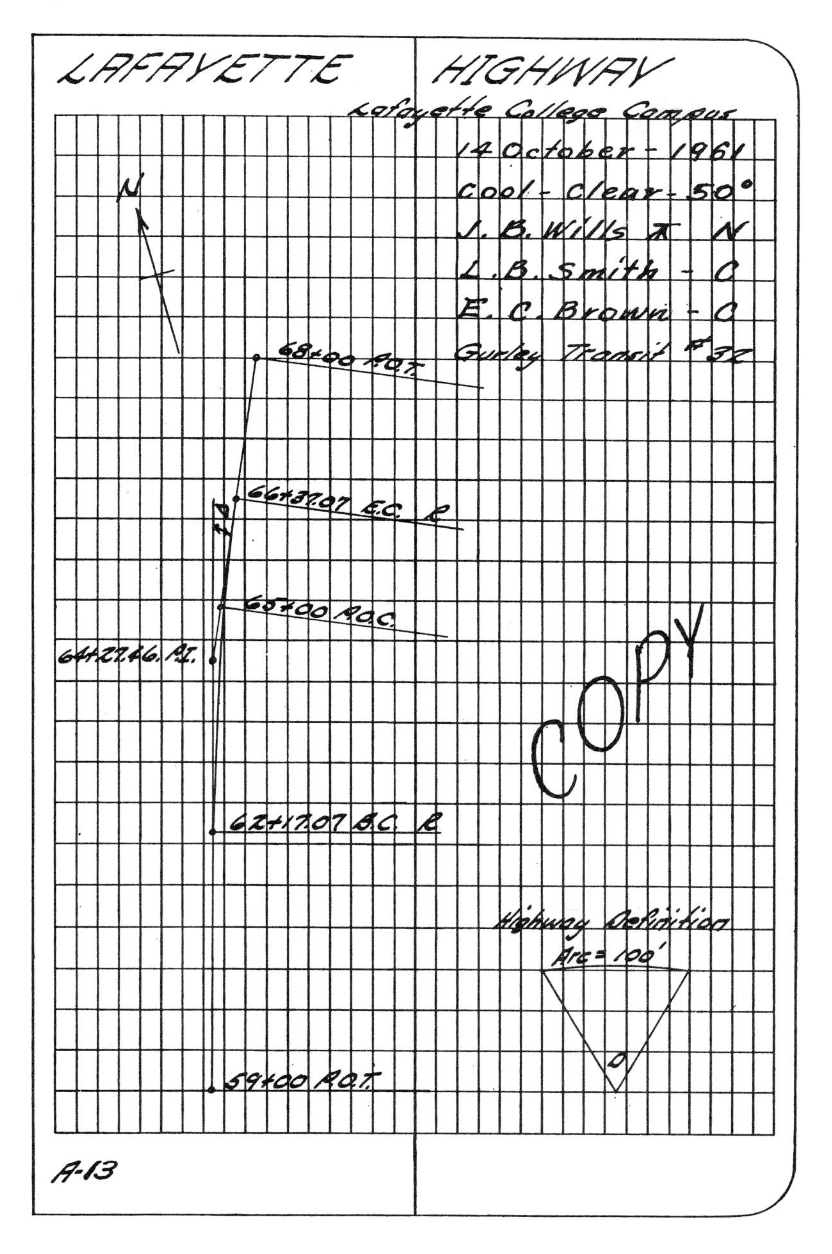
LAFAYETTE HIGHWAY
Lafayette College Campus
14 October - 1961
Cool - Clear - 50°
J. B. Wills ⊼ N
L. B. Smith - C
E. C. Brown - C
Gurley Transit #32
N
68+00 P.O.T.
66+37.07 E.C. R
Δ
65+00 P.O.C.
64+27.46 P.I.
COPY
62+17.07 B.C. R
Highway Definition
Arc = 100'
D
59+00 P.O.T.
A-13

appendix B

TABLE I. SAMPLE PAGES FROM TRAVERSE TABLE*

Distance.	15°.		15¼°.		15½°.		15¾°.		Distance.
	Lat.	Dep.	Lat.	Dep.	Lat.	Dep.	Lat.	Dep.	
1	0.97	0.26	0.96	0.26	0.96	0.27	0.96	0.27	1
2	1.93	0.52	1.93	0.53	1.93	0.53	1.92	0.54	2
3	2.90	0.78	2.89	0.79	2.89	0.80	2.89	0.81	3
4	3.86	1.04	3.86	1.05	3.85	1.07	3.85	1.09	4
5	4.83	1.29	4.82	1.32	4.82	1.34	4.81	1.36	5
6	5.80	1.55	5.79	1.58	5.78	1.60	5.77	1.63	6
7	6.76	1.81	6.75	1.84	6.75	1.87	6.74	1.90	7
8	7.73	2.07	7.72	2.10	7.71	2.14	7.70	2.17	8
9	8.69	2.33	8.68	2.37	8.67	2.41	8.66	2.44	9
10	9.66	2.59	9.65	2.63	9.64	2.67	9.62	2.71	10
11	10.63	2.85	10.61	2.89	10.60	2.94	10.59	2.99	11
12	11.59	3.11	11.58	3.16	11.56	3.21	11.55	3.26	12
13	12.56	3.36	12.54	3.42	12.53	3.47	12.51	3.53	13
14	13.52	3.62	13.51	3.68	13.49	3.74	13.47	3.80	14
15	14.49	3.88	14.47	3.95	14.45	4.01	14.44	4.07	15
16	15.45	4.14	15.44	4.21	15.42	4.28	15.40	4.34	16
17	16.42	4.40	16.40	4.47	16.38	4.54	16.36	4.61	17
18	17.39	4.66	17.37	4.73	17.35	4.81	17.32	4.89	18
19	18.35	4.92	18.33	5.00	18.31	5.08	18.29	5.16	19
20	19.32	5.18	19.30	5.26	19.27	5.34	19.25	5.43	20
21	20.28	5.44	20.26	5.52	20.24	5.61	20.21	5.70	21
22	21.25	5.69	21.23	5.79	21.20	5.88	21.17	5.97	22
23	22.22	5.95	22.19	6.05	22.16	6.15	22.14	6.24	23
24	23.18	6.21	23.15	6.31	23.13	6.41	23.10	6.51	24
25	24.15	6.47	24.12	6.58	24.09	6.68	24.06	6.79	25
26	25.11	6.73	25.08	6.84	25.05	6.95	25.02	7.06	26
27	26.08	6.99	26.05	7.10	26.02	7.22	25.99	7.33	27
28	27.05	7.25	27.01	7.36	26.98	7.48	26.95	7.60	28
29	28.01	7.51	27.98	7.63	27.95	7.75	27.91	7.87	29
30	28.98	7.76	28.94	7.89	28.91	8.02	28.87	8.14	30
31	29.94	8.02	29.91	8.15	29.87	8.28	29.84	8.41	31
32	30.91	8.28	30.87	8.42	30.84	8.55	30.80	8.69	32
33	31.88	8.54	31.84	8.68	31.80	8.82	31.76	8.96	33
34	32.84	8.80	32.80	8.94	32.76	9.09	32.72	9.23	34
35	33.81	9.06	33.77	9.21	33.73	9.35	33.69	9.50	35
36	34.77	9.32	34.73	9.47	34.69	9.62	34.65	9.77	36
37	35.74	9.58	35.70	9.73	35.65	9.89	35.61	10.04	37
38	36.71	9.84	36.66	10.00	36.62	10.16	36.57	10.31	38
39	37.67	10.09	37.63	10.26	37.58	10.42	37.54	10.59	39
40	38.64	10.35	38.59	10.52	38.55	10.69	38.50	10.86	40
41	39.60	10.61	39.56	10.78	39.51	10.96	39.46	11.13	41
42	40.57	10.87	40.52	11.05	40.47	11.22	40.42	11.40	42
43	41.53	11.13	41.49	11.31	41.44	11.49	41.39	11.67	43
44	42.50	11.39	42.45	11.57	42.40	11.76	42.35	11.94	44
45	43.47	11.65	43.42	11.84	43.36	12.03	43.31	12.21	45
46	44.43	11.91	44.38	12.10	44.33	12.29	44.27	12.49	46
47	45.40	12.16	45.35	12.36	45.29	12.56	45.24	12.76	47
48	46.36	12.42	46.31	12.63	46.25	12.83	46.20	13.03	48
49	47.33	12.68	47.27	12.89	47.22	13.09	47.16	13.30	49
50	48.30	12.94	48.24	13.15	48.18	13.36	48.12	13.57	50
Distance.	Dep.	Lat.	Dep.	Lat.	Dep.	Lat.	Dep.	Lat.	Distance.
	75°.		74¾°.		74½°.		74¼°.		

* From U.S. Department of the Interior, *Standard Field Tables and Trigonometric Formulas*, 8th ed. (GPO, 1950).

TABLE I. SAMPLE PAGES FROM TRAVERSE TABLE

Distance.	15°.		15¼°.		15½°.		15¾°.		Distance.
	Lat.	Dep.	Lat.	Dep.	Lat.	Dep.	Lat.	Dep.	
51	49.26	13.20	49.20	13.41	49.15	13.63	49.09	13.84	51
52	50.23	13.46	50.17	13.68	50.11	13.90	50.05	14.11	52
53	51.19	13.72	51.13	13.94	51.07	14.16	51.01	14.39	53
54	52.16	13.98	52.10	14.20	52.04	14.43	51.97	14.66	54
55	53.13	14.24	53.06	14.47	53.00	14.70	52.94	14.93	55
56	54.09	14.49	54.03	14.73	53.96	14.97	53.90	15.20	56
57	55.06	14.75	54.99	14.99	54.93	15.23	54.86	15.47	57
58	56.02	15.01	55.96	15.26	55.89	15.50	55.82	15.74	58
59	56.99	15.27	56.92	15.52	56.85	15.77	56.78	16.01	59
60	57.96	15.53	57.89	15.78	57.82	16.03	57.75	16.29	60
61	58.92	15.79	58.85	16.04	58.78	16.30	58.71	16.56	61
62	59.89	16.05	59.82	16.31	59.75	16.57	59.67	16.83	62
63	60.85	16.31	60.78	16.57	60.71	16.84	60.63	17.10	63
64	61.82	16.56	61.75	16.83	61.67	17.10	61.60	17.37	64
65	62.79	16.82	62.71	17.10	62.64	17.37	62.56	17.64	65
66	63.75	17.08	63.68	17.36	63.60	17.64	63.52	17.92	66
67	64.72	17.34	64.64	17.62	64.56	17.90	64.48	18.19	67
68	65.68	17.60	65.61	17.89	65.53	18.17	65.45	18.46	68
69	66.65	17.86	66.57	18.15	66.49	18.44	66.41	18.73	69
70	67.61	18.12	67.54	18.41	67.45	18.71	67.37	19.00	70
71	68.58	18.38	68.50	18.68	68.42	18.97	68.33	19.27	71
72	69.55	18.63	69.46	18.94	69.38	19.24	69.30	19.54	72
73	70.51	18.89	70.43	19.20	70.35	19.51	70.26	19.82	73
74	71.48	19.15	71.39	19.46	71.31	19.78	71.22	20.09	74
75	72.44	19.41	72.36	19.73	72.27	20.04	72.18	20.36	75
76	73.41	19.67	73.32	19.99	73.24	20.31	73.15	20.63	76
77	74.38	19.93	74.29	20.25	74.20	20.58	74.11	20.90	77
78	75.34	20.19	75.25	20.52	75.16	20.84	75.07	21.17	78
79	76.31	20.45	76.22	20.78	76.13	21.11	76.03	21.44	79
80	77.27	20.71	77.18	21.04	77.09	21.38	77.00	21.72	80
81	78.24	20.96	78.15	21.31	78.05	21.65	77.96	21.99	81
82	79.21	21.22	79.11	21.57	79.02	21.91	78.92	22.26	82
83	80.17	21.48	80.08	21.83	79.98	22.18	79.88	22.53	83
84	81.14	21.74	81.04	22.09	80.94	22.45	80.85	22.80	84
85	82.10	22.00	82.01	22.36	81.91	22.72	81.81	23.07	85
86	83.07	22.26	82.97	22.62	82.87	22.98	82.77	23.34	86
87	84.04	22.52	83.94	22.88	83.84	23.25	83.73	23.62	87
88	85.00	22.78	84.90	23.15	84.80	23.52	84.70	23.89	88
89	85.97	23.03	85.87	23.41	85.76	23.78	85.66	24.16	89
90	86.93	23.29	86.83	23.67	86.73	24.05	86.62	24.43	90
91	87.90	23.55	87.80	23.94	87.69	24.32	87.58	24.70	91
92	88.87	23.81	88.76	24.20	88.65	24.59	88.55	24.97	92
93	89.83	24.07	89.73	24.46	89.62	24.85	89.51	25.24	93
94	90.80	24.33	90.69	24.72	90.58	25.12	90.47	25.52	94
95	91.76	24.59	91.65	24.99	91.54	25.39	91.43	25.79	95
96	92.73	24.85	92.62	25.25	92.51	25.65	92.40	26.06	96
97	93.69	25.11	93.58	25.51	93.47	25.92	93.36	26.33	97
98	94.66	25.36	94.55	25.78	94.44	26.19	94.32	26.60	98
99	95.63	25.62	95.51	26.04	95.40	26.46	95.28	26.87	99
100	96.59	25.88	96.48	26.30	96.36	26.72	96.25	27.14	100
Distance.	Dep.	Lat.	Dep.	Lat.	Dep.	Lat.	Dep.	Lat.	Distance.
	75°.		74¾°.		74½°.		74¼°.		

TABLE II. STADIA REDUCTIONS

Minutes	0°		1°		2°		3°	
	Hor. Dist.	Diff. Elev.	Hor. Dist.	Diff. Elev.	Hor. Dist.	Diff. Elev.	Hor. Dist.	Diff. Elev.
0	100.00	.00	99.97	1.74	99.88	3.49	99.73	5.23
2	100.00	.06	99.97	1.80	99.87	3.55	99.72	5.28
4	100.00	.12	99.97	1.86	99.87	3.60	99.71	5.34
6	100.00	.17	99.96	1.92	99.87	3.66	99.71	5.40
8	100.00	.23	99.96	1.98	99.86	3.72	99.70	5.46
10	100.00	.29	99.96	2.04	99.86	3.78	99.69	5.52
12	100.00	.35	99.96	2.09	99.85	3.84	99.69	5.57
14	100.00	.41	99.95	2.15	99.85	3.89	99.68	5.63
16	100.00	.47	99.95	2.21	99.84	3.95	99.68	5.69
18	100.00	.52	99.95	2.27	99.84	4.01	99.67	5.75
20	100.00	.58	99.95	2.33	99.83	4.07	99.66	5.80
22	100.00	.64	99.94	2.38	99.83	4.13	99.66	5.86
24	100.00	.70	99.94	2.44	99.82	4.18	99.65	5.92
26	99.99	.76	99.94	2.50	99.82	4.24	99.64	5.98
28	99.99	.81	99.93	2.56	99.81	4.30	99.63	6.04
30	99.99	.87	99.93	2.62	99.81	4.36	99.63	6.09
32	99.99	.93	99.93	2.67	99.80	4.42	99.62	6.15
34	99.99	.99	99.93	2.73	99.80	4.47	99.61	6.21
36	99.99	1.05	99.92	2.79	99.79	4.53	99.61	6.27
38	99.99	1.11	99.92	2.85	99.79	4.59	99.60	6.32
40	99.99	1.16	99.92	2.91	99.78	4.65	99.59	6.38
42	99.99	1.22	99.91	2.97	99.78	4.71	99.58	6.44
44	99.98	1.28	99.91	3.02	99.77	4.76	99.58	6.50
46	99.98	1.34	99.90	3.08	99.77	4.82	99.57	6.56
48	99.98	1.40	99.90	3.14	99.76	4.88	99.56	6.61
50	99.98	1.45	99.90	3.20	99.76	4.94	99.55	6.67
52	99.98	1.51	99.89	3.26	99.75	4.99	99.55	6.73
54	99.98	1.57	99.89	3.31	99.74	5.05	99.54	6.79
56	99.97	1.63	99.89	3.37	99.74	5.11	99.53	6.84
58	99.97	1.69	99.88	3.43	99.73	5.17	99.52	6.90
60	99.97	1.74	99.88	3.49	99.73	5.23	99.51	6.96
$C =$ .75	.75	.01	.75	.02	.75	.03	.75	.05
$C =$ 1.00	1.00	.01	1.00	.03	1.00	.04	1.00	.06
$C =$ 1.25	1.25	.02	1.25	.03	1.25	.05	1.25	.08

TABLE II. STADIA REDUCTIONS

Minutes	4°		5°		6°		7°	
	Hor. Dist.	Diff. Elev.	Hor. Dist.	Diff. Elev.	Hor. Dist.	Diff. Elev.	Hor. Dist.	Diff. Elev.
0	99.51	6.96	99.24	8.68	98.91	10.40	98.51	12.10
2	99.51	7.02	99.23	8.74	98.90	10.45	98.50	12.15
4	99.50	7.07	99.22	8.80	98.88	10.51	98.49	12.21
6	99.49	7.13	99.21	8.85	98.87	10.57	98.47	12.27
8	99.48	7.19	99.20	8.91	98.86	10.62	98.46	12.32
10	99.47	7.25	99.19	8.97	98.85	10.68	98.44	12.38
12	99.46	7.30	99.18	9.03	98.83	10.74	98.43	12.43
14	99.46	7.36	99.17	9.08	98.82	10.79	98.41	12.49
16	99.45	7.42	99.16	9.14	98.81	10.85	98.40	12.55
18	99.44	7.48	99.15	9.20	98.80	10.91	98.39	12.60
20	99.43	7.53	99.14	9.25	98.78	10.96	98.37	12.66
22	99.42	7.59	99.13	9.31	98.77	11.02	98.36	12.72
24	99.41	7.65	99.11	9.37	98.76	11.08	98.34	12.77
26	99.40	7.71	99.10	9.43	98.74	11.13	98.33	12.83
28	99.39	7.76	99.09	9.48	98.73	11.19	98.31	12.88
30	99.38	7.82	99.08	9.54	98.72	11.25	98.30	12.94
32	99.38	7.88	99.07	9.60	98.71	11.30	98.28	13.00
34	99.37	7.94	99.06	9.65	98.69	11.36	98.27	13.05
36	99.36	7.99	99.05	9.71	98.68	11.42	98.25	13.11
38	99.35	8.05	99.04	9.77	98.67	11.47	98.24	13.17
40	99.34	8.11	99.03	9.83	98.65	11.53	98.22	13.22
42	99.33	8.17	99.01	9.88	98.64	11.59	98.20	13.28
44	99.32	8.22	99.00	9.94	98.63	11.64	98.19	13.33
46	99.31	8.28	98.99	10.00	98.61	11.70	98.17	13.39
48	99.30	8.34	98.98	10.05	98.60	11.76	98.16	13.45
50	99.29	8.40	98.97	10.11	98.58	11.81	98.14	13.50
52	99.28	8.45	98.96	10.17	98.57	11.87	98.13	13.56
54	99.27	8.51	98.94	10.22	98.56	11.93	98.11	13.61
56	99.26	8.57	98.93	10.28	98.54	11.98	98.10	13.67
58	99.25	8.63	98.92	10.34	98.53	12.04	98.08	13.73
60	99.24	8.68	98.91	10.40	98.51	12.10	98.06	13.78
$C =$ .75	.75	.06	.75	.07	.75	.08	.74	.10
$C =$ 1.00	1.00	.08	1.00	.10	.99	.11	.99	.13
$C =$ 1.25	1.25	.10	1.24	.12	1.24	.14	1.24	.16

TABLE II. STADIA REDUCTIONS

Minutes	8°		9°		10°		11°	
	Hor. Dist.	Diff. Elev.	Hor. Dist.	Diff. Elev.	Hor. Dist.	Diff. Elev.	Hor. Dist.	Diff. Elev.
0	98.06	13.78	97.55	15.45	96.98	17.10	96.36	18.73
2	98.05	13.84	97.53	15.51	96.96	17.16	96.34	18.78
4	98.03	13.89	97.52	15.56	96.94	17.21	96.32	18.84
6	98.01	13.95	97.50	15.62	96.92	17.26	96.29	18.89
8	98.00	14.01	97.48	15.67	96.90	17.32	96.27	18.95
10	97.98	14.06	97.46	15.73	96.88	17.37	96.25	19.00
12	97.97	14.12	97.44	15.78	96.86	17.43	96.23	19.05
14	97.95	14.17	97.43	15.84	96.84	17.48	96.21	19.11
16	97.93	14.23	97.41	15.89	96.82	17.54	96.18	19.16
18	97.92	14.28	97.39	15.95	96.80	17.59	96.16	19.21
20	97.90	14.34	97.37	16.00	96.78	17.65	96.14	19.27
22	97.88	14.40	97.35	16.06	96.76	17.70	96.12	19.32
24	97.87	14.45	97.33	16.11	96.74	17.76	96.09	19.38
26	97.85	14.51	97.31	16.17	96.72	17.81	96.07	19.43
28	97.83	14.56	97.29	16.22	96.70	17.86	96.05	19.48
30	97.82	14.62	97.28	16.28	96.68	17.92	96.03	19.54
32	97.80	14.67	97.26	16.33	96.66	17.97	96.00	19.59
34	97.78	14.73	97.24	16.39	96.64	18.03	95.98	19.64
36	97.76	14.79	97.22	16.44	96.62	18.08	95.96	19.70
38	97.75	14.84	97.20	16.50	96.60	18.14	95.93	19.75
40	97.73	14.90	97.18	16.55	96.57	18.19	95.91	19.80
42	97.71	14.95	97.16	16.61	96.55	18.24	95.89	19.86
44	97.69	15.01	97.14	16.66	96.53	18.30	95.86	19.91
46	97.68	15.06	97.12	16.72	96.51	18.35	95.84	19.96
48	97.66	15.12	97.10	16.77	96.49	18.41	95.82	20.02
50	97.64	15.17	97.08	16.83	96.47	18.46	95.79	20.07
52	97.62	15.23	97.06	16.88	96.45	18.51	95.77	20.12
54	97.61	15.28	97.04	16.94	96.42	18.57	95.75	20.18
56	97.59	15.34	97.02	16.99	96.40	18.62	95.72	20.23
58	97.57	15.40	97.00	17.05	96.38	18.68	95.70	20.28
60	97.55	15.45	96.98	17.10	96.36	18.73	95.68	20.34
$C =$ .75	.74	.11	.74	.12	.74	.14	.73	.15
$C =$ 1.00	.99	.15	.99	.17	.98	.18	.98	.20
$C =$ 1.25	1.24	.18	1.23	.21	1.23	.23	1.22	.25

TABLE II. STADIA REDUCTIONS

Minutes	12°		13°		14°		15°	
	Hor. Dist.	Diff. Elev.	Hor. Dist.	Diff. Elev.	Hor. Dist.	Diff. Elev.	Hor. Dist.	Diff. Elev.
0	95.68	20.34	94.94	21.92	94.15	23.47	93.30	25.00
2	95.65	20.39	94.91	21.97	94.12	23.52	93.27	25.05
4	95.63	20.44	94.89	22.02	94.09	23.58	93.24	25.10
6	95.61	20.50	94.86	22.08	94.07	23.63	93.21	25.15
8	95.58	20.55	94.84	22.13	94.04	23.68	93.18	25.20
10	95.56	20.60	94.81	22.18	94.01	23.73	93.16	25.25
12	95.53	20.66	94.79	22.23	93.98	23.78	93.13	25.30
14	95.51	20.71	94.76	22.28	93.95	23.83	93.10	25.35
16	95.49	20.76	94.73	22.34	93.93	23.88	93.07	25.40
18	95.46	20.81	94.71	22.39	93.90	23.93	93.04	25.45
20	95.44	20.87	94.68	22.44	93.87	23.99	93.01	25.50
22	95.41	20.92	94.66	22.49	93.84	24.04	92.98	25.55
24	95.39	20.97	94.63	22.54	93.82	24.09	92.95	25.60
26	95.36	21.03	94.60	22.60	93.79	24.14	92.92	25.65
28	95.34	21.08	94.58	22.65	93.76	24.19	92.89	25.70
30	95.32	21.13	94.55	22.70	93.73	24.24	92.86	25.75
32	95.29	21.18	94.52	22.75	93.70	24.29	92.83	25.80
34	95.27	21.24	94.50	22.80	93.67	24.34	92.80	25.85
36	95.24	21.29	94.47	22.85	93.65	24.39	92.77	25.90
38	95.22	21.34	94.44	22.91	93.62	24.44	92.74	25.95
40	95.19	21.39	94.42	22.96	93.59	24.49	92.71	26.00
42	95.17	21.45	94.39	23.01	93.56	24.55	92.68	26.05
44	95.14	21.50	94.36	23.06	93.53	24.60	92.65	26.10
46	95.12	21.55	94.34	23.11	93.50	24.65	92.62	26.15
48	95.09	21.60	94.31	23.16	93.47	24.70	92.59	26.20
50	95.07	21.66	94.28	23.22	93.45	24.75	92.56	26.25
52	95.04	21.71	94.26	23.27	93.42	24.80	92.53	26.30
54	95.02	21.76	94.23	23.32	93.39	24.85	92.49	26.35
56	94.99	21.81	94.20	23.37	93.36	24.90	92.46	26.40
58	94.97	21.87	94.17	23.42	93.33	24.95	92.43	26.45
60	94.94	21.92	94.15	23.47	93.30	25.00	92.40	26.50
$C =$ .75	.73	.16	.73	.18	.73	.19	.72	.20
$C =$ 1.00	.98	.22	.97	.23	.97	.25	.96	.27
$C =$ 1.25	1.22	.27	1.22	.29	1.21	.31	1.20	.33

TABLE II. STADIA REDUCTIONS

Minutes	16°		17°		18°		19°	
	Hor. Dist.	Diff. Elev.	Hor. Dist.	Diff. Elev.	Hor. Dist.	Diff. Elev.	Hor. Dist.	Diff. Elev.
0	92.40	26.50	91.45	27.96	90.45	29.39	89.40	30.78
2	92.37	26.55	91.42	28.01	90.42	29.44	89.36	30.83
4	92.34	26.59	91.39	28.06	90.38	29.48	89.33	30.87
6	92.31	26.64	91.35	28.10	90.35	29.53	89.29	30.92
8	92.28	26.69	91.32	28.15	90.31	29.58	89.26	30.97
10	92.25	26.74	91.29	28.20	90.28	29.62	89.22	31.01
12	92.22	26.79	91.26	28.25	90.24	29.67	89.18	31.06
14	92.19	26.84	91.22	28.30	90.21	29.72	89.15	31.10
16	92.15	26.89	91.19	28.34	90.18	29.76	89.11	31.15
18	92.12	26.94	91.16	28.39	90.14	29.81	89.08	31.19
20	92.09	26.99	91.12	28.44	90.11	29.86	89.04	31.24
22	92.06	27.04	91.09	28.49	90.07	29.90	89.00	31.28
24	92.03	27.09	91.06	28.54	90.04	29.95	88.97	31.33
26	92.00	27.13	91.02	28.58	90.00	30.00	88.93	31.38
28	91.97	27.18	90.99	28.63	89.97	30.04	88.89	31.42
30	91.93	27.23	90.96	28.68	89.93	30.09	88.86	31.47
32	91.90	27.28	90.92	28.73	89.90	30.14	88.82	31.51
34	91.87	27.33	90.89	28.77	89.86	30.18	88.78	31.56
36	91.84	27.38	90.86	28.82	89.83	30.23	88.75	31.60
38	91.81	27.43	90.82	28.87	89.79	30.28	88.71	31.65
40	91.77	27.48	90.79	28.92	89.76	30.32	88.67	31.69
42	91.74	27.52	90.76	28.96	89.72	30.37	88.64	31.74
44	91.71	27.57	90.72	29.01	89.69	30.41	88.60	31.78
46	91.68	27.62	90.69	29.06	89.65	30.46	88.56	31.83
48	91.65	27.67	90.66	29.11	89.61	30.51	88.53	31.87
50	91.61	27.72	90.62	29.15	89.58	30.55	88.49	31.92
52	91.58	27.77	90.59	29.20	89.54	30.60	88.45	31.96
54	91.55	27.81	90.55	29.25	89.51	30.65	88.41	32.01
56	91.52	27.86	90.52	29.30	89.47	30.69	88.38	32.05
58	91.48	27.91	90.49	29.34	89.44	30.74	88.34	32.09
60	91.45	27.96	90.45	29.39	89.40	30.78	88.30	32.14
$C = .75$	.72	.21	.72	.23	.71	.24	.71	.25
$C = 1.00$	.96	.28	.95	.30	.95	.32	.94	.33
$C = 1.25$	1.20	.36	1.19	.38	1.19	.40	1.18	.42

TABLE II. STADIA REDUCTIONS

Minutes	20°		21°		22°		23°	
	Hor. Dist.	Diff. Elev.	Hor. Dist.	Diff. Elev.	Hor. Dist.	Diff. Elev.	Hor. Dist.	Diff. Elev.
0	88.30	32.14	87.16	33.46	85.97	34.73	84.73	35.97
2	88.26	32.18	87.12	33.50	85.93	34.77	84.69	36.01
4	88.23	32.23	87.08	33.54	85.89	34.82	84.65	36.05
6	88.19	32.27	87.04	33.59	85.85	34.86	84.61	36.09
8	88.15	32.32	87.00	33.63	85.80	34.90	84.57	36.13
10	88.11	32.36	86.96	33.67	85.76	34.94	84.52	36.17
12	88.08	32.41	86.92	33.72	85.72	34.98	84.48	36.21
14	88.04	32.45	86.88	33.76	85.68	35.02	84.44	36.25
16	88.00	32.49	86.84	33.80	85.64	35.07	84.40	36.29
18	87.96	32.54	86.80	33.84	85.60	35.11	84.35	36.33
20	87.93	32.58	86.77	33.89	85.56	35.15	84.31	36.37
22	87.89	32.63	86.73	33.93	85.52	35.19	84.27	36.41
24	87.85	32.67	86.69	33.97	85.48	35.23	84.23	36.45
26	87.81	32.72	86.65	34.01	85.44	35.27	84.18	36.49
28	87.77	32.76	86.61	34.06	85.40	35.31	84.14	36.53
30	87.74	32.80	86.57	34.10	85.36	35.36	84.10	36.57
32	87.70	32.85	86.53	34.14	85.31	35.40	84.06	36.61
34	87.66	32.89	86.49	34.18	85.27	35.44	84.01	36.65
36	87.62	32.93	86.45	34.23	85.23	35.48	83.97	36.69
38	87.58	32.98	86.41	34.27	85.19	35.52	83.93	36.73
40	87.54	33.02	86.37	34.31	85.15	35.56	83.89	36.77
42	87.51	33.07	86.33	34.35	85.11	35.60	83.84	36.80
44	87.47	33.11	86.29	34.40	85.07	35.64	83.80	36.84
46	87.43	33.15	86.25	34.44	85.02	35.68	83.76	36.88
48	87.39	33.20	86.21	34.48	84.98	35.72	83.72	36.92
50	87.35	33.24	86.17	34.52	84.94	35.76	83.67	36.96
52	87.31	33.28	86.13	34.57	84.90	35.80	83.63	37.00
54	87.27	33.33	86.09	34.61	84.86	35.85	83.59	37.04
56	87.24	33.37	86.05	34.65	84.82	35.89	83.54	37.08
58	87.20	33.41	86.01	34.69	84.77	35.93	83.50	37.12
60	87.16	33.46	85.97	34.73	84.73	35.97	83.46	37.16
$C =$ .75	.70	.26	.70	.27	.69	.29	.69	.30
$C =$ 1.00	.94	.35	.93	.37	.92	.38	.92	.40
$C =$ 1.25	1.17	.44	1.16	.46	1.15	.48	1.15	.50

TABLE II. STADIA REDUCTIONS

Minutes	24°		25°		26°		27°	
	Hor. Dist.	Diff. Elev.	Hor. Dist.	Diff. Elev.	Hor. Dist.	Diff. Elev.	Hor. Dist.	Diff. Elev.
0	83.46	37.16	82.14	38.30	80.78	39.40	79.39	40.45
2	83.41	37.20	82.09	38.34	80.74	39.44	79.34	40.49
4	83.37	37.23	82.05	38.38	80.69	39.47	79.30	40.52
6	83.33	37.27	82.01	38.41	80.65	39.51	79.25	40.55
8	83.28	37.31	81.96	38.45	80.60	39.54	79.20	40.59
10	83.24	37.35	81.92	38.49	80.55	39.58	79.15	40.62
12	83.20	37.39	81.87	38.53	80.51	39.61	79.11	40.66
14	83.15	37.43	81.83	38.56	80.46	39.65	79.06	40.69
16	83.11	37.47	81.78	38.60	80.41	39.69	79.01	40.72
18	83.07	37.51	81.74	38.64	80.37	39.72	78.96	40.76
20	83.02	37.54	81.69	38.67	80.32	39.76	78.92	40.79
22	82.98	37.58	81.65	38.71	80.28	39.79	78.87	40.82
24	82.93	37.62	81.60	38.75	80.23	39.83	78.82	40.86
26	82.89	37.66	81.56	38.78	80.18	39.86	78.77	40.89
28	82.85	37.70	81.51	38.82	80.14	39.90	78.73	40.92
30	82.80	37.74	81.47	38.86	80.09	39.93	78.68	40.96
32	82.76	37.77	81.42	38.89	80.04	39.97	78.63	40.99
34	82.72	37.81	81.38	38.93	80.00	40.00	78.58	41.02
36	82.67	37.85	81.33	38.97	79.95	40.04	78.54	41.06
38	82.63	37.89	81.28	39.00	79.90	40.07	78.49	41.09
40	82.58	37.93	81.24	39.04	79.86	40.11	78.44	41.12
42	82.54	37.96	81.19	39.08	79.81	40.14	78.39	41.16
44	82.49	38.00	81.15	39.11	79.76	40.18	78.34	41.19
46	82.45	38.04	81.10	39.15	79.72	40.21	78.30	41.22
48	82.41	38.08	81.06	39.18	79.67	40.24	78.25	41.26
50	82.36	38.11	81.01	39.22	79.62	40.28	78.20	41.29
52	82.32	38.15	80.97	39.26	79.58	40.31	78.15	41.32
54	82.27	38.19	80.92	39.29	79.53	40.35	78.10	41.35
56	82.23	38.23	80.87	39.33	79.48	40.38	78.06	41.39
58	82.18	38.26	80.83	39.36	79.44	40.42	78.01	41.42
60	82.14	38.30	80.78	39.40	79.39	40.45	77.96	41.45
$C =$ .75	.68	.31	.68	.32	.67	.33	.67	.35
$C =$ 1.00	.91	.41	.90	.43	.89	.45	.89	.46
$C =$ 1.25	1.14	.52	1.13	.54	1.12	.56	1.11	.58

TABLE II. STADIA REDUCTIONS

Minutes	28°		29°		30°	
	Hor. Dist.	Diff. Elev.	Hor. Dist.	Diff. Elev.	Hor. Dist.	Diff. Elev.
0	77.96	41.45	76.50	42.40	75.00	43.30
2	77.91	41.48	76.45	42.43	74.95	43.33
4	77.86	41.52	76.40	42.46	74.90	43.36
6	77.81	41.55	76.35	42.49	74.85	43.39
8	77.77	41.58	76.30	42.53	74.80	43.42
10	77.72	41.61	76.25	42.56	74.75	43.45
12	77.67	41.65	76.20	42.59	74.70	43.47
14	77.62	41.68	76.15	42.62	74.65	43.50
16	77.57	41.71	76.10	42.65	74.60	43.53
18	77.52	41.74	76.05	42.68	74.55	43.56
20	77.48	41.77	76.00	42.71	74.49	43.59
22	77.42	41.81	75.95	42.74	74.44	43.62
24	77.38	41.84	75.90	42.77	74.39	43.65
26	77.33	41.87	75.85	42.80	74.34	43.67
28	77.28	41.90	75.80	42.83	74.29	43.70
30	77.23	41.93	75.75	42.86	74.24	43.73
32	77.18	41.97	75.70	42.89	74.19	43.76
34	77.13	42.00	75.65	42.92	74.14	43.79
36	77.09	42.03	75.60	42.95	74.09	43.82
38	77.04	42.06	75.55	42.98	74.04	43.84
40	76.99	42.09	75.50	43.01	73.99	43.87
42	76.94	42.12	75.45	43.04	73.93	43.90
44	76.89	42.15	75.40	43.07	73.88	43.93
46	76.84	42.19	75.35	43.10	73.83	43.95
48	76.79	42.22	75.30	43.13	73.78	43.98
50	76.74	42.25	75.25	43.16	73.73	44.01
52	76.69	42.28	75.20	43.18	73.68	44.04
54	76.64	42.31	75.15	43.21	73.63	44.07
56	76.59	42.34	75.10	43.24	73.58	44.09
58	76.55	42.37	75.05	43.27	73.52	44.12
60	76.50	42.40	75.00	43.30	73.47	44.15
$C =$ .75	.66	.36	.65	.37	.65	.38
$C =$ 1.00	.88	.48	.87	.49	.86	.51
$C =$ 1.25	1.10	.60	1.09	.62	1.08	.63

TABLE III. TABLE OF DISTANCES FOR SUBTENSE BAR (IN FEET)*
(Sample Pages, for Interval 0° 30′ to 0° 38′ Only)

0°	0″	1″	2″	3″	4″	5″	6″	7″	8″	9″	0°
30′00″	**751.9**	**751.5**	**751.1**	**750.7**	**750.2**	**749.8**	**749.4**	**749.0**	**748.6**	**748.2**	**30′00″**
10	747.7	747.3	746.9	746.5	746.1	745.7	745.3	744.9	744.5	744.0	10
20	743.6	743.2	742.8	742.4	742.0	741.6	741.2	740.8	740.4	740.0	20
30	739.6	739.2	738.8	738.4	738.0	737.6	737.2	736.8	736.4	736.0	30
40	735.6	735.2	734.8	734.4	734.0	733.6	733.2	732.8	732.4	732.0	40
50	731.6	731.2	730.8	730.4	730.0	729.6	729.2	728.9	728.5	728.1	50
31′00″	**727.7**	**727.3**	**726.9**	**726.5**	**726.1**	**725.7**	**725.3**	**724.9**	**724.6**	**724.2**	**31′00″**
10	723.8	723.4	723.0	722.6	722.2	721.8	721.5	721.1	720.7	720.3	10
20	719.9	719.5	719.2	718.8	718.4	718.0	717.6	717.2	716.9	716.5	20
30	716.1	715.7	715.4	715.0	714.6	714.2	713.9	713.5	713.1	712.7	30
40	712.3	712.0	711.6	711.2	710.8	710.5	710.1	709.7	709.4	709.0	40
50	708.6	708.2	707.9	707.5	707.1	706.8	706.4	706.0	705.7	705.3	50
32′00″	**704.9**	**704.5**	**704.2**	**703.8**	**703.5**	**703.1**	**702.7**	**702.4**	**702.0**	**701.6**	**32′00″**
10	701.3	700.9	700.5	700.2	699.8	699.5	699.1	698.7	698.4	698.0	10
20	697.7	697.3	696.9	696.6	696.2	695.9	695.5	695.2	694.8	694.4	20
30	694.1	693.7	693.4	693.0	692.7	692.3	691.9	691.6	691.3	690.9	30
40	690.5	690.2	689.8	689.5	689.1	688.8	688.4	688.1	687.7	687.4	40
50	687.0	686.7	686.3	686.0	685.6	685.3	685.0	684.6	684.3	683.9	50
33′00″	**683.6**	**683.2**	**682.9**	**682.5**	**682.2**	**681.8**	**681.5**	**681.1**	**680.8**	**680.5**	**33′00″**
10	680.1	679.8	679.4	679.1	678.8	678.4	678.1	677.7	677.4	677.0	10
20	676.7	676.4	676.0	675.7	675.3	675.0	674.7	674.4	674.0	673.7	20
30	673.4	673.0	672.7	672.4	672.0	671.7	671.4	671.0	670.7	670.4	30
40	670.0	669.7	699.4	669.0	668.7	668.4	668.1	667.7	667.4	667.0	40
50	666.7	666.4	666.1	665.7	665.4	665.1	664.8	664.4	664.1	663.8	50

0°	0″	1″	2″	3″	4″	5″	6″	7″	8″	9″	0°
34′00″	**663.5**	**663.1**	**662.8**	**662.5**	**662.2**	**661.8**	**661.5**	**661.2**	**660.9**	**660.5**	**34′00″**
10	660.2	659.9	659.6	659.3	658.9	658.6	658.3	658.0	657.7	657.3	10
20	657.0	656.7	656.4	656.1	655.8	655.4	655.1	654.8	654.5	654.2	20
30	653.9	653.5	653.2	652.9	652.6	652.3	651.9	651.6	651.3	651.0	30
40	650.7	650.4	650.1	649.7	649.4	649.1	648.8	648.5	648.2	647.9	40
50	647.6	647.3	647.0	646.6	646.3	646.0	645.7	645.4	645.1	644.8	50
35′00″	**644.5**	**644.2**	**643.9**	**643.6**	**643.3**	**643.0**	**642.7**	**642.4**	**642.1**	**641.7**	**35′00″**
10	641.4	641.1	640.8	640.5	640.2	639.9	639.6	639.3	639.0	638.7	10
20	638.4	638.1	637.8	637.5	637.2	636.9	636.6	636.3	636.0	635.7	20
30	635.4	635.1	634.8	634.5	634.2	633.9	633.6	633.3	633.0	632.8	30
40	632.5	632.2	631.9	631.6	631.3	631.0	630.7	630.4	630.1	629.8	40
50	629.5	629.2	628.9	628.6	628.4	628.1	627.8	627.5	627.2	626.9	50
36′00″	**626.6**	**626.3**	**626.0**	**625.7**	**625.4**	**625.1**	**624.9**	**624.6**	**624.3**	**624.0**	**36′00″**
10	623.7	623.4	623.1	622.8	622.6	622.3	622.0	621.7	621.4	621.1	10
20	620.8	620.5	620.3	620.0	619.7	619.4	619.1	618.8	618.6	618.3	20
30	618.0	617.7	617.5	617.2	616.9	616.6	616.3	616.1	615.8	615.5	30
40	615.2	614.9	614.6	614.3	614.1	613.8	613.5	613.2	613.0	612.7	40
50	612.4	612.1	611.9	611.6	611.3	611.0	610.8	610.5	610.2	609.9	50
37′00″	**609.7**	**609.4**	**609.1**	**608.8**	**608.6**	**608.3**	**608.0**	**607.8**	**607.5**	**607.2**	**37′00″**
10	606.9	606.6	606.4	606.1	605.8	605.6	605.3	605.0	604.8	604.5	10
20	604.2	603.9	603.7	603.4	603.1	602.9	602.6	602.3	602.1	601.8	20
30	601.5	601.3	601.0	600.7	600.5	600.2	599.9	599.7	599.4	599.1	30
40	598.9	598.6	598.3	598.1	597.8	597.6	597.3	597.0	596.8	596.5	40
50	596.2	596.0	595.7	595.5	595.2	594.9	594.7	594.4	594.1	593.9	50
38′00″	**593.6**	**593.4**	**593.1**	**592.8**	**592.6**	**592.3**	**592.1**	**591.8**	**591.5**	**591.3**	**38′00″**

* Courtesy of Kern Instruments, Inc. The original of these two pages also includes values for 39′, and 40′.

TABLE IV. CONVERGENCY OF RANGE LINES

Latitude Degrees	Difference Between South and North Boundaries of Township Links	Angle of Convergency of Adjacent Range Lines ′ ″	Difference of Longitude per Range		Difference of Latitude, in Minutes of Arc, for	
			Arc ′ ″	Time Seconds	1 Mile in Arc	6 Miles in Arc
25	33.9	2 25	5 44.34	22.96		
26	35.4	2 32	5 47.20	23.15		
27	37.0	2 39	5 50.22	23.35	0.871	5.229
28	38.6	2 46	5 53.40	23.56		
29	40.2	2 53	5 56.74	23.78		
30	41.9	3 0	6 0.36	24.02		
31	43.6	3 7	6 4.02	24.27		
32	45.4	3 15	6 7.93	24.53	0.871	5.225
33	47.2	3 23	6 12.00	24.80		
34	49.1	3 30	6 16.31	25.09		
35	50.9	3 38	6 20.95	25.40		
36	52.7	3 46	6 25.60	25.71		
37	54.7	3 55	6 30.59	26.04	0.870	5.221
38	56.8	4 4	6 35.81	26.39		
39	58.8	4 13	6 41.34	26.76		
40	60.9	4 22	6 47.13	27.14		
41	63.1	4 31	6 53.22	27.55		
42	65.4	4 41	6 59.62	27.97	0.869	5.217
43	67.7	4 51	7 6.27	28.42		
44	70.1	5 1	7 13.44	28.90		
45	72.6	5 12	7 20.93	29.39		
46	75.2	5 23	7 28.81	29.92		
47	77.8	5 34	7 37.10	30.47	0.869	5.212
48	80.6	5 46	7 45.79	31.05		
49	83.5	5 59	7 55.12	31.67		
50	86.4	6 12	8 4.83	32.32		
51	89.6	6 25	8 15.17	33.01		
52	92.8	6 39	8 26.13	33.74	0.868	5.207
53	96.2	6 54	8 37.75	34.52		
54	99.8	7 9	8 50.07	35.34		
55	103.5	7 25	9 3.18	36.22		
56	107.5	7 42	9 17.12	37.14		
57	111.6	8 0	9 31.97	38.13	0.867	5.202
58	116.0	8 19	9 47.83	39.19		
59	120.6	8 38	10 4.78	40.32		
60	125.5	8 59	10 22.94	41.52		
61	130.8	9 22	10 42.42	42.83		
62	136.3	9 46	11 3.38	44.22	0.866	5.198
63	142.2	10 11	11 25.97	45.73		
64	148.6	10 38	11 50.37	47.36		
65	155.0	11 8	12 16.82	49.12		
66	162.8	11 39	12 45.55	51.04		
67	170.7	12 13	13 16.88	53.12	0.866	5.195
68	179.3	12 51	13 51.15	55.41		
69	188.7	13 31	14 28.77	57.92		
70	199.1	14 15	15 10.26	60.68	0.866	5.193

TABLE V. AZIMUTHS OF THE SECANT*

Lat.	0 mi.	1 mi.	2 mi.	3 mi.	Deflection angle 6 mi.
°	° ′	° ′	° ′		′ ″
25	89 58.8	89 59.2	89 59.6	90°	2 25
26	58.7	59.2	59.6	E or W.	2 32
27	58.7	59.1	59.6	" " "	2 39
28	58.6	59.1	59.5	" " "	2 46
29	58.6	59.0	59.5	" " "	2 53
30	58.5	59.0	59.5	" " "	3 0
31	58.4	59.0	59.5	" " "	3 7
32	58.4	58.9	59.5	" " "	3 15
33	58.3	58.9	59.4	" " "	3 23
34	58.2	58.8	59.4	" " "	3 30
35	58.2	58.8	59.4	" " "	3 38
36	58.1	58.7	59.4	" " "	3 46
37	58.0	58.7	59.3	" " "	3 55
38	58.0	58.6	59.3	" " "	4 4
39	57.9	58.6	59.3	" " "	4 13
40	57.8	58.5	59.3	" " "	4 22
41	57.7	58.5	59.2	" " "	4 31
42	57.7	58.4	59.2	" " "	4 41
43	57.6	58.4	59.2	" " "	4 51
44	57.5	58.3	59.2	" " "	5 1
45	57.4	58.3	59.1	" " "	5 12
46	57.3	58.2	59.1	" " "	5 23
47	57.2	58.1	59.1	" " "	5 34
48	57.1	58.1	59.0	" " "	5 46
49	57.0	58.0	59.0	" " "	5 59
50	56.9	57.9	59.0	" " "	6 12
51	56.8	57.9	58.9	" " "	6 25
52	56.7	57.8	58.9	" " "	6 39
53	56.6	57.7	58.8	" " "	6 54
54	56.4	57.6	58.8	" " "	7 9
55	56.3	57.5	58.8	" " "	7 25
56	56.2	57.4	58.7	" " "	7 42
57	56.0	57.3	58.7	" " "	8 0
58	55.8	57.2	58.6	" " "	8 19
59	55.7	57.1	58.6	" " "	8 38
60	55.5	57.0	58.5	" " "	8 59
61	55.3	56.9	58.4	" " "	9 22
62	55.1	56.7	58.4	" " "	9 46
63	54.9	56.6	58.3	" " "	10 11
64	54.7	56.5	58.2	" " "	10 38
65	54.4	56.3	58.1	" " "	11 8
66	54.2	56.1	58.1	" " "	11 39
67	53.9	55.9	58.0	" " "	12 13
68	53.6	55.7	57.9	" " "	12 51
69	53.2	55.5	57.8	" " "	13 31
70	89 52.9	89 55.3	89 57.6	" " "	14 15
	6 mi.	5 mi.	4 mi.	3 mi.	

* From U.S. Department of the Interior, *Standard Field Tables and Trigonometric Formulas*, 8th ed. (GPO, 1950).

TABLE VI. OFFSETS, IN LINKS, FROM THE SECANT TO THE PARALLEL*

Lat.	0 mi.	½ mi.	1 mi.	1½ mi.	2 mi.	2½ mi.	3 mi.
°							
25	2 N.	1 N.	0	1 S.	1 S.	2 S.	2 S.
26	2	1	0	1	1	2	2
27	3	1	0	1	2	2	2
28	3	1	0	1	2	2	2
29	3	1	0	1	2	2	2
30	3	1	0	1	2	2	2
31	3	1	0	1	2	2	2
32	3	1	0	1	2	2	3
33	3	1	0	1	2	2	3
34	3	2	0	1	2	3	3
35	4	2	0	1	2	3	3
36	4	2	0	1	2	3	3
37	4	2	0	1	2	3	3
38	4	2	0	1	2	3	3
39	4	2	0	1	2	3	3
40	4	2	0	1	3	3	3
41	4	2	0	2	3	3	4
42	5	2	0	2	3	3	4
43	5	2	0	2	3	4	4
44	5	2	0	2	3	4	4
45	5	2	0	2	3	4	4
46	5	2	0	2	3	4	4
47	5	2	0	2	3	4	4
48	6	3	0	2	3	4	4
49	6	3	0	2	3	4	5
50	6	3	0	2	4	4	5
51	6	3	0	2	4	5	5
52	6	3	0	2	4	5	5
53	7	3	0	2	4	5	5
54	7	3	0	2	4	5	6
55	7	3	0	3	4	5	6
56	7	3	0	3	4	6	6
57	8	3	0	3	5	6	6
58	8	4	0	3	5	6	6
59	8	4	0	3	5	6	7
60	9	4	0	3	5	7	7
61	9	4	0	3	5	7	7
62	9	4	0	3	6	7	8
63	10	4	0	3	6	7	8
64	10	5	0	4	6	8	8
65	11	5	0	4	6	8	9
66	11	5	0	4	7	8	9
67	12	5	0	4	7	9	9
68	12	6	0	4	7	9	10
69	13	6	0	5	8	10	10
70	14 N.	6 N.	0	5 S.	8 S.	10 S.	11 S.
	6 mi.	5½ mi.	5 mi.	4½ mi.	4 mi.	3½ mi.	3 mi.

* From U.S. Department of the Interior, *Standard Field Tables and Trigonometric Formulas*, 8th ed. (GPO, 1950).

TABLE VII. FUNCTIONS OF CIRCULAR CURVES

Degree of Curve D	Defl. Per Ft of Sta. (Min)	Chord Definition			Arc Definition	
		Radius R	log R	C.O. 1 Sta.	Radius R	log R
0° 0'		Infinite	Infin.		Infinite	Infin.
1'	0.005	343775.	5.536274	0.03	343775.	5.536274
2'	0.01	171887.	5.235244	0.06	171887.	5.235244
3'	0.015	114592.	5.059153	0.09	114592.	5.059153
4'	0.02	85943.7	4.934214	0.12	85943.7	4.934214
5'	0.025	68754.9	4.837304	0.15	68754.9	4.837304
6'	0.03	57295.8	4.758123	0.17	57295.8	4.758123
7'	0.035	49110.7	4.691176	0.20	49110.7	4.691176
8'	0.04	42971.8	4.633184	0.23	42971.8	4.633184
9'	0.045	38197.2	4.582031	0.26	38197.2	4.582031
10'	0.05	34377.5	4.536274	0.29	34377.5	4.536274
11'	0.055	31252.3	4.494881	0.32	31252.2	4.494881
12'	0.06	28647.8	4.457093	0.35	28647.8	4.457093
13'	0.065	26444.2	4.422331	0.38	26444.2	4.422331
14'	0.07	24555.4	4.390146	0.41	24555.3	4.390146
15'	0.075	22918.3	4.360183	0.44	22918.3	4.360183
16'	0.08	21485.9	4.332154	0.47	21485.9	4.332154
17'	0.085	20222.1	4.305825	0.49	20222.0	4.305825
18'	0.09	19098.6	4.281002	0.52	19098.6	4.281001
19'	0.095	18093.4	4.257521	0.55	18093.4	4.257520
20'	0.1	17188.8	4.235244	0.58	17188.7	4.235244
21'	0.105	16370.2	4.214055	0.61	16370.2	4.214055
22'	0.11	15626.1	4.193852	0.64	15626.1	4.193851
23'	0.115	14946.8	4.174547	0.67	14946.7	4.174546
24'	0.12	14324.0	4.156064	0.70	14323.9	4.156063
25'	0.125	13751.0	4.138335	0.73	13751.0	4.138334
26'	0.13	13222.1	4.121302	0.76	13222.1	4.121300
27'	0.135	12732.4	4.104911	0.79	12732.4	4.104910
28'	0.14	12277.7	4.089117	0.81	12277.7	4.089116
29'	0.145	11854.3	4.073877	0.84	11854.3	4.073876
30'	0.15	11459.2	4.059154	0.87	11459.2	4.059153
31'	0.155	11089.6	4.044914	0.90	11089.5	4.044912
32'	0.16	10743.0	4.031125	0.93	10743.0	4.031124
33'	0.165	10417.5	4.017762	0.96	10417.4	4.017760
34'	0.17	10111.1	4.004797	0.99	10111.0	4.004795
35'	0.175	9822.18	3.992208	1.02	9822.13	3.992206
36'	0.18	9549.34	3.979973	1.05	9549.29	3.979971
37'	0.185	9291.25	3.968074	1.07	9291.21	3.968072
38'	0.19	9046.75	3.956493	1.11	9046.70	3.956490
39'	0.195	8814.78	3.945212	1.13	8814.73	3.945209
40'	0.2	8594.42	3.934216	1.16	8594.37	3.934214
41'	0.205	8384.80	3.923493	1.19	8384.75	3.923490
42'	0.21	8185.16	3.913027	1.21	8185.11	3.913025
43'	0.215	7994.81	3.902808	1.25	7994.76	3.902805
44'	0.22	7813.11	3.892824	1.28	7813.06	3.892821
45'	0.225	7639.49	3.883065	1.31	7639.44	3.883061
46'	0.23	7473.42	3.873519	1.34	7473.36	3.873516
47'	0.235	7314.41	3.864179	1.37	7314.35	3.864176
48'	0.24	7162.03	3.855036	1.39	7161.97	3.855033
49'	0.245	7015.87	3.846082	1.43	7015.81	3.846078
50'	0.25	6875.55	3.837308	1.45	6875.49	3.837304
51'	0.255	6740.74	3.828708	1.48	6740.68	3.828704
52'	0.26	6611.12	3.820275	1.51	6611.05	3.820270
53'	0.265	6486.38	3.812002	1.54	6486.31	3.811998
54'	0.27	6366.26	3.803885	1.57	6366.20	3.803880
55'	0.275	6250.51	3.795916	1.60	6250.45	3.795911
56'	0.28	6138.90	3.788091	1.63	6138.83	3.788086
57'	0.285	6031.20	3.780404	1.66	6031.14	3.780399
58'	0.29	5927.22	3.772851	1.69	5927.15	3.772846
59'	0.295	5826.76	3.765427	1.72	5826.69	3.765422

TABLE VII. FUNCTIONS OF CIRCULAR CURVES

Degree of Curve D	Defl. Per Ft of Sta. (Min)	Chord Definition			Arc Definition	
		Radius R	log R	C.O. 1 Sta.	Radius R	log R
1° 0'	0.3	5729.65	3.758128	1.75	5729.58	3.758123
1'	0.305	5635.72	3.750950	1.77	5635.65	3.750944
2'	0.31	5544.83	3.743888	1.80	5544.75	3.743882
3'	0.315	5456.82	3.736939	1.83	5456.74	3.736933
4'	0.32	5371.56	3.730100	1.86	5371.48	3.730094
5'	0.325	5288.92	3.723367	1.89	5288.84	3.723360
6'	0.33	5208.79	3.716737	1.92	5208.71	3.716730
7'	0.335	5131.05	3.710206	1.95	5130.97	3.710199
8'	0.34	5055.59	3.703772	1.98	5055.51	3.703765
9'	0.345	4982.33	3.697432	2.01	4982.24	3.697425
10'	0.35	4911.15	3.691183	2.03	4911.07	3.691176
11'	0.355	4841.98	3.685023	2.07	4841.90	3.685015
12'	0.36	4774.74	3.678949	2.09	4774.65	3.678941
13'	0.365	4709.33	3.672959	2.12	4709.24	3.672951
14'	0.37	4645.69	3.667051	2.15	4645.60	3.667042
15'	0.375	4583.75	3.661221	2.18	4583.66	3.661213
16'	0.38	4523.44	3.655469	2.21	4523.35	3.655460
17'	0.385	4464.70	3.649792	2.24	4464.61	3.649783
18'	0.39	4407.46	3.644189	2.27	4407.37	3.644179
19'	0.395	4351.67	3.638656	2.30	4351.58	3.638647
20'	0.4	4297.28	3.633194	2.33	4297.18	3.633184
21'	0.405	4244.23	3.627799	2.35	4244.13	3.627789
22'	0.41	4192.47	3.622470	2.39	4192.37	3.622460
23'	0.415	4141.96	3.617206	2.41	4141.86	3.617196
24'	0.42	4092.66	3.612005	2.44	4092.56	3.611995
25'	0.425	4044.51	3.606866	2.47	4044.41	3.606855
26'	0.43	3997.49	3.601787	2.50	3997.38	3.601775
27'	0.435	3951.54	3.596766	2.53	3951.43	3.596755
28'	0.44	3906.64	3.591803	2.56	3906.53	3.591791
29'	0.445	3862.74	3.586896	2.59	3862.64	3.586884
30'	0.45	3819.83	3.582044	2.62	3819.71	3.582031
31'	0.455	3777.85	3.577245	2.65	3777.74	3.577232
32'	0.46	3736.79	3.572499	2.68	3736.68	3.572486
33'	0.465	3696.61	3.567804	2.71	3696.50	3.567791
34'	0.47	3657.29	3.563160	2.73	3657.18	3.563146
35'	0.475	3618.80	3.558564	2.76	3618.68	3.558550
36'	0.48	3581.10	3.554017	2.79	3580.99	3.554003
37'	0.485	3544.19	3.549517	2.82	3544.07	3.549502
38'	0.49	3508.02	3.545063	2.85	3507.91	3.545048
39'	0.495	3472,59	3.540654	2.88	3472.47	3.540638
40'	0.5	3437.87	3.536289	2.91	3437.75	3.536274
41'	0.505	3403.83	3.531968	2.94	3403.71	3.531952
42'	0.51	3370.46	3.527690	2.97	3370.34	3.527673
43'	0.515	3337.74	3.523453	3.00	3337.62	3.523437
44'	0.52	3305.65	3.519257	3.03	3305.53	3.519241
45'	0.525	3274.17	3.515101	3.05	3274.04	3.515085
46'	0.53	3243.29	3.510985	3.08	3243.16	3.510968
47'	0.535	3212.98	3.506908	3.11	3212.85	3.506890
48'	0.54	3183.23	3.502868	3.14	3183.10	3.502850
49'	0.545	3154.03	3.498866	3.17	3153.90	3.498847
50'	0.55	3125.36	3.494900	3.20	3125.22	3.494881
51'	0.555	3097.20	3.490970	3.23	3097.07	3.490951
52'	0.56	3069.55	3.487075	3.26	3069.42	3.487056
53'	0.565	3042.39	3.483215	3.29	3042.25	3.483195
54'	0.57	3015.71	3.479389	3.32	3015.57	3.479369
55'	0.575	2989.48	3.475596	3.35	2989.34	3.475576
56'	0.58	2963.72	3.471836	3.37	2963.58	3.471816
57'	0.585	2938.39	3.468109	3.40	2938.25	3.468087
58'	0.59	2913.49	3.464413	3.43	2913.34	3.464392
59'	0.595	2889.01	3.460749	3.46	2888.86	3.460727

TABLE VII. FUNCTIONS OF CIRCULAR CURVES

Degree of Curve D	Defl. Per Ft of Sta. (Min)	Chord Definition			Arc Definition	
		Radius R	log R	C.O. 1 Sta.	Radius R	log R
2° 0'	0.6	2864.93	3.457115	3.49	2864.79	3.457093
1'	0.605	2841.26	3.453511	3.52	2841.11	3.453488
2'	0.61	2817.97	3.449937	3.55	2817.83	3.449914
3'	0.615	2795.06	3.446392	3.58	2794.92	3.446369
4'	0.62	2772.53	3.442876	3.61	2772.38	3.442852
5'	0.625	2750.35	3.439388	3.64	2750.20	3.439364
6'	0.63	2728.52	3.435928	3.66	2728.37	3.435903
7'	0.635	2707.04	3.432495	3.69	2706.89	3.432470
8'	0.64	2685.89	3.429089	3.72	2685.74	3.429064
9'	0.645	2665.08	3.425710	3.75	2664.92	3.425684
10'	0.65	2644.58	3.422356	3.78	2644.42	3.422330
11'	0.655	2624.39	3.419029	3.81	2624.23	3.419002
12'	0.66	2604.51	3.415727	3.84	2604.35	3.415700
13'	0.665	2584.93	3.412449	3.87	2584.77	3.412422
14'	0.67	2565.65	3.409197	3.90	2565.48	3.409169
15'	0.675	2546.64	3.405968	3.93	2546.48	3.405940
16'	0.68	2527.92	3.402763	3.96	2527.75	3.402735
17'	0.685	2509.47	3.399582	3.98	2509.30	3.399553
18'	0.69	2491.29	3.396424	4.01	2491.12	3.396395
19'	0.695	2473.37	3.393289	4.04	2473.20	3.393259
20'	0.7	2455.70	3.390176	4.07	2455.53	3.390145
21'	0.705	2438.29	3.387085	4.10	2438.12	3.387055
22'	0.71	2421.12	3.384016	4.13	2420.95	3.383985
23'	0.715	2404.19	3.380969	4.16	2404.02	3.380938
24'	0.72	2387.50	3.377943	4.19	2387.32	3.377911
25'	0.725	2371.04	3.374938	4.22	2370.86	3.374905
26'	0.73	2354.80	3.371954	4.25	2354.62	3.371921
27'	0.735	2338.78	3.368990	4.28	2338.60	3.368956
28'	0.74	2322.98	3.366046	4.30	2322.80	3.366012
29'	0.745	2307.39	3.363122	4.33	2307.21	3.363087
30'	0.75	2292.01	3.360217	4.36	2291.83	3.360183
31'	0.755	2276.84	3.357332	4.39	2276.65	3.357297
32'	0.76	2261.86	3.354466	4.42	2261.68	3.354430
33'	0.765	2247.08	3.351618	4.45	2246.89	3.351582
34'	0.77	2232.49	3.348789	4.48	2232.30	3.348753
35'	0.775	2218.09	3.345797	4.51	2217.90	3.345942
36'	0.78	2203.87	3.343187	4.54	2203.68	3.343149
37'	0.785	2189.84	3.340412	4.57	2189.65	3.340374
38'	0.79	2175.98	3.337655	4.60	2175.79	3.337617
39'	0.795	2162.30	3.334916	4.62	2162.10	3.334877
40'	0.8	2148.79	3.332193	4.65	2148.59	3.332154
41'	0.805	2135.44	3.329488	4.68	2135.25	3.329448
42'	0.81	2122.26	3.326799	4.71	2122.07	3.326759
43'	0.815	2109.24	3.324127	4.74	2109.05	3.324086
44'	0.82	2096.39	3.321471	4.77	2096.19	3.321430
45'	0.825	2083.68	3.318832	4.80	2083.48	3.318790
46'	0.83	2071.13	3.316208	4.83	2070.93	3.316166
47'	0.835	2058.73	3.313600	4.86	2058.53	3.313557
48'	0.84	2046.48	3.311008	4.89	2046.28	3.310964
49'	0.845	2034.37	3.308431	4.92	2034.17	3.308387
50'	0.85	2022.41	3.305869	4.94	2022.20	3.305825
51'	0.855	2010.59	3.303323	4.97	2010.38	3.303278
52'	0.86	1998.90	3.300791	5.00	1998.69	3.300745
53'	0.865	1987.35	3.298274	5.03	1987.14	3.298228
54'	0.87	1975.93	3.295771	5.06	1975.72	3.295725
55'	0.875	1964.64	3.293283	5.09	1964.43	3.293236
56'	0.88	1953.48	3.290809	5.12	1953.27	3.290761
57'	0.885	1942.44	3.288349	5.15	1942.23	3.288301
58'	0.89	1931.53	3.285902	5.18	1931.32	3.285854
59'	0.895	1920.75	3.283470	5.21	1920.53	3.283421

TABLE VII. FUNCTIONS OF CIRCULAR CURVES

Degree of Curve D	Defl. Per Ft of Sta. (Min)	Chord Definition			Arc Definition	
		Radius R	log R	C.O. 1 Sta.	Radius R	log R
3° 0'	0.9	1910.08	3.281051	5.24	1909.86	3.281001
1'	0.905	1899.53	3.278646	5.26	1899.31	3.278595
2'	0.91	1889.09	3.276253	5.29	1888.87	3.276203
3'	0.915	1878.77	3.273874	5.32	1878.55	3.273823
4'	0.92	1868.56	3.271508	5.35	1868.34	3.271456
5'	0.925	1858.47	3.269155	5.38	1858.24	3.269102
6'	0.93	1848.48	3.266814	5.41	1848.25	3.266761
7'	0.935	1838.59	3.264486	5.44	1838.37	3.264432
8'	0.94	1828.82	3.262170	5.47	1828.59	3.262116
9'	0.945	1819.14	3.259867	5.50	1818.91	3.259812
10'	0.95	1809.57	3.257576	5.53	1809.34	3.257520
11'	0.955	1800.10	3.255296	5.56	1799.87	3.255240
12'	0.96	1790.73	3.253029	5.58	1790.49	3.252973
13'	0.965	1781.45	3.250774	5.61	1781.22	3.250716
14'	0.97	1772.27	3.248530	5.64	1772.03	3.248472
15'	0.975	1763.18	3.246297	5.67	1762.95	3.246239
16'	0.98	1754.19	3.244077	5.70	1753.95	3.244018
17'	0.985	1745.29	3.241867	5.73	1745.05	3.241808
18'	0.99	1736.48	3.239669	5.76	1736.24	3.239609
19'	0.995	1727.75	3.237481	5.79	1727.51	3.237421
20'	1.0	1719.12	3.235305	5.82	1718.87	3.235244
21'	1.005	1710.57	3.233140	5.85	1710.32	3.233078
22'	1.01	1702.10	3.230985	5.88	1701.85	3.230922
23'	1.015	1693.72	3.228841	5.90	1693.47	3.228778
24'	1.02	1685.42	3.226707	5.93	1685.17	3.226644
25'	1.025	1677.20	3.224584	5.96	1676.95	3.224520
26'	1.03	1669.06	3.222472	5.99	1668.81	3.222407
27'	1.035	1661.00	3.220369	6.02	1660.75	3.220303
28'	1.04	1653.01	3.218277	6.05	1652.76	3.218210
29'	1.045	1645.11	3.216195	6.08	1644.85	3.216128
30'	1.05	1637.28	3.214122	6.11	1637.02	3.214055
31'	1.055	1629.52	3.212060	6.14	1629.26	3.211991
32'	1.06	1621.84	3.210007	6.17	1621.58	3.209938
33'	1.065	1614.22	3.207964	6.19	1613.96	3.207894
34'	1.07	1606.68	3.205930	6.22	1606.42	3.205860
35'	1.075	1599.21	3.203906	6.25	1598.95	3.203835
36'	1.08	1591.81	3.201892	6.28	1591.55	3.201820
37'	1.085	1584.48	3.199886	6.31	1584.21	3.199814
38'	1.09	1577.21	3.197890	6.34	1576.95	3.197817
39'	1.095	1570.01	3.195903	6.37	1569.75	3.195830
40'	1.1	1562.88	3.193925	6.40	1562.61	3.193851
41'	1.105	1555.81	3.191956	6.43	1555.54	3.191881
42'	1.11	1548.80	3.189996	6.46	1548.53	3.189921
43'	1.115	1541.86	3.188045	6.49	1541.59	3.187969
44'	1.12	1534.98	3.186103	6.51	1534.71	3.186026
45'	1.125	1528.16	3.184169	6.54	1527.89	3.184091
46'	1.13	1521.40	3.182244	6.57	1521.13	3.182165
47'	1.135	1514.17	3.180327	6.60	1514.43	3.180248
48'	1.14	1508.06	3.178419	6.63	1507.78	3.178339
49'	1.145	1501.48	3.176519	6.66	1501.20	3.176438
50'	1.15	1494.95	3.174627	6.69	1494.67	3.174546
51'	1.155	1488.48	3.172744	6.72	1488.20	3.172661
52'	1.16	1482.07	3.170868	6.75	1481.79	3.170786
53'	1.165	1475.71	3.169001	6.78	1475.43	3.168918
54'	1.17	1469.41	3.167142	6.81	1469.12	3.167058
55'	1.175	1463.16	3.165291	6.83	1462.87	3.165206
56'	1.18	1456.96	3.163447	6.86	1456.67	3.163362
57'	1.185	1450.81	3.161612	6.89	1450.53	3.161526
58'	1.19	1444.72	3.159784	6.92	1444.43	3.159697
59'	1.195	1438.68	3.157963	6.95	1438.39	3.157876

TABLE VII. FUNCTIONS OF CIRCULAR CURVES

Degree of Curve D	Defl. Per Ft of Sta. (Min)	Chord Definition			Arc Definition	
		Radius R	log R	C.O. 1 Sta.	Radius R	log R
4° 0'	1.2	1432.69	3.156151	6.98	1432.39	3.156063
1'	1.205	1426.74	3.154346	7.01	1426.45	3.154257
2'	1.21	1420.85	3.152548	7.04	1420.56	3.152458
3'	1.215	1415.01	3.150758	7.07	1414.71	3.150667
4'	1.22	1409.21	3.148975	7.10	1408.91	3.148884
5'	1.225	1403.46	3.147200	7.13	1403.16	3.147108
6'	1.23	1397.76	3.145431	7.15	1397.46	3.145339
7'	1.235	1392.10	3.143670	7.18	1391.80	3.143577
8'	1.24	1386.49	3.141916	7.21	1386.19	3.141822
9'	1.245	1380.92	3.140170	7.24	1380.62	3.140074
10'	1.25	1375.40	3.138430	7.27	1375.10	3.138334
11'	1.255	1369.92	3.136697	7.30	1369.62	3.136600
12'	1.26	1364.49	3.134971	7.33	1364.19	3.134873
13'	1.265	1359.10	3.133251	7.36	1358.79	3.133153
14'	1.27	1353.75	3.131539	7.39	1353.44	3.131440
15'	1.275	1348.45	3.129833	7.42	1348.14	3.129734
16'	1.28	1343.18	3.128134	7.45	1342.87	3.128034
17'	1.285	1337.96	3.126442	7.47	1337.64	3.126341
18'	1.29	1332.77	3.124756	7.50	1332.46	3.124654
19'	1.295	1327.63	3.123077	7.53	1327.32	3.122974
20'	1.3	1322.53	3.121404	7.56	1322.21	3.121300
21'	1.305	1317.46	3.119738	7.59	1317.14	3.119633
22'	1.31	1312.43	3.118078	7.62	1312.12	3.117972
23'	1.315	1307.45	3.116424	7.65	1307.13	3.116318
24'	1.32	1302.50	3.114777	7.68	1302.18	3.114669
25'	1.325	1297.58	3.113136	7.71	1297.26	3.113028
26'	1.33	1292.71	3.111501	7.74	1292.39	3.111392
27'	1.335	1287.87	3.109872	7.76	1287.55	3.109762
28'	1.34	1283.07	3.108249	7.79	1282.74	3.108139
29'	1.345	1278.30	3.106632	7.82	1277.97	3.106521
30'	1.35	1273.57	3.105022	7.85	1273.24	3.104910
31'	1.355	1268.87	3.103417	7.88	1268.54	3.103304
32'	1.36	1264.21	3.101818	7.91	1263.88	3.101705
33'	1.365	1259.58	3.100225	7.94	1259.25	3.100111
34'	1.37	1254.98	3.098638	7.97	1254.65	3.098523
35'	1.375	1250.42	3.097057	8.00	1250.09	3.096941
36'	1.38	1245.89	3.095481	8.03	1245.56	3.095365
37'	1.385	1241.40	3.093912	8.06	1241.06	3.093974
38'	1.39	1236.94	3.092347	8.08	1236.60	3.092229
39'	1.395	1232.51	3.090789	8.11	1232.17	3.090670
40'	1.4	1228.11	3.089236	8.14	1227.77	3.089116
41'	1.405	1223.74	3.087689	8.17	1223.40	3.087566
42'	1.41	1219.40	3.086147	8.20	1219.06	3.086025
43'	1.415	1215.09	3.084610	8.23	1214.75	3.084487
44'	1.42	1210.82	3.083079	8.26	1210.47	3.082955
45'	1.425	1206.57	3.081553	8.29	1206.23	3.081429
46'	1.43	1202.36	3.080033	8.32	1202.01	3.079908
47'	1.435	1198.17	3.078518	8.35	1197.82	3.078392
48'	1.44	1194.01	3.077008	8.38	1193.66	3.076881
49'	1.445	1189.88	3.075504	8.40	1189.53	3.075376
50'	1.45	1185.78	3.074005	8.43	1185.43	3.073876
51'	1.455	1181.71	3.072511	8.46	1181.36	3.072381
52'	1.46	1177.66	3.071022	8.49	1177.31	3.070891
53'	1.465	1173.65	3.069538	8.52	1173.29	3.069406
54'	1.47	1169.66	3.068059	8.55	1169.30	3.067927
55'	1.475	1165.70	3.066585	8.58	1165.34	3.066452
56'	1.48	1161.76	3.065116	8.61	1161.40	3.064982
57'	1.485	1157.85	3.063653	8.64	1157.49	3.063517
58'	1.49	1153.97	3.062194	8.67	1153.61	3.062057
59'	1.495	1150.11	3.060740	8.69	1149.75	3.060603

TABLE VII. FUNCTIONS OF CIRCULAR CURVES

Degree of Curve D	Defl. Per Ft of Sta. (Min)	Chord Definition			Arc Definition	
		Radius R	log R	C.O. 1 Sta.	Radius R	log R
5° 0'	1.5	1146.28	3.059290	8.72	1145.92	3.059153
1'	1.505	1142.47	3.057846	8.75	1142.11	3.057707
2'	1.51	1138.69	3.056407	8.78	1138.33	3.056267
3'	1.515	1134.94	3.054972	8.81	1134.57	3.054831
4'	1.52	1131.21	3.053542	8.84	1130.84	3.053400
5'	1.525	1127.50	3.052116	8.87	1127.13	3.051974
6'	1.53	1123.82	3.050696	8.90	1123.45	3.050553
7'	1.535	1120.16	3.049280	8.93	1119.79	3.049135
8'	1.54	1116.52	3.047868	8.96	1116.15	3.047723
9'	1.545	1112.91	3.046462	8.99	1112.54	3.046315
10'	1.55	1109.33	3.045059	9.01	1108.95	3.044912
11'	1.555	1105.76	3.043662	9.04	1105.38	3.043514
12'	1.56	1102.22	3.042268	9.07	1101.84	3.042119
13'	1.565	1098.70	3.040880	9.10	1098.32	3.040729
14'	1.57	1095.20	3.039495	9.13	1094.82	3.039343
15'	1.575	1091.73	3.038115	9.16	1091.35	3.037963
16'	1.58	1088.28	3.036740	9.19	1087.89	3.036587
17'	1.585	1084.85	3.035368	9.22	1084.46	3.035215
18'	1.59	1081.44	3.034002	9.25	1081.05	3.033847
19'	1.595	1078.05	3.032639	9.28	1077.66	3.032483
20'	1.6	1074.68	3.031281	9.31	1074.30	3.031124
21'	1.605	1071.34	3.029927	9.33	1070.95	3.029769
22'	1.61	1068.01	3.028577	9.36	1067.62	3.028418
23'	1.615	1064.71	3.027231	9.39	1064.32	3.027071
24'	1.62	1061.43	3.025890	9.42	1061.03	3.025729
25'	1.625	1058.16	3.024552	9.45	1057.77	3.024390
26'	1.63	1054.92	3.023219	9.48	1054.52	3.023056
27'	1.635	1051.70	3.021890	9.51	1051.30	3.021726
28'	1.64	1048.49	3.020565	9.54	1048.09	3.020400
29'	1.645	1045.31	3.019244	9.57	1044.91	3.019078
30'	1.65	1042.14	3.017927	9.60	1041.74	3.017760
31'	1.655	1039.00	3.016614	9.62	1038.59	3.016446
32'	1.66	1035.87	3.015305	9.65	1035.47	3.015136
33'	1.665	1032.76	3.013999	9.68	1032.36	3.013829
34'	1.67	1029.67	3.012698	9.71	1029.27	3.012527
35'	1.675	1026.60	3.011401	9.74	1026.19	3.011229
36'	1.68	1023.55	3.010107	9.77	1023.14	3.009935
37'	1.685	1020.51	3.008818	9.80	1020.10	3.008644
38'	1.69	1017.49	3.007532	9.83	1017.08	3.007357
39'	1.695	1014.50	3.006250	9.86	1014.08	3.006074
40'	1.7	1011.51	3.004972	9.89	1011.10	3.004795
41'	1.705	1008.55	3.003698	9.92	1008.14	3.003520
42'	1.71	1005.60	3.002427	9.94	1005.19	3.002248
43'	1.715	1002.67	3.001160	9.97	1002.26	3.000980
44'	1.72	999.762	2.999897	10.00	999.345	2.999715
45'	1.725	996.867	2.998637	10.03	996.448	2.998455
46'	1.73	993.988	2.997381	10.06	993.568	2.997198
47'	1.735	991.126	2.996129	10.09	990.705	2.995944
48'	1.74	988.280	2.994880	10.12	987.858	2.994695
49'	1.745	985.451	2.993635	10.15	985.028	2.993448
50'	1.75	982.638	2.992393	10.18	982.213	2.992206
51'	1.755	979.840	2.991155	10.21	979.415	2.990967
52'	1.76	977.060	2.989921	10.23	976.632	2.989731
53'	1.765	974.294	2.988690	10.26	973.866	2.988499
54'	1.77	971.544	2.987463	10.29	971.115	2.987271
55'	1.775	968.810	2.986239	10.32	968.379	2.986045
56'	1.78	966.091	2.985018	10.35	965.659	2.984824
57'	1.785	963.387	2.983801	10.38	962.954	2.983606
58'	1.79	960.698	2.982587	10.41	960.264	2.982391
59'	1.795	958.025	2.981377	10.44	957.590	2.981179

TABLE VIII. LENGTHS OF CIRCULAR ARCS FOR RADIUS = 1

Deg.	Length	Deg.	Length	Min.	Length	Sec.	Length
1	0.017 45 329	**61**	1.064 65 084	**1**	.000 29 089	**1**	.000 00 485
2	.034 90 659	62	.082 10 414	2	0 58 178	2	00 970
3	.052 35 988	63	.099 55 743	3	0 87 266	3	01 454
4	.069 81 317	64	.117 01 072	4	1 16 355	4	01 939
5	0.087 26 646	**65**	1.134 46 401	**5**	.001 45 444	**5**	.000 02 424
6	.104 71 976	66	.151 91 731	6	1 74 533	6	02 909
7	.122 17 305	67	.169 37 060	7	2 03 622	7	03 394
8	.139 62 634	68	.186 82 389	8	2 32 711	8	03 879
9	.157 07 963	69	.204 27 718	9	2 61 799	9	04 363
10	0.174 53 293	**70**	1.221 73 048	**10**	.002 90 888	**10**	.000 04 848
11	.191 98 622	71	.239 18 377	11	3 19 977	11	05 333
12	.209 43 951	72	.256 63 706	12	3 49 066	12	05 818
13	.226 89 280	73	.274 09 035	13	3 78 155	13	06 303
14	.244 34 610	74	.291 54 365	14	4 07 243	14	06 787
15	0.261 79 939	**75**	1.308 99 694	**15**	.004 36 332	**15**	.000 07 272
16	.279 25 268	76	.326 45 023	16	4 65 421	16	07 757
17	.296 70 597	77	.343 90 352	17	4 94 510	17	08 242
18	.314 15 927	78	.361 35 682	18	5 23 599	18	08 727
19	.331 61 256	79	.378 81 011	19	5 52 688	19	09 211
20	0.349 06 585	**80**	1.396 26 340	**20**	.005 81 776	**20**	.000 09 696
21	.366 51 914	81	.413 71 669	21	6 10 865	21	10 181
22	.383 97 244	82	.431 16 999	22	6 39 954	22	10 666
23	.401 42 573	83	.448 62 328	23	6 69 043	23	11 151
24	.418 87 902	84	.466 07 657	24	6 98 132	24	11 636
25	0.436 33 231	**85**	1.483 52 986	**25**	.007 27 221	**25**	.000 12 120
26	.453 78 561	86	.500 98 316	26	7 56 309	26	12 605
27	.471 23 890	87	.518 43 645	27	7 85 398	27	13 090
28	.488 69 219	88	.535 88 974	28	8 14 487	28	13 575
29	.506 14 548	89	.553 34 303	29	8 43 576	29	14 060
30	0.523 59 878	**90**	1.570 79 633	**30**	.008 72 665	**30**	.000 14 544
31	.541 05 207	91	.588 24 962	31	9 01 753	31	15 029
32	.558 50 536	92	.605 70 291	32	9 30 842	32	15 514
33	.575 95 865	93	.623 15 620	33	9 59 931	33	15 999
34	.593 41 195	94	.640 60 950	34	9 89 020	34	16 484
35	0.610 86 524	**95**	1.658 06 279	**35**	.010 18 109	**35**	.000 16 969
36	.628 31 853	96	.675 51 608	36	10 47 198	36	17 453
37	.645 77 182	97	.692 96 937	37	10 76 286	37	17 938
38	.663 22 512	98	.710 42 267	38	11 05 375	38	18 423
39	.680 67 841	99	.727 87 596	39	11 34 464	39	18 908
40	0.698 13 170	**100**	1.745 32 925	**40**	.011 63 553	**40**	.000 19 393
41	.715 58 499	101	.762 78 254	41	11 92 642	41	19 877
42	.733 03 829	102	.780 23 584	42	12 21 730	42	20 362
43	.750 49 158	103	.797 68 913	43	12 50 819	43	20 847
44	.767 94 487	104	.815 14 242	44	12 79 908	44	21 332
45	0.785 39 816	**105**	1.832 59 571	**45**	.013 08 997	**45**	.000 21 817
46	.802 85 146	106	.850 04 901	46	13 38 086	46	22 301
47	.820 30 475	107	.867 50 230	47	13 67 175	47	22 786
48	.837 75 804	108	.884 95 559	48	13 96 263	48	23 271
49	.855 21 133	109	.902 40 888	49	14 25 352	49	23 756
50	0.872 66 463	**110**	1.919 86 218	**50**	.014 54 441	**50**	.000 24 241
51	.890 11 792	111	.937 31 547	51	14 83 530	51	24 726
52	.907 57 121	112	.954 76 876	52	15 12 619	52	25 210
53	.925 02 450	113	.972 22 205	53	15 41 708	53	25 695
54	.942 47 780	114	.989 67 535	54	15 70 796	54	26 180
55	0.959 93 109	**115**	2.007 12 864	**55**	.015 99 885	**55**	.000 26 665
56	0.977 38 438	116	.024 58 193	56	16 28 974	56	27 150
57	0.994 83 767	117	.042 03 522	57	16 58 063	57	27 634
58	1.012 29 097	118	.059 48 852	58	16 87 152	58	28 119
59	1.029 74 426	119	.076 94 181	59	17 16 240	59	28 604
60	1.047 19 755	120	.094 39 510	60	17 45 329	60	29 089

TABLE IX. TRIGONOMETRIC FORMULAS FOR THE SOLUTION OF RIGHT TRIANGLES

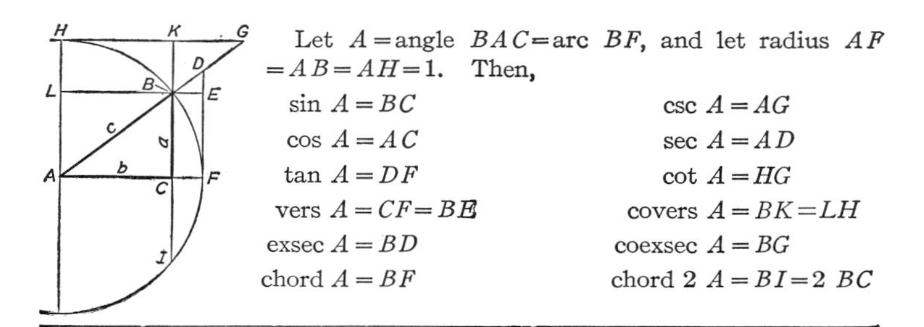

Let A = angle BAC = arc BF, and let radius $AF = AB = AH = 1$. Then,

$\sin A = BC$	$\csc A = AG$
$\cos A = AC$	$\sec A = AD$
$\tan A = DF$	$\cot A = HG$
vers $A = CF = BE$	covers $A = BK = LH$
exsec $A = BD$	coexsec $A = BG$
chord $A = BF$	chord $2A = BI = 2BC$

In the right-angled triangle ABC, let $AB = c$, $BC = a$, $CA = b$. Then,

1. $\sin A = \frac{a}{c}$	**11.** $a = c \sin A = b \tan A$
2. $\cos A = \frac{b}{c}$	**12.** $b = c \cos A = a \cot A$
3. $\tan A = \frac{a}{b}$	**13.** $c = \frac{a}{\sin A} = \frac{b}{\cos A}$
4. $\cot A = \frac{b}{a}$	**14.** $a = c \cos B = b \cot B$
5. $\sec A = \frac{c}{b}$	**15.** $b = c \sin B = a \tan B$
6. $\csc A = \frac{c}{a}$	**16.** $c = \frac{a}{\cos B} = \frac{b}{\sin B}$
7. vers $A = 1 - \cos A = \frac{c-b}{c}$ = covers B	**17.** $a = \sqrt{c^2 - b^2} = \sqrt{(c-b)(c+b)}$
8. exsec $A = \sec A - 1 = \frac{c-b}{b}$ = coexsec B	**18.** $b = \sqrt{c^2 - a^2} = \sqrt{(c-a)(c+a)}$
9. covers $A = \frac{c-a}{c}$ = vers B	**19.** $c = \sqrt{a^2 + b^2}$
10. coexsec $A = \frac{c-a}{a}$ = exsec B	**20.** $C = 90° = A + B$

21. Area $= \frac{1}{2}ab$

TABLE X. TRIGONOMETRIC FORMULAS FOR THE SOLUTION OF OBLIQUE TRIANGLES

No.	Given	Sought	Formula
22	A, B, a	C, b, c	$C=180°-(A+B)$
			$b=\frac{a}{\sin A}\times \sin B$
			$c=\frac{a}{\sin A}\times \sin (A+B)=\frac{a}{\sin A}\times \sin C$
		Area	$\text{Area}=\frac{1}{2}ab \sin C=\frac{a^2 \sin B \sin C}{2 \sin A}$
23	A, a, b	B, C, c	$\sin B=\frac{\sin A}{a}\times b$
			$C=180°-(A+B)$
			$c=\frac{a}{\sin A}\times \sin C$
		Area	$\text{Area}=\frac{1}{2}ab \sin C$
24	$C, a, b,$	c	$c=\sqrt{a^2+b^2-2ab \cos C}$
25		$\frac{1}{2}(A+B)$	$\frac{1}{2}(A+B)=90°-\frac{1}{2}C$
26		$\frac{1}{2}(A-B)$	$\tan \frac{1}{2}(A-B)=\frac{a-b}{a+b}\times \tan \frac{1}{2}(A+B)$
27		A, B	$A=\frac{1}{2}(A+B)+\frac{1}{2}(A-B)$
			$B=\frac{1}{2}(A+B)-\frac{1}{2}(A-B)$
28		c	$c=(a+b)\times\frac{\cos \frac{1}{2}(A+B)}{\cos \frac{1}{2}(A-B)}=(a-b)\times\frac{\sin \frac{1}{2}(A+B)}{\sin \frac{1}{2}(A-B)}$
29		Area	$\text{Area}=\frac{1}{2}ab \sin C$
30	a, b, c	A	Let $s=\frac{a+b+c}{2}$
31			$\sin \frac{1}{2} A=\sqrt{\frac{(s-b)(s-c)}{bc}}$
			$\cos \frac{1}{2} A=\sqrt{\frac{s(s-a)}{bc}}$
			$\tan \frac{1}{2} A=\sqrt{\frac{(s-b)(s-c)}{s(s-a)}}$
32			$\sin A=\frac{2\sqrt{s(s-a)(s-b)(s-c)}}{bc}$
			$\cos A=\frac{b^2+c^2-a^2}{2bc}$
33		Area	$\text{Area}=\sqrt{s(s-a)(s-b)(s-c)}$

TABLE XI. RELATIONS BETWEEN LINEAR AND ANGULAR ERRORS

Allowable Angular Error for Given Linear Precision		Allowable Linear Error for Given Angular Precision					
Precision of Linear Measurements	Allowable Angular Error	Least Reading in Angular Measurements	Allowable Linear Error in 100′	500′	1000′	5000′	Ratio
$\frac{1}{500}$	6′53″	5′	.145	.727	1.454	7.272	$\frac{1}{688}$
$\frac{1}{1000}$	3 26	1′	.029	.145	.291	1.454	$\frac{1}{3440}$
$\frac{1}{5000}$	0 41	30″	.015	.073	.145	.727	$\frac{1}{6880}$
$\frac{1}{10,000}$	0 21	20″	.010	.049	.097	.485	$\frac{1}{10,300}$
$\frac{1}{50,000}$	0 04	10″	.005	.024	.049	.242	$\frac{1}{20,600}$
$\frac{1}{100,000}$	0 02	5″	.002	.012	.024	.121	$\frac{1}{41,200}$
$\frac{1}{1,000,000}$	0 00.2	2″	.001	.005	.010	.048	$\frac{1}{103,100}$
		1″		.002	.005	.024	$\frac{1}{206,300}$

TABLE XII. PRECISION OF COMPUTED VALUES

Size of Angle and Function		Angular Error				
		1′	30″	20″	10″	5″
		Precision of computed value using sine or cosine				
sin 5°	or cos 85°	$\frac{1}{300}$	$\frac{1}{600}$	$\frac{1}{900}$	$\frac{1}{1800}$	$\frac{1}{3600}$
10	80	$\frac{1}{610}$	$\frac{1}{1210}$	$\frac{1}{1820}$	$\frac{1}{3640}$	$\frac{1}{7280}$
20	70	$\frac{1}{1250}$	$\frac{1}{2500}$	$\frac{1}{3750}$	$\frac{1}{7500}$	$\frac{1}{15,000}$
30	60	$\frac{1}{1990}$	$\frac{1}{3970}$	$\frac{1}{5960}$	$\frac{1}{11,970}$	$\frac{1}{23,940}$
40	50	$\frac{1}{2890}$	$\frac{1}{5770}$	$\frac{1}{8660}$	$\frac{1}{17,310}$	$\frac{1}{34,620}$
50	40	$\frac{1}{4100}$	$\frac{1}{8190}$	$\frac{1}{12,290}$	$\frac{1}{24,580}$	$\frac{1}{49,160}$
60	30	$\frac{1}{5950}$	$\frac{1}{11,900}$	$\frac{1}{17,860}$	$\frac{1}{35,720}$	$\frac{1}{71,440}$
70	20	$\frac{1}{9450}$	$\frac{1}{18,900}$	$\frac{1}{28,330}$	$\frac{1}{56,670}$	$\frac{1}{113,340}$
80	10	$\frac{1}{19,500}$	$\frac{1}{39,000}$	$\frac{1}{58,500}$	$\frac{1}{117,000}$	$\frac{1}{234,000}$
		Precision of computed value using tan or cot				
tan or cot	5°	$\frac{1}{300}$	$\frac{1}{600}$	$\frac{1}{900}$	$\frac{1}{1790}$	$\frac{1}{3580}$
	10	$\frac{1}{590}$	$\frac{1}{1180}$	$\frac{1}{1760}$	$\frac{1}{3530}$	$\frac{1}{7050}$
	20	$\frac{1}{1100}$	$\frac{1}{2210}$	$\frac{1}{3310}$	$\frac{1}{6620}$	$\frac{1}{13,250}$
	30	$\frac{1}{1490}$	$\frac{1}{2980}$	$\frac{1}{4470}$	$\frac{1}{8930}$	$\frac{1}{17,870}$
	40	$\frac{1}{1690}$	$\frac{1}{3390}$	$\frac{1}{5080}$	$\frac{1}{10,160}$	$\frac{1}{20,320}$
	45	$\frac{1}{1720}$	$\frac{1}{3440}$	$\frac{1}{5160}$	$\frac{1}{10,310}$	$\frac{1}{20,630}$
	50	$\frac{1}{1690}$	$\frac{1}{3390}$	$\frac{1}{5080}$	$\frac{1}{10,160}$	$\frac{1}{20,320}$
	60	$\frac{1}{1490}$	$\frac{1}{2980}$	$\frac{1}{4470}$	$\frac{1}{8930}$	$\frac{1}{17,870}$
	70	$\frac{1}{1100}$	$\frac{1}{2210}$	$\frac{1}{3310}$	$\frac{1}{6620}$	$\frac{1}{13,250}$
	80	$\frac{1}{590}$	$\frac{1}{1180}$	$\frac{1}{1760}$	$\frac{1}{3530}$	$\frac{1}{7050}$
	85	$\frac{1}{300}$	$\frac{1}{600}$	$\frac{1}{900}$	$\frac{1}{1790}$	$\frac{1}{3580}$

TABLE XIII. NATURAL SINES AND COSINES

′	0°		1°		2°		′
	Sine	Cosine	Sine	Cosine	Sine	Cosine	
0	0.0000000	1.0000000	0.0174524	0.9998477	0.0348995	0.9993908	60
1	.0002909	1.0000000	.0177432	.9998426	.0351902	.9993806	59
2	.0005818	.9999998	.0180341	.9998374	.0354809	.9993704	58
3	.0008727	.9999996	.0183249	.9998321	.0357716	.9993600	57
4	.0011636	.9999993	.0186158	.9998267	.0360623	.9993495	56
5	.0014544	.9999989	.0189066	.9998213	.0363530	.9993390	55
6	.0017453	.9999985	.0191974	.9998157	.0366437	.9993284	54
7	.0020362	.9999979	.0194883	.9998101	.0369344	.9993177	53
8	.0023271	.9999973	.0197791	.9998044	.0372251	.9993069	52
9	.0026180	.9999966	.0200699	.9997986	.0375158	.9992960	51
10	0.0029089	0.9999958	0.0203608	0.9997927	0.0378065	0.9992851	50
11	.0031998	.9999949	.0206516	.9997867	.0380971	.9992740	49
12	.0034907	.9999939	.0209424	.9997807	.0383878	.9992629	48
13	.0037815	.9999928	.0212332	.9997745	.0386785	.9992517	47
14	.0040724	.9999917	.0215241	.9997683	.0389692	.9992404	46
15	.0043633	.9999905	.0218149	.9997620	.0392598	.9992290	45
16	.0046542	.9999892	.0221057	.9997556	.0395505	.9992176	44
17	.0049451	.9999878	.0223965	.9997492	.0398411	.9992060	43
18	.0052360	.9999863	.0226873	.9997426	.0401318	.9991944	42
19	.0055268	.9999847	.0229781	.9997360	.0404224	.9991827	41
20	0.0058177	0.9999831	0.0232690	0.9997292	0.0407131	0.9991709	40
21	.0061086	.9999813	.0235598	.9997224	.0410037	.9991590	39
22	.0063995	.9999795	.0238506	.9997155	.0412944	.9991470	38
23	.0066904	.9999776	.0241414	.9997086	.0415850	.9991350	37
24	.0069813	.9999756	.0244322	.9997015	.0418757	.9991228	36
25	.0072721	.9999736	.0247230	.9996943	.0421663	.9991106	35
26	.0075630	.9999714	.0250138	.9996871	.0424569	.9990983	34
27	.0078539	.9999692	.0253046	.9996798	.0427475	.9990859	33
28	.0081448	.9999668	.0255954	.9996724	.0430382	.9990734	32
29	.0084357	.9999644	.0258862	.9996649	.0433288	.9990609	31
30	0.0087265	0.9999619	0.0261769	0.9996573	0.0436194	0.9990482	30
31	.0090174	.9999593	.0264677	.9996497	.0439100	.9990355	29
32	.0093083	.9999567	.0267585	.9996419	.0442006	.9990227	28
33	.0095992	.9999539	.0270493	.9996341	.0444912	.9990098	27
34	.0098900	.9999511	.0273401	.9996262	.0447818	.9989968	26
35	.0101809	.9999482	.0276309	.9996182	.0450724	.9989837	25
36	.0104718	.9999452	.0279216	.9996101	.0453630	.9989706	24
37	.0107627	.9999421	.0282124	.9996020	.0456536	.9989573	23
38	.0110535	.9999389	.0285032	.9995937	.0459442	.9989440	22
39	.0113444	.9999357	.0287940	.9995854	.0462347	.9989306	21
40	0.0116353	0.9999323	0.0290847	0.9995770	0.0465253	0.9989171	20
41	.0119261	.9999289	.0293755	.9995684	.0468159	.9989035	19
42	.0122170	.9999254	.0296662	.9995599	.0471065	.9988899	18
43	.0125079	.9999218	.0299570	.9995512	.0473970	.9988761	17
44	.0127987	.9999181	.0302478	.9995424	.0476876	.9988623	16
45	.0130896	.9999143	.0305385	.9995336	.0479781	.9988484	15
46	.0133805	.9999105	.0308293	.9995247	.0482687	.9988344	14
47	.0136713	.9999065	.0311200	.9995157	.0485592	.9988203	13
48	.0139622	.9999025	.0314108	.9995066	.0488498	.9988061	12
49	.0142530	.9998981	.0317015	.9994974	.0491403	.9987919	11
50	0.0145439	0.9998942	0.0319922	0.9994881	0.0494308	0.9987775	10
51	.0148348	.9998900	.0322830	.9994788	.0497214	.9987631	9
52	.0151256	.9998853	.0325737	.9994693	.0500119	.9987486	8
53	.0154165	.9998812	.0328644	.9994598	.0503024	.9987340	7
54	.0157073	.9998766	.0331552	.9994502	.0505929	.9987194	6
55	.0159982	.9998720	.0334459	.9994405	.0508835	.9987046	5
56	.0162890	.9998673	.0337366	.9994308	.0511740	.9986898	4
57	.0165799	.9998625	.0340274	.9994209	.0514645	.9986748	3
58	.0168707	.9998577	.0343181	.9994110	.0517550	.9986598	2
59	.0171616	.9998527	.0346088	.9994009	.0520455	.9986447	1
60	0.0174524	0.9998477	0.0348995	0.9993908	0.0523360	0.9986295	0
	Cosine	Sine	Cosine	Sine	Cosine	Sine	
′	89°		88°		87°		′

TABLE XIII. NATURAL SINES AND COSINES

′	3°		4°		5°		′
	Sine	Cosine	Sine	Cosine	Sine	Cosine	
0	0.0523360	0.9986295	0.0697565	0 9975641	0.0871557	0.9961947	60
1	.0526264	.9986143	.0700467	.9975437	.0874455	.9961693	59
2	.0529169	.9985989	.0703368	.9975233	.0877353	.9961438	58
3	.0532074	.9985835	.0706270	.9975028	.0880251	.9961183	57
4	.0534979	.9985680	.0709171	.9974822	.0883148	.9960926	56
5	.0537883	.9985524	.0712073	.9974615	.0886046	.9960669	55
6	.0540788	.9985367	.0714974	.9974408	.0888943	.9960411	54
7	.0543693	.9985209	.0717876	.9974199	.0891840	.9960152	53
8	.0546597	.9985050	.0720777	.9973990	.0894738	.9959892	52
9	.0549502	.9984891	.0723678	.9973780	.0897635	.9959631	51
10	0.0552406	0.9984731	0.0726580	0.9973569	0.0900532	0.9959370	50
11	.0555311	.9984570	.0729481	.9973357	.0903429	.9959107	49
12	.0558215	.9984408	.0732382	.9973145	.0906326	.9958844	48
13	.0561119	.9984245	.0735283	.9972931	.0909223	.9958580	47
14	.0564024	.9984081	.0738184	.9972717	.0912119	.9958315	46
15	.0566928	.9983917	.0741085	.9972502	.0915016	.9958049	45
16	.0569832	.9983751	.0743986	.9972286	.0917913	.9957783	44
17	.0572736	.9983585	.0746887	.9972069	.0920809	.9957515	43
18	.0575640	.9983418	.0749787	.9971851	.0923706	.9957247	42
19	.0578544	.9983250	.0752688	.9971633	.0926602	.9956978	41
20	0.0581448	0.9983082	0.0755589	0.9971413	0.0929499	0.9956708	40
21	.0584352	.9982912	.0758489	.9971193	.0932395	.9956437	39
22	.0587256	.9982742	.0761390	.9970972	.0935291	.9956165	38
23	.0590160	.9982570	.0764290	.9970750	.0938187	.9955893	37
24	.0593064	.9982398	.0767190	.9970528	.0941083	.9955620	36
25	.0595967	.9982225	.0770091	.9970304	.0943979	.9955345	35
26	.0598871	.9982052	.0772991	.9970080	.0946875	.9955070	34
27	.0601775	.9981877	.0775891	.9969854	.0949771	.9954795	33
28	.0604678	.9981701	.0778791	.9969628	.0952666	.9954518	32
29	.0607582	.9981525	.0781691	.9969401	.0955562	.9954240	31
30	0.0610485	0.9981348	0.0784591	0.9969173	0.0958458	0.9953962	30
31	.0613389	.9981170	.0787491	.9968945	.0961353	.9953683	29
32	.0616292	.9980991	.0790391	.9968715	.0964248	.9953403	28
33	.0619196	.9980811	.0793290	.9968485	.0967144	.9953122	27
34	.0622099	.9980631	.0796190	.9968254	.0970039	.9952840	26
35	.0625002	.9980450	.0799090	.9968022	.0972934	.9952557	25
36	.0627905	.9980267	.0801989	.9967789	.0975829	.9952274	24
37	.0630808	.9980084	.0804889	.9967555	.0978724	.9951990	23
38	.0633711	.9979900	.0807788	.9967321	.0981619	.9951705	22
39	.0636614	.9979716	.0810687	.9967085	.0984514	.9951419	21
40	0.0639517	0.9979530	0.0813587	0.9966849	0.0987408	0.9951132	20
41	.0642420	.9979343	.0816486	.9966612	.0990303	.9950844	19
42	.0645323	.9979156	.0819385	.9966374	.0993197	.9950556	18
43	.0648226	.9978968	.0822284	.9966135	.0996092	.9950266	17
44	.0651129	.9978779	.0825183	.9965895	.0998986	.9949976	16
45	.0654031	.9978589	.0828082	.9965655	.1001881	.9949685	15
46	.0656934	.9978399	.0830981	.9965414	.1004775	.9949393	14
47	.0659836	.9978207	.0833880	.9965172	.1007669	.9949101	13
48	.0662739	.9978015	.0836778	.9964929	.1010563	.9948807	12
49	.0665641	.9977821	.0839677	.9964685	.1013457	.9948513	11
50	0.0668544	0.9977627	0.0842576	0.9964440	0.1016351	0.9948217	10
51	.0671446	.9977433	.0845474	.9964195	.1019245	.9947921	9
52	.0674349	.9977237	.0848373	.9963948	.1022138	.9947625	8
53	.0677251	.9977040	.0851271	.9963701	.1025032	.9947327	7
54	.0680153	.9976843	.0854169	.9963453	.1027925	.9947028	6
55	.0683055	.9976645	.0857067	.9963204	.1030819	.9946729	5
56	.0685957	.9976445	.0859966	.9962954	.1033712	.9946428	4
57	.0688859	.9976245	.0862864	.9962704	.1036605	.9946127	3
58	.0691761	.9976045	.0865762	.9962452	.1039499	.9945825	2
59	.0694663	.9975843	.0868660	.9962200	.1042392	.9945523	1
60	0.0697565	0.9975641	0.0871557	0.9961947	0.1045285	0.9945219	0
	Cosine	Sine	Cosine	Sine	Cosine	Sine	
′	86°		85°		84°		′

TABLE XIII. NATURAL SINES AND COSINES

′	6°		7°		8°		′
	Sine	Cosine	Sine	Cosine	Sine	Cosine	
0	0.1045285	0.9945219	0.1218693	0.9925462	0.1391731	0.9902681	60
1	.1048178	.9944914	.1221581	.9925107	.1394612	.9902275	59
2	.1051070	.9944609	.1224468	.9924751	.1397492	.9901869	58
3	.1053963	.9944303	.1227355	.9924394	.1400372	.9901462	57
4	.1056856	.9943996	.1230241	.9924037	.1403252	.9901055	56
5	.1059748	.9943688	.1233128	.9923679	.1406132	.9900646	55
6	.1062641	.9943379	.1236015	.9923319	.1409012	.9900237	54
7	.1065533	.9943070	.1238901	.9922959	.1411892	.9899826	53
8	.1068425	.9942760	.1241788	.9922599	.1414772	.9899415	52
9	.1071318	.9942448	.1244674	.9922237	.1417651	.9899003	51
10	0.1074210	0.9942136	0.1247560	0.9921874	0.1420531	0.9898590	50
11	.1077102	.9941823	.1250446	.9921511	.1423410	.9898177	49
12	.1079994	.9941510	.1253332	.9921147	.1426289	.9897762	48
13	.1082885	.9941195	.1256218	.9920782	.1429168	.9897347	47
14	.1085777	.9940880	.1259104	.9920416	.1432047	.9896931	46
15	.1088669	.9940563	.1261990	.9920049	.1434926	.9896514	45
16	.1091560	.9940246	.1264875	.9919682	.1437805	.9896096	44
17	.1094452	.9939928	.1267761	.9919314	.1440684	.9895677	43
18	.1097343	.9939610	.1270646	.9918944	.1443562	.9895258	42
19	.1100234	.9939290	.1273531	.9918574	.1446440	.9894838	41
20	0.1103126	0.9938969	0.1276416	0.9918204	0.1449319	0.9894416	40
21	.1106017	.9938648	.1279302	.9917832	.1452197	.9893994	39
22	.1108908	.9938326	.1282186	.9917459	.1455075	.9893572	38
23	.1111799	.9938003	.1285071	.9917086	.1457953	.9893148	37
24	.1114689	.9937679	.1287956	.9916712	.1460830	.9892723	36
25	.1117580	.9937355	.1290841	.9916337	.1463708	.9892298	35
26	.1120471	.9937029	.1293725	.9915961	.1466585	.9891872	34
27	.1123361	.9936703	.1296609	.9915584	.1469463	.9891445	33
28	.1126252	.9936375	.1299494	.9915206	.1472340	.9891017	32
29	.1129142	.9936047	.1302378	.9914828	.1475217	.9890588	31
30	0.1132032	0.9935719	0.1305262	0.9914449	0.1478094	0.9890159	30
31	.1134922	.9935389	.1308146	.9914069	.1480971	.9889728	29
32	.1137812	.9935058	.1311030	.9913688	.1483848	.9889297	28
33	.1140702	.9934727	.1313913	.9913306	.1486724	.9888865	27
34	.1143592	.9934395	.1316797	.9912923	.1489601	.9888432	26
35	.1146482	.9934062	.1319681	.9912540	.1492477	.9887998	25
36	.1149372	.9933728	.1322564	.9912155	.1495353	.9887564	24
37	.1152261	.9933393	.1325447	.9911770	.1498230	.9887128	23
38	.1155151	.9933057	.1328330	.9911384	.1501106	.9886692	22
39	.1158040	.9932721	.1331213	.9910997	.1503981	.9886255	21
40	0.1160929	0.9932384	0.1334096	0.9910610	0.1506857	0.9885817	20
41	.1163818	.9932045	.1336979	.9910221	.1509733	.9885378	19
42	.1166707	.9931706	.1339862	.9909832	.1512608	.9884939	18
43	.1169596	.9931367	.1342744	.9909442	.1515484	.9884498	17
44	.1172485	.9931026	.1345627	.9909051	.1518359	.9884057	16
45	.1175374	.9930685	.1348509	.9908659	.1521234	.9883615	15
46	.1178263	.9930342	.1351392	.9908266	.1524109	.9883172	14
47	.1181151	.9929999	.1354274	.9907873	.1526984	.9882728	13
48	.1184040	.9929655	.1357156	.9907478	.1529858	.9882284	12
49	.1186928	.9929310	.1360038	.9907083	.1532733	.9881838	11
50	0.1189816	0.9928965	0.1362919	0.9906687	0.1535607	0.9881392	10
51	.1192704	.9928618	.1365801	.9906290	.1538482	.9880945	9
52	.1195593	.9928271	.1368683	.9905893	.1541356	.9880497	8
53	.1198481	.9927922	.1371564	.9905494	.1544230	.9880048	7
54	.1201368	.9927573	.1374445	.9905095	.1547104	.9879599	6
55	.1204256	.9927224	.1377327	.9904694	.1549978	.9879148	5
56	.1207144	.9926873	.1380208	.9904293	.1552851	.9878697	4
57	.1210031	.9926521	.1383089	.9903891	.1555725	.9878245	3
58	.1212919	.9926169	.1385970	.9903489	.1558598	.9877792	2
59	.1215806	.9925816	.1388850	.9903085	.1561472	.9877338	1
60	0.1218693	0.9925462	0.1391731	0.9902681	0.1564345	0.9876883	0
′	Cosine	Sine	Cosine	Sine	Cosine	Sine	′
	83°		82°		81°		

TABLE XIII. NATURAL SINES AND COSINES

′	9°		10°		11°		′
	Sine	Cosine	Sine	Cosine	Sine	Cosine	
0	0.1564345	0.9876883	0.1736482	0.9848078	0.1908090	0.9816272	60
1	.1567218	.9876428	.1739346	.9847572	.1910945	.9815716	59
2	.1570091	.9875972	.1742211	.9847066	.1913801	.9815160	58
3	.1572963	.9875514	.1745075	.9846558	.1916656	.9814603	57
4	.1575836	.9875057	.1747939	.9846050	.1919510	.9814045	56
5	.1578708	.9874598	.1750803	.9845542	.1922365	.9813486	55
6	.1581581	.9874138	.1753667	.9845032	.1925220	.9812927	54
7	.1584453	.9873678	.1756531	.9844521	.1928074	.9812366	53
8	.1587325	.9873216	.1759395	.9844010	.1930928	.9811805	52
9	.1590197	.9872754	.1762258	.9843498	.1933782	.9811243	51
10	0.1593069	0.9872291	0.1765121	0.9842985	0.1936636	0.9810680	50
11	.1595940	.9871827	.1767984	.9842471	.1939490	.9810116	49
12	.1598812	.9871363	.1770847	.9841956	.1942344	.9809552	48
13	.1601683	.9870897	.1773710	.9841441	.1945197	.9808986	47
14	.1604555	.9870431	.1776573	.9840924	.1948050	.9808420	46
15	.1607426	.9869964	.1779435	.9840407	.1950903	.9807853	45
16	.1610297	.9869496	.1782298	.9839889	.1953756	.9807285	44
17	.1613167	.9869027	.1785160	.9839370	.1956609	.9806716	43
18	.1616038	.9868557	.1788022	.9838850	.1959461	.9806147	42
19	.1618909	.9868087	.1790834	.9838330	.1962314	.9805576	41
20	0.1621779	0.9867615	0.1793746	0.9837808	0.1965166	0.9805005	40
21	.1624650	.9867143	.1796607	.9837286	.1968018	.9804433	39
22	.1627520	.9866670	.1799469	.9836763	.1970870	.9803860	38
23	.1630390	.9866196	.1802330	.9836239	.1973722	.9803286	37
24	.1633260	.9865722	.1805191	.9835715	.1976573	.9802712	36
25	.1636129	.9865246	.1808052	.9835189	.1979425	.9802136	35
26	.1638999	.9864770	.1810913	.9834663	.1982276	.9801560	34
27	.1641868	.9864293	.1813774	.9834136	.1985127	.9800983	33
28	.1644738	.9863815	.1816635	.9833608	.1987978	.9800405	32
29	.1647607	.9863336	.1819495	.9833079	.1990829	.9799827	31
30	0.1650476	0.9862856	0.1822355	0.9832549	0.1993679	0.9799247	30
31	.1653345	.9862375	.1825215	.9832019	.1996530	.9798667	29
32	.1656214	.9861894	.1828075	.9831487	.1999380	.9798086	28
33	.1659082	.9861412	.1830935	.9830955	.2002230	.9797504	27
34	.1661951	.9860929	.1833795	.9830422	.2005080	.9796921	26
35	.1664819	.9860445	.1836654	.9829888	2007930	.9796337	25
36	.1667687	.9859960	.1839514	.9829353	.2010779	.9795752	24
37	.1670556	.9859475	.1842373	.9828818	.2013629	.9795167	23
38	.1673423	.9858988	.1845232	.9828282	.2016478	.9794581	22
39	.1676291	.9858501	.1848091	.9827744	.2019327	.9793994	21
40	0.1679159	0.9858013	0.1850949	0.9827206	0.2022176	0.9793406	20
41	.1682026	.9857524	.1853808	.9826668	.2025024	.9792818	19
42	.1684894	.9857035	.1856666	.9826128	.2027873	.9792228	18
43	.1687761	.9856544	.1859524	.9825587	.2030721	.9791638	17
44	.1690628	.9856053	.1862382	.9825046	.2033569	.9791047	16
45	.1693495	.9855561	.1865240	.9824504	.2036418	.9790455	15
46	.1696362	.9855068	.1868098	.9823961	.2039265	.9789862	14
47	.1699228	.9854574	.1870956	.9823417	.2042113	.9789268	13
48	.1702095	.9854079	.1873813	.9822873	.2044961	.9788674	12
49	.1704961	.9853583	.1876670	.9822327	.2047808	.9788079	11
50	0.1707828	0.9853087	0.1879528	0.9821781	0.2050655	0.9787483	10
51	.1710694	.9852590	.1882385	.9821234	.2053502	.9786886	9
52	.1713560	.9852092	.1885241	.9820686	.2056349	.9786288	8
53	.1716425	.9851593	.1888098	.9820137	.2059195	.9785689	7
54	.1719291	.9851093	.1890954	.9819587	.2062042	.9785090	6
55	.1722156	.9850593	.1893811	.9819037	.2064888	.9784490	5
56	.1725022	.9850091	.1896667	.9818485	.2067734	.9783889	4
57	.1727887	.9849589	.1899523	.9817933	.2070580	.9783287	3
58	.1730752	.9849086	.1902379	.9817380	.2073426	.9782684	2
59	.1733617	.9848582	.1905234	.9816826	.2076272	.9782080	1
60	0.1736482	0.9848078	0.1908090	0.9816272	0.2079117	0.9781476	0
	Cosine	Sine	Cosine	Sine	Cosine	Sine	
′	80°		79°		78°		′

TABLE XIII. NATURAL SINES AND COSINES

′	12° Sine	12° Cosine	13° Sine	13° Cosine	14° Sine	14° Cosine	′
0	0.2079117	0.9781476	0.2249511	0.9743701	0.2419219	0.9702957	60
1	.2081962	.9780871	.2252345	.9743046	.2422041	.9702253	59
2	.2084807	.9780265	.2255179	.9742390	.2424863	.9701548	58
3	.2087652	.9779658	.2258013	.9741734	.2427685	.9700842	57
4	.2090497	.9779050	.2260846	.9741077	.2430507	.9700136	56
5	.2093341	.9778442	.2263680	.9740419	.2433329	.9699428	55
6	.2096186	.9777832	.2266513	.9739760	.2436150	.9698720	54
7	.2099030	.9777222	.2269346	.9739100	.2438971	.9698011	53
8	.2101874	.9776611	.2272179	.9738439	.2441792	.9697301	52
9	.2104718	.9775999	.2275012	.9737778	.2444613	.9696591	51
10	0.2107561	0.9775387	0.2277844	0.9737116	0.2447433	0.9695879	50
11	.2110405	.9774773	.2280677	.9736453	.2450254	.9695167	49
12	2113248	.9774159	.2283509	.9735789	.2453074	.9694453	48
13	.2116091	.9773544	.2286341	.9735124	.2455894	.9693740	47
14	.2118934	.9772928	.2289172	.9734459	.2458713	.9693025	46
15	.2121777	.9772311	.2292004	.9733793	.2461533	.9692309	45
16	.2124619	.9771693	.2294835	.9733125	.2464352	.9691593	44
17	.2127462	.9771075	.2297666	.9732458	.2467171	.9690875	43
18	.2130304	.9770456	.2300497	.9731789	.2469990	.9690157	42
19	.2133146	.9769836	.2303328	.9731119	.2472809	.9689438	41
20	0.2135988	0.9769215	0.2306159	0.9730449	0.2475627	0.9688719	40
21	.2138829	.9768593	.2308989	.9729777	.2478445	.9687998	39
22	.2141671	.9767970	.2311819	.9729105	.2481263	.9687277	38
23	.2144512	.9767347	.2314649	.9728432	.2484081	.9686555	37
24	.2147353	.9766723	.2317479	.9727759	.2486899	.9685832	36
25	.2150194	.9766098	.2320309	.9727084	.2489716	.9685108	35
26	.2153035	.9765472	.2323138	.9726409	.2492533	.9684383	34
27	.2155876	.9764845	.2325967	.9725733	.2495350	.9683658	33
28	.2158716	.9764218	.2328796	.9725056	.2498167	.9682931	32
29	.2161556	.9763589	.2331625	.9724378	.2500984	.9682204	31
30	0.2164396	0.9762960	0.2334454	0.9723699	0.2503800	0.9681476	30
31	.2167236	.9762330	.2337282	.9723020	.2506616	.9680748	29
32	.2170076	.9761699	.2340110	.9722339	.2509432	.9680018	28
33	.2172915	.9761068	.2342938	.9721658	.2512248	.9679288	27
34	.2175754	.9760435	.2345766	.9720976	.2515063	.9678557	26
35	.2178593	.9759802	.2348594	.9720294	.2517879	.9677825	25
36	.2181432	.9759168	.2351421	.9719610	.2520694	.9677092	24
37	.2184271	.9758533	.2354248	.9718926	.2523508	.9676358	23
38	.2187110	.9757897	.2357075	.9718240	.2526323	.9675624	22
39	.2189948	.9757260	.2359902	.9717554	.2529137	.9674888	21
40	0.2192786	0.9756623	0.2362729	0.9716867	0.2531952	0.9674152	20
41	.2195624	.9755985	.2365555	.9716180	.2534766	.9673415	19
42	.2198462	.9755345	.2368381	.9715491	.2537579	.9672678	18
43	.2201300	.9754706	.2371207	.9714802	.2540393	.9671939	17
44	.2204137	.9754065	.2374033	.9714112	.2543206	.9671200	16
45	.2206974	.9753423	.2376859	.9713421	.2546019	.9670459	15
46	.2209811	.9752781	.2379684	.9712729	.2548832	.9669718	14
47	.2212648	.9752138	.2382510	.9712036	.2551645	.9668977	13
48	.2215485	.9751494	.2385335	.9711343	.2554458	.9668234	12
49	.2218321	.9750849	.2388159	.9710649	.2557270	.9667490	11
50	0.2221158	0.9750203	0.2390984	0.9709953	0.2560082	0.9666746	10
51	.2223994	.9749556	.2393808	.9709258	.2562894	.9666001	9
52	.2226830	.9748909	.2396633	.9708561	.2565705	.9665255	8
53	.2229666	.9748261	.2399457	.9707863	.2568517	.9664508	7
54	.2232501	.9747612	.2402280	.9707165	.2571328	.9663761	6
55	.2235337	.9746962	.2405104	.9706466	.2574139	.9663012	5
56	.2238172	.9746311	.2407927	.9705766	.2576950	.9662263	4
57	.2241007	.9745660	.2410751	.9705065	.2579760	.9661513	3
58	.2243842	.9745008	.2413574	.9704363	.2582570	.9660762	2
59	.2246676	.9744355	.2416396	.9703661	.2585381	.9660011	1
60	0.2249511	0.9743701	0.2419219	0.9702957	0.2588190	0.9659258	0
′	Cosine	Sine	Cosine	Sine	Cosine	Sine	′
	77°		76°		75°		

TABLE XIII. NATURAL SINES AND COSINES

′	15°		16°		17°		′
	Sine	Cosine	Sine	Cosine	Sine	Cosine	
0	0.2588190	0.9659258	0.2756374	0.9612617	0.2923717	0.9563048	60
1	.2591000	.9658505	.2759170	.9611815	.2926499	.9562197	59
2	.2593810	.9657751	.2761965	.9611012	.2929280	.9561345	58
3	.2596619	.9656996	.2764761	.9610208	.2932061	.9560492	57
4	.2599428	.9656240	.2767556	.9609403	.2934842	.9559639	56
5	.2602237	.9655484	.2770352	.9608598	.2937623	.9558785	55
6	.2605045	.9654726	.2773147	.9607792	.2940403	.9557930	54
7	.2607853	.9653968	.2775941	.9606984	.2943183	.9557074	53
8	.2610662	.9653209	.2778736	.9606177	.2945963	.9556218	52
9	.2613469	.9652449	.2781530	.9605368	.2948743	.9555361	51
10	0.2616277	0.9651689	0.2784324	0.9604558	0.2951522	0.9554502	50
11	.2619085	.9650927	.2787118	.9603748	.2954302	.9553643	49
12	.2621892	.9650165	.2789911	.9602937	.2957081	.9552784	48
13	.2624699	.9649402	.2792704	.9602125	.2959859	.9551923	47
14	.2627506	.9648638	.2795497	.9601312	.2962638	.9551062	46
15	.2630312	.9647873	.2798290	.9600499	.2965416	.9550199	45
16	.2633118	.9647108	.2801083	.9599684	.2968194	.9549336	44
17	.2635925	.9646341	.2803875	.9598869	.2970971	.9548473	43
18	.2638730	.9645574	.2806667	.9598053	.2973749	.9547608	42
19	.2641536	.9644806	.2809459	.9597233	.2976526	.9546743	41
20	0.2644342	0.9644037	0.2812251	0.9596418	0.2979303	0.9545876	40
21	.2647147	.9643268	.2815042	.9595600	.2982079	.9545009	39
22	.2649952	.9642497	.2817833	.9594781	.2984856	.9544141	38
23	.2652757	.9641726	.2820624	.9593961	.2987632	.9543273	37
24	.2655561	.9640954	.2823415	.9593140	.2990408	.9542403	36
25	.2658366	.9640181	.2826205	.9592318	.2993184	.9541533	35
26	.2661170	.9639407	.2828995	.9591496	.2995959	.9540662	34
27	.2663973	.9638633	.2831785	.9590672	.2998734	.9539790	33
28	.2666777	.9637858	.2834575	.9589848	.3001509	.9538917	32
29	.2669581	.9637081	.2837364	.9589023	.3004284	.9538044	31
30	0.2672384	0.9636305	0.2840153	0.9588197	0.3007058	0.9537170	30
31	.2675187	.9635527	.2842942	.9587371	.3009832	.9536294	29
32	.2677989	.9634748	.2845731	.9586543	.3012606	.9535418	28
33	.2680792	.9633969	.2848520	.9585715	.3015380	.9534542	27
34	.2683594	.9633189	.2851308	.9584886	.3018153	.9533664	26
35	.2686396	.9632408	.2854096	.9584056	.3020926	.9532786	25
36	.2689198	.9631626	.2856884	.9583226	.3023699	.9531907	24
37	.2692000	.9630843	.2859671	.9582394	.3026471	.9531027	23
38	.2694801	.9630060	.2862458	.9581562	.3029244	.9530146	22
39	.2697602	.9629275	.2865246	.9580729	.3032016	.9529264	21
40	0.2700403	0.9628490	0.2868032	0.9579895	0.3034788	0.9528382	20
41	.2703204	.9627704	.2870819	.9579060	.3037559	.9527499	19
42	.2706004	.9626917	.2873605	.9578225	.3040331	.9526615	18
43	.2708805	.9626130	.2876391	.9577389	.3043102	.9525730	17
44	.2711605	.9625342	.2879177	.9576552	.3045872	.9524844	16
45	.2714404	.9624552	.2881963	.9575714	.3048643	.9523958	15
46	.2717204	.9623762	.2884748	.9574875	.3051413	.9523071	14
47	.2720003	.9622972	.2887533	.9574035	.3054183	.9522183	13
48	.2722802	.9622180	.2890318	.9573195	.3056953	.9521294	12
49	.2725601	.9621387	.2893103	.9572354	.3059723	.9520404	11
50	0.2728400	0.9620594	0.2895887	0.9571512	0.3062492	0.9519514	10
51	.2731198	.9619800	.2898671	.9570669	.3065261	.9518623	9
52	.2733997	.9619005	.2901455	.9569825	.3068030	.9517731	8
53	.2736794	.9618210	.2904239	.9568981	.3070798	.9516838	7
54	.2739592	.9617413	.2907022	.9568136	.3073566	.9515944	6
55	.2742390	.9616616	.2909805	.9567290	.3076334	.9515050	5
56	.2745187	.9615818	.2912588	.9566443	.3079102	.9514154	4
57	.2747984	.9615019	.2915371	.9565595	.3081869	.9513258	3
58	.2750781	.9614219	.2918153	.9564747	.3084636	.9512361	2
59	.2753577	.9613418	.2920935	.9563898	.3087403	.9511464	1
60	0.2756374	0.9612617	0.2923717	0.9563048	0.3090170	0.9510565	0
	Cosine	Sine	Cosine	Sine	Cosine	Sine	
′	74°		73°		72°		′

TABLE XIII. NATURAL SINES AND COSINES

′	18° Sine	18° Cosine	19° Sine	19° Cosine	20° Sine	20° Cosine	′
0	0.3090170	0.9510565	0.3255682	0.9455186	0.3420201	0.9396926	60
1	.3092936	.9509666	.3258432	.9454238	.3422935	.9395931	59
2	.3095702	.9508766	.3261182	.9453290	.3425668	.9394935	58
3	.3098468	.9507865	.3263932	.9452341	.3428400	.9393938	57
4	.3101234	.9506963	.3266681	.9451391	.3431133	.9392940	56
5	.3103999	.9506061	.3269430	.9450441	.3433865	.9391942	55
6	.3106764	.9505157	.3272179	.9449489	.3436597	.9390943	54
7	.3109529	.9504253	.3274928	.9448537	.3439329	.9389942	53
8	.3112294	.9503348	.3277676	.9447584	.3442060	.9388942	52
9	.3115058	.9502443	.3280424	.9446630	.3444791	.9387940	51
10	0.3117822	0.9501536	0.3283172	0.9445675	0.3447521	0.9386938	50
11	.3120586	.9500629	.3285919	.9444720	.3450252	.9385934	49
12	.3123349	.9499721	.3288666	.9443764	.3452982	.9384930	48
13	.3126112	.9498812	.3291413	.9442807	.3455712	.9383925	47
14	.3128875	.9497902	.3294160	.9441849	.3458441	.9382920	46
15	.3131638	.9496991	.3296906	.9440890	.3461171	.9381913	45
16	.3134400	.9496080	.3299653	.9439931	.3463900	.9380906	44
17	.3137163	.9495168	.3302398	.9438971	.3466628	.9379898	43
18	.3139925	.9494255	.3305144	.9438010	.3469357	.9378889	42
19	.3142686	.9493341	.3307889	.9437048	.3472085	.9377880	41
20	0.3145448	0.9492426	0.3310634	0.9436085	0.3474812	0.9376869	40
21	.3148209	.9491511	.3313379	.9435122	.3477540	.9375858	39
22	.3150969	.9490595	.3316123	.9434157	.3480267	.9374846	38
23	.3153730	.9489678	.3318867	.9433192	.3482994	.9373833	37
24	.3156490	.9488760	.3321611	.9432227	.3485720	.9372820	36
25	.3159250	.9487842	.3324355	.9431260	.3488447	.9371806	35
26	.3162010	.9486922	.3327098	.9430293	.3491173	.9370790	34
27	.3164770	.9486002	.3329841	.9429324	.3493898	.9369774	33
28	.3167529	.9485081	.3332584	.9428355	.3496624	.9368758	32
29	.3170288	.9484159	.3335326	.9427386	.3499349	.9367740	31
30	0.3173047	0.9483237	0.3338069	0.9426415	0.3502074	0.9366722	30
31	.3175805	.9482313	.3340810	.9425444	.3504798	.9365703	29
32	.3178563	.9481389	.3343552	.9424471	.3507523	.9364683	28
33	.3181321	.9480464	.3346293	.9423498	.3510246	.9363662	27
34	.3184079	.9479538	.3349034	.9422525	.3512970	.9362641	26
35	.3186836	.9478612	.3351775	.9421550	.3515693	.9361618	25
36	.3189593	.9477684	.3354516	.9420575	.3518416	.9360595	24
37	.3192350	.9476756	.3357256	.9419598	.3521139	.9359571	23
38	.3195106	.9475827	.3359996	.9418621	.3523862	.9358547	22
39	.3197863	.9474897	.3362735	.9417644	.3526584	.9357521	21
40	0.3200619	0.9473966	0.3365475	0.9416665	0.3529306	0.9356495	20
41	.3203374	.9473035	.3368214	.9415686	.3532027	.9355468	19
42	.3206130	.9472103	.3370953	.9414705	.3534748	.9354440	18
43	.3208885	.9471170	.3373691	.9413724	.3537469	.9353412	17
44	.3211640	.9470236	.3376429	.9412743	.3540190	.9352382	16
45	.3214395	.9469301	.3379167	.9411760	.3542910	.9351352	15
46	.3217149	.9468366	.3381905	.9410777	.3545630	.9350321	14
47	.3219903	.9467430	.3384642	.9409793	.3548350	.9349289	13
48	.3222657	.9466493	.3387379	.9408808	.3551070	.9348257	12
49	.3225411	.9465555	.3390116	.9407822	.3553789	.9347223	11
50	0.3228164	0.9464616	0.3392852	0.9406835	0.3556508	0.9346189	10
51	.3230917	.9463677	.3395589	.9405848	.3559226	.9345154	9
52	.3233670	.9462736	.3398325	.9404860	.3561944	.9344119	8
53	.3236422	.9461795	.3401060	.9403871	.3564662	.9343082	7
54	.3239174	.9460854	.3403796	.9402881	.3567380	.9342045	6
55	.3241926	.9459911	.3406531	.9401891	.3570097	.9341007	5
56	.3244678	.9458968	.3409265	.9400899	.3572814	.9339968	4
57	.3247429	.9458023	.3412000	.9399907	.3575531	.9338928	3
58	.3250180	.9457078	.3414734	.9398914	.3578248	.9337888	2
59	.3252931	.9456132	.3417468	.9397921	.3580964	.9336846	1
60	0.3255682	0.9455186	0.3420201	0.9396926	0.3583679	0.9335804	0
′	71° Cosine	71° Sine	70° Cosine	70° Sine	69° Cosine	69° Sine	′

TABLE XIII. NATURAL SINES AND COSINES

′	21° Sine	21° Cosine	22° Sine	22° Cosine	23° Sine	23° Cosine	′
0	0.3583679	0.9335804	0.3746066	0.9271839	0.3907311	0.9205049	60
1	.3586395	.9334761	.3748763	.9270748	.3909989	.9203912	59
2	.3589110	.9333718	.3751459	.9269658	.3912666	.9202774	58
3	.3591825	.9332673	.3754156	.9268566	.3915343	.9201635	57
4	.3594540	.9331628	.3756852	.9267474	.3918019	.9200496	56
5	.3597254	.9330582	.3759547	.9266380	.3920695	.9199356	55
6	.3599968	.9329535	.3762243	.9265286	.3923371	.9198215	54
7	.3602682	.9328488	.3764938	.9264192	.3926047	.9197073	53
8	.3605395	.9327439	.3767632	.9263096	.3928722	.9195931	52
9	.3608108	.9326390	.3770327	.9262000	.3931397	.9194788	51
10	0.3610821	0.9325340	0.3773021	0.9260902	0.3934071	0.9193644	50
11	.3613534	.9324290	.3775714	.9259805	.3936745	.9192499	49
12	.3616246	.9323238	.3778408	.9258706	.3939419	.9191353	48
13	.3618958	.9322186	.3781101	.9257606	.3942093	.9190207	47
14	.3621669	.9321133	.3783794	.9256506	.3944766	.9189060	46
15	.3624380	.9320079	.3786486	.9255405	.3947439	.9187912	45
16	.3627091	.9319024	.3789178	.9254303	.3950111	.9186763	44
17	.3629802	.9317969	.3791870	.9253201	.3952783	.9185614	43
18	.3632512	.9316912	.3794562	.9252097	.3955455	.9184464	42
19	.3635222	.9315855	.3797253	.9250993	.3958127	.9183313	41
20	0.3637932	0.9314797	0.3799944	0.9249888	0.3960798	0.9182161	40
21	.3640641	.9313739	.3802634	.9248782	.3963468	.9181009	39
22	.3643351	.9312679	.3805324	.9247676	.3966139	.9179855	38
23	.3646059	.9311619	.3808014	.9246568	.3968809	.9178701	37
24	.3648768	.9310558	.3810704	.9245460	.3971479	.9177546	36
25	.3651476	.9309496	.3813393	.9244351	.3974148	.9176391	35
26	.3654184	.9308434	.3816082	.9243242	.3976818	.9175234	34
27	.3656891	.9307370	.3818770	.9242131	.3979486	.9174077	33
28	.3659599	.9306306	.3821459	.9241020	.3982155	.9172919	32
29	.3662306	.9305241	.3824147	.9239908	.3984823	.9171760	31
30	0.3665012	0.9304176	0.3826834	0.9238795	0.3987491	0.9170601	30
31	.3667719	.9303109	.3829522	.9237682	.3990158	.9169440	29
32	.3670425	.9302042	.3832209	.9236567	.3992825	.9168279	28
33	.3673130	.9300974	.3834895	.9235452	.3995492	.9167118	27
34	.3675836	.9299905	.3837582	.9234336	.3998158	.9165955	26
35	.3678541	.9298835	.3840268	.9233220	.4000825	.9164791	25
36	.3681246	.9297765	.3842953	.9232102	.4003490	.9163627	24
37	.3683950	.9296694	.3845639	.9230984	.4006156	.9162462	23
38	.3686654	.9295622	.3848324	.9229865	.4008821	.9161297	22
39	.3689358	.9294549	.3851008	.9228745	.4011486	.9160130	21
40	0.3692061	0.9293475	0.3853693	0.9227624	0.4014150	0.9158963	20
41	.3694765	.9292401	.3856377	.9226503	.4016814	.9157795	19
42	.3697468	.9291326	.3859060	.9225381	.4019478	.9156626	18
43	.3700170	.9290250	.3861744	.9224258	.4022141	.9155456	17
44	.3702872	.9289173	.3864427	.9223134	.4024804	.9154286	16
45	.3705574	.9288096	.3867110	.9222010	.4027467	.9153115	15
46	.3708276	.9287017	.3869792	.9220884	.4030129	.9151943	14
47	.3710977	.9285938	.3872474	.9219758	.4032791	.9150770	13
48	.3713678	.9284858	.3875156	.9218632	.4035453	.9149597	12
49	.3716379	.9283778	.3877837	.9217504	.4038114	.9148422	11
50	0.3719079	0.9282696	0.3880518	0.9216375	0.4040775	0.9147247	10
51	.3721780	.9281614	.3883199	.9215246	.4043436	.9146072	9
52	.3724479	.9280531	.3885880	.9214116	.4046096	.9144895	8
53	.3727179	.9279447	.3888560	.9212986	.4048756	.9143718	7
54	.3729878	.9278363	.3891240	.9211854	.4051416	.9142540	6
55	.3732577	.9277277	.3893919	.9210722	.4054075	.9141361	5
56	.3735275	.9276191	.3896598	.9209589	.4056734	.9140181	4
57	.3737973	.9275104	.3899277	.9208455	.4059393	.9139001	3
58	.3740671	.9274016	.3901955	.9207320	.4062051	.9137819	2
59	.3743369	.9272928	.3904633	.9206185	.4064709	.9136637	1
60	0.3746066	0.9271839	0.3907311	0.9205049	0.4067366	0.9135455	0
′	68° Cosine	68° Sine	67° Cosine	67° Sine	66° Cosine	66° Sine	′

TABLE XIII. NATURAL SINES AND COSINES

′	24°		25°		26°		′
	Sine	Cosine	Sine	Cosine	Sine	Cosine	
0	0.4067366	0.9135455	0.4226183	0.9063078	0.4383711	0.8987940	60
1	.4070024	.9134271	.4228819	.9061848	.4386326	.8986665	59
2	.4072681	.9133087	.4231455	.9060618	.4388940	.8985389	58
3	.4075337	.9131902	.4234090	.9059386	.4391553	.8984112	57
4	.4077993	.9130716	.4236725	.9058154	.4394166	.8982834	56
5	.4080649	.9129529	.4239360	.9056922	.4396779	.8981555	55
6	.4083305	.9128342	.4241994	.9055688	.4399392	.8980276	54
7	.4085960	.9127154	.4244628	.9054454	.4402004	.8978996	53
8	.4088615	.9125965	.4247262	.9053219	.4404615	.8977715	52
9	.4091269	.9124775	.4249895	.9051983	.4407227	.8976433	51
10	0.4093923	0.9123584	0.4252528	0.9050746	0.4409838	0.8975151	50
11	.4096577	.9122393	.4255161	.9049509	.4412448	.8973868	49
12	.4099230	.9121201	.4257793	.9048271	.4415059	.8972584	48
13	.4101883	.9120008	.4260425	.9047032	.4417668	.8971299	47
14	.4104536	.9118815	.4263056	.9045792	.4420278	.8970014	46
15	.4107189	.9117620	.4265687	.9044551	.4422887	.8968727	45
16	.4109841	.9116425	.4268318	.9043310	.4425496	.8967440	44
17	.4112492	.9115229	.4270949	.9042068	.4428104	.8966153	43
18	.4115144	.9114033	.4273579	.9040825	.4430712	.8964864	42
19	.4117795	.9112835	.4276208	.9039582	.4433319	.8963575	41
20	0.4120445	0.9111637	0.4278838	0.9038338	0.4435927	0.8962285	40
21	.4123096	.9110438	.4281467	.9037093	.4438534	.8960994	39
22	.4125745	.9109238	.4284095	.9035847	.4441140	.8959703	38
23	.4128395	.9108038	.4286723	.9034600	.4443746	.8958411	37
24	.4131044	.9106837	.4289351	.9033353	.4446352	.8957118	36
25	.4133693	.9105635	.4291979	.9032105	.4448957	.8955824	35
26	.4136342	.9104432	.4294606	.9030856	.4451562	.8954529	34
27	.4138990	.9103228	.4297233	.9029606	.4454167	.8953234	33
28	.4141638	.9102024	.4299859	.9028356	.4456771	.8951938	32
29	.4144285	.9100819	.4302485	.9027105	.4459375	.8950641	31
30	0.4146932	0.9099613	0.4305111	0.9025853	0.4461978	0.8949344	30
31	.4149579	.9098406	.4307736	.9024600	.4464581	.8948045	29
32	.4152226	.9097199	.4310361	.9023347	.4467184	.8946746	28
33	.4154872	.9095990	.4312986	.9022092	.4469786	.8945446	27
34	.4157517	.9094781	.4315610	.9020838	.4472388	.8944146	26
35	.4160163	.9093572	.4318234	.9019582	.4474990	.8942844	25
36	.4162808	.9092361	.4320857	.9018325	.4477591	.8941542	24
37	.4165453	.9091150	.4323481	.9017068	.4480192	.8940240	23
38	.4168097	.9089938	.4326103	.9015810	.4482792	.8938936	22
39	.4170741	.9088725	.4328726	.9014551	.4485392	.8937632	21
40	0.4173385	0.9087511	0.4331348	0.9013292	0.4487992	0.8936326	20
41	.4176028	.9086297	.4333970	.9012031	.4490591	.8935021	19
42	.4178671	.9085082	.4336591	.9010770	.4493190	.8933714	18
43	.4181313	.9083866	.4339212	.9009508	.4495789	.8932406	17
44	.4183956	.9082649	.4341832	.9008246	.4498387	.8931098	16
45	.4186597	.9081432	.4344453	.9006982	.4500984	.8929789	15
46	.4189239	.9080214	.4347072	.9005718	.4503582	.8928480	14
47	.4191880	.9078995	.4349692	.9004453	.4506179	.8927169	13
48	.4194521	.9077775	.4352311	.9003188	.4508775	.8925858	12
49	.4197161	.9076554	.4354930	.9001921	.4511372	.8924546	11
50	0.4199801	0.9075333	0.4357548	0.9000654	0.4513967	0.8923234	10
51	.4202441	.9074111	.4360166	.8999386	.4516563	.8921920	9
52	.4205080	.9072888	.4362784	.8998117	.4519158	.8920606	8
53	.4207719	.9071665	.4365401	.8996848	.4521753	.8919291	7
54	.4210358	.9070440	.4368018	.8995578	.4524347	.8917975	6
55	.4212996	.9069215	.4370634	.8994307	.4526941	.8916659	5
56	.4215634	.9067989	.4373251	.8993035	.4529535	.8915342	4
57	.4218272	.9066762	.4375866	.8991763	.4532128	.8914024	3
58	.4220909	.9065535	.4378482	.8990489	.4534721	.8912705	2
59	.4223546	.9064307	.4381097	.8989215	.4537313	.8911385	1
60	0.4226183	0.9063078	0.4383711	0.8987940	0.4539905	0.8910065	0
	Cosine	Sine	Cosine	Sine	Cosine	Sine	
′	65°		64°		63°		′

TABLE XIII. NATURAL SINES AND COSINES

′	27°		28°		29°		′
	Sine	Cosine	Sine	Cosine	Sine	Cosine	
0	0.4539905	0.8910065	0.4694716	0.8829476	0.4848096	0.8746197	60
1	.4542497	.8908744	.4697284	.8828110	.4850640	.8744786	59
2	.4545088	.8907423	.4699852	.8826743	.4853184	.8743375	58
3	.4547679	.8906100	.4702419	.8825376	.4855727	.8741963	57
4	.4550269	.8904777	.4704986	.8824007	.4858270	.8740550	56
5	.4552859	.8903453	.4707553	.8822638	.4860812	.8739137	55
6	.4555449	.8902128	.4710119	.8821269	.4863354	.8737722	54
7	.4558038	.8900803	.4712685	.8819898	.4865895	.8736307	53
8	.4560627	.8899476	.4715250	.8818527	.4868436	.8734891	52
9	.4563216	.8898149	.4717815	.8817155	.4870977	.8733475	51
10	0.4565804	0.8896822	0.4720380	0.8815782	0.4873517	0.8732058	50
11	.4568392	.8895493	.4722944	.8814409	.4876057	.8730640	49
12	.4570979	.8894164	.4725508	.8813035	.4878597	.8729221	48
13	.4573566	.8892834	.4728071	.8811660	.4881136	.8727801	47
14	.4576153	.8891503	.4730634	.8810284	.4883674	.8726381	46
15	.4578739	.8890171	.4733197	.8808907	.4886212	.8724960	45
16	.4581325	.8888839	.4735759	.8807530	.4888750	.8723538	44
17	.4583910	.8887506	.4738321	.8806152	.4891288	.8722116	43
18	.4586496	.8886172	.4740882	.8804774	.4893825	.8720693	42
19	.4589080	.8884838	.4743443	.8803394	.4896361	.8719269	41
20	0.4591665	0.8883503	0.4746004	0.8802014	0.4898897	0.8717844	40
21	.4594248	.8882166	.4748564	.8800633	.4901433	.8716419	39
22	.4596832	.8880830	.4751124	.8799251	.4903968	.8714993	38
23	.4599415	.8879492	.4753683	.8797869	.4906503	.8713566	37
24	.4601998	.8878154	.4756242	.8796486	.4909038	.8712138	36
25	.4604580	.8876815	.4758801	.8795102	.4911572	.8710710	35
26	.4607162	.8875475	.4761359	.8793717	.4914105	.8709281	34
27	.4609744	.8874134	.4763917	.8792332	.4916638	.8707851	33
28	.4612325	.8872793	.4766474	.8790946	.4919171	.8706420	32
29	.4614906	.8871451	.4769031	.8789559	.4921704	.8704989	31
30	0.4617486	0.8870108	0.4771588	0.8788171	0.4924236	0.8703557	30
31	.4620066	.8868765	.4774144	.8786783	.4926767	.8702124	29
32	.4622646	.8867420	.4776700	.8785394	.4929298	.8700691	28
33	.4625225	.8866075	.4779255	.8784004	.4931829	.8699256	27
34	.4627804	.8864730	.4781810	.8782613	.4934359	.8697821	26
35	.4630382	.8863383	.4784364	.8781222	.4936889	.8696386	25
36	.4632960	.8862036	.4786919	.8779830	.4939419	.8694949	24
37	.4635538	.8860688	.4789472	.8778437	.4941948	.8693512	23
38	.4638115	.8859339	.4792026	.8777043	.4944476	.8692074	22
39	.4640692	.8857989	.4794579	.8775649	.4947005	.8690636	21
40	0.4643269	0.8856639	0.4797131	0.8774254	0.4949532	0.8689196	20
41	.4645845	.8855288	.4799683	.8772858	.4952060	.8687756	19
42	.4648420	.8853936	.4802235	.8771462	.4954587	.8686315	18
43	.4650996	.8852584	.4804786	.8770064	.4957113	.8684874	17
44	.4653571	.8851230	.4807337	.8768666	.4959639	.8683431	16
45	.4656145	.8849876	.4809888	.8767268	.4962165	.8681988	15
46	.4658719	.8848522	.4812438	.8765868	.4964690	.8680544	14
47	.4661293	.8847166	.4814987	.8764468	.4967215	.8679100	13
48	.4663866	.8845810	.4817537	.8763067	.4969740	.8677655	12
49	.4666439	.8844453	.4820086	.8761665	.4972264	.8676209	11
50	0.4669012	0.8843095	0.4822634	0.8760263	0.4974787	0.8674762	10
51	.4671584	.8841736	.4825182	.8758859	.4977310	.8673314	9
52	.4674156	.8840377	.4827730	.8757455	.4979833	.8671866	8
53	.4676727	.8839017	.4830277	.8756051	.4982355	.8670417	7
54	.4679298	.8837656	.4832824	.8754645	.4984877	.8668967	6
55	.4681869	.8836295	.4835370	.8753239	.4987399	.8667517	5
56	.4684439	.8834933	.4837916	.8751832	.4989920	.8666066	4
57	.4687009	.8833569	.4840462	.8750425	.4992441	.8664614	3
58	.4689578	.8832206	.4843007	.8749016	.4994961	.8663161	2
59	.4692147	.8830841	.4845552	.8747607	.4997481	.8661708	1
60	0.4694716	0.8829476	0.4848096	0.8746197	0.5000000	0.8660254	0
	Cosine	Sine	Cosine	Sine	Cosine	Sine	
′	62°		61°		60°		′

TABLE XIII. NATURAL SINES AND COSINES

′	30°		31°		32°		′
	Sine	Cosine	Sine	Cosine	Sine	Cosine	
0	0.5000000	0.8660254	0.5150381	0.8571673	0.5299193	0.8480481	60
1	.5002519	.8658799	.5152874	.8570174	.5301659	.8478939	59
2	.5005037	.8657344	.5155367	.8568675	.5304125	.8477397	58
3	.5007556	.8655887	.5157859	.8567175	.5306591	.8475853	57
4	.5010073	.8654430	.5160351	.8565674	.5309057	.8474309	56
5	.5012591	.8652973	.5162842	.8564173	.5311521	.8472765	55
6	.5015107	.8651514	.5165333	.8562671	.5313986	.8471219	54
7	.5017624	.8650055	.5167824	.8561168	.5316450	.8469673	53
8	.5020140	.8648595	.5170314	.8559664	.5318913	.8468126	52
9	.5022655	.8647134	.5172804	.8558160	.5321376	.8466579	51
10	0.5025170	0.8645673	0.5175293	0.8556655	0.5323839	0.8465030	50
11	.5027685	.8644211	.5177782	.8555149	.5326301	.8463481	49
12	.5030199	.8642748	.5180270	.8553643	.5328763	.8461932	48
13	.5032713	.8641284	.5182758	.8552135	.5331224	.8460381	47
14	.5035227	.8639820	.5185246	.8550627	.5333685	.8458830	46
15	.5037740	.8638355	.5187733	.8549119	.5336145	.8457278	45
16	.5040252	.8636889	.5190219	.8547609	.5338605	.8455726	44
17	.5042765	.8635423	.5192705	.8546099	.5341065	.8454172	43
18	.5045276	.8633956	.5195191	.8544588	.5343523	.8452618	42
19	.5047788	.8632488	.5197676	.8543077	.5345982	.8451064	41
20	0.5050298	0.8631019	0.5200161	0.8541564	0.5348440	0.8449508	40
21	.5052809	.8629549	.5202646	.8540051	.5350898	.8447952	39
22	.5055319	.8628079	.5205130	.8538538	.5353355	.8446395	38
23	.5057828	.8626608	.5207613	.8537023	.5355812	.8444838	37
24	.5060338	.8625137	.5210096	.8535508	.5358268	.8443279	36
25	.5062846	.8623664	.5212579	.8533992	.5360724	.8441720	35
26	.5065355	.8622191	.5215061	.8532475	.5363179	.8440161	34
27	.5067863	.8620717	.5217543	.8530958	.5365634	.8438600	33
28	.5070370	.8619243	.5220024	.8529440	.5368089	.8437039	32
29	.5072877	.8617768	.5222505	.8527921	.5370543	.8435477	31
30	0.5075384	0.8616292	0.5224986	0.8526402	0.5372996	0.8433914	30
31	.5077890	.8614815	.5227466	.8524881	.5375449	.8432351	29
32	.5080396	.8613337	.5229945	.8523360	.5377902	.8430787	28
33	.5082901	.8611859	.5232424	.8521839	.5380354	.8429222	27
34	.5085406	.8610380	.5234903	.8520316	.5382806	.8427657	26
35	.5087910	.8608901	.5237381	.8518793	.5385257	.8426091	25
36	.5090414	.8607420	.5239859	.8517269	.5387708	.8424524	24
37	.5092918	.8605939	.5242336	.8515745	.5390158	.8422956	23
38	.5095421	.8604457	.5244813	.8514219	.5392608	.8421388	22
39	.5097924	.8602975	.5247290	.8512693	.5395058	.8419818	21
40	0.5100426	0.8601491	0.5249766	0.8511167	0.5397507	0.8418249	20
41	.5102928	.8600007	.5252241	.8509639	.5399955	.8416679	19
42	.5105429	.8598523	.5254717	.8508111	.5402403	.8415108	18
43	.5107930	.8597037	.5257191	.8506582	.5404851	.8413536	17
44	.5110431	.8595551	.5259665	.8505053	.5407298	.8411963	16
45	.5112931	.8594064	.5262139	.8503522	.5409745	.8410390	15
46	.5115431	.8592576	.5264613	.8501991	.5412191	.8408816	14
47	.5117930	.8591088	.5267085	.8500459	.5414637	.8407241	13
48	.5120429	.8589599	.5269558	.8498927	.5417082	.8405666	12
49	.5122927	.8588109	.5272030	.8497394	.5419527	.8404090	11
50	0.5125425	0.8586619	0.5274502	0.8495860	0.5421971	0.8402513	10
51	.5127923	.8585127	.5276973	.8494325	.5424415	.8400936	9
52	.5130420	.8583635	.5279443	.8492790	.5426859	.8399357	8
53	.5132916	.8582143	.5281914	.8491254	.5429302	.8397778	7
54	.5135413	.8580649	.5284383	.8489717	.5431744	.8396199	6
55	.5137908	.8579155	.5286853	.8488179	.5434187	.8394618	5
56	.5140404	.8577660	.5289322	.8486641	.5436628	.8393037	4
57	.5142899	.8576164	.5291790	.8485102	.5439069	.8391455	3
58	.5145393	.8574668	.5294258	.8483562	.5441510	.8389873	2
59	.5147887	.8573171	.5296726	.8482022	.5443951	.8388290	1
60	0.5150381	0.8571673	0.5299193	0.8480481	0.5446390	0.8386706	0
	Cosine	Sine	Cosine	Sine	Cosine	Sine	
′	59°		58°		57°		′

TABLE XIII. NATURAL SINES AND COSINES

′	33°		34°		35°		′
	Sine	Cosine	Sine	Cosine	Sine	Cosine	
0	0.5446390	0.8386706	0.5591929	0.8290376	0.5735764	0.8191520	60
1	.5448830	.8385121	.5594340	.8288749	.5738147	.8189852	59
2	.5451269	.8383536	.5596751	.8287121	.5740529	.8188182	58
3	.5453707	.8381950	.5599162	.8285493	.5742911	.8186512	57
4	.5456145	.8380363	.5601572	.8283864	.5745292	.8184841	56
5	.5458583	.8378775	.5603981	.8282234	.5747672	.8183169	55
6	.5461020	.8377187	.5606390	.8280603	.5750053	.8181497	54
7	.5463456	.8375598	.5608798	.8278972	.5752432	.8179824	53
8	.5465892	.8374009	.5611206	.8277340	.5754811	.8178151	52
9	.5468328	.8372418	.5613614	.8275708	.5757190	.8176476	51
10	0.5470763	0.8370827	0.5616021	0.8274074	0.5759568	0.8174801	50
11	.5473198	.8369236	.5618428	.8272440	.5761946	.8173125	49
12	.5475632	.8367643	.5620834	.8270806	.5764323	.8171449	48
13	.5478066	.8366050	.5623239	.8269170	.5766700	.8169772	47
14	.5480499	.8364456	.5625645	.8267534	.5769076	.8168094	46
15	.5482932	.8362862	.5628049	.8265897	.5771452	.8166416	45
16	.5485365	.8361266	.5630453	.8264260	.5773827	.8164736	44
17	.5487797	.8359670	.5632857	.8262622	.5776202	.8163056	43
18	.5490228	.8358074	.5635260	.8260983	.5778576	.8161376	42
19	.5492659	.8356476	.5637663	.8259343	.5780950	.8159695	41
20	0.5495090	0.8354878	0.5640066	0.8257703	0.5783323	0.8158013	40
21	.5497520	.8353279	.5642467	.8256062	.5785696	.8156330	39
22	.5499950	.8351680	.5644869	.8254420	.5788069	.8154647	38
23	.5502379	.8350080	.5647270	.8252778	.5790440	.8152963	37
24	.5504807	.8348479	.5649670	.8251135	.5792812	.8151278	36
25	.5507236	.8346877	.5652070	.8249491	.5795183	.8149593	35
26	.5509663	.8345275	.5654469	.8247847	.5797553	.8147906	34
27	.5512091	.8343672	.5656868	.8246202	.5799923	.8146220	33
28	.5514518	.8342068	.5659267	.8244556	.5802292	.8144532	32
29	.5516944	.8340463	.5661665	.8242909	.5804661	.8142844	31
30	0.5519370	0.8338858	0.5664062	0.8241262	0.5807030	0.8141155	30
31	.5521795	.8337252	.5666459	.8239614	.5809397	.8139466	29
32	.5524220	.8335646	.5668856	.8237965	.5811765	.8137775	28
33	.5526645	.8334038	.5671252	.8236316	.5814132	.8136084	27
34	.5529069	.8332430	.5673648	.8234666	.5816498	.8134393	26
35	.5531492	.8330822	.5676043	.8233015	.5818864	.8132701	25
36	.5533915	.8329212	.5678437	.8231364	.5821230	.8131008	24
37	.5536338	.8327602	.5680832	.8229712	.5823595	.8129314	23
38	.5538760	.8325991	.5683225	.8228059	.5825959	.8127620	22
39	.5541182	.8324380	.5685619	.8226405	.5828323	.8125925	21
40	0.5543603	0.8322768	0.5688011	0.8224751	0.5830687	0.8124229	20
41	.5546024	.8321155	.5690403	.8223096	.5833050	.8122532	19
42	.5548444	.8319541	.5692795	.8221440	.5835412	.8120835	18
43	.5550864	.8317927	.5695187	.8219784	.5837774	.8119137	17
44	.5553283	.8316312	.5697577	.8218127	.5840136	.8117439	16
45	.5555702	.8314696	.5699968	.8216469	.5842497	.8115740	15
46	.5558121	.8313080	.5702357	.8214811	.5844857	.8114040	14
47	.5560539	.8311463	.5704747	.8213152	.5847217	.8112339	13
48	.5562956	.8309845	.5707136	.8211492	.5849577	.8110638	12
49	.5565373	.8308226	.5709524	.8209832	.5851936	.8108936	11
50	0.5567790	0.8306607	0.5711912	0.8208170	0.5854294	0.8107234	10
51	.5570206	.8304987	.5714299	.8206509	.5856652	.8105530	9
52	.5572621	.8303366	.5716686	.8204846	.5859010	.8103826	8
53	.5575036	.8301745	.5719073	.8203183	.5861367	.8102122	7
54	.5577451	.8300123	.5721459	.8201519	.5863724	.8100416	6
55	.5579865	.8298500	.5723844	.8199854	.5866080	.8098710	5
56	.5582279	.8296877	.5726229	.8198189	.5868435	.8097004	4
57	.5584692	.8295252	.5728614	.8196523	.5870790	.8095296	3
58	.5587105	.8293628	.5730998	.8194856	.5873145	.8093588	2
59	.5589517	.8292002	.5733381	.8193189	.5875499	.8091879	1
60	0.5591929	0.8290376	0.5735764	0.8191520	0.5877853	0.8090170	0
	Cosine	Sine	Cosine	Sine	Cosine	Sine	
′	56°		55°		54°		′

TABLE XIII. NATURAL SINES AND COSINES

′	36°		37°		38°		′
	Sine	Cosine	Sine	Cosine	Sine	Cosine	
0	0.5877853	0.8090170	0.6018150	0.7986355	0.6156615	0.7880108	60
1	.5880206	.8088460	.6020473	.7984604	.6158907	.7878316	59
2	.5882558	.8086749	.6022795	.7982853	.6161198	.7876524	58
3	.5884910	.8085037	.6025117	.7981100	.6163489	.7874732	57
4	.5887262	.8083325	.6027439	.7979347	.6165780	.7872939	56
5	.5889613	.8081612	.6029760	.7977594	.6168069	.7871145	55
6	.5891964	.8079899	.6032080	.7975839	.6170359	.7869350	54
7	.5894314	.8078185	.6034400	.7974084	.6172648	.7867555	53
8	.5896663	.8076470	.6036719	.7972329	.6174936	.7865759	52
9	.5899012	.8074754	.6039038	.7970572	.6177224	.7863963	51
10	0.5901361	0.8073038	0.6041356	0.7968815	0.6179511	0.7862165	50
11	.5903709	.8071321	.6043674	.7967058	.6181798	.7860367	49
12	.5906057	.8069603	.6045991	.7965299	.6184084	.7858569	48
13	.5908404	.8067885	.6048303	.7963540	.6186370	.7856770	47
14	.5910750	.8066166	.6050624	.7961780	.6188655	.7854970	46
15	.5913096	.8064446	.6052940	.7960020	.6190939	.7853169	45
16	.5915442	.8062726	.6055255	.7958259	.6193224	.7851368	44
17	.5917787	.8061005	.6057570	.7956497	.6195507	.7849566	43
18	.5920132	.8059283	.6059884	.7954735	.6197790	.7847764	42
19	.5922476	.8057560	.6062198	.7952972	.6200073	.7845961	41
20	0.5924819	0.8055837	0.6064511	0.7951208	0.6202355	0.7844157	40
21	.5927163	.8054113	.6066824	.7949444	.6204636	.7842352	39
22	.5929505	.8052389	.6069136	.7947678	.6206917	.7840547	38
23	.5931847	.8050664	.6071447	.7945913	.6209198	.7838741	37
24	.5934189	.8048938	.6073758	.7944146	.6211478	.7836935	36
25	.5936530	.8047211	.6076069	.7942379	.6213757	.7835127	35
26	.5938871	.8045484	.6078379	.7940611	.6216036	.7833320	34
27	.5941211	.8043756	.6080689	.7938843	.6218314	.7831511	33
28	.5943550	.8042028	.6082998	.7937074	.6220592	.7829702	32
29	.5945889	.8040299	.6085306	.7935304	.6222870	.7827892	31
30	0.5948228	0.8038569	0.6087614	0.7933533	0.6225146	0.7826082	30
31	.5950566	.8036838	.6089922	.7931762	.6227423	.7824270	29
32	.5952904	.8035107	.6092229	.7929990	.6229698	.7822459	28
33	.5955241	.8033375	.6094535	.7928218	.6231974	.7820646	27
34	.5957577	.8031642	.6096841	.7926445	.6234248	.7818833	26
35	.5959913	.8029909	.6099147	.7924671	.6236522	.7817019	25
36	.5962249	.8028175	.6101452	.7922896	.6238796	.7815205	24
37	.5964584	.8026440	.6103756	.7921121	.6241069	.7813390	23
38	.5966918	.8024705	.6106060	.7919345	.6243342	.7811574	22
39	.5969252	.8022969	.6108363	.7917569	.6245614	.7809757	21
40	0.5971586	0.8021232	0.6110666	0.7915792	0.6247885	0.7807940	20
41	.5973919	.8019495	.6112969	.7914014	.6250156	.7806123	19
42	.5976251	.8017756	.6115270	.7912235	.6252427	.7804304	18
43	.5978583	.8016018	.6117572	.7910456	.6254696	.7802485	17
44	.5980915	.8014278	.6119873	.7908676	.6256966	.7800665	16
45	.5983246	.8012538	.6122173	.7906896	.6259235	.7798845	15
46	.5985577	.8010797	.6124473	.7905115	.6261503	.7797024	14
47	.5987906	.8009056	.6126772	.7903333	.6263771	.7795202	13
48	.5990236	.8007314	.6129071	.7901550	.6266038	.7793380	12
49	.5992565	.8005571	.6131369	.7899767	.6268305	.7791557	11
50	0.5994893	0.8003827	0.6133666	0.7897983	0.6270571	0.7789733	10
51	.5997221	.8002083	.6135964	.7896198	.6272837	.7787909	9
52	.5999549	.8000338	.6138260	.7894413	.6275102	.7786084	8
53	.6001876	.7998593	.6140556	.7892627	.6277366	.7784258	7
54	.6004202	.7996847	.6142852	.7890841	.6279631	.7782431	6
55	.6006528	.7995100	.6145147	.7889054	.6281894	.7780604	5
56	.6008854	.7993352	.6147442	.7887266	.6284157	.7778777	4
57	.6011179	.7991604	.6149736	.7885477	.6286420	.7776949	3
58	.6013503	.7989855	.6152029	.7883688	.6288682	.7775120	2
59	.6015827	.7988105	.6154322	.7881898	.6290943	.7773290	1
60	0.6018150	0.7986355	0.6156615	0.7880108	0.6293204	0.7771460	0
	Cosine	Sine	Cosine	Sine	Cosine	Sine	
′	53°		52°		51°		′

TABLE XIII. NATURAL SINES AND COSINES

′	39° Sine	39° Cosine	40° Sine	40° Cosine	41° Sine	41° Cosine	′
0	0.6293204	0.7771460	0.6427876	0.7660444	0.6560590	0.7547096	60
1	.6295464	.7769629	.6430104	.7658574	.6562785	.7545187	59
2	.6297724	.7767797	.6432332	.7656704	.6564980	.7543278	58
3	.6299983	.7765965	.6434559	.7654832	.6567174	.7541368	57
4	.6302242	.7764132	.6436785	.7652960	.6569367	.7539457	56
5	.6304500	.7762298	.6439011	.7651087	.6571560	.7537546	55
6	.6306758	.7760464	.6441236	.7649214	.6573752	.7535634	54
7	.6309015	.7758629	.6443461	.7647340	.6575944	.7533721	53
8	.6311272	.7756794	.6445685	.7645465	.6578135	.7531808	52
9	.6313528	.7754957	.6447909	.7643590	.6580326	.7529894	51
10	0.6315784	0.7753121	0.6450132	0.7641714	0.6582516	0.7527980	50
11	.6318039	.7751283	.6452355	.7639838	.6584706	.7526065	49
12	.6320293	.7749445	.6454577	.7637960	.6586895	.7524149	48
13	.6322547	.7747606	.6456798	.7636082	.6589083	.7522233	47
14	.6324800	.7745767	.6459019	.7634204	.6591271	.7520316	46
15	.6327053	.7743926	.6461240	.7632325	.6593458	.7518398	45
16	.6329306	.7742086	.6463460	.7630445	.6595645	.7516480	44
17	.6331557	.7740244	.6465679	.7628564	.6597831	.7514561	43
18	.6333809	.7738402	.6467898	.7626683	.6600017	.7512641	42
19	.6336059	.7736559	.6470116	.7624802	.6602202	.7510721	41
20	0.6338310	0.7734716	0.6472334	0.7622919	0.6604383	0.7508800	40
21	.6340559	.7732872	.6474551	.7621036	.6606570	.7506879	39
22	.6342808	.7731027	.6476767	.7619152	.6608754	.7504957	38
23	.6345057	.7729182	.6478984	.7617268	.6610936	.7503034	37
24	.6347305	.7727336	.6481199	.7615383	.6613119	.7501111	36
25	.6349553	.7725489	.6483414	.7613497	.6615300	.7499187	35
26	.6351800	.7723642	.6485628	.7611611	.6617482	.7497262	34
27	.6354046	.7721794	.6487842	.7609724	.6619662	.7495337	33
28	.6356292	.7719945	.6490056	.7607837	.6621842	.7493411	32
29	.6358537	.7718096	.6492268	.7605949	.6624022	.7491484	31
30	0.6360782	0.7716246	0.6494480	0.7604060	0.6626200	0.7489557	30
31	.6363026	.7714395	.6496692	.7602170	.6628379	.7487629	29
32	.6365270	.7712544	.6498903	.7600280	.6630557	.7485701	28
33	.6367513	.7710692	.6501114	.7598389	.6632734	.7483772	27
34	.6369756	.7708840	.6503324	.7596498	.6634910	.7481842	26
35	.6371998	.7706986	.6505533	.7594606	.6637087	.7479912	25
36	.6374240	.7705132	.6507742	.7592713	.6639262	.7477981	24
37	.6376481	.7703278	.6509951	.7590820	.6641437	.7476049	23
38	.6378721	.7701423	.6512158	.7588926	.6643612	.7474117	22
39	.6380961	.7699567	.6514366	.7587031	.6645785	.7472184	21
40	0.6383201	0.7697710	0.6516572	0.7585136	0.6647959	0.7470251	20
41	.6385440	.7695853	.6518778	.7583240	.6650131	.7468317	19
42	.6387678	.7693996	.6520984	.7581343	.6652304	.7466382	18
43	.6389916	.7692137	.6523189	.7579446	.6654475	.7464446	17
44	.6392153	.7690278	.6525394	.7577548	.6656646	.7462510	16
45	.6394390	.7688418	.6527598	.7575650	.6658817	.7460574	15
46	.6396626	.7686558	.6529801	.7573751	.6660987	.7458636	14
47	.6398862	.7684697	.6532004	.7571851	.6663156	.7456699	13
48	.6401097	.7682835	.6534206	.7569951	.6665325	.7454760	12
49	.6403332	.7680973	.6536408	.7568050	.6667493	.7452821	11
50	0.6405566	0.7679110	0.6538609	0.7566148	0.6669661	0.7450881	10
51	.6407799	.7677246	.6540810	.7564246	.6671828	.7448941	9
52	.6410032	.7675382	.6543010	.7562343	.6673994	.7446999	8
53	.6412264	.7673517	.6545209	.7560439	.6676160	.7445058	7
54	.6414496	.7671652	.6547408	.7558535	.6678326	.7443115	6
55	.6416728	.7669785	.6549607	.7556630	.6680490	.7441173	5
56	.6418958	.7667918	.6551804	.7554724	.6682655	.7439229	4
57	.6421189	.7666051	.6554002	.7552818	.6684818	.7437285	3
58	.6423418	.7664183	.6556198	.7550911	.6686981	.7435340	2
59	.6425647	.7662314	.6558395	.7549004	.6689144	.7433394	1
60	0.6427876	0.7660444	0.6560590	0.7547096	0.6691306	0.7431448	0
′	Cosine 50°	Sine 50°	Cosine 49°	Sine 49°	Cosine 48°	Sine 48°	′

TABLE XIII. NATURAL SINES AND COSINES

′	42°		43°		44°		′
	Sine	Cosine	Sine	Cosine	Sine	Cosine	
0	0.6691306	0.7431448	0.6819984	0.7313537	0.6946584	0.7193398	60
1	.6693468	.7429502	.6822111	.7311553	.6948676	.7191377	59
2	.6695628	.7427554	.6824237	.7309568	.6950767	.7189355	58
3	.6697789	.7425606	.6826363	.7307583	.6952858	.7187333	57
4	.6699948	.7423658	.6828489	.7305597	.6954949	.7185310	56
5	.6702108	.7421708	.6830613	.7303610	.6957039	.7183287	55
6	.6704266	.7419758	.6832738	.7301623	.6959128	.7181263	54
7	.6706424	.7417808	.6834861	.7299635	.6961217	.7179238	53
8	.6708582	.7415857	.6836984	.7297646	.6963305	.7177213	52
9	.6710739	.7413905	.6839107	.7295657	.6965392	.7175187	51
10	0.6712895	0.7411953	0.6841229	0.7293668	0.6967479	0.7173161	50
11	.6715051	.7410000	.6843350	.7291677	.6969565	.7171134	49
12	.6717206	.7408046	.6845471	.7289686	.6971651	.7169106	48
13	.6719361	.7406092	.6847591	.7287695	.6973736	.7167078	47
14	.6721515	.7404137	.6849711	.7285703	.6975821	.7165049	46
15	.6723668	.7402181	.6851830	.7283710	.6977905	.7163019	45
16	.6725821	.7400225	.6853948	.7281716	.6979988	.7160989	44
17	.6727973	.7398268	.6856066	.7279722	.6982071	.7158959	43
18	.6730125	.7396311	.6858184	.7277728	.6984153	.7156927	42
19	.6732276	.7394353	.6860300	.7275732	.6986234	.7154895	41
20	0.6734427	0.7392394	0.6862416	0.7273736	0.6988315	0.7152863	40
21	.6736577	.7390435	.6864532	.7271740	.6990396	.7150830	39
22	.6738727	.7388475	.6866647	.7269743	.6992476	.7148796	38
23	.6740876	.7386515	.6868761	.7267745	.6994555	.7146762	37
24	.6743024	.7384553	.6870875	.7265747	.6996633	.7144727	36
25	.6745172	.7382592	.6872988	.7263748	.6998711	.7142691	35
26	.6747319	.7380629	.6875101	.7261748	.7000789	.7140655	34
27	.6749466	.7378666	.6877213	.7259748	.7002866	.7138618	33
28	.6751612	.7376703	.6879325	.7257747	.7004942	.7136581	32
29	.6753757	.7374738	.6881435	.7255746	.7007018	.7134543	31
30	0.6755902	0.7372773	0.6883546	0.7253744	0.7009093	0.7132504	30
31	.6758046	.7370808	.6885655	.7251741	.7011167	.7130465	29
32	.6760190	.7368842	.6887765	.7249738	.7013241	.7128426	28
33	.6762333	.7366875	.6889873	.7247734	.7015314	.7126385	27
34	.6764476	.7364908	.6891981	.7245729	.7017387	.7124344	26
35	.6766618	.7362940	.6894089	.7243724	.7019459	.7122303	25
36	.6768760	.7360971	.6896195	.7241719	.7021531	.7120260	24
37	.6770901	.7359002	.6898302	.7239712	.7023601	.7118218	23
38	.6773041	.7357032	.6900407	.7237705	.7025672	.7116174	22
39	.6775181	.7355061	.6902512	.7235698	.7027741	.7114130	21
40	0.6777320	0.7353090	0.6904617	0.7233690	0.7029811	0.7112086	20
41	.6779459	.7351118	.6906721	.7231681	.7031879	.7110041	19
42	6781597	.7349146	.6908824	.7229671	.7033947	.7107995	18
43	.6783734	.7347173	.6910927	.7227661	.7036014	.7105948	17
44	.6785871	.7345199	.6913029	.7225651	.7038081	.7103901	16
45	.6788007	.7343225	.6915131	.7223640	.7040147	.7101854	15
46	.6790143	.7341250	.6917232	.7221628	.7042213	.7099806	14
47	.6792278	.7339275	.6919332	.7219615	.7044278	.7097757	13
48	.6794413	.7337299	.6921432	.7217602	.7046342	.7095707	12
49	.6796547	.7335322	.6923531	.7215589	.7048406	.7093657	11
50	0.6798681	0.7333345	0.6925630	0.7213574	0.7050469	0.7091607	10
51	.6800813	.7331367	.6927728	.7211559	.7052532	.7089556	9
52	.6802946	.7329388	.6929825	.7209544	.7054594	.7087504	8
53	.6805078	.7327409	.6931922	.7207528	.7056655	.7085451	7
54	.6807209	.7325429	.6934018	.7205511	.7058716	.7083398	6
55	.6809339	.7323449	.6936114	.7203494	.7060776	.7081345	5
56	.6811469	.7321467	.6938209	.7201476	.7062835	.7079291	4
57	.6813599	.7319486	.6940304	.7199457	.7064894	.7077236	3
58	.6815728	.7317503	.6942398	.7197438	.7066953	.7075180	2
59	.6817856	.7315521	.6944491	.7195418	.7069011	.7073124	1
60	0.6819984	0.7313537	0.6946584	0.7193398	0.7071068	0.7071068	0
	Cosine	Sine	Cosine	Sine	Cosine	Sine	
′	47°		46°		45°		′

TABLE XIV. NATURAL TANGENTS AND COTANGENTS

′	0°		1°		2°		′
	Tangent	Cotangent	Tangent	Cotangent	Tangent	Cotangent	
0	0.0000000	Infinite	0.0174551	57.289962	0.0349208	28.636253	60
1	.0002909	3437.7467	.0177460	56.350590	.0352120	28.399397	59
2	.0005818	1718.8732	.0180370	55.441517	.0355033	28.166422	58
3	.0008727	1145.9153	.0183280	54.561300	.0357945	27.937233	57
4	.0011636	859.43630	.0186190	53.708588	.0360858	27.711740	56
5	.0014544	687.54887	.0189100	52.882109	.0363771	27.489853	55
6	.0017453	572.95721	.0192010	52.080673	.0366683	27.271486	54
7	.0020362	491.10600	.0194920	51.303157	.0369596	27.056557	53
8	.0023271	429.71757	.0197830	50.548506	.0372509	26.844984	52
9	.0026180	381.97099	.0200740	49.815726	.0375422	26.636690	51
10	0.0029089	343.77371	0.0203650	49.103881	0.0378335	26.431600	50
11	.0031998	312.52137	.0206560	48.412084	.0381248	26.229638	49
12	.0034907	286.47773	.0209470	47.739501	.0384161	26.030736	48
13	.0037816	264.44080	.0212380	47.085343	.0387074	25.834823	47
14	.0040725	245.55198	.0215291	46.448862	.0389988	25.641832	46
15	.0043634	229.18166	.0218201	45.829351	.0392901	25.451700	45
16	.0046542	214.85762	.0221111	45.226141	.0395814	25.264361	44
17	.0049451	202.21875	.0224021	44.638596	.0398728	25.079757	43
18	.0052360	190.98419	.0226932	44.066113	.0401641	24.897826	42
19	.0055269	180.93220	.0229842	43.508122	.0404555	24.718512	41
20	0.0058178	171.88540	0.0232753	42.964077	0.0407469	24.541758	40
21	.0061087	163.70019	.0235663	42.433464	.0410383	24.367509	39
22	.0063996	156.25908	.0238574	41.915790	.0413296	24.195714	38
23	.0066905	149.46502	.0241484	41.410588	.0416210	24.026320	37
24	.0069814	143.23712	.0244395	40.917412	.0419124	23.859277	36
25	.0072723	137.50745	.0247305	40.435837	.0422038	23.694537	35
26	.0075632	132.21851	.0250216	39.965461	.0424952	23.532052	34
27	.0078541	127.32134	.0253127	39.505895	.0427866	23.371777	33
28	.0081450	122.77396	.0256038	39.056771	.0430781	23.213666	32
29	.0084360	118.54018	.0258948	38.617738	.0433695	23.057677	31
30	0.0087269	114.58865	0.0261859	38.188459	0.0436609	22.903766	30
31	.0090178	110.89205	.0264770	37.768613	.0439524	22.751892	29
32	.0093087	107.42648	.0267681	37.357892	.0442438	22.602015	28
33	.0095996	104.17094	.0270592	36.956001	.0445353	22.454096	27
34	.0098905	101.10690	.0273503	36.562659	.0448268	22.308097	26
35	.0101814	98.217943	.0276414	36.177596	.0451183	22.163980	25
36	.0104724	95.489475	.0279325	35.800553	.0454097	22.021710	24
37	.0107633	92.908487	.0282236	35.431282	.0457012	21.881251	23
38	.0110542	90.463336	.0285148	35.069546	.0459927	21.742569	22
39	.0113451	88.143572	.0288059	34.715115	.0462842	21.605630	21
40	0.0116361	85.939791	0.0290970	34.367771	0.0465757	21.470401	20
41	.0119270	83.843507	.0293882	34.027303	.0468673	21.336851	19
42	.0122179	81.847041	.0296793	33.693509	.0471588	21.204949	18
43	.0125088	79.943430	.0299705	33.366194	.0474503	21.074664	17
44	.0127998	78.126342	.0302616	33.045173	.0477419	20.945966	16
45	.0130907	76.390009	.0305528	32.730264	.0480334	20.818828	15
46	.0133817	74.729165	.0308439	32.421295	.0483250	20.693220	14
47	.0136726	73.138991	.0311351	32.118099	.0486166	20.569115	13
48	.0139635	71.615070	.0314263	31.820516	.0489082	20.446486	12
49	.0142545	70.153346	.0317174	31.528392	.0491997	20.325308	11
50	0.0145454	68.750087	0.0320086	31.241577	0.0494913	20.205553	10
51	.0148364	67.401854	.0322998	30.959928	.0497829	20.087199	9
52	.0151273	66.105473	.0325910	30.683307	.0500746	19.970219	8
53	.0154183	64.858008	.0328822	30.411580	.0503662	19.854591	7
54	.0157093	63.656741	.0331734	30.144619	.0506578	19.740291	6
55	.0160002	62.499154	.0334646	29.882299	.0509495	19.627296	5
56	.0162912	61.382905	.0337558	29.624499	.0512411	19.515584	4
57	.0165821	60.305820	.0340471	29.371106	.0515328	19.405133	3
58	.0168731	59.265872	.0343383	29.122005	.0518244	19.295922	2
59	.0171641	58.261174	.0346295	28.877089	.0521161	19.187930	1
60	0.0174551	57.289962	0.0349208	28.636253	0.0524078	19.081137	0
	Cotangent	Tangent	Cotangent	Tangent	Cotangent	Tangent	
′	89°		88°		87°		′

TABLE XIV. NATURAL TANGENTS AND COTANGENTS

′	3°		4°		5°		′
	Tangent	Cotangent	Tangent	Cotangent	Tangent	Cotangent	
0	0.0524078	19.081137	0.0699268	14.300666	0.0874887	11.430052	60
1	.0526995	18.975523	.0702191	14.241134	.0877818	11.391885	59
2	.0529912	18.871068	.0705115	14.182092	.0880749	11.353970	58
3	.0532829	18.767754	.0708038	14.123536	.0883681	11.316304	57
4	.0535746	18.665562	.0710961	14.065459	.0886612	11.278885	56
5	.0538663	18.564473	.0713885	14.007856	.0889544	11.241712	55
6	.0541581	18.464471	.0716809	13.950719	.0892476	11.204780	54
7	.0544498	18.365537	.0719733	13.894045	.0895408	11.168089	53
8	.0547416	18.267654	.0722657	13.837827	.0898341	11.131635	52
9	.0550333	18.170807	.0725581	13.782060	.0901273	11.095416	51
10	0.0553251	18.074977	0.0728505	13.726738	0.0904206	11.059431	50
11	.0556169	17.980150	.0731430	13.671856	.0907138	11.023676	49
12	.0559087	17.886310	.0734354	13.617409	.0910071	10.988150	48
13	.0562005	17.793442	.0737279	13.563391	.0913004	10.952850	47
14	.0564923	17.701529	.0740203	13.509799	.0915938	10.917775	46
15	.0567841	17.610559	.0743128	13.456625	.0918871	10.882921	45
16	.0570759	17.520516	.0746053	13.403867	.0921804	10.848288	44
17	.0573678	17.431385	.0748979	13.351518	.0924738	10.813872	43
18	.0576596	17.343155	.0751904	13.299574	.0927672	10.779673	42
19	.0579515	17.255810	.0754829	13.248031	.0930606	10.745687	41
20	0.0582434	17.169337	0.0757755	13.196883	0.0933540	10.711913	40
21	.0585352	17.083724	.0760680	13.146127	.0936474	10.678348	39
22	.0588271	16.998957	.0763606	13.095757	.0939409	10.644992	38
23	.0591190	16.915025	.0766532	13.045769	.0942344	10.611841	37
24	.0594109	16.831915	.0769458	12.996160	.0945278	10.578895	36
25	.0597029	16.749614	.0772384	12.946924	.0948213	10.546151	35
26	.0599948	16.668112	.0775311	12.898058	.0951148	10.513607	34
27	.0602867	16.587396	.0778237	12.849557	.0954084	10.481261	33
28	.0605787	16.507456	.0781164	12.801417	.0957019	10.449112	32
29	.0608706	16.428279	.0784090	12.753634	.0959955	10.417158	31
30	0.0611626	16.349855	0.0787017	12.706205	0.0962890	10.385397	30
31	.0614546	16.272174	.0789944	12.659125	.0965826	10.353827	29
32	.0617466	16.195225	.0792871	12.612390	.0968763	10.322447	28
33	.0620386	16.118998	.0795798	12.565997	.0971699	10.291255	27
34	.0623306	16.043482	.0798726	12.519942	.0974635	10.260249	26
35	.0626226	15.968667	.0801653	12.474221	.0977572	10.229428	25
36	.0629147	15.894545	.0804581	12.428831	.0980509	10.198789	24
37	.0632067	15.821105	.0807509	12.383768	.0983446	10.168332	23
38	.0634988	15.748337	.0810437	12.339028	.0986383	10.138054	22
39	.0637908	15.676233	.0813365	12.294609	.0989320	10.107954	21
40	0.0640829	15.604784	0.0816293	12.250505	0.0992257	10.078031	20
41	.0643750	15.533981	.0819221	12.206716	.0995195	10.048283	19
42	.0646671	15.463814	.0822150	12.163236	.0998133	10.018708	18
43	.0649592	15.394276	.0825078	12.120062	.1001071	9.9893050	17
44	.0652513	15.325358	.0828007	12.077192	.1004009	9.9600724	16
45	.0655435	15.257052	.0830936	12.034622	.1006947	9.9310088	15
46	.0658356	15.189349	.0833865	11.992349	.1009886	9.9021125	14
47	.0661278	15.122242	.0836794	11.950371	.1012824	9.8733823	13
48	.0664199	15.055723	.0839723	11.908682	.1015763	9.8448166	12
49	.0667121	14.989784	.0842653	11.867282	.1018702	9.8164140	11
50	0.0670043	14.924417	0.0845583	11.826167	0.1021641	9.7881732	10
51	.0672965	14.859616	.0848512	11.785333	.1024580	9.7600927	9
52	.0675887	14.795372	.0851442	11.744779	.1027520	9.7321713	8
53	.0678809	14.731679	.0854372	11.704500	.1030460	9.7044075	7
54	.0681732	14.668529	.0857302	11.664495	.1033400	9.6768000	6
55	.0684654	14.605916	.0860233	11.624761	.1036340	9.6493475	5
56	.0687577	14.543833	.0863163	11.585294	.1039280	9.6220486	4
57	.0690499	14.482273	.0866094	11.546093	.1042220	9.5949022	3
58	.0693422	14.421230	.0869025	11.507154	.1045161	9.5679068	2
59	.0696345	14.360696	.0871956	11.468474	.1048101	9.5410613	1
60	0.0699268	14.300666	0.0874887	11.430052	0.1051042	9.5143645	0
	Cotangent	Tangent	Cotangent	Tangent	Cotangent	Tangent	
′	86°		85°		84°		′

TABLE XIV. NATURAL TANGENTS AND COTANGENTS

′	6° Tangent	6° Cotangent	7° Tangent	7° Cotangent	8° Tangent	8° Cotangent	′
0	0.1051042	9.5143645	0.1227846	8.1443464	0.1405408	7.1153697	60
1	.1053983	9.4878149	.1230798	8.1248071	.1408375	7.1003826	59
2	.1056925	9.4614116	.1233752	8.1053599	.1411342	7.0854573	58
3	.1059866	9.4351531	.1236705	8.0860042	.1414308	7.0705934	57
4	.1062808	9.4090384	.1239658	8.0667394	.1417276	7.0557905	56
5	.1065750	9.3830663	.1242612	8.0475647	.1420243	7.0410482	55
6	.1068692	9.3572355	.1245566	8.0284796	.1423211	7.0263662	54
7	.1071634	9.3315450	.1248520	8.0094835	.1426179	7.0117441	53
8	.1074576	9.3059936	.1251474	7.9905756	.1429147	6.9971816	52
9	.1077519	9.2805802	.1254429	7.9717555	.1432115	6.9826781	51
10	0.1080462	9.2553035	0.1257384	7.9530224	0.1435084	6.9682335	50
11	.1083405	9.2301627	.1260339	7.9343758	.1438053	6.9538473	49
12	.1086348	9.2051564	.1263294	7.9158151	.1441022	6.9395192	48
13	.1089291	9.1802838	.1266249	7.8973396	.1443991	6.9252489	47
14	.1092234	9.1555436	.1269205	7.8789489	.1446961	6.9110359	46
15	.1095178	9.1309348	.1272161	7.8606423	.1449931	6.8968799	45
16	.1098122	9.1064564	.1275117	7.8424191	.1452901	6.8827807	44
17	.1101066	9.0821074	.1278073	7.8242790	.1455872	6.8687378	43
18	.1104010	9.0578867	.1281030	7.8062212	.1458842	6.8547508	42
19	.1106955	9.0337933	.1283986	7.7882453	.1461813	6.8408196	41
20	0.1109899	9.0098261	0.1286943	7.7703506	0.1464784	6.8269437	40
21	.1112844	8.9859843	.1289900	7.7525366	.1467756	6.8131227	39
22	.1115789	8.9622668	.1292858	7.7348028	.1470727	6.7993565	38
23	.1118734	8.9386726	.1295815	7.7171486	.1473699	6.7856446	37
24	.1121680	8.9152009	.1298773	7.6995735	.1476672	6.7719867	36
25	.1124625	8.8918505	.1301731	7.6820769	.1479644	6.7583826	35
26	.1127571	8.8686206	.1304690	7.6646584	.1482617	6.7448318	34
27	.1130517	8.8455103	.1307648	7.6473174	.1485590	6.7313341	33
28	.1133463	8.8225186	.1310607	7.6300533	.1488563	6.7178891	32
29	.1136410	8.7996446	.1313566	7.6128657	.1491536	6.7044966	31
30	0.1139356	8.7768874	0.1316525	7.5957541	0.1494510	6.6911562	30
31	.1142303	8.7542461	.1319484	7.5787179	.1497484	6.6778677	29
32	.1145250	8.7317198	.1322444	7.5617567	.1500458	6.6646307	28
33	.1148197	8.7093077	.1325404	7.5448699	.1503433	6.6514449	27
34	.1151144	8.6870088	.1328364	7.5280571	.1506408	6.6383100	26
35	.1154092	8.6648223	.1331324	7.5113178	.1509383	6.6252258	25
36	.1157039	8.6427475	.1334285	7.4946514	.1512358	6.6121919	24
37	.1159987	8.6207833	.1337246	7.4780576	.1515333	6.5992080	23
38	.1162936	8.5989290	.1340207	7.4615357	.1518309	6.5862739	22
39	.1165884	8.5771838	.1343168	7.4450855	.1521285	6.5733892	21
40	0.1168832	8.5555468	0.1346129	7.4287064	0.1524262	6.5605538	20
41	.1171781	8.5340172	.1349091	7.4123978	.1527238	6.5477672	19
42	.1174730	8.5125943	.1352053	7.3961595	.1530215	6.5350293	18
43	.1177679	8.4912772	.1355015	7.3799909	.1533192	6.5223396	17
44	.1180628	8.4700651	.1357978	7.3638916	.1536170	6.5096981	16
45	.1183578	8.4489573	.1360940	7.3478610	.1539147	6.4971043	15
46	.1186528	8.4279531	.1363903	7.3318989	.1542125	6.4845581	14
47	.1189478	8.4070511	.1366866	7.3160047	.1545103	6.4720591	13
48	.1192428	8.3862519	.1369830	7.3001780	.1548082	6.4596070	12
49	.1195378	8.3655536	.1372793	7.2844184	.1551061	6.4472017	11
50	0.1198329	8.3449558	0.1375757	7.2687255	0.1554040	6.4348428	10
51	.1201279	8.3244577	.1378721	7.2530987	.1557019	6.4225301	9
52	.1204230	8.3040586	.1381685	7.2375378	.1559998	6.4102633	8
53	.1207182	8.2837579	.1384650	7.2220422	.1562978	6.3980422	7
54	.1210133	8.2635547	.1387615	7.2066116	.1565958	6.3858665	6
55	.1213085	8.2434485	.1390580	7.1912456	.1568939	6.3737359	5
56	.1216036	8.2234384	.1393545	7.1759437	.1571919	6.3616502	4
57	.1218988	8.2035239	.1396510	7.1607056	.1574900	6.3496092	3
58	.1221941	8.1837041	.1399476	7.1455308	.1577881	6.3376126	2
59	.1224893	8.1639786	.1402442	7.1304190	.1580863	6.3256601	1
60	0.1227846	8.1443464	0.1405408	7.1153697	0.1583844	6.3137515	0
′	Cotangent	Tangent	Cotangent	Tangent	Cotangent	Tangent	′
	83°		82°		81°		

TABLE XIV. NATURAL TANGENTS AND COTANGENTS

′	9°		10°		11°		′
	Tangent	Cotangent	Tangent	Cotangent	Tangent	Cotangent	
0	0.1583844	6.3137515	0.1763270	5.6712818	0.1943803	5.1445540	60
1	.1586826	6.3018866	.1766269	5.6616509	.1946822	5.1365763	59
2	.1589809	6.2900651	.1769269	5.6520516	.1949841	5.1286224	58
3	.1592791	6.2782868	.1772269	5.6424838	.1952861	5.1206921	57
4	.1595774	6.2665515	.1775270	5.6329474	.1955881	5.1127855	56
5	.1598757	6.2548588	.1778270	5.6234421	.1958901	5.1049024	55
6	.1601740	6.2432086	.1781271	5.6139680	.1961922	5.0970426	54
7	.1604724	6.2316007	.1784273	5.6045247	.1964943	5.0892061	53
8	.1607708	6.2200347	.1787274	5.5951121	.1967964	5.0813928	52
9	.1610692	6.2085106	.1790276	5.5857302	.1970986	5.0736025	51
10	0.1613677	6.1970279	0.1793279	5.5763786	0.1974008	5.0658352	50
11	.1616662	6.1855867	.1796281	5.5670574	.1977031	5.0580907	49
12	.1619647	6.1741865	.1799284	5.5577663	.1980053	5.0503690	48
13	.1622632	6.1628272	.1802287	5.5485052	.1983076	5.0426700	47
14	.1625618	6.1515085	.1805291	5.5392740	.1986100	5.0349935	46
15	.1628603	6.1402303	.1808295	5.5300724	.1989124	5.0273395	45
16	.1631590	6.1289923	.1811299	5.5209005	.1992148	5.0197078	44
17	.1634576	6.1177943	.1814303	5.5117579	.1995172	5.0120984	43
18	.1637563	6.1066360	.1817308	5.5026446	.1998197	5.0045111	42
19	.1640550	6.0955174	.1820313	5.4935604	.2001222	4.9969459	41
20	0.1643537	6.0844381	0.1823319	5.4845052	0.2004248	4.9894027	40
21	.1646525	6.0733979	.1826324	5.4754788	.2007274	4.9818813	39
22	.1649513	6.0623967	.1829330	5.4664812	.2010300	4.9743817	38
23	.1652501	6.0514343	.1832337	5.4575121	.2013327	4.9669037	37
24	.1655489	6.0405103	.1835343	5.4485715	.2016354	4.9594474	36
25	.1658478	6.0296247	.1838350	5.4396592	.2019381	4.9520125	35
26	.1661467	6.0187772	.1841358	5.4307750	.2022409	4.9445990	34
27	.1664456	6.0079676	.1844365	5.4219188	.2025437	4.9372068	33
28	.1667446	5.9971957	.1847373	5.4130906	.2028465	4.9298358	32
29	.1670436	5.9864614	.1850382	5.4042901	.2031494	4.9224859	31
30	0.1673426	5.9757644	0.1853390	5.3955172	0.2034523	4.9151570	30
31	.1676417	5.9651045	.1856399	5.3867718	.2037552	4.9078491	29
32	.1679407	5.9544815	.1859409	5.3780538	.2040582	4.9005620	28
33	.1682398	5.9438952	.1862418	5.3693630	.2043612	4.8932956	27
34	.1685390	5.9333455	.1865428	5.3606993	.2046643	4.8860499	26
35	.1688381	5.9228322	.1868439	5.3520626	.2049674	4.8788248	25
36	.1691373	5.9123550	.1871449	5.3434527	.2052705	4.8716201	24
37	.1694366	5.9019138	.1874460	5.3348696	.2055737	4.8644359	23
38	.1697358	5.8915084	.1877471	5.3263131	.2058769	4.8572719	22
39	.1700351	5.8811386	.1880483	5.3177830	.2061802	4.8501282	21
40	0.1703344	5.8708042	0.1883495	5.3092793	0.2064834	4.8430045	20
41	.1706338	5.8605051	.1886507	5.3008018	.2067867	4.8359010	19
42	.1709331	5.8502410	.1889520	5.2923505	.2070900	4.8288174	18
43	.1712325	5.8400117	.1892533	5.2839251	.2073934	4.8217536	17
44	.1715320	5.8298172	.1895546	5.2755255	.2076968	4.8147096	16
45	.1718314	5.8196572	.1898559	5.2671517	.2080003	4.8076854	15
46	.1721309	5.8095315	.1901573	5.2588035	.2083038	4.8006808	14
47	.1724304	5.7994400	.1904587	5.2504809	.2086073	4.7936957	13
48	.1727300	5.7893825	.1907602	5.2421836	.2089109	4.7867300	12
49	.1730296	5.7793588	.1910617	5.2339116	.2092145	4.7797837	11
50	0.1733292	5.7693688	0.1913632	5.2256647	0.2095181	4.7728568	10
51	.1736288	5.7594122	.1916648	5.2174428	.2098218	4.7659490	9
52	.1739285	5.7494889	.1919664	5.2092459	.2101255	4.7590603	8
53	.1742282	5.7395988	.1922680	5.2010738	.2104293	4.7521907	7
54	.1745279	5.7297416	.1925696	5.1929264	.2107331	4.7453401	6
55	.1748277	5.7199173	.1928713	5.1848035	.2110369	4.7385083	5
56	.1751275	5.7101256	.1931731	5.1767051	.2113407	4.7316954	4
57	.1754273	5.7003663	.1934748	5.1686311	.2116446	4.7249012	3
58	.1757272	5.6906394	.1937766	5.1605813	.2119486	4.7181256	2
59	.1760271	5.6809446	.1940784	5.1525557	.2122525	4.7113686	1
60	0.1763270	5.6712818	0.1943803	5.1445540	0.2125566	4.7046301	0
	Cotangent	Tangent	Cotangent	Tangent	Cotangent	Tangent	
′	80°		79°		78°		′

TABLE XIV. NATURAL TANGENTS AND COTANGENTS

′	12°		13°		14°		′
	Tangent	Cotangent	Tangent	Cotangent	Tangent	Cotangent	
0	0.2125566	4.7046301	0.2308682	4.3314759	0.2493280	4.0107809	60
1	.2128606	4.6979100	.2311746	4.3257347	.2496370	4.0058164	59
2	.2131647	4.6912083	.2314811	4.3200079	.2499460	4.0008636	58
3	.2134688	4.6845248	.2317876	4.3142955	.2502551	3.9959223	57
4	.2137730	4.6778595	.2320941	4.3085974	.2505642	3.9909924	56
5	.2140772	4.6712124	.2324007	4.3029136	.2508734	3.9860739	55
6	.2143814	4.6645832	.2327073	4.2972440	.2511826	3.9811669	54
7	.2146857	4.6579721	.2330140	4.2915885	.2514919	3.9762712	53
8	.2149900	4.6513788	.2333207	4.2859472	.2518012	3.9713868	52
9	.2152944	4.6448034	.2336274	4.2803199	.2521106	3.9665137	51
10	0.2155988	4.6382457	0.2339342	4.2747066	0.2524200	3.9616518	50
11	.2159032	4.6317056	.2342411	4.2691072	.2527294	3.9568011	49
12	.2162077	4.6251832	.2345479	4.2635218	.2530389	3.9519615	48
13	.2165122	4.6186783	.2348548	4.2579501	.2533484	3.9471331	47
14	.2168167	4.6121908	.2351617	4.2523923	.2536580	3.9423157	46
15	.2171213	4.6057207	.2354687	4.2468482	.2539676	3.9375094	45
16	.2174259	4.5992680	.2357758	4.2413177	.2542773	3.9327141	44
17	.2177306	4.5928325	.2360829	4.2358009	.2545870	3.9279297	43
18	.2180353	4.5864141	.2363900	4.2302977	.2548968	3.9231563	42
19	.2183400	4.5800129	.2366971	4.2248080	.2552066	3.9183937	41
20	0.2186448	4.5736287	0.2370044	4.2193318	0.2555165	3.9136420	40
21	.2189496	4.5672615	.2373116	4.2138690	.2558264	3.9089011	39
22	.2192544	4.5609111	.2376189	4.2084196	.2561363	3.9041710	38
23	.2195593	4.5545776	.2379262	4.2029835	.2564463	3.8994516	37
24	.2198643	4.5482608	.2382336	4.1975606	.2567564	3.8947429	36
25	.2201692	4.5419608	.2385410	4.1921510	.2570664	3.8900448	35
26	.2204742	4.5356773	.2388485	4.1867546	.2573766	3.8853574	34
27	.2207793	4.5294105	.2391560	4.1813713	.2576868	3.8806805	33
28	.2210844	4.5231601	.2394635	4.1760011	.2579970	3.8760142	32
29	.2213895	4.5169261	.2397711	4.1706440	.2583073	3.8713584	31
30	0.2216947	4.5107085	0.2400788	4.1652998	0.2586176	3.8667131	30
31	.2219999	4.5045072	.2403864	4.1599685	.2589280	3.8620782	29
32	.2223051	4.4983221	.2406942	4.1546501	.2592384	3.8574537	28
33	.2226104	4.4921532	.2410019	4.1493446	.2595488	3.8528396	27
34	.2229157	4.4860004	.2413097	4.1440519	.2598593	3.8482358	26
35	.2232211	4.4798636	.2416176	4.1387719	.2601699	3.8436424	25
36	.2235265	4.4737428	.2419255	4.1335046	.2604805	3.8390591	24
37	.2238319	4.4676379	.2422334	4.1282499	.2607911	3.8344861	23
38	.2241374	4.4615489	.2425414	4.1230079	.2611018	3.8299233	22
39	.2244429	4.4554756	.2428494	4.1177784	.2614126	3.8253707	21
40	0.2247485	4.4494181	0.2431575	4.1125614	0.2617234	3.8208281	20
41	.2250541	4.4433762	.2434656	4.1073569	.2620342	3.8162957	19
42	.2253597	4.4373500	.2437737	4.1021649	.2623451	3.8117733	18
43	.2256654	4.4313392	.2440819	4.0969852	.2626560	3.8072609	17
44	.2259711	4.4253439	.2443902	4.0918178	.2629670	3.8027585	16
45	.2262769	4.4193641	.2446984	4.0866627	.2632780	3.7982661	15
46	.2265827	4.4133996	.2450068	4.0815199	.2635891	3.7937835	14
47	.2268885	4.4074504	.2453151	4.0763892	.2639002	3.7893109	13
48	.2271944	4.4015164	.2456236	4.0712707	.2642114	3.7848481	12
49	.2275003	4.3955977	.2459320	4.0661643	.2645226	3.7803951	11
50	0.2278063	4.3896940	0.2462405	4.0610700	0.2648339	3.7759519	10
51	.2281123	4.3838054	.2465491	4.0559877	.2651452	3.7715185	9
52	.2284184	4.3779317	.2468577	4.0509174	.2654566	3.7670947	8
53	.2287244	4.3720731	.2471663	4.0458590	.2657680	3.7626807	7
54	.2290306	4.3662293	.2474750	4.0408125	.2660794	3.7582763	6
55	.2293367	4.3604003	.2477837	4.0357779	.2663909	3.7538815	5
56	.2296429	4.3545861	.2480925	4.0307550	.2667025	3.7494963	4
57	.2299492	4.3487866	.2484013	4.0257440	.2670141	3.7451207	3
58	.2302555	4.3430018	.2487102	4.0207446	.2673257	3.7407546	2
59	.2305618	4.3372316	.2490191	4.0157570	.2676374	3.7363980	1
60	0.2308682	4.3314759	0.2493280	4.0107809	0.2679492	3.7320508	0
	Cotangent	Tangent	Cotangent	Tangent	Cotangent	Tangent	
′	77°		76°		75°		′

TABLE XIV. NATURAL TANGENTS AND COTANGENTS

′	15°		16°		17°		′
	Tangent	Cotangent	Tangent	Cotangent	Tangent	Cotangent	
0	0.2679492	3.7320508	0.2867454	3.4874144	0.3057307	3.2708526	60
1	.2682610	3.7277131	.2870602	3.4835896	.3060488	3.2674529	59
2	.2685728	3.7233847	.2873751	3.4797726	.3063670	3.2640596	58
3	.2688847	3.7190658	.2876900	3.4759632	.3066852	3.2606728	57
4	.2691967	3.7147561	.2880050	3.4721616	.3070034	3.2572924	56
5	.2695087	3.7104558	.2883201	3.4683676	.3073218	3.2539184	55
6	.2698207	3.7061648	.2886352	3.4645813	.3076402	3.2505508	54
7	.2701328	3.7018830	.2889503	3.4608026	.3079586	3.2471895	53
8	.2704449	3.6976104	.2892655	3.4570315	.3082771	3.2438346	52
9	.2707571	3.6933469	.2895808	3.4532679	.3085957	3.2404861	51
10	0.2710694	3.6890927	0.2898961	3.4495120	0.3089143	3.2371438	50
11	.2713817	3.6848475	.2902114	3.4457635	.3092330	3.2338078	49
12	.2716940	3.6806115	.2905269	3.4420226	.3095517	3.2304780	48
13	.2720064	3.6763845	.2908423	3.4382891	.3098705	3.2271546	47
14	.2723188	3.6721665	.2911578	3.4345631	.3101893	3.2238373	46
15	.2726313	3.6679575	.2914734	3.4308446	.3105083	3.2205263	45
16	.2729438	3.6637575	.2917890	3.4271334	.3108272	3.2172215	44
17	.2732564	3.6595665	.2921047	3.4234297	.3111462	3.2139228	43
18	.2735690	3.6553844	.2924205	3.4197333	.3114653	3.2106304	42
19	.2738817	3.6512111	.2927363	3.4160443	.3117845	3.2073440	41
20	0.2741945	3.6470467	0.2930521	3.4123626	0.3121036	3.2040638	40
21	.2745072	3.6428911	.2933680	3.4086882	.3124229	3.2007897	39
22	.2748201	3.6387444	.2936839	3.4050210	.3127422	3.1975217	38
23	.2751330	3.6346064	.2939999	3.4013612	.3130616	3.1942597	37
24	.2754459	3.6304771	.2943160	3.3977085	.3133810	3.1910039	36
25	.2757589	3.6263566	.2946321	3.3940631	.3137005	3.1877540	35
26	.2760719	3.6222447	.2949483	3.3904249	.3140200	3.1845102	34
27	.2763850	3.6181415	.2952645	3.3867938	.3143396	3.1812724	33
28	.2766981	3.6140469	.2955808	3.3831699	.3146593	3.1780406	32
29	.2770113	3.6099609	.2958971	3.3795531	.3149790	3.1748147	31
30	0.2773245	3.6058835	0.2962135	3.3759434	0.3152988	3.1715948	30
31	.2776378	3.6018146	.2965299	3.3723408	.3156186	3.1683808	29
32	.2779512	3.5977543	.2968464	3.3687453	.3159385	3.1651728	28
33	.2782646	3.5937024	.2971630	3.3651568	.3162585	3.1619706	27
34	.2785780	3.5896590	.2974796	3.3615753	.3165785	3.1587744	26
35	.2788915	3.5856241	.2977962	3.3580008	.3168986	3.1555840	25
36	.2792050	3.5815975	.2981129	3.3544333	.3172187	3.1523994	24
37	.2795186	3.5775794	.2984297	3.3508728	.3175389	3.1492207	23
38	.2798322	3.5735696	.2987465	3.3473191	.3178591	3.1460478	22
39	.2801459	3.5695681	.2990634	3.3437724	.3181794	3.1428807	21
40	0.2804597	3.5655749	0.2993803	3.3402326	0.3184998	3.1397194	20
41	.2807735	3.5615900	.2996973	3.3366997	.3188202	3.1365639	19
42	.2810873	3.5576133	.3000144	3.3331736	.3191407	3.1334141	18
43	.2814012	3.5536449	.3003315	3.3296543	.3194613	3.1302701	17
44	.2817152	3.5496846	.3006486	3.3261419	.3197819	3.1271317	16
45	.2820292	3.5457325	.3009658	3.3226362	.3201025	3.1239991	15
46	.2823432	3.5417886	.3012831	3.3191373	.3204232	3.1208722	14
47	.2826573	3.5378528	.3016004	3.3156452	.3207440	3.1177509	13
48	.2829715	3.5339251	.3019178	3.3121598	.3210649	3.1146353	12
49	.2832857	3.5300054	.3022352	3.3086811	.3213858	3.1115254	11
50	0.2835999	3.5260938	0.3025527	3.3052091	0.3217067	3.1084210	10
51	.2839143	3.5221902	.3028703	3.3017438	.3220278	3.1053223	9
52	.2842286	3.5182946	.3031879	3.2982851	.3223489	3.1022291	8
53	.2845430	3.5144070	.3035055	3.2948330	.3226700	3.0991416	7
54	.2848575	3.5105273	.3038232	3.2913876	.3229912	3.0960596	6
55	.2851720	3.5066555	.3041410	3.2879487	.3233125	3.0929831	5
56	.2854866	3.5027916	.3044588	3.2845164	.3236338	3.0899122	4
57	.2858012	3.4989356	.3047767	3.2810907	.3239552	3.0868468	3
58	.2861159	3.4950874	.3050946	3.2776715	.3242766	3.0837869	2
59	.2864306	3.4912470	.3054126	3.2742588	.3245981	3.0807325	1
60	0.2867454	3.4874144	0.3057307	3.2708526	0.3249197	3.0776835	0
′	Cotangent	Tangent	Cotangent	Tangent	Cotangent	Tangent	′
	74°		73°		72°		

TABLE XIV. NATURAL TANGENTS AND COTANGENTS

′	18°		19°		20°		′
	Tangent	Cotangent	Tangent	Cotangent	Tangent	Cotangent	
0	0.3249197	3.0776835	0.3443276	2.9042109	0.3639702	2.7474774	60
1	.3252413	3.0746400	.3446530	2.9014688	.3642997	2.7449927	59
2	.3255630	3.0716020	.3449785	2.8987314	.3646292	2.7425120	58
3	.3258848	3.0685694	.3453040	2.8959986	.3649588	2.7400352	57
4	.3262066	3.0655421	.3456296	2.8932704	.3652885	2.7375623	56
5	.3265284	3.0625203	.3459553	2.8905467	.3656182	2.7350934	55
6	.3268504	3.0595038	.3462810	2.8878277	.3659480	2.7326284	54
7	.3271724	3.0564928	.3466068	2.8851132	.3662779	2.7301674	53
8	.3274944	3.0534870	.3469327	2.8824033	.3666079	2.7277102	52
9	.3278165	3.0504866	.3472586	2.8796979	.3669379	2.7252569	51
10	0.3281387	3.0474915	0.3475846	2.8769970	0.3672680	2.7228076	50
11	.3284610	3.0445018	.3479107	2.8743007	.3675981	2.7203620	49
12	.3287833	3.0415173	.3482368	2.8716088	.3679284	2.7179204	48
13	.3291056	3.0385381	.3485630	2.8689215	.3682587	2.7154826	47
14	.3294281	3.0355641	.3488893	2.8662386	.3685890	2.7130487	46
15	.3297505	3.0325954	.3492156	2.8635602	.3689195	2.7106186	45
16	.3300731	3.0296320	.3495420	2.8608863	.3692500	2.7081923	44
17	.3303957	3.0266737	.3498685	2.8582168	.3695806	2.7057699	43
18	.3307184	3.0237207	.3501950	2.8555517	.3699112	2.7033513	42
19	.3310411	3.0207728	.3505216	2.8528911	.3702420	2.7009364	41
20	0.3313639	3.0178301	0.3508483	2.8502349	0.3705728	2.6985254	40
21	.3316868	3.0148926	.3511750	2.8475831	.3709036	2.6961181	39
22	.3320097	3.0119603	.3515018	2.8449356	.3712346	2.6937147	38
23	.3323327	3.0090330	.3518287	2.8422926	.3715656	2.6913149	37
24	.3326557	3.0061109	.3521556	2.8396539	.3718967	2.6889190	36
25	.3329788	3.0031939	.3524826	2.8370196	.3722278	2.6865267	35
26	.3333020	3.0002820	.3528096	2.8343896	.3725590	2.6841383	34
27	.3336252	2.9973751	.3531368	2.8317640	.3728903	2.6817535	33
28	.3339485	2.9944734	.3534640	2.8291426	.3732217	2.6793725	32
29	.3342719	2.9915766	.3537912	2.8265256	.3735532	2.6769951	31
30	0.3345953	2.9886850	0.3541186	2.8239129	0.3738847	2.6746215	30
31	.3349188	2.9857983	.3544460	2.8213045	.3742163	2.6722516	29
32	.3352424	2.9829167	.3547734	2.8187003	.3745479	2.6698853	28
33	.3355660	2.9800400	.3551010	2.8161004	.3748797	2.6675227	27
34	.3358896	2.9771683	.3554286	2.8135048	.3752115	2.6651638	26
35	.3362134	2.9743016	.3557562	2.8109134	.3755433	2.6628085	25
36	.3365372	2.9714399	.3560840	2.8083263	.3758753	2.6604569	24
37	.3368610	2.9685831	.3564118	2.8057433	.3762073	2.6581089	23
38	.3371850	2.9657312	.3567397	2.8031646	.3765394	2.6557645	22
39	.3375090	2.9628842	.3570676	2.8005901	.3768716	2.6534238	21
40	0.3378330	2.9600422	0.3573956	2.7980198	0.3772038	2.6510867	20
41	.3381571	2.9572050	.3577237	2.7954537	.3775361	2.6487531	19
42	.3384813	2.9543727	.3580518	2.7928917	.3778685	2.6464232	18
43	.3388056	2.9515453	.3583801	2.7903339	.3782010	2.6440969	17
44	.3391299	2.9487227	.3587083	2.7877802	.3785335	2.6417741	16
45	.3394543	2.9459050	.3590367	2.7852307	.3788661	2.6394549	15
46	.3397787	2.9430921	.3593651	2.7826853	.3791988	2.6371392	14
47	.3401032	2.9402840	.3596936	2.7801440	.3795315	2.6348271	13
48	.3404278	2.9374807	.3600222	2.7776069	.3798644	2.6325186	12
49	.3407524	2.9346822	.3603508	2.7750738	.3801973	2.6302136	11
50	0.3410771	2.9318885	0.3606795	2.7725448	0.3805302	2.6279121	10
51	.3414019	2.9290995	.3610082	2.7700199	.3808633	2.6256141	9
52	.3417267	2.9263152	.3613371	2.7674991	.3811964	2.6233196	8
53	.3420516	2.9235358	.3616660	2.7649822	.3815296	2.6210286	7
54	.3423765	2.9207610	.3619949	2.7624695	.3818629	2.6187411	6
55	.3427015	2.9179909	.3623240	2.7599608	.3821962	2.6164571	5
56	.3430266	2.9152256	.3626531	2.7574561	.3825296	2.6141766	4
57	.3433518	2.9124649	.3629823	2.7549554	.3828631	2.6118995	3
58	.3436770	2.9097089	.3633115	2.7524588	.3831967	2.6096259	2
59	.3440023	2.9069576	.3636408	2.7499661	.3835303	2.6073558	1
60	0.3443276	2.9042109	0.3639702	2.7474774	0.3838640	2.6050891	0
	Cotangent	Tangent	Cotangent	Tangent	Cotangent	Tangent	
′	71°		70°		69°		′

TABLE XIV. NATURAL TANGENTS AND COTANGENTS

′	21°		22°		23°		′
	Tangent	Cotangent	Tangent	Cotangent	Tangent	Cotangent	
0	0.3838640	2.6050891	0.4040262	2.4750869	0.4244748	2.3558524	60
1	.3841978	2.6028258	.4043646	2.4730155	.4248182	2.3539483	59
2	.3845317	2.6005659	.4047031	2.4709470	.4251616	2.3520469	58
3	.3848656	2.5983095	.4050417	2.4688816	.4255051	2.3501481	57
4	.3851996	2.5960564	.4053804	2.4668191	.4258487	2.3482519	56
5	.3855337	2.5938068	.4057191	2.4647596	.4261924	2.3463582	55
6	.3858679	2.5915606	.4060579	2.4627030	.4265361	2.3444672	54
7	.3862021	2.5893177	.4063968	2.4606494	.4268800	2.3425787	53
8	.3865364	2.5870782	.4067358	2.4585987	.4272239	2.3406928	52
9	.3868708	2.5848421	.4070748	2.4565510	.4275680	2.3388095	51
10	0.3872053	2.5826094	0.4074139	2.4545061	0.4279121	2.3369287	50
11	.3875398	2.5803800	.4077531	2.4524642	.4282563	2.3350505	49
12	.3878744	2.5781539	.4080924	2.4504252	.4286005	2.3331748	48
13	.3882091	2.5759312	.4084318	2.4483891	.4289449	2.3313017	47
14	.3885439	2.5737118	.4087713	2.4463559	.4292894	2.3294311	46
15	.3888787	2.5714957	.4091108	2.4443256	.4296339	2.3275630	45
16	.3892136	2.5692830	.4094504	2.4422982	.4299785	2.3256975	44
17	.3895486	2.5670735	.4097901	2.4402736	.4303232	2.3238345	43
18	.3898837	2.5648674	.4101299	2.4382519	.4306680	2.3219740	42
19	.3902189	2.5626645	.4104697	2.4362331	.4310129	2.3201160	41
20	0.3905541	2.5604649	0.4108097	2.4342172	0.4313579	2.3182606	40
21	.3908894	2.5582686	.4111497	2.4322041	.4317030	2.3164076	39
22	.3912247	2.5560756	.4114898	2.4301938	.4320481	2.3145571	38
23	.3915602	2.5538858	.4118300	2.4281864	.4323933	2.3127092	37
24	.3918957	2.5516992	.4121703	2.4261819	.4327386	2.3108637	36
25	.3922313	2.5495160	.4125106	2.4241801	.4330840	2.3090206	35
26	.3925670	2.5473359	.4128510	2.4221812	.4334295	2.3071801	34
27	.3929027	2.5451591	.4131915	2.4201851	.4337751	2.3053420	33
28	.3932386	2.5429855	.4135321	2.4181918	.4341208	2.3035064	32
29	.3935745	2.5408151	.4138728	2.4162013	.4344665	2.3016732	31
30	0.3939105	2.5386479	0.4142136	2.4142136	0.4348124	2.2998425	30
31	.3942465	2.5864839	.4145544	2.4122286	.4351583	2.2980143	29
32	.3945827	2.5343231	.4148953	2.4102465	.4355043	2.2961885	28
33	.3949189	2.5321655	.4152363	2.4082672	.4358504	2.2943651	27
34	.3952552	2.5300111	.4155774	2.4062906	.4361966	2.2925442	26
35	.3955916	2.5278598	.4159186	2.4043168	.4365429	2.2907257	25
36	.3959280	2.5257117	.4162598	2.4023457	.4368893	2.2889096	24
37	.3962645	2.5235667	.4166012	2.4003774	.4372357	2.2870959	23
38	.3966011	2.5214249	.4169426	2.3984118	.4375823	2.2852846	22
39	.3969378	2.5192863	.4172841	2.3964490	.4379289	2.2834758	21
40	0.3972746	2.5171507	0.4176257	2.3944889	0.4382756	2.2816693	20
41	.3976114	2.5150183	.4179673	2.3925316	.4386224	2.2798653	19
42	.3979483	2.5128890	.4183091	2.3905769	.4389693	2.2780636	18
43	.3982853	2.5107629	.4186509	2.3886250	.4393163	2.2762643	17
44	.3986224	2.5086398	.4189928	2.3866758	.4396634	2.2744674	16
45	.3989595	2.5065198	.4193348	2.3847293	.4400105	2.2726729	15
46	.3992968	2.5044029	.4196769	2.3827855	.4403578	2.2708807	14
47	.3996341	2.5022891	.4200190	2.3808444	.4407051	2.2690909	13
48	.3999715	2.5001784	.4203613	2.3789060	.4410526	2.2673035	12
49	.4003089	2.4980707	.4207036	2.3769703	.4414001	2.2655184	11
50	0.4006465	2.4959661	0.4210460	2.3750372	0.4417477	2.2637357	10
51	.4009841	2.4938645	.4213885	2.3731068	.4420954	2.2619554	9
52	.4013218	2.4917660	.4217311	2.3711791	.4424432	2.2601773	8
53	.4016596	2.4896706	.4220738	2.3692540	.4427910	2.2584016	7
54	.4019974	2.4875781	.4224165	2.3673316	.4431390	2.2566283	6
55	.4023354	2.4854887	.4227594	2.3654118	.4434871	2.2548572	5
56	.4026734	2.4834023	.4231023	2.3634946	.4438352	2.2530885	4
57	.4030115	2.4813190	.4234453	2.3615801	.4441834	2.2513221	3
58	.4033496	2.4792386	.4237884	2.3596683	.4445318	2.2495580	2
59	.4036879	2.4771612	.4241316	2.3577590	.4448802	2.2477962	1
60	0.4040262	2.4750869	0.4244748	2.3558524	0.4452287	2.2460368	0
	Cotangent	Tangent	Cotangent	Tangent	Cotangent	Tangent	
′	68°		67°		66°		′

TABLE XIV. NATURAL TANGENTS AND COTANGENTS

′	24°		25°		26°		′
	Tangent	Cotangent	Tangent	Cotangent	Tangent	Cotangent	
0	0.4452287	2.2460368	0.4663077	2.1445069	0.4877326	2.0503038	60
1	.4455773	2.2442796	.4666618	2.1428793	.4880927	2.0487910	59
2	.4459260	2.2425247	.4670161	2.1412537	.4884530	2.0472800	58
3	.4462747	2.2407721	.4673705	2.1396301	.4888133	2.0457708	57
4	.4466236	2.2390218	.4677250	2.1380085	.4891737	2.0442634	56
5	.4469726	2.2372738	.4680796	2.1363890	.4895343	2.0427578	55
6	.4473216	2.2355280	.4684342	2.1347714	.4898949	2.0412540	54
7	.4476708	2.2337845	.4687890	2.1331559	.4902557	2.0397519	53
8	.4480200	2.2320433	.4691439	2.1315423	.4906166	2.0382517	52
9	.4483693	2.2303043	.4694988	2.1299308	.4909775	2.0367532	51
10	0.4487187	2.2285676	0.4698539	2.1283213	0.4913386	2.0352565	50
11	.4490682	2.2268331	.4702090	2.1267137	.4916997	2.0337615	49
12	.4494178	2.2251009	.4705643	2.1251082	.4920610	2.0322683	48
13	.4497675	2.2233709	.4709196	2.1235046	.4924224	2.0307769	47
14	.4501173	2.2216432	.4712751	2.1219030	.4927838	2.0292873	46
15	.4504672	2.2199177	.4716306	2.1203034	.4931454	2.0277994	45
16	.4508171	2.2181944	.4719863	2.1187057	.4935071	2.0263133	44
17	.4511672	2.2164733	.4723420	2.1171101	.4938689	2.0248289	43
18	.4515173	2.2147545	.4726978	2.1155164	.4942308	2.0233462	42
19	.4518676	2.2130379	.4730538	2.1139246	.4945928	2.0218654	41
20	0.4522179	2.2113234	0.4734098	2.1123348	0.4949549	2.0203862	40
21	.4525683	2.2096112	.4737659	2.1107470	.4953171	2.0189088	39
22	.4529188	2.2079012	.4741222	2.1091611	.4956794	2.0174331	38
23	.4532694	2.2061934	.4744785	2.1075771	.4960418	2.0159592	37
24	.4536201	2.2044878	.4748349	2.1059951	.4964043	2.0144869	36
25	.4539709	2.2027843	.4751914	2.1044150	.4967669	2.0130164	35
26	.4543218	2.2010831	.4755481	2.1028369	.4971297	2.0115477	34
27	.4546728	2.1993840	.4759048	2.1012607	.4974925	2.0100806	33
28	.4550238	2.1976871	.4762616	2.0996864	.4978554	2.0086153	32
29	.4553750	2.1959923	.4766185	2.0981140	.4982185	2.0071516	31
30	0.4557263	2.1942997	0.4769755	2.0965436	0.4985816	2.0056897	30
31	.4560776	2.1926093	.4773326	2.0949751	.4989449	2.0042295	29
32	.4564290	2.1909210	.4776899	2.0934085	.4993082	2.0027710	28
33	.4567806	2.1892349	.4780472	2.0918437	.4996717	2.0013142	27
34	.4571322	2.1875510	.4784046	2.0902809	.5000352	1.9998590	26
35	.4574839	2.1858691	.4787621	2.0887200	.5003989	1.9984056	25
36	.4578357	2.1841894	.4791197	2.0871610	.5007627	1.9969539	24
37	.4581877	2.1825119	.4794774	2.0856039	.5011266	1.9955038	23
38	.4585397	2.1808364	.4798352	2.0840487	.5014906	1.9940554	22
39	.4588918	2.1791631	.4801932	2.0824953	.5018547	1.9926087	21
40	0.4592439	2.1774920	0.4805512	2.0809438	0.5022189	1.9911637	20
41	.4595962	2.1758229	.4809093	2.0793942	.5025832	1.9897204	19
42	.4599486	2.1741559	.4812675	2.0778465	.5029476	1.9882787	18
43	.4603011	2.1724911	.4816258	2.0763007	.5033121	1.9868387	17
44	.4606537	2.1708283	.4819842	2.0747567	.5036768	1.9854003	16
45	.4610063	2.1691677	.4823427	2.0732146	.5040415	1.9839636	15
46	.4613591	2.1675091	.4827014	2.0716743	.5044063	1.9825286	14
47	.4617119	2.1658527	.4830601	2.0701359	.5047713	1.9810952	13
48	.4620649	2.1641983	.4834189	2.0685994	.5051363	1.9796635	12
49	.4624179	2.1625460	.4837778	2.0670646	.5055015	1.9782334	11
50	0.4627710	2.1608958	0.4841368	2.0655318	0.5058668	1.9768050	10
51	.4631243	2.1592476	.4844959	2.0640008	.5062322	1.9753782	9
52	.4634776	2.1576015	.4848552	2.0624716	.5065977	1.9739531	8
53	.4638310	2.1559575	.4852145	2.0609442	.5069633	1.9725296	7
54	.4641845	2.1543156	.4855739	2.0594187	.5073290	1.9711077	6
55	.4645382	2.1526757	.4859334	2.0578950	.5076948	1.9696874	5
56	.4648919	2.1510378	.4862931	2.0563732	.5080607	1.9682688	4
57	.4652457	2.1494021	.4866528	2.0548531	.5084267	1.9668518	3
58	.4655996	2.1477683	.4870126	2.0533349	.5087929	1.9654364	2
59	.4659536	2.1461366	.4873726	2.0518185	.5091591	1.9640227	1
60	0.4663077	2.1445069	0.4877326	2.0503038	0.5095254	1.9626105	0
	Cotangent	Tangent	Cotangent	Tangent	Cotangent	Tangent	
′	65°		64°		63°		′

TABLE XIV. NATURAL TANGENTS AND COTANGENTS

′	27°		28°		29°		′
	Tangent	Cotangent	Tangent	Cotangent	Tangent	Cotangent	
0	0.5095254	1.9626105	0.5317094	1.8807265	0.5543091	1.8040478	60
1	.5098919	1.9612000	.5320826	1.8794074	.5546894	1.8028108	59
2	.5102585	1.9597910	.5324559	1.8780898	.5550698	1.8015751	58
3	.5106252	1.9583837	.5328293	1.8767736	.5554504	1.8003408	57
4	.5109919	1.9569780	.5332029	1.8754588	.5558311	1.7991077	56
5	.5113588	1.9555739	.5335765	1.8741455	.5562119	1.7978759	55
6	.5117259	1.9541713	.5339503	1.8728336	.5565929	1.7966454	54
7	.5120930	1.9527704	.5343242	1.8715231	.5569739	1.7954162	53
8	.5124602	1.9513711	.5346981	1.8702141	.5573551	1.7941883	52
9	.5128275	1.9499733	.5350723	1.8689065	.5577364	1.7929616	51
10	0.5131950	1.9485772	0.5354465	1.8676003	0.5581179	1.7917362	50
11	.5135625	1.9471826	.5358208	1.8662955	.5584994	1.7905121	49
12	.5139302	1.9457896	.5361953	1.8649921	.5588811	1.7892893	48
13	.5142980	1.9443981	.5365699	1.8636902	.5592629	1.7880678	47
14	.5146658	1.9430083	.5369446	1.8623896	.5596449	1.7868475	46
15	.5150338	1.9416200	.5373194	1.8610905	.5600269	1.7856285	45
16	.5154019	1.9402333	.5376943	1.8597928	.5604091	1.7844107	44
17	.5157702	1.9388481	.5380694	1.8584965	.5607914	1.7831943	43
18	.5161385	1.9374645	.5384445	1.8572015	.5611738	1.7819790	42
19	.5165069	1.9360825	.5388198	1.8559080	.5615564	1.7807651	41
20	0.5168755	1.9347020	0.5391952	1.8546159	0.5619391	1.7795524	40
21	.5172441	1.9333231	.5395707	1.8533252	.5623219	1.7783409	39
22	.5176129	1.9319457	.5399464	1.8520358	.5627048	1.7771307	38
23	.5179818	1.9305699	.5403221	1.8507479	.5630879	1.7759218	37
24	.5183508	1.9291956	.5406980	1.8494613	.5634711	1.7747141	36
25	.5187199	1.9278228	.5410740	1.8481761	.5638543	1.7735076	35
26	.5190891	1.9264516	.5414501	1.8468923	.5642378	1.7723024	34
27	.5194584	1.9250819	.5418263	1.8456099	.5646213	1.7710985	33
28	.5198278	1.9237138	.5422027	1.8443289	.5650050	1.7698958	32
29	.5201974	1.9223472	.5425791	1.8430492	.5653888	1.7686943	31
30	0.5205671	1.9209821	0.5429557	1.8417709	0.5657728	1.7674940	30
31	.5209368	1.9196185	.5433324	1.8404940	.5661568	1.7662950	29
32	.5213067	1.9182565	.5437092	1.8392184	.5665410	1.7650972	28
33	.5216767	1.9168960	.5440862	1.8379442	.5669254	1.7639007	27
34	.5220468	1.9155370	.5444632	1.8366713	.5673098	1.7627053	26
35	.5224170	1.9141795	.5448404	1.8353999	.5676944	1.7615112	25
36	.5227874	1.9128236	.5452177	1.8341297	.5680791	1.7603183	24
37	.5231578	1.9114691	.5455951	1.8328610	.5684639	1.7591267	23
38	.5235284	1.9101162	.5459727	1.8315936	.5688488	1.7579362	22
39	.5238990	1.9087647	.5463503	1.8303275	.5692339	1.7567470	21
40	0.5242698	1.9074147	0.5467281	1.8290628	0.5696191	1.7555590	20
41	.5246407	1.9060663	.5471060	1.8277994	.5700045	1.7543722	19
42	.5250117	1.9047193	.5474840	1.8265374	.5703899	1.7531866	18
43	.5253829	1.9033738	.5478621	1.8252767	.5707755	1.7520023	17
44	.5257541	1.9020299	.5482404	1.8240173	.5711612	1.7508191	16
45	.5261255	1.9006874	.5486188	1.8227593	.5715471	1.7496371	15
46	.5264969	1.8993464	.5489973	1.8215026	.5719331	1.7484564	14
47	.5268685	1.8980068	.5493759	1.8202473	.5723192	1.7472768	13
48	.5272402	1.8966688	.5497547	1.8189932	.5727054	1.7460984	12
49	.5276120	1.8953322	.5501335	1.8177405	.5730918	1.7449213	11
50	0.5279839	1.8939971	0.5505125	1.8164892	0.5734783	1.7437453	10
51	.5283560	1.8926635	.5508916	1.8152391	.5738649	1.7425705	9
52	.5287281	1.8913313	.5512708	1.8139904	.5742516	1.7413969	8
53	.5291004	1.8900006	.5516502	1.8127430	.5746385	1.7402245	7
54	.5294727	1.8886713	.5520297	1.8114969	.5750255	1.7390533	6
55	.5298452	1.8873436	.5524093	1.8102521	.5754126	1.7378833	5
56	.5302178	1.8860172	.5527890	1.8090086	.5757999	1.7367144	4
57	.5305906	1.8846924	.5531688	1.8077664	.5761873	1.7355468	3
58	.5309634	1.8833690	.5535488	1.8065256	.5765748	1.7343803	2
59	.5313364	1.8820470	.5539288	1.8052860	.5769625	1.7332149	1
60	0.5317094	1.8807265	0.5543091	1.8040478	0.5773503	1.7320508	0
	Cotangent	Tangent	Cotangent	Tangent	Cotangent	Tangent	
′	62°		61°		60°		′

TABLE XIV. NATURAL TANGENTS AND COTANGENTS

′	30°		31°		32°		′
	Tangent	Cotangent	Tangent	Cotangent	Tangent	Cotangent	
0	0.5773503	1.7320508	0.6008606	1.6642795	0.6248694	1.6003345	60
1	.5777382	1.7308878	.6012536	1.6631834	.6252739	1.5992991	59
2	.5781262	1.7297260	.6016527	1.6620884	.6256786	1.5982647	58
3	.5785144	1.7285654	.6020490	1.6609945	.6260834	1.5972312	57
4	.5789027	1.7274060	.6024454	1.6599016	.6264884	1.5961987	56
5	.5792912	1.7262477	.6028419	1.6588097	.6268935	1.5951672	55
6	.5796797	1.7250905	.6032386	1.6577189	.6272988	1.5941366	54
7	.5800684	1.7239346	.6036354	1.6566292	.6277042	1.5931070	53
8	.5804573	1.7227797	.6040323	1.6555405	.6281098	1.5920783	52
9	.5808462	1.7216261	.6044294	1.6544529	.6285155	1.5910505	51
10	0.5812353	1.7204736	0.6048266	1.6533663	0.6289215	1.5900238	50
11	.5816245	1.7193222	.6052240	1.6522808	.6293274	1.5889979	49
12	.5820139	1.7181720	.6056215	1.6511963	.6297336	1.5879731	48
13	.5824034	1.7170230	.6060192	1.6501128	.6301399	1.5869491	47
14	.5827930	1.7158751	.6064170	1.6490304	.6305464	1.5859261	46
15	.5831828	1.7147283	.6068149	1.6479490	.6309530	1.5849041	45
16	.5835726	1.7135827	.6072130	1.6468687	.6313598	1.5838830	44
17	.5839627	1.7124382	.6076112	1 6457893	.6317667	1.5828628	43
18	.5843528	1.7112949	.6080095	1.6447111	.6321738	1.5818436	42
19	.5847431	1.7101527	.6084080	1.6436338	.6325810	1.5808253	41
20	0.5851335	1.7090116	0.6088067	1.6425576	0.6329883	1.5798079	40
21	.5855241	1.7078717	.6092054	1.6414824	.6333959	1.5787915	39
22	.5859148	1.7067329	.6096043	1.6404082	.6338035	1.5777760	38
23	.5863056	1.7055953	.6100034	1.6393351	.6342113	1.5767615	37
24	.5866965	1.7044587	.6104026	1.6382630	.6346193	1.5757479	36
25	.5870876	1.7033233	.6108019	1.6371919	.6350274	1.5747352	35
26	.5874788	1.7021890	.6112014	1.6361218	.6354357	1.5737234	34
27	.5878702	1.7010599	.6116011	1.6350528	.6358441	1.5727126	33
28	.5882616	1.6999238	.6120008	1.6339847	.6362527	1.5717026	32
29	.5886533	1.6987929	.6124007	1.6329177	.6366614	1.5706936	31
30	0.5890450	1.6976631	0.6128008	1.6318517	0.6370703	1.5696856	30
31	.5894369	1.6965344	.6132010	1.6307867	.6374793	1.5686784	29
32	.5898289	1.6954069	.6136014	1.6297227	.6378885	1.5676722	28
33	.5902211	1.6942804	.6140018	1.6286597	.6382978	1.5666669	27
34	.5906134	1.6931550	.6144024	1.6275977	.6387073	1.5656625	26
35	.5910058	1.6920308	.6148032	1.6265368	.6391169	1.5646590	25
36	.5913984	1.6909077	.6152041	1.6254768	.6395267	1.5636564	24
37	.5917910	1.6897856	.6156052	1.6244178	.6399366	1.5626548	23
38	.5921839	1.6886647	.6160064	1.6233599	.6403467	1.5616540	22
39	.5925768	1.6875449	.6164077	1.6223029	.6407569	1.5606542	21
40	0.5929699	1.6864261	0.6168092	1.6212469	0.6411673	1.5596552	20
41	.5933632	1.6853085	.6172108	1.6201920	.6415779	1.5586572	19
42	.5937565	1.6841919	.6176126	1.6191380	.6419883	1.5576601	18
43	.5941501	1.6830765	.6180145	1.6180850	.6423994	1.5566639	17
44	.5945437	1.6819621	.6184166	1.6170330	.6428105	1.5556685	16
45	.5949375	1.6808489	.6188188	1.6159820	.6432216	1.5546741	15
46	.5953314	1.6797367	.6192211	1.6149320	.6436329	1.5536806	14
47	.5957255	1.6786256	.6196236	1.6138829	.6440444	1.5526880	13
48	.5961196	1.6775156	.6200263	1.6128349	.6444560	1.5516963	12
49	.5965140	1.6764067	.6204291	1.6117878	.6448678	1.5507054	11
50	0.5969084	1.6752988	0.6208320	1.6107417	0.6452797	1.5497155	10
51	.5973030	1.6741921	.6212351	1.6096966	.6456918	1.5487264	9
52	.5976978	1.6730864	.6216383	1.6086525	.6461041	1.5477383	8
53	.5980926	1.6719818	.6220417	1.6076094	.6465165	1.5467510	7
54	.5984877	1.6708782	.6224452	1.6065672	.6469290	1.5457647	6
55	.5988828	1.6697758	.6228488	1.6055260	.6473417	1.5447792	5
56	.5992781	1.6686744	.6232527	1.6044858	.6477546	1.5437946	4
57	.5996735	1.6675741	.6236566	1.6034465	.6481676	1.5428108	3
58	.6000691	1.6664748	.6240607	1.6024082	.6485808	1.5418280	2
59	.6004648	1.6653766	.6244650	1.6013709	.6489941	1.5408460	1
60	0.6008606	1.6642795	0.6248694	1.6003345	0.6494076	1.5398650	0
	Cotangent	Tangent	Cotangent	Tangent	Cotangent	Tangent	
′	59°		58°		57°		′

TABLE XIV. NATURAL TANGENTS AND COTANGENTS

′	33°		34°		35°		′
	Tangent	Cotangent	Tangent	Cotangent	Tangent	Cotangent	
0	0.6494076	1.5398650	0.6745085	1.4825610	0.7002075	1.4281480	60
1	.6498212	1.5388848	.6749318	1.4816311	.7006411	1.4272642	59
2	.6502350	1.5379054	.6753553	1.4807021	.7010749	1.4263811	58
3	.6506490	1.5369270	.6757790	1.4797738	.7015089	1.4254988	57
4	.6510631	1.5359494	.6762028	1.4788463	.7019430	1.4246171	56
5	.6514774	1.5349727	.6766268	1.4779197	.7023773	1.4237362	55
6	.6518918	1.5339969	.6770509	1.4769938	.7028118	1.4228561	54
7	.6523064	1.5330219	.6774752	1.4760688	.7032464	1.4219766	53
8	.6527211	1.5320479	.6778997	1.4751445	.7036813	1.4210979	52
9	.6531360	1.5310746	.6783243	1.4742210	.7041163	1.4202200	51
10	0.6535511	1.5301023	0.6787492	1.4732983	0.7045515	1.4193427	50
11	.6539663	1.5291308	.6791741	1.4723764	.7049869	1.4184662	49
12	.6543817	1.5281602	.6795993	1.4714553	.7054224	1.4175904	48
13	.6547972	1.5271904	.6800246	1.4705350	.7058581	1.4167153	47
14	.6552129	1.5262215	.6804501	1.4696155	.7062940	1.4158409	46
15	.6556287	1.5252535	.6808758	1.4686967	.7067301	1.4149673	45
16	.6560447	1.5242863	.6813016	1.4677788	.7071664	1.4140943	44
17	.6564609	1.5233200	.6817276	1.4668616	.7076028	1.4132221	43
18	.6568772	1.5223545	.6821537	1.4659452	.7080395	1.4123506	42
19	.6572937	1.5213899	.6825801	1.4650296	.7084763	1.4114799	41
20	0.6577103	1.5204261	0.6830066	1.4641147	0.7089133	1.4106098	40
21	.6581271	1.5194632	.6834333	1.4632007	.7093504	1.4097405	39
22	.6585441	1.5185012	.6838601	1.4622874	.7097878	1.4088718	38
23	.6589612	1.5175400	.6842871	1.4613749	.7102253	1.4080039	37
24	.6593785	1.5165796	.6847143	1.4604632	.7106630	1.4071367	36
25	.6597960	1.5156201	.6851416	1.4595522	.7111009	1.4062702	35
26	.6602136	1.5146614	.6855692	1.4586420	.7115390	1.4054044	34
27	.6606313	1.5137036	.6859969	1.4577326	.7119772	1.4045393	33
28	.6610492	1.5127466	.6864247	1.4568240	.7124157	1.4036749	32
29	.6614673	1.5117905	.6868528	1.4559161	.7128543	1.4028113	31
30	0.6618856	1.5108352	0.6872810	1.4550090	0.7132931	1.4019483	30
31	.6623040	1.5098807	.6877093	1.4541027	.7137320	1.4010860	29
32	.6627225	1.5089271	.6881379	1.4531971	.7141712	1.4002245	28
33	.6631413	1.5079743	.6885666	1.4522923	.7146106	1.3993636	27
34	.6635601	1.5070224	.6889955	1.4513883	.7150501	1.3985034	26
35	.6639792	1.5060713	.6894246	1.4504850	.7154898	1.3976440	25
36	.6643984	1.5051210	.6898538	1.4495825	.7159297	1.3967852	24
37	.6648178	1.5041716	.6902832	1.4486808	.7163698	1.3959272	23
38	.6652373	1.5032229	.6907128	1.4477798	.7168100	1.3950698	22
39	.6656570	1.5022751	.6911425	1.4468796	.7172505	1.3942131	21
40	0.6660769	1.5013282	0.6915725	1.4459801	0.7176911	1.3933571	20
41	.6664969	1.5003821	.6920026	1.4450814	.7181319	1.3925019	19
42	.6669171	1.4994367	.6924328	1.4441834	.7185729	1.3916473	18
43	.6673374	1.4984923	.6928633	1.4432862	.7190140	1.3907934	17
44	.6677580	1.4975486	.6932939	1.4423897	.7194554	1.3899401	16
45	.6681786	1.4966058	.6937247	1.4414940	.7198970	1.3890876	15
46	.6685995	1.4956637	.6941557	1.4405991	.7203387	1.3882358	14
47	.6690205	1.4947225	.6945868	1.4397049	.7207806	1.3873847	13
48	.6694417	1.4937822	.6950181	1.4388114	.7212227	1.3865342	12
49	.6698630	1.4928426	.6954496	1.4379187	.7216650	1.3856844	11
50	0.6702845	1.4919039	0.6958813	1.4370268	0.7221075	1.3848353	10
51	.6707061	1.4909659	.6963131	1.4361356	.7225502	1.3839869	9
52	.6711280	1.4900288	.6967451	1.4352451	.7229930	1.3831392	8
53	.6715500	1.4890925	.6971773	1.4343554	.7234361	1.3822922	7
54	.6719721	1.4881570	.6976097	1.4334664	.7238793	1.3814458	6
55	.6723944	1.4872223	.6980422	1.4325781	.7243227	1.3806001	5
56	.6728169	1.4862884	.6984749	1.4316906	.7247663	1.3797551	4
57	.6732396	1.4853554	.6989078	1.4308039	.7252101	1.3789108	3
58	.6736624	1.4844231	.6993409	1.4299178	.7256540	1.3780672	2
59	.6740854	1.4834916	.6997741	1.4290326	.7260982	1.3772242	1
60	0.6745085	1.4825610	0.7002075	1.4281480	0.7265425	1.3763819	0
	Cotangent	Tangent	Cotangent	Tangent	Cotangent	Tangent	
′	56°		55°		54°		′

TABLE XIV. NATURAL TANGENTS AND COTANGENTS

′	36°		37°		38°		
	Tangent	Cotangent	Tangent	Cotangent	Tangent	Cotangent	
0	0.7265425	1.3763819	0.7535541	1.3270448	0.7812856	1.2799416	60
1	.7269871	1.3755403	.7540102	1.3262420	.7817542	1.2791745	59
2	.7274318	1.3746994	.7544666	1.3254397	.7822229	1.2784079	58
3	.7278767	1.3738591	.7549232	1.3246381	.7826919	1.2776419	57
4	.7283218	1.3730195	.7553799	1.3238371	.7831611	1.2768765	56
5	.7287671	1.3721806	.7558369	1.3230368	.7836305	1.2761116	55
6	.7292125	1.3713423	.7562941	1.3222370	.7841002	1.2753473	54
7	.7296582	1.3705047	.7567514	1.3214379	.7845700	1.2745835	53
8	.7301041	1.3696678	.7572090	1.3206393	.7850400	1.2738204	52
9	.7305501	1.3688315	.7576668	1.3198414	.7855103	1.2730578	51
10	0.7309963	1.3679959	0.7581248	1.3190441	0.7859808	1.2722957	50
11	.7314428	1.3671610	.7585829	1.3182474	.7864515	1.2715342	49
12	.7318894	1.3663267	.7590413	1.3174513	.7869224	1.2707733	48
13	.7323362	1.3654931	.7594999	1.3166559	.7873935	1.2700130	47
14	.7327832	1.3646602	.7599587	1.3158610	.7878649	1.2692532	46
15	.7332303	1.3638279	.7604177	1.3150668	.7883364	1.2684940	45
16	.7336777	1.3629963	.7608769	1.3142731	.7888082	1.2677353	44
17	.7341253	1.3621653	.7613363	1.3134801	.7892802	1.2669772	43
18	.7345730	1.3613350	.7617959	1.3126876	.7897524	1.2662196	42
19	.7350210	1.3605054	.7622557	1.3118958	.7902248	1.2654626	41
20	0.7354691	1.3596764	0.7627157	1.3111046	0.7906975	1.2647062	40
21	.7359174	1.3588481	.7631759	1.3103140	.7911703	1.2639503	39
22	.7363660	1.3580204	.7636363	1.3095239	.7916434	1.2631950	38
23	.7368147	1.3571934	.7640969	1.3087345	.7921167	1.2624402	37
24	.7372636	1.3563670	.7645577	1.3079457	.7925902	1.2616860	36
25	.7377127	1.3555413	.7650188	1.3071575	.7930640	1.2609323	35
26	.7381620	1.3547162	.7654800	1.3063699	.7935379	1.2601792	34
27	.7386115	1.3538918	.7659414	1.3055828	.7940121	1.2594267	33
28	.7390611	1.3530680	.7664031	1.3047964	.7944865	1.2586747	32
29	.7395110	1.3522449	.7668649	1.3040106	.7949611	1.2579232	31
30	0.7399611	1.3514224	0.7673270	1.3032254	0.7954359	1.2571723	30
31	.7404113	1.3506006	.7677893	1.3024408	.7959110	1.2564219	29
32	.7408618	1.3497794	.7682517	1.3016567	.7963862	1.2556721	28
33	.7413124	1.3489589	.7687144	1.3008733	.7968617	1.2549229	27
34	.7417633	1.3481390	.7691773	1.3000904	.7973374	1.2541742	26
35	.7422143	1.3473198	.7696404	1.2993081	.7978134	1.2534260	25
36	.7426655	1.3465011	.7701037	1.2985265	.7982895	1.2526784	24
37	.7431170	1.3456832	.7705672	1.2977454	.7987659	1.2519313	23
38	.7435686	1.3448658	.7710309	1.2969649	.7992425	1.2511848	22
39	.7440204	1.3440492	.7714948	1.2961850	.7997193	1.2504388	21
40	0.7444724	1.3432331	0.7719589	1.2954057	0.8001963	1.2496933	20
41	.7449246	1.3424177	.7724233	1.2946270	.8006736	1.2489484	19
42	.7453770	1.3416029	.7728878	1.2938488	.8011511	1.2482040	18
43	.7458296	1.3407888	.7733526	1.2930713	.8016288	1.2474602	17
44	.7462824	1.3399753	.7738176	1.2922943	.8021067	1.2467169	16
45	.7467354	1.3391624	.7742827	1.2915179	.8025849	1.2459742	15
46	.7471886	1.3383502	.7747481	1.2907421	.8030632	1.2452320	14
47	.7476420	1.3375386	.7752137	1.2899669	.8035418	1.2444903	13
48	.7480956	1.3367276	.7756795	1.2891922	.8040206	1.2437492	12
49	.7485494	1.3359172	.7761455	1.2884182	.8044997	1.2430086	11
50	0.7490033	1.3351075	0.7766118	1.2876447	0.8049790	1.2422685	10
51	.7494575	1.3342984	.7770782	1.2868718	.8054584	1.2415290	9
52	.7499119	1.3334900	.7775448	1.2860995	.8059382	1.2407900	8
53	.7503665	1.3326822	.7780117	1.2853277	.8064181	1.2400515	7
54	.7508212	1.3318750	.7784788	1.2845566	.8068983	1.2393136	6
55	.7512762	1.3310684	.7789460	1.2837860	.8073787	1.2385762	5
56	.7517314	1.3302624	.7794135	1.2830160	.8078593	1.2378393	4
57	.7521867	1.3294571	.7798812	1.2822466	.8083401	1.2371030	3
58	.7526423	1.3286524	.7803492	1.2814776	.8088212	1.2363672	2
59	.7530981	1.3278483	.7808173	1.2807094	.8093025	1.2356319	1
60	0.7535541	1.3270448	0.7812856	1.2799416	0.8097840	1.2348972	0
	Cotangent	Tangent	Cotangent	Tangent	Cotangent	Tangent	
′	53°		52°		51°		′

TABLE XIV. NATURAL TANGENTS AND COTANGENTS

′	39°		40°		41°		′
	Tangent	Cotangent	Tangent	Cotangent	Tangent	Cotangent	
0	0.8097840	1.2348972	0.8390996	1.1917536	0.8692867	1.1503684	60
1	.8102658	1.2341629	.8395955	1.1910498	.8697976	1.1496928	59
2	.8107478	1.2334292	.8400915	1.1903465	.8703087	1.1490176	58
3	.8112300	1.2326961	.8405878	1.1896437	.8708200	1.1483429	57
4	.8117124	1.2319634	.8410844	1.1889414	.8713316	1.1476687	56
5	.8121951	1.2312313	.8415812	1.1882395	.8718435	1.1469949	55
6	.8126780	1.2304997	.8420782	1.1875382	.8723556	1.1463215	54
7	.8131611	1.2297687	.8425755	1.1868373	.8728680	1.1456486	53
8	.8136444	1.2290381	.8430730	1.1861369	.8733806	1.1449762	52
9	.8141280	1.2283081	.8435708	1.1854370	.8738935	1.1443041	51
10	0.8146118	1.2275786	0.8440688	1.1847376	0.8744067	1.1436326	50
11	.8150958	1.2268496	.8445670	1.1840387	.8749201	1.1429615	49
12	.8155801	1.2261211	.8450655	1.1833402	.8754338	1.1422908	48
13	.8160646	1.2253932	.8455643	1.1826422	.8759478	1.1416206	47
14	.8165493	1.2246658	.8460633	1.1819447	.8764620	1.1409508	46
15	.8170343	1.2239389	.8465625	1.1812477	.8769765	1.1402815	45
16	.8175195	1.2232125	.8470620	1.1805512	.8774912	1.1396126	44
17	.8180049	1.2224866	.8475617	1.1798551	.8780062	1.1389441	43
18	.8184905	1.2217613	.8480617	1.1791595	.8785215	1.1382761	42
19	.8189764	1.2210364	.8485619	1.1784644	.8790370	1.1376086	41
20	0.8194625	1.2203121	0.8490624	1.1777698	0.8795528	1.1369414	40
21	.8199488	1.2195883	.8495631	1.1770756	.8800688	1.1362747	39
22	.8204354	1.2188650	.8500640	1.1763820	.8805852	1.1356085	38
23	.8209222	1.2181422	.8505653	1.1756888	.8811017	1.1349427	37
24	.8214093	1.2174199	.8510667	1.1749960	.8816186	1.1342773	36
25	.8218965	1.2166982	.8515684	1.1743038	.8821357	1.1336124	35
26	.8223840	1.2159769	.8520704	1.1736120	.8826531	1.1329479	34
27	.8228718	1.2152562	.8525726	1.1729207	.8831707	1.1322839	33
28	.8233597	1.2145359	.8530750	1.1722298	.8836883	1.1316203	32
29	.8238479	1.2138162	.8535777	1.1715395	.8842068	1.1309571	31
30	0.8243364	1.2130970	0.8540807	1.1708496	0.8847253	1.1302944	30
31	.8248251	1.2123783	.8545839	1.1701601	.8852440	1.1296321	29
32	.8253140	1.2116601	.8550873	1.1694712	.8857630	1.1289702	28
33	.8258031	1.2109424	.8555910	1.1687827	.8862822	1.1283088	27
34	.8262925	1.2102252	.8560950	1.1680947	.8868017	1.1276478	26
35	.8267821	1.2095085	.8565992	1.1674071	.8873215	1.1269872	25
36	.8272719	1.2087924	.8571037	1.1667200	.8878415	1.1263271	24
37	.8277620	1.2080767	.8576084	1.1660334	.8883619	1.1256674	23
38	.8282523	1.2073615	.8581133	1.1653472	.8888825	1.1250081	22
39	.8287429	1.2066468	.8586185	1.1646615	.8894033	1.1243493	21
40	0.8292337	1.2059327	0.8591240	1.1639763	0.8899244	1.1236909	20
41	.8297247	1.2052190	.8596297	1.1632916	.8904458	1.1230329	19
42	.8302160	1.2045058	.8601357	1.1626073	.8909675	1.1223754	18
43	.8307075	1.2037932	.8606419	1.1619234	.8914894	1.1217183	17
44	.8311992	1.2030810	.8611484	1.1612400	.8920116	1.1210616	16
45	.8316912	1.2023693	.8616551	1.1605571	.8925341	1.1204053	15
46	.8321834	1.2016581	.8621621	1.1598747	.8930569	1.1197495	14
47	.8326759	1.2009475	.8626694	1.1591927	.8935799	1.1190941	13
48	.8331686	1.2002373	.8631768	1.1585112	.8941032	1.1184391	12
49	.8336615	1.1995276	.8636846	1.1578301	.8946268	1.1177846	11
50	0.8341547	1.1988184	0.8641926	1.1571495	0.8951506	1.1171305	10
51	.8346481	1.1981097	.8647009	1.1564693	.8956747	1.1164768	9
52	.8351418	1.1974015	.8652094	1.1557896	.8961991	1.1158235	8
53	.8356357	1.1966938	.8657181	1.1551104	.8967238	1.1151706	7
54	.8361298	1.1959866	.8662272	1.1544316	.8972487	1.1145182	6
55	.8366242	1.1952799	.8667365	1.1537532	.8977739	1.1138662	5
56	.8371188	1.1945736	.8672460	1.1530754	.8982994	1.1132146	4
57	.8376136	1.1938679	.8677558	1.1523979	.8988251	1.1125635	3
58	.8381087	1.1931626	.8682659	1.1517210	.8993512	1.1119127	2
59	.8386041	1.1924579	.8687762	1.1510445	.8998775	1.1112624	1
60	0.8390996	1.1917536	0.8692867	1.1503684	0.9004040	1.1106125	0
	Cotangent	Tangent	Cotangent	Tangent	Cotangent	Tangent	
′	50°		49°		48°		′

TABLE XIV. NATURAL TANGENTS AND COTANGENTS

′	42°		43°		44°		′
	Tangent	Cotangent	Tangent	Cotangent	Tangent	Cotangent	
0	0.9004040	1.1106125	0.9325151	1.0723687	0.9656888	1.0355303	60
1	.9009309	1.1099630	.9330591	1.0717435	.9662511	1.0349277	59
2	.9014580	1.1093140	.9336034	1.0711187	.9668137	1.0343254	58
3	.9019854	1.1086653	.9341479	1.0704943	.9673767	1.0337235	57
4	.9025131	1.1080171	.9346928	1.0698702	.9679399	1.0331220	56
5	.9030411	1.1073693	.9352380	1.0692466	.9685035	1.0325208	55
6	.9035693	1.1067219	.9357834	1.0686233	.9690674	1.0319199	54
7	.9040979	1.1060750	.9363292	1.0680004	.9696316	1.0313195	53
8	.9046267	1.1054284	.9368753	1.0673779	.9701962	1.0307194	52
9	.9051557	1.1047823	.9374216	1.0667558	.9707610	1.0301196	51
10	0.9056851	1.1041365	0.9379683	1.0661341	0.9713262	1.0295203	50
11	.9062147	1.1034913	.9385153	1.0655128	.9718917	1.0289212	49
12	.9067446	1.1028463	.9390625	1.0648918	.9724575	1.0283226	48
13	.9072748	1.1022019	.9396101	1.0642713	.9730236	1.0277243	47
14	.9078053	1.1015578	.9401579	1.0636511	.9735901	1.0271263	46
15	.9083360	1.1009141	.9407061	1.0630313	.9741569	1.0265287	45
16	.9088671	1.1002709	.9412545	1.0624119	.9747240	1.0259315	44
17	.9093984	1.0996281	.9418033	1.0617929	.9752914	1.0253346	43
18	.9099300	1.0989857	.9423523	1.0611742	.9758591	1.0247381	42
19	.9104619	1.0983436	.9429017	1.0605560	.9764272	1.0241419	41
20	0.9109940	1.0977020	0.9434513	1.0599381	0.9769956	1.0235461	40
21	.9115265	1.0970609	.9440013	1.0593206	.9775643	1.0229506	39
22	.9120592	1.0964201	.9445516	1.0587035	.9781333	1.0223555	38
23	.9125922	1.0957797	.9451021	1.0580867	.9787027	1.0217608	37
24	.9131255	1.0951397	.9456530	1.0574704	.9792724	1.0211664	36
25	.9136591	1.0945002	.9462042	1.0568544	.9798424	1.0205723	35
26	.9141929	1.0938610	.9467556	1.0562388	.9804127	1.0199786	34
27	.9147270	1.0932223	.9473074	1.0556235	.9809833	1.0193853	33
28	.9152615	1.0925840	.9478595	1.0550087	.9815543	1.0187923	32
29	.9157962	1.0919460	.9484119	1.0543942	.9821256	1.0181997	31
30	0.9163312	1.0913085	0.9489646	1.0537801	0.9826973	1.0176074	30
31	.9168665	1.0906714	.9495176	1.0531664	.9832692	1.0170155	29
32	.9174020	1.0900347	.9500709	1.0525531	.9838415	1.0164239	28
33	.9179379	1.0893984	.9506245	1.0519401	.9844141	1.0158326	27
34	.9184740	1.0887624	.9511784	1.0513275	.9849871	1.0152418	26
35	.9190104	1.0881269	.9517326	1.0507153	.9855603	1.0146512	25
36	.9195471	1.0874918	.9522871	1.0501034	.9861339	1.0140610	24
37	.9200841	1.0868571	.9528420	1.0494920	.9867079	1.0134712	23
38	.9206214	1.0862228	.9533971	1.0488809	.9872821	1.0128817	22
39	.9211590	1.0855889	.9539526	1.0482702	.9878567	1.0122925	21
40	0.9216969	1.0849554	0.9545083	1.0476598	0.9884316	1.0117038	20
41	.9222350	1.0843223	.9550644	1.0470498	.9890069	1.0111153	19
42	.9227734	1.0836896	.9556208	1.0464402	.9895825	1.0105272	18
43	.9233122	1.0830573	.9561774	1.0458310	.9901584	1.0099394	17
44	.9238512	1.0824254	.9567344	1.0452221	.9907346	1.0093520	16
45	.9243905	1.0817939	.9572917	1.0446136	.9913112	1.0087649	15
46	.9249301	1.0811628	.9578494	1.0440055	.9918881	1.0081782	14
47	.9254700	1.0805321	.9584073	1.0433977	.9924654	1.0075918	13
48	.9260102	1.0799018	.9589655	1.0427904	.9930429	1.0070058	12
49	.9265506	1.0792718	.9595241	1.0421833	.9936208	1.0064201	11
50	0.9270914	1.0786423	0.9600829	1.0415767	0.9941991	1.0058348	10
51	.9276324	1.0780132	.9606421	1.0409704	.9947777	1.0052497	9
52	.9281738	1.0773845	.9612016	1.0403645	.9953566	1.0046651	8
53	.9287154	1.0767561	.9617614	1.0397589	.9959358	1.0040808	7
54	.9292573	1.0761282	.9623215	1.0391538	.9965154	1.0034968	6
55	.9297996	1.0755006	.9628819	1.0385489	.9970953	1.0029131	5
56	.9303421	1.0748734	.9634427	1.0379445	.9976756	1.0023298	4
57	.9308849	1.0742467	.9640037	1.0373404	.9982562	1.0017469	3
58	.9314280	1.0736203	.9645651	1.0367367	.9988371	1.0011642	2
59	.9319714	1.0729943	.9651268	1.0361333	.9994184	1.0005819	1
60	0.9325151	1.0723687	0.9656888	1.0355303	1.0000000	1.0000000	0
	Cotangent	Tangent	Cotangent	Tangent	Cotangent	Tangent	
′	47°		46°		45°		′

REFERENCES

AMERICAN SOCIETY OF CIVIL ENGINEERS, Committee of the Surveying and Mapping Division on Definition of Surveying Terms. *Definitions of Surveying, Mapping, and Related Terms.* Manual No. 34, 1954.

BOUCHARD, HARRY and FRANCIS H. MOFFITT. Surveying, 5th ed. Scranton: International Textbook Company, 1965.

BRINKER, RUSSELL C. 4300 *Review Questions for Surveyors,* 9th ed. Published by the author, Box 399, Sun City, Arizona, 85351, 1971.

BRINKER, RUSSELL C., and BROTHER B. AUSTIN BARRY. *Noteforms for Surveying Measurements.* Scranton: International Textbook Company, 1957.

BROWN, CURTIS M. *Boundary Control and Legal Principles.* New York: John Wiley & Sons, Inc., 1957.

BROWN, CURTIS M., and WINFIELD H. ELDRIDGE. *Evidence and Procedures for Boundary Location.* New York: John Wiley & Sons, Inc., 1962.

CLARK, FRANK E. *Law of Surveying and Boundaries.* Indianapolis: The Bobbs-Merrill Company, 1939.

DAVIS, RAYMOND E., FRANCIS S. FOOTE, and JOE W. KELLY. *Surveying: Theory and Practice,* 5th ed. New York: McGraw-Hill Book Company, Inc., 1966.

EICHLER, JOHN O., and HARRY TUBIS. *Photogrammetry Laboratory Kit,* 1953.

Ephemeris. (Ephemerides are published annually by W. & L. E. Gurley, by Keuffel and Esser Company, and by C. L. Berger and Sons, Inc.)

LOW, JULIAN W. *Plane Table Mapping.* New York: Harper & Brothers, 1952.

MEYER, CARL F. *Route Surveying,* 4th ed. Scranton: International Textbook Company, 1969.

MOFFITT, FRANCIS H. *Photogrammetry,* 2d ed. Scranton: International Textbook Company, 1967.

Optical Alignment Equipment. Keuffel and Esser Company.

Surveying and Mapping. [The quarterly journal of the American Congress on Surveying and Mapping.]

UNITED STATES BUREAU OF LAND MANAGEMENT. *Manual of Instructions for the Survey of the Public Lands of the United States.* Washington, D.C.: Government Printing Office, 1947.

UNITED STATES GENERAL LAND OFFICE. *Standard Field Tables.* Washington, D.C.: Government Printing Office, 1942.

index

Index